Algebraic Geometry

Algebraic Geometry

A Volume in Memory of Paolo Francia

Editors
M. C. Beltrametti
F. Catanese
C. Ciliberto
A. Lanteri
C. Pedrini

Walter de Gruyter · Berlin · New York 2002

Editors

Mauro C. Beltrametti
Dipartimento di Matematica
Università di Genova
Via Dodecaneso 35
16146 Genova
Italy

Fabrizio Catanese
Lehrstuhl Mathematik VIII
Universität Bayreuth
95440 Bayreuth
Germany

Ciro Ciliberto
Dipartimento di Matematica
Università di Roma Tor Vergata
Via della Ricerca Scientifica
00133 Roma
Italy

Antonio Lanteri
Dipartimento di Matematica
„F. Enriques"
Via C. Saldini 50
20133 Milano
Italy

Claudio Pedrini
Dipartimento di Matematica
Università di Genova
Via Dodecaneso 35
16146 Genova
Italy

Mathematics Subject Classification 2000:
14-06; 14J29, 14E30

Keywords:
Surfaces of general type, minimal model program (Mori theory, extremal rays), special varieties, motives theory, numerical algebraic geometry

⊚ Printed on acid-free paper which falls within the guidelines of the ANSI to ensure permanence and durability.

Library of Congress – Cataloging-in-Publication Data

```
Algebraic geometry : a volume in memory of Paolo Francia /
editors, M. C. Beltrametti ... [et al.].
   p. cm.
Includes bibliographical references.
ISBN 3 11 017180 5 (cloth : alk. paper)
   1. Geometry, Algebraic.    I. Francia, Paolo, 1951–    II.
Beltrametti, Mauro, 1948–
QA564 .A19   2002
516.3'5–dc21
                                            2002074088
```

Die Deutsche Bibliothek – Cataloging-in-Publication Data

```
Algebraic geometry : a volume in memory of Paolo Francia /
ed. M. C. Beltrametti ... . – Berlin ; New York : de Gruyter,
2002
ISBN 3-11-017180-5
```

© Copyright 2002 by Walter de Gruyter GmbH & Co. KG, 10785 Berlin, Germany.
All rights reserved, including those of translation into foreign languages. No part of this book may be reproduced or transmitted in any form or by any means, electronic or mechanical, including photocopy, recording or any information storage and retrieval system, without permission in writing from the publisher.
Printed in Germany.
Cover design: Thomas Bonnie, Hamburg.
Typeset using the authors' T$_E$X files: I. Zimmermann, Freiburg.
Printing and binding: Hubert & Co. GmbH & Co. KG, Göttingen.

Preface

Paolo Francia was born in Torino, on September 4, 1951. He got his Laurea at the University of Genoa in 1975, under the supervision of Prof. D. Gallarati. Paolo Francia spent most of his academic life at the University of Genoa. He was appointed a full professor at the University of Messina in 1987. He moved back to Genoa the next academic year. After a shorter period at the Facoltà di Informatica (Faculty of Computer Science) and a longer one (1988 to 1994) spent at the Faculty of Engineering, Paolo Francia held a chair of Geometry at the Mathematics Department. He was also a member of the "Accademia Ligure di Scienze e Lettere".

As far as the field of algebraic geometry is concerned, Paolo Francia has been perhaps the most important and brilliant mathematician the University of Genoa has ever had.

His scientific work was mainly devoted to classification problems of higher dimensional varieties, and to surfaces of general type. On both topics he obtained fundamental and beautiful results, most of them properly commented upon and illustrated in some of the contributions in the present volume.

Paolo left Andreina, his wife, and his children Lucia, Pietro and Laura. We all lost a dear friend. We remember him with deep and true admiration, and with sincere love.

Publications of Paolo Francia

[1] Su alcuni esempi di Lascu e Moishezon, Atti Sem. Mat. Fis. Univ. Modena 25 (1976), 27–35.

[2] (with M. C. Beltrametti and F. Odetti) Su un problema di razionalità posto da K. Ueno, Rend. Sem. Mat. Univ. e Politec. Torino 35 (1976/77), 215–217.

[3] (with M. C. Beltrametti) Some remarks on the cohomology of an invertible sheaf, Proc. Konink. Nederl. Akad. Wetensch. Ser. A 82 (1979), 473–480 = Indag. Math. 41 (1979), 473–480.

[4] Some remarks on minimal models I, Compositio Math. 40 (1980), 301–313.

[5] (with M. C. Beltrametti) Conic bundles on non-rational surfaces, Algebraic Geometry–Open Problems, Proceedings Ravello 1982, Lecture Notes in Math. 997, Springer-Verlag, Berlin 1983, 34–89.

[6] (with M. C. Beltrametti) Threefolds with negative Kodaira dimension and positive irregularity, Nagoya Math. J. 91 (1983), 163–172.

[7] Pluricanonical mappings for surfaces of general type, Geometry Seminars, 1982–1983, Univ. di Bologna 1984.

[8] (with M. C. Beltrametti) A property of regular morphisms, Proc. Konink. Nederl. Akad. Wetensch. Ser. A 87 (1984), 361–368 = Indag. Math. 46 (1984), 361–368.

[9] (with M. C. Beltrametti and A. J. Sommese) On Reider's method and higher order embeddings, Duke Math. J. 58 (1989), 425–439.

[10] On the base points of the bicanonical system, in: 1988 Cortona Proceedings on Projective Surfaces and their Classification (F. Catanese and C. Ciliberto, eds.), Sympos. Math. 32, INDAM, Academic Press, London 1992, 141–150.

[11] (with L. Bădescu and M. C. Beltrametti) Positive curves in polarized manifolds, Manuscripta Math. 92 (1997), 369–388.

[12] (with C. Ciliberto and M. Mendes Lopes) Remarks on the bicanonical map for surfaces of general type, Math. Z. 224 (1997), 137–166.

Paolo Francia, September 4, 1951 – July 6, 2000

Table of Contents

Preface v

Lucian Bădescu and Michael Schneider
Formal functions, connectivity and homogeneous spaces 1

Luca Barbieri-Viale
On algebraic 1-motives related to Hodge cycles 25

Arnaud Beauville
The Szpiro inequality for higher genus fibrations 61

Giuseppe Borrelli
On regular surfaces of general type with $p_g = 2$ and non-birational
bicanonical map... 65

Fabrizio Catanese and Frank-Olaf Schreyer
Canonical projections of irregular algebraic surfaces 79

Ciro Ciliberto and Margarida Mendes Lopes
On surfaces with $p_g = 2$, $q = 1$ and non-birational bicanonical map 117

Alberto Conte, Marina Marchisio and Jacob Murre
On unirationality of double covers of fixed degree and large dimension;
a method of Ciliberto 127

Alessio Corti and Miles Reid
Weighted Grassmannians 141

Tommaso de Fernex and Lawrence Ein
Resolution of indeterminacy of pairs 165

Vladimir Guletskiĭ and Claudio Pedrini
The Chow motive of the Godeaux surface 179

Yujiro Kawamata
Francia's flip and derived categories................................. 197

Kazuhiro Konno
On the quadric hull of a canonical surface 217

Adrian Langer
A note on Bogomolov's instability and Higgs sheaves 237

Antonio Lanteri and Raquel Mallavibarrena
Jets of antimulticanonical bundles on Del Pezzo surfaces of degree ≤ 2 . . . 257

Margarida Mendes Lopes and Rita Pardini
A survey on the bicanonical map of surfaces with $p_g = 0$ and $K^2 \geq 2$ 277

Francesco Russo
The antibirational involutions of the plane and the classification of real del Pezzo surfaces 289

Vyacheslav V. Shokurov
Letters of a bi-rationalist: IV. Geometry of log flips 313

Andrew J. Sommese, Jean Verschelde and Charles Wampler
A method for tracking singular paths with application to the numerical irreducible decomposition 329

Lectures of the Conference 347

List of Contributors 349

List of Participants 353

Formal functions, connectivity and homogeneous spaces*

Lucian Bădescu and Michael Schneider

To the memory of Paolo Francia

1. Introduction

Fix an algebraically closed ground field k and let X be an irreducible projective variety over k. Then $K(X)$ will denote the field of rational functions of X, and $K(X_{/Y})$ will denote the ring of formal-rational functions of X along a closed subscheme Y of X. For X normal and Y connected of positive dimension $K(X_{/Y})$ is a field. The field $K(X)$ can be considered naturally as a subfield of $K(X_{/Y})$, and the extension problem of formal-rational functions asks when can any formal-rational function on X along Y be extended to a rational function on X, i.e. when do we have $K(X) = K(X_{/Y})$?

The formal functions have been introduced and studied systematically in algebraic geometry by Zariski in [Z]. In particular, Zariski used them to prove his famous connectedness theorem. Later on this theory has been considerably extended and deepened by Grothendieck in the framework of formal schemes, by providing powerful tools (such as Grothendieck's existence theorem, see [EGAIII], theorem (5.1.4)) to study related questions (see [SGA1], [SGA2]).

There is an analogous problem in the complex-analytic setting. Given a pair (X, Y) consisting of a complex irreducible projective variety X and a closed subvariety Y of X, let U be a connected open neighbourhood of Y in X (in the complex topology), and denote by $\mathcal{M}(U)$ the $\mathbb{C}$-algebra of all meromorphic functions on U. We have the field extensions $K(X) \subseteq \mathcal{M}(U) \subseteq K(X_{/Y})$, and therefore the extension problem for meromorphic functions can be solved if it can be solved for formal-rational functions, i.e. the equality $K(X) = \mathcal{M}(U)$ is a consequence of the equality $K(X) = K(X_{/Y})$.

Extension of analytic or formal objects like functions or subvarieties is a rather classical subject. The earliest reference we are aware of is a paper of Severi [Se], in which he proved that any meromorphic function defined in a small complex connected neighbourhood of a smooth hypersurface in $\mathbb{P}^n(\mathbb{C})$ ($n \geq 2$) can be extended to all of $\mathbb{P}^n(\mathbb{C})$. After some partial extension of Severi's result by Van de Ven and Remmert (see [RVdV]), Barth [Ba] generalized the above-mentioned result of Severi to arbitrary

---

*This paper was mostly done in the Spring 1997 when the first-named author was visiting Max Planck Institut in Bonn for six months and University of Bayreuth for two short visits; unfortunately, a couple of months later (August, 29, 1997) Michael Schneider passed away in a tragic accident when climbing in mountains.

positive-dimensional connected subvarieties of $\mathbb{P}^n(\mathbb{C})$. On the other hand, Hironaka [H] extended Severi's result to arbitrary smooth projective varieties X and smooth hypersurfaces Y in X with ample normal bundle. His approach is algebraic and he considers the extension of formal-rational functions along Y. Further important contributions to the extension problem of meromorphic functions or analytic objects are due to Griffiths [Gr], Chow [Ch1], Rossi [Ro] and others.

In a basic paper Hironaka and Matsumura [HM] showed that $K(\mathbb{P}^n) = K(\mathbb{P}^n_{/Y})$ for all positive-dimensional connected subvarieties $Y \subset \mathbb{P}^n$ and proved that for a subvariety $Y \subset A$, A an abelian variety, the extension problem can be solved precisely if Y generates A as a group and the canonical map $\mathrm{Alb}(Y) \to A$ has connected fibers. Further important contributions to the extension problem of formal-rational functions are due to Hartshorne [Ha2], Chow [Ch2], Faltings [F1], [F2], and others. The above mentioned result of Hironaka–Matsumura on formal-rational functions implies the result of Barth on meromorphic functions.

Here is the contents of this paper. In the second section some algebraicity criteria for formal-rational functions are proved. Specifically, let $f : X' \to X$ be a proper surjective morphism of projective algebraic varieties, Y' a closed subvariety of X' and Y a closed subvariety of X such that $f(Y') \subseteq Y$. Assume that $K(X_{/Y})$ is a field, and let $\zeta \in K(X_{/Y})$ be a formal-rational function. If the pull-back function $\tilde{f}^*(\zeta)$ is a rational function on X' then ζ is algebraic over the field $K(X)$ of rational functions of X (see Theorem (2.5)). As a consequence one proves the following result (which emphasizes in particular the close relationship between formal geometry and connectedness theorems in algebraic geometry, see Theorem (2.7)): Let X be a projective irreducible algebraic variety and $Y \subset X$ a closed subvariety of X. Then the following two conditions are equivalent:

a) For every proper surjective morphism $f : X' \to X$ from an irreducible variety X', $f^{-1}(Y)$ is connected, and

b) $K(X_{/Y})$ is a field and $K(X)$ is algebraically closed in $K(X_{/Y})$.

The third section relates – via GAGA – the problem of extending formal-rational functions with the analogous problem in Complex Geometry of extending meromorphic functions defined in a small complex neighbourhood of a closed subvariety.

In the last section we make some remarks concerning the problem of extending formal-rational functions on projective homogeneous spaces. After recalling a fundamental result of Chow [Ch2] (Theorem (4.5), (i)), we show that Theorem (2.5) becomes a lot more precise in the case of homogeneous varieties (see Theorem (4.5), (ii)). Then we apply these facts to recover a result of Hironaka–Matsumura [HM] (Corollary (4.6)) concerning abelian varieties. Among the other applications of Chow's result we mention Proposition (4.8) which answers affirmatively a problem raised by Hartshorne in [H1] when the ambient variety is a homogeneous space. Finally, using a general observation (see Lemma (4.14)), we show that Chow's result implies a result of Faltings [F1], asserting that if Y is a "small-codimensional" subvariety of a rational projective homogeneous space X, then $K(X) = K(X_{/Y})$.

Acknowledgement. The first named author is grateful to the referee for supplying a list of small corrections (mostly of typographical nature) and for suggesting some improvements of the presentation.

2. Algebraicity criteria for formal-rational functions

(2.1). Fix a normal irreducible variety X and a closed connected subvariety Y, and let $\alpha : K(X) \to K(X_{/Y})$ be the canonical homomorphism of k-algebras (see [HM]). Then by [HM], lemma (1.4) the ring $K(X_{/Y})$ is actually a field, so that α makes $K(X_{/Y})$ a field extension of $K(X)$. We want to characterize those formal-rational functions $\zeta \in K(X_{/Y})$ that are algebraic over $\alpha(K(X))$, or simply over $K(X)$.

Let $f : X' \to X$ be a proper surjective morphism of irreducible varieties, and let Y be a closed subvariety of X. Then by [HM] there is a canonical morphism $\hat{f} : X'_{/f^{-1}(Y)} \to X_{/Y}$ of formal schemes which is compatible with f, and which induces a unique map of k-algebras $\hat{f}^* : K(X_{/Y}) \to K(X'_{/f^{-1}(Y)})$ making the following diagram

$$\begin{array}{ccc} K(X) & \xrightarrow{f^*} & K(X') \\ \alpha \downarrow & & \downarrow \alpha' \\ K(X_{/Y}) & \xrightarrow{\hat{f}^*} & K(X'_{/f^{-1}(Y)}) \end{array}$$

commutative. Assume now that Y' is a closed subvariety of X' such that $f(Y') \subseteq Y$. Then $Y' \subseteq f^{-1}(Y)$, whence there is the canonical flat (restriction) morphism $\psi : X'_{/Y'} \to X'_{/f^{-1}(Y)}$, which induces the canonical map $\psi^* : K(X'_{/f^{-1}(Y)}) \to K(X'_{/Y'})$. Therefore we get the map $\tilde{f} := \hat{f} \circ \psi : X'_{/Y'} \to X_{/Y}$ which induces the unique map of k-algebras $\tilde{f}^* := \psi^* \circ \hat{f}^* : K(X_{/Y}) \to K(X'_{/Y'})$ making the following diagram commutative

$$\begin{array}{ccc} K(X) & \xrightarrow{f^*} & K(X') \\ \alpha \downarrow & & \downarrow \alpha' \\ K(X_{/Y}) & \xrightarrow{\tilde{f}^*} & K(X'_{/Y'}) \end{array}$$

With these notations one has the following important result of Hironaka and Matsumura (see [HM], theorem (2.7)), whose proof is based essentially on Grothendieck's existence theorem (see [EGAIII], theorem (5.1.4)):

Theorem (2.2) (Hironaka–Matsumura). *Let $f : X' \to X$ be a proper surjective morphism of irreducible varieties. Then for every closed subvariety Y of X there* is a canonical isomorphism $K(X'_{/f^{-1}(Y)}) \cong [K(X_{/Y}) \otimes_{K(X)} K(X')]_0$, where $[A]_0$ denotes the total ring of fractions of a commutative unitary ring A.

Corollary (2.3) ([HM]). *Let X be an irreducible algebraic variety, and let Y be closed subvariety of X. Let $u : X' \to X$ be the (birational) normalization of X. Then $K(X_{/Y})$ is a field if and only if $u^{-1}(Y)$ is connected. If Y is a closed subvariety of an irreducible variety X then $K(X_{/Y})$ is a finite product of fields.*

Proof. By Theorem (2.2), $K(X'_{/u^{-1}(Y)}) \cong [K(X_{/Y}) \otimes_{K(X)} K(X')]_0 \cong K(X_{/Y})$ (since $K(X') \cong K(X)$). Hence by lemma (1.4) of [HM], $K(X'_{/u^{-1}(Y)})$ is a field, since $u^{-1}(Y)$ is connected. Conversely, if $u^{-1}(Y)$ is the disjoint union of two non-empty closed subsets Y_1 and Y_2 then clearly

$$K(X'_{/u^{-1}(Y)}) \cong K(X'_{/Y_1}) \times K(X'_{/Y_2}),$$

whence $K(X'_{/u^{-1}(Y)})$ cannot be a field.

For the last statement, by what we said above, we may replace X by its normalization X' and Y by $u^{-1}(Y)$. Therefore we may assume X normal. Let $Y_1,\ldots,Y_m$ be the connected components of Y. Then

$$K(X_{/Y}) \cong K(X_{/Y_1}) \times \cdots \times K(X_{/Y_m}),$$

and by the first part, $K(X_{/Y_i})$ is a field for every $i = 1, \ldots, m$. $\square$

Corollary (2.4). *Let Y be a closed subvariety of the irreducible variety X such that $K(X_{/Y})$ is a field. Then every finitely generated field subextension $E|K(X)$ of the field extension $K(X_{/Y})|K(X)$ is separable, i.e., there exists a transcendence basis $\{t_1, \ldots, t_r\}$ of E over $K(X)$ such that E is a finite separable field extension of $K(X)(t_1, \ldots, t_r)$.*

Proof. By [J], théorème 3.3, the statement is equivalent to the fact that for every purely inseparable finite extension L of $K(X)$, the ring $K(X_{/Y}) \otimes_{K(X)} L$ is reduced. To check this latter condition, let $f : X' \to X$ be the normalization of X in the field L (note that since L is a finite field extension of $K(X)$, L is finitely generated over k). Then f is a finite morphism and $K(X') = L$, so by Theorem (2.2) there is a canonical isomorphism

$$K(X'_{/f^{-1}(Y)}) \cong [K(X_{/Y}) \otimes_{K(X)} L]_0.$$

By Corollary (2.3), $K(X'_{/f^{-1}(Y)})$ is a finite product of fields, and in particular, is a reduced ring. It follows that $[K(X_{/Y}) \otimes_{K(X)} L]_0$ is reduced, whence the subring $K(X_{/Y}) \otimes_{K(X)} L$ is also reduced, as required. $\square$

Next we prove the following algebraicity criterion, which will play an important role in the sequel:

Theorem (2.5). *Let $\zeta \in K(X_{/Y})$ be a formal-rational function of an irreducible variety X along a closed subvariety Y of X such that $K(X_{/Y})$ is a field (e.g. if X is normal and Y is connected, see Corollary (2.3) above). Then the following three conditions are equivalent:*

(i) *ζ is algebraic over $K(X)$.*

(ii) *There is a proper surjective morphism $f : X' \to X$ from an irreducible variety X' and a closed subvariety Y' of X' such that $f(Y') \subseteq Y$ and $\tilde{f}^*(\zeta)$ is algebraic over $K(X')$.*

(iii) *There is a proper surjective morphism $f : X' \to X$ from an irreducible variety X' and a closed subvariety Y' of X' such that $f(Y') \subseteq Y$ and $\tilde{f}^*(\zeta) \in K(X')$.*

Consider also the following condition:

(iv) *There is a finite surjective morphism $f : X' \to X$, with X' a normal irreducible variety, and a closed subvariety Y' of X' such that $f(Y') \subseteq Y$, $f|Y' : Y' \to Y$ is an isomorphism and f is étale at every point of Y' (i.e. an étale neighbourhood of (X, Y), in the terminology of [Gi]), with the property that $\tilde{f}^*(\zeta) \in K(X')$.*

Then (iv) implies any of (i)–(iii). If moreover X is normal then these four conditions are all equivalent.

Proof. Let us first prove that (iii)$\Longrightarrow$(i). Assume that ζ is transcendental over $K(X)$. Since $K(X_{/Y})$ is a field, the map $\tilde{f}^*$ is injective. It follows that $\tilde{f}^*(\zeta)$ is transcendental over $K(X)$. If we set $\zeta' = \tilde{f}^*(\zeta)$, we get the purely transcendental field subextension $K(X) \subset E := K(X)(\zeta')$ of $K(X')$. Set $V := X \times \mathbb{P}^1$. Then V is identified with a variety whose field of rational functions is E such that the map $g^* : K(X) \to K(V) = E$ induced by the projection $g : V = X \times \mathbb{P}^1 \to X$ is just the subextension $K(X) \subset E$. Let $v : \tilde{X} \to X$ be the (birational) normalizarion morphism. By Corollary (2.3) the hypothesis that $K(X_{/Y})$ is a field is equivalent to the fact that $v^{-1}(Y)$ is connected. It follows that $v^{-1}(Y) \times \mathbb{P}^1$ is also connected. Since $v \times \mathrm{id}_{\mathbb{P}^1} : \tilde{X} \times \mathbb{P}^1 \to V = X \times \mathbb{P}^1$ is the normalization of V and $(v \times \mathrm{id}_{\mathbb{P}^1})^{-1}(g^{-1}(Y)) = v^{-1}(Y) \times \mathbb{P}^1$ is connected, Corollary (2.3) again implies that $K(V_{/g^{-1}(Y)})$ is a field.

On the other hand, by (iii), there is a rational function $t \in K(X')$ such that $\tilde{f}^*(\zeta) = \alpha'(t)$. Since $K(X_{/Y})$ is a field, the map $\tilde{f}^*$ is injective, and therefore ζ transcendental over $K(X)$ implies that t is also transcendental over $K(X)$. These observations show that the field extension $f^* : K(X) \to K(X')$ factors as $K(X) \subset K(X)(\zeta) = K(V) \subseteq K(X')$, where the second field extension is induced by the inclusion $K(X) \subset K(X')$ (via f^*) by mapping ζ to t (this makes sense because t is transcendental over $K(X)$). It follows that there exists a unique dominant rational map $h : X' \dashrightarrow V$ such that $f = g \circ h$ (so that the extension $K(V) \subseteq K(X')$ coincides with $h^* : K(V) \to K(X')$). Let X'' be the normalization of the closure (in $X' \times V$) of the graph of h, let $h' : X'' \to V$ be the corresponding projection onto V (which is a proper surjective morphism), and let $u : X'' \to X'$ be the projection onto X'. Clearly, u is a birational proper morphism. Therefore, if we denote by $Y'' := u^{-1}(Y')$, the induced maps $u^* : K(X') \to K(X'')$ and $\hat{u}^* : K(X'_{/Y'}) \to K(X''_{/Y''})$ are both isomorphisms. The fact that u^* is an isomorphism is obvious, while the fact that $\hat{u}^*$ is an isomorphism is an immediate consequence of Theorem (2.2) of Hironaka and Matsumura. So, without loss of generality we may assume that the rational dominant map h is a proper surjective morphism. In other words, we showed that we may assume that f factors as $f = g \circ h$, with $g : V \to X$ and $h : X' \to V$ proper surjective morphisms such that $K(V) = K(X)(\zeta')$, with $\zeta' \in K(X')$ transcendental over $K(X)$ (and of the form $\hat{f}^*(\zeta)$). Thus we get the following commutative diagram:

$$\begin{array}{ccccc} K(X) & \xrightarrow{g^*} & K(V) & \xrightarrow{h^*} & K(X') \\ \alpha \downarrow & & \downarrow \alpha'' & & \downarrow \alpha' \\ K(X_{/Y}) & \xrightarrow{\hat{g}^*} & K(V_{/g^{-1}(Y)}) & \xrightarrow{\tilde{h}^*} & K(X'_{/Y'}) \end{array}$$

At this point we can apply Theorem (2.2) above to deduce that $K(V_{/g^{-1}(Y)})$ is canonically isomorphic to $[K(X_{/Y}) \otimes_{K(X)} K(V)]_0$. In particular, $\hat{g}^*(\zeta)$ is transcendental over $K(V)$ (because ζ is transcendental over $K(X)$). Since we saw above that $K(V_{/g^{-1}(Y)})$ is a field, the map $\tilde{h}^*$ is injective. It follows that $\tilde{h}^*(\hat{g}^*(\zeta)) = \hat{f}^*(\zeta)$ is transcendental over $h^*(K(V))$. However, this is not possible because by the construction of V, $\hat{f}^*(\zeta) = \alpha'(t)$ and $t = h^*(\zeta)$. This shows that the assumption that ζ was transcendental over $K(X)$ is absurd, proving that (iii)$\Longrightarrow$(i).

Obviously, (iv)$\Longrightarrow$(iii) and (iii)$\Longrightarrow$(ii).

We prove now that (i)$\Longrightarrow$(iv) (if X is normal). Since ζ is algebraic over $K(X)$, $K(X) \subset K(X)(\zeta)$ is a finite field extension; in particular, $K(X)(\zeta)$ is a function field. Then take as $f : X' \to X$ the normalization of X in $K(X)(\zeta)$. By construction, $\zeta \in K(X')$. It remains to check that the immersion $Y \subset X$ lifts to X', and that, if Y' is the image of Y in X' then f is étale at each point of Y'. Although not stated explicitly, this is done in [Gi], theorem 4.3, whose proof works in our situation perfectly well. Indeed, the only change which has to be made in the proof of theorem 4.3 of [Gi] is to take $K' = K(X)(\zeta)$ (instead of $K' = K(X')$). So with this minor change Gieseker's proof of theorem 4.3 of [Gi] yields exactly what we need.

Note that the proof of the implication (i)$\Longrightarrow$(iv) yields the fact that the finite surjective morphism $f : X' \to X$ from (iv) can be chosen such that its degree equals to the degree of the minimal polynomial of ζ over $K(X)$.

Therefore the conditions (i), (iii) and (iv) are all equivalent (if X is normal), and (iii)$\Longrightarrow$(ii). However, if X is not normal and (i) holds, let $u : \tilde{X} \to X$ be the morphism of (birational) normalization. By Theorem (2.2), the birational morphism u yields the $K(X_{/Y}) \cong K(\tilde{X}_{/u^{-1}(Y)})$. By the implication (i)$\Longrightarrow$(iv) just proved (applied to $\zeta \in K(\tilde{X}_{/u^{-1}(Y)})$), there is a finite surjective morphism $g : X' \to \tilde{X}$ and a closed subvariety $Y' \subset X'$ such that $g(Y') \subseteq u^{-1}(Y)$ and $\tilde{g}^*(\zeta) \in K(X')$.

Then taking $f := u \circ g$, we get a finite surjective morphism $f : X' \to X$ such that $f(Y') \subseteq Y$ and $\tilde{f}^*(\zeta) \in K(X')$. In other words, we showed that (i)$\Longrightarrow$(iii) for any arbitrary irreducible (not necessarily normal) variety X. Therefore (i)$\Longleftrightarrow$(iii) and (iii)$\Longrightarrow$(ii).

It remains therefore to prove that (ii)$\Longrightarrow$(iii). By the remarks made at the beginning of the proof we may replace X' by its normalization, so that we may assume X' also normal. Then by the implication (i)$\Longrightarrow$(iv) (applied to X' and $\tilde{f}^*(\zeta)$), the fact that $\tilde{f}^*(\zeta)$ is algebraic over $K(X')$ yields the existence of a finite surjective morphism $u : Z \to X'$ and a closed subset W of Z such that $u(W) = Y'$ such that $\tilde{u}^*(\tilde{f}^*(\zeta)) \in K(Z)$. Then taking the proper surjective morphism $g := f \circ u : Z \to X$ we get $g(W) \subseteq Y$ and $\tilde{g}^*(\zeta) \in K(Z)$, whence (ii)$\Longrightarrow$(iii). $\square$

Remark (2.6). Let $x \in X$ be a normal point of the irreducible variety X. Denote by $\hat{\mathcal{O}}_{X,x}$ the completion of the local ring $\mathcal{O}_{X,x}$ with respect to its maximal ideal $m_{X,x}$. Then Theorem (2.5) asserts in this case that a formal-rational function $\zeta \in K(X_{/\{x\}}) = [\hat{\mathcal{O}}_{X,x}]_0$ is algebraic over $K(X)$ if and only if there exists a finite morphism $f : X' \to X$ and a point $x' \in X$ such that $f(x') = x$, at which f is étale (i.e. an étale neighbourhood (X', x') of (X, x)) such that $\hat{f}^*(\zeta) \in K(X')$. On the other hand, let $\mathcal{O}^h_{X,x}$ be the henselization of the local ring $\mathcal{O}_{X,x}$. By the definition of $\mathcal{O}^h_{X,x}$ we have $\mathcal{O}^h_{X,x} = \mathrm{dir\,lim}_{(X',x')} \mathcal{O}_{X',x'}$, where (X', x') runs over the set of all étale neighbourhoods of (X, x). Therefore Theorem (2.5) implies the following well known facts:

$$\mathcal{O}^h_{X,x} = \{\zeta \in \hat{\mathcal{O}}_{X,x} \mid \zeta \text{ algebraic over } K(X)\}, \text{ and}$$

$$[\mathcal{O}^h_{X,x}]_0 = \{\zeta \in K(X'_{/\{x'\}}) \mid \zeta \text{ algebraic over } K(X)\}.$$

Using Theorem (2.5) we prove the following result (which was proved independently also by E. Szabó):

Theorem (2.7). *Let X be an irreducible variety, and let Y be a closed subvariety of X. The following two conditions are equivalent:*

(i) *For every proper surjective morphism $f : X' \to X$ from an irreducible variety X', $f^{-1}(Y)$ is connected.*

(ii) *$K(X_{/Y})$ is a field and $K(X)$ is algebraically closed in $K(X_{/Y})$.*

(iii) *$K(X_{/Y})$ is a field and the algebraic closure of $K(X)$ in $K(X_{/Y})$ is purely inseparable over $K(X)$ (i.e. the field extension $K(X_{/Y}) | K(X)$ is primary).*

Proof. Let $u : \tilde{X} \to X$ be the canonical morphism from the normalization of X in $K(X)$. Observe first that since u is birational and finite, by Theorem (2.2), $K(\tilde{X}) \cong K(X)$ and $K(\tilde{X}_{/u^{-1}(Y)}) \cong K(X_{/Y})$. Thus the conditions (ii) and (iii) are stable under passing to the normalization of X. On the other hand, if $f : X' \to X$ is an arbitrary proper surjective morphism from an irreducible variety X' and if $v : \tilde{X}' \to X'$ the normalization morphism of X' in $K(X')$, by the universal property of the normalization one gets a commutative diagram

$$\begin{array}{ccc} \tilde{X}' & \xrightarrow{v} & X' \\ \tilde{f} \downarrow & & \downarrow f \\ \tilde{X} & \xrightarrow{u} & X \end{array}$$

where $\tilde{f} : \tilde{X}' \to \tilde{X}$ is also proper and surjective. From this it clearly follows that the condition (i) holds for the pair (X, Y) if it does for $(\tilde{X}, u^{-1}(Y))$. In other words, there is no loss of generality in assuming that X is normal.

Assume that (i) holds. Since X is normal and Y is connected (apply (i) to the identity morphism id_X), [HM], lemma (1.4) implies that $K(X_{/Y})$ is a field. On the other hand, if there would exist an algebraic formal-rational function $\zeta \in K(X_{/Y})$ over $K(X)$ which does not belong to $K(X)$ then by Theorem (2.5) there is a finite morphism $f : X' \to X$ of degree > 1 (equal to the degree of the minimal polynomial of ζ over $K(X)$) satisfying the condition (iv) of Theorem (2.5). Then we claim that Y' is a connected component of $f^{-1}(Y)$, and $Y' \neq f^{-1}(Y)$. Indeed, if we assume that $Y' = f^{-1}(Y)$ it would follow that $\deg(f) = 1$ because f is étale in an open neighbourhood U of Y' in X'. Thus $Y' \neq f^{-1}(Y)$. Write $f^{-1}(Y) = Y' \cup Y^*$, with $\emptyset \neq Y^* \not\subseteq Y'$. Since $Y' \subseteq U$, Y' is a section of $f|U$ and $f|U$ is étale, a simple general fact (see [SGA1], éxposé I, corollaire 5.3) implies that Y' is a connected component of $f^{-1}(Y)$, and since $\emptyset \neq Y^* \not\subseteq Y'$, it follows that $f^{-1}(Y)$ is not connected, contradicting (i).

Clearly (ii)$\Longrightarrow$(iii).

Now assume that (iii) holds true and let $f : X' \to X$ be an arbitrary proper surjective morphism. Using (iii) and a well known result (see [EGAIV], (4.3.1)), we infer that the ring $K(X_{/Y}) \otimes_{K(X)} K(X')$ has a unique minimal prime ideal p. Let S be the (multiplicative) set of all non-zero divisors of $K(X_{/Y}) \otimes_{K(X)} K(X')$. Then $S \cap p = \emptyset$ (in fact, S is just the complement of p in $K(X_{/Y}) \otimes_{K(X)} K(X')$), whence by Theorem (2.2) we have

$$K(X'_{/f^{-1}(Y)}) \cong [K(X_{/Y}) \otimes_{K(X)} K(X')]_0 = S^{-1}(K(X_{/Y}) \otimes_{K(X)} K(X')).$$

It follows that the ring $K(X'_{/f^{-1}(Y)})$ has also a unique minimal prime ideal (namely $pS^{-1}(K(X_{/Y}) \otimes_{K(X)} K(X'))$. If $f^{-1}(Y)$ would not be connected then $K(X'_{/f^{-1}(Y)})$ would be, by Corollary (2.3), isomorphic to the product of at least two fields. In particular, $K(X'_{/f^{-1}(Y)})$ would have more than one minimal prime ideal. Thus we proved that (iii)$\Longrightarrow$(i). □

To illustrate Theorem (2.7), recall Fulton–Hansen's connectedness theorem in a slight generalized form for weighted projective spaces (see [FH], or [FL1], and [B]):

Theorem (2.8) (Fulton–Hansen [FH]). *Let $f : X' \to \mathbb{P}^n(e) \times \mathbb{P}^n(e)$ be a proper morphism from the irreducible variety X' to the product of the weighted projective space $\mathbb{P}^n(e)$ of weights $e = (e_0, e_1, \ldots, e_n)$ by itself. If $\dim(f(X')) > n$ then $f^{-1}(\Delta)$ is connected, where Δ is the diagonal of $\mathbb{P}^n(e) \times \mathbb{P}^n(e)$.*

Then by Theorem (2.7), Theorem (2.8) can be restated in terms of formal-rational functions as:

Corollary (2.9). *Let X be a closed irreducible subvariety of $\mathbb{P}^n(e) \times \mathbb{P}^n(e)$ of dimension $> n$, and let Y denote the intersection $X \cap \Delta$. Then $K(X_{/Y})$ is a field and $K(X)$ is algebraically closed in $K(X_{/Y})$ (via the field extension $\alpha : K(X) \to K(X_{/Y})$).*

In fact, the following much stronger result holds:

Theorem (2.10) (Bădescu [B]). *In the hypotheses of Corollary (2.9), Y is G3 in X, i.e. the canonical map $\alpha : K(X) \to K(X_{/Y})$ is an isomorphism. In particular, for every proper morphism $f : X' \to \mathbb{P}^n(e) \times \mathbb{P}^n(e)$ from an irreducible variety X' with $\dim(f(X')) > n$, $f^{-1}(\Delta)$ is G3 in X'.*

3. Formal-rational functions and meromorphic functions

Throughout this section the ground field will be the field $\mathbb{C}$ of complex numbers.

3.1. Let X be a complex projective algebraic variety and let Y be a closed irreducible subvariety of X. For an arbitrary complex algebraic variety Z we shall denote by Z^{an} the corresponding analytic space. Then in the analytic category it makes sense to consider the formal completion $X^{\mathrm{an}}_{/Y^{\mathrm{an}}}$ (see e.g. [BSt]). By projectivity assumptions we can apply GAGA to get a canonical isomorphism

$$K(X_{/Y}) \cong K(X^{\mathrm{an}}_{/Y^{\mathrm{an}}}),$$

where the second ring is the ring of formal-rational functions in the analytic category (defined in a completely analogous way as in the algebraic setting). Let U be a connected neighbourhood of Y^{an} in X^{an}. Clearly $\mathcal{M}(U) \subseteq K(X^{\mathrm{an}}_{/Y^{\mathrm{an}}})$ and therefore we get a natural inclusion $\mathcal{M}(U) \subseteq K(X_{/Y})$.

The Gi conditions ($i = 1, 2, 3$) have the following analytic counterparts:

Definition (3.2). Let Y be a connected closed subvariety of a projective irreducible algebraic variety X over $\mathbb{C}$.

(i) Y is said to be *analytically $G1$* if for any connected complex neighbourhood U of Y in X we have $H^0(U, \mathcal{O}_U^{\mathrm{an}}) = \mathbb{C}$.

(ii) Y is said to be *analytically $G2$* if for any connected complex neighbourhood U of Y in X the restriction map $\mathcal{M}(X) \to \mathcal{M}(U)$ defines a finite field extension.

(iii) Y is said to be *analytically G3* if for any connected complex neighbourhood U of Y in X the restriction map $\mathcal{M}(X) \to \mathcal{M}(U)$ is an isomorphism.

As in the algebraic case "analytic Gi" implies "analytic $G(i-1)$" for $i = 2, 3$. On the other hand by our remarks above "algebraic Gi" implies "analytic Gi" for $i = 1, 2, 3$. In general the converse is not true. We illustrate this by the following example due to Serre.

Example (3.3) (Serre). There is a complex projective geometrically ruled surface $p : X \to B$ over an elliptic curve B and a curve Y in X, which is a section of p, with the following properties (see [Ha1], page 231, example 3.2):

(i) $Y^2 = 0$ (and in particular, $X \setminus Y$ is not affine).

(ii) $(X \setminus Y)^{\mathrm{an}}$ is analytically isomorphic to $\mathbb{C}^* \times \mathbb{C}^*$, and therefore is a Stein manifold.

By condition (i), Y cannot be $G2$, see [HM] or also [Ha1]. We are going to show however that Y is analytically $G3$ in X. In particular $K(X) = \mathcal{M}(U) \subsetneq K(X_{/Y})$.

More generally, let Y be a closed irreducible subvariety of a smooth complex projective variety X of dimension ≥ 2 such that $(X \setminus Y)^{\mathrm{an}}$ is a Stein manifold. Then it is well known that there exists a strictly plurisubharmonic exhaustion function $\varphi : (X \setminus Y)^{\mathrm{an}} \to \mathbb{R}$, whence for all $a \in \mathbb{R}$ the open subsets $\{\varphi < a\}$ are relatively compact in $(X \setminus Y)^{\mathrm{an}}$. The unions of Y^{an} and the complements of the adherence (in $(X \setminus Y)^{\mathrm{an}}$) of $\{\varphi < a\}$, for all $a \in \mathbb{R}$, form a fundamental system of strictly pseudoconcave neighbourhoods of Y^{an} in X^{an}. Then one can apply some of the well known results of Andreotti [A] to deduce that for every such connected pseudoconcave neighbourhood U of Y^{an} in X^{an}, $\mathcal{M}(X) = K(X) \to \mathcal{M}(U)$ is a finite algebraic field extension, i.e. Y is analytically $G2$ in X.

In fact one can say more. Since every $\zeta \in \mathcal{M}(U)$ can be extended to a meromorphic function to the whole X^{an} because on a Stein manifold Z of dimension ≥ 2 every meromorphic function defined in the complement of a compact set K such that $Z \setminus K$ is connected, can be extended to a unique meromorphic function on Z (see [KaSa]).

Thus, in the above situation one deduces that the canonical map $\mathcal{M}(X) = K(X) \to \mathcal{M}(U)$ is an isomorphism for every connected complex neighbourhood U of Y in X, i.e. Y is even analytically $G3$ in X.

3.4. We want to study the relationship between $\mathcal{M}(U)$ and $K(X_{/Y})$ more closely. Observe that the natural map $\alpha : K(X) \to K(X_{/Y})$ decomposes as $\alpha = \gamma \circ \beta$, where

$$\beta : K(X) = \mathcal{M}(X) \to \varinjlim_U \mathcal{M}(U)$$

is the canonical map into the direct limit, and

$$\gamma : \varinjlim_U \mathcal{M}(U) \to K(X_{/Y})$$

is induced by the inclusions $\mathcal{M}(U) \subseteq K(X_{/Y})$ (and where U runs over all connected complex neighbourhoods of Y in X). Clearly, the maps β and γ are both injective.

Proposition (3.5). *Let Y be a closed connected subvariety of the normal projective variety X over $\mathbb{C}$. In the notations of (3.4) one has*

$$\overline{K(X)} \subseteq \gamma(\operatorname*{dir\,lim}_{U} \mathcal{M}(U)),$$

where $\overline{K(X)}$ denotes the algebraic closure of $K(X)$ in $K(X_{/Y})$.

Proof. Since X is normal and Y is connected $K(X_{/Y})$ is a field by [HM], lemma (1.4). Let $\zeta \in K(X_{/Y})$ be an arbitrary formal-rational function which is algebraic over $K(X)$. By Theorem (2.5) there exists a finite surjective morphism $f : X' \to X$ and a closed subvariety Y' of X' with the following properties: $f(Y') \subseteq Y$, $f|Y' : Y' \to Y$ is an isomorphism, f is étale along Y', and $\tilde{f}^*(\zeta) \in K(X')$. In particular, for every connected complex open neighbourhood U' of Y' in X', $\tilde{f}^*(\zeta) \in \mathcal{M}(U')$. Since f is étale along Y' and $f|Y' : Y' \to Y$ is an isomorphism, for sufficiently small U' the restriction $f|U' : U' \to U := f(U')$ is an isomorphism. This implies that $\zeta \in \mathcal{M}(U)$. $\square$

Corollary (3.6). *In the hypotheses of Proposition (3.5), assume furthermore that Y is analytically G3 in X. Then $K(X)$ is algebraically closed in $K(X_{/Y})$. Moreover, for every proper surjective morphism $f : X' \to X$ from an irreducible variety X', $f^{-1}(Y)$ is connected.*

Proof. Indeed, since Y is analytically G3 in X, $K(X) = \mathcal{M}(U)$ for all U, whence $K(X) = \operatorname{dir\,lim}_U \mathcal{M}(U)$. The last part follows from the first one and from Theorem (2.7). $\square$

Corollary (3.7). *If X is the geometrically ruled surface over B, and Y is the curve on X given in Serre's example (see (3.3)), then $K(X)$ is algebraically closed in $K(X_{/Y})$ and $\mathcal{M}(U) \neq K(X_{/Y})$ for every connected complex open neighbourhood U of Y in X. Moreover, $f^{-1}(Y)$ is connected for every proper surjective morphism $f : X' \to X$ from an irreducible variety X'.*

Proof. Everything follows from Example (3.3) and Corollary (3.6). $\square$

4. Formal-rational functions on homogeneous spaces

Throughout this section k will be an algebraically closed field of characteristic zero.

Let X be a projective homogeneous space over k, acted on transitively by the connected algebraic group G, and let Y be a closed irreducible subvariety of X. According to Chow [Ch1] we shall use the following notations and terminology. Fix a point $p \in Y$ and consider the morphism $\varphi : G \to X$ defined by $\varphi(g) = g \cdot p$. Then the subset $G_{Y,p} := \varphi^{-1}(Y) = \{g \in G | g \cdot p \in Y\}$ is closed in G. Denote by G_Y the subgroup of G generated by $G_{Y,p}$. It is easily seen that G_Y depends only on Y, and not of the reference point $p \in Y$. A basic elementary fact is that G_Y is a closed subgroup of G (see [Ch1], or also (4.2) below for the case when G is a linear algebraic group).

Definition (4.1) (Chow [Ch1]). One says that the subvariety Y generates the homogeneous space X if $G_Y = G$, i.e. if $G_{Y,p}$ generates G (in the group-theoretical sense).

Let us first remark that if Y generates X then $g \cdot Y$ also generates X for every $g \in G$. Indeed, if we fix a point $p \in Y$ then $G_{g \cdot Y, g \cdot p} = g G_{Y,p} g^{-1}$ as is easily seen, whence $G_{g \cdot Y} = g G_Y g^{-1}$, i.e. the subgroups G_Y and $G_{g \cdot Y}$ are conjugate in G. In particular, $G_{Y,p}$ generates G if and only if $G_{g \cdot Y, g \cdot p}$ generates G.

4.2. From now on we shall fix a smooth point p of Y as the reference point. Since for all $x \in X$ the fiber $\varphi^{-1}(x)$ is isomorphic to the isotropy group G_x of x, we infer that the morphism $\varphi | G_{Y,p} : G_{Y,p} \to Y$ is smooth, and in particular is open. Moreover, $G_{Y,p}$ is irreducible if G acts on X by connected isotropy groups, because the fibers of $\varphi | G_{Y,p}$ are all irreducible, Y is irreducible and the morphism $\varphi | G_{Y,p}$ is open. This is the case when G is a linear algebraic group. Indeed, the isotropy group $P = G_p$ of the reference point p is a parabolic subgroup of G (since $X \cong G/G_p$ is projective), whence P contains a Borel subgroup B. If P^0 is the connected component of P containing the unit element e of G, then P^0 still contains B (since B is connected). In particular, P^0 is also a parabolic subgroup of G and P is contained in the normalizer $N_G(P^0)$ of P^0 in G. But by a theorem of Chevalley (see [Bo], theorem 11.15), $P^0 = N_G(P^0)$, whence $P = P^0$, i.e. P is connected. (Alternatively, the canonical morphism $\pi : G/P^0 \to G/P = X$ induced by the inclusion $P^0 \subseteq P$ is a finite étale morphism of degree equal to the index of P^0 in P. Since X is a rational variety, it is algebraically simply connected (see e.g. [SGA1]), whence π is an isomorphism, i.e. $P^0 = P$.)

In the case when G is an arbitrary connected algebraic group, $p = \varphi(e)$ is a smooth point of Y and the morphism $\varphi | G_{Y,p}$ is smooth, whence e is a smooth point of $G_{Y,p}$. In particular, there is a unique irreducible component, call it H, of $G_{Y,p}$ passing through the point e. By using general dimensionality arguments, Chow observes in [Ch1] that Y generates X if and only if H generates G (in the group-theoretical sense).

In the above notations, fix a natural number $i \geq 1$ and consider the morphism $\varphi_i : H^{2i} \to G$ (where $H^{2i} = H \times \cdots \times H$ ($2i$ times)) defined by

$$\varphi_i(g_1, \ldots, g_{2i}) = g_{2i}^{-1} g_{2i-1} \cdots g_t^{(-1)^{t+1}} \cdots g_2^{-1} g_1.$$

Since $\varphi_i(g_1, \ldots, g_{2i}) = \varphi_{i+1}(g_1, \ldots, g_{2i}, e, e)$, we get $\varphi_i(H^{2i}) \subseteq \varphi_{i+1}(H^{2(i+1)})$ for every $i \geq 1$. Moreover, $H \subseteq \varphi_1(H^2)$ because $g = \varphi_1(e, g)$ for all $g \in G$. From the definition of φ_i we get that $g^{-1} \in \varphi_i(H^{2i})$, for every $g \in \varphi_i(H^{2i})$, and

$\varphi_i(H^{2i}) \cdot \varphi_j(H^{2j}) \subseteq \varphi_{i+j}(H^{2(i+j)})$ for all $i, j \in \mathbb{N}$. It follows that the subgroup $\langle H \rangle$ generated by H coincides with $\bigcup_{i=1}^{\infty} \varphi_i(H^{2i})$.

On the other hand, the ascending sequence

$$\overline{\varphi_1(H^2)} \subseteq \overline{\varphi_2(H^4)} \subseteq \cdots \subseteq \overline{\varphi_i(H^{2i})} \subseteq \cdots$$

of closed irreducible subsets (where "overline" means taking closure) has to become stationary, i.e. there is an $n \geq 1$ such that $\overline{\varphi_n(H^{2n})} = \overline{\varphi_{n+p}(H^{2(n+p)})}$ for every $p \geq 1$. Set $F := \overline{\varphi_n(H^{2n})}$. Then F is the closure in G of the subgroup $\langle H \rangle$, and so F is a closed (connected) subgroup of G. Since by a general result of Chevalley $\varphi_n(H^{2n})$ contains a non-empty open subset U of F, we infer that

$$\langle H \rangle = \bigcup_{g \in \langle H \rangle} gU,$$

and in particular, $\langle H \rangle$ is an open subgroup of F, whence $\langle H \rangle = F$. In other words, we proved the well known fact that the subgroup $\langle H \rangle$ generated by a closed irreducible subvariety H passing through e is a closed connected group of G. Note that we also showed that there exists an $n \geq 1$ such that $\varphi_n(H^{2n})$ generates the subgroup $\langle H \rangle$ and the morphism $\varphi_n : H^{2n} \to \langle H \rangle$ is dominant.

Proposition (4.3). *Let X be a projective homogeneous space, and let Y be a closed irreducible subvariety of X. Then Y generates X if and only if Y intersects every irreducible hypersurface of X.*

Proof. This result was proved by Speiser in the case when X is an abelian variety (see [Sp]), and by Debarre in general in [De1], (2.1). The proof which follows is an adaptation of Speiser's proof to the general case. Assume that Y generates X and that there is an irreducible hypersurface D of X such that $Y \cap D = \emptyset$. If $\varphi : G \to X$ is the morphism $\varphi(g) = g \cdot p$ from the beginning of this section, set $D' := \varphi^{-1}(D)$.

Claim 1. For every $g \in G$ we have either $gG_{Y,p} \cap D' = \emptyset$, or $gG_{Y,p} \subseteq D'$.

Indeed, since $G_{Y,p} = \varphi^{-1}(Y)$ claim 1 is a consequence of the following statement: either $gY \cap D = \emptyset$, or $gY \subseteq D$. Assume that this statement does not hold, i.e. $gY \cap D \neq \emptyset$ and $gY \not\subseteq D$. It follows that $Y \cap g^{-1}D \neq \emptyset$ and $Y \not\subseteq g^{-1}D$. Hence the cycles $Y \cdot D = 0$ and $Y \cdot g^{-1}D > 0$ are numerically equivalent on Y, but this is impossible since Y is a projective variety. This proves claim 1.

Claim 2. For every $g, h \in G_{Y,p}$ one has $D'h^{-1}g \subseteq D'$.

To prove claim 2, let $g, h \in G_{Y,p}$ and $\delta \in D'$ be arbitrary elements. Then $\delta \in \delta h^{-1}G_{Y,p} \cap D'$ (because $h \in G_{Y,p}$), and so, by claim 1 we get $\delta h^{-1}G_{Y,p} \subseteq D'$. In particular, $\delta h^{-1}g \in D'$, and since δ was arbitrary in D', we get $D'h^{-1}g \subseteq D'$.

By applying claim 2 repeatedly we get that for any $n \geq 1$ and for every $g_1, g_2, \ldots, g_{2n} \in G_{Y,p}$ one has

$$D'g_{2n}^{-1}g_{2n-1} \cdots g_2^{-1}g_1 \subseteq D',$$

from which, using (4.2), we get $D' \cdot G_Y \subseteq D'$. So far we didn't use the hypothesis that Y generates X. This means that $G = G_Y$. Therefore $D' \cdot G \subseteq D'$, which is obviously impossible because D' is a hypersurface of G.

The converse is immediate. Indeed, if Y does not generate X, set $X = G/P$, with P the isotropy group of the reference point $p \in Y$, and denote by P' the (closed) subgroup of G_Y, and by $\pi : X = G/P \to X' := G/P'$ the canonical (surjective) morphism. Then $\pi(Y)$ is a point of X' and $\dim(X') \geq 1$ (because Y does not generate X), whence there exists an irreducible hypersurface H of X' such that $\pi(Y) \not\in H$. Then any component of $\pi^{-1}(H)$ is an irreducible hypersurface of X which does not meet Y. □

Corollary (4.4). *Let X be a projective homogeneous space, and let Y be a closed irreducible subvariety of X of dimension ≥ 2. Let $\mathcal{O}_X(1)$ be a very ample line bundle on X. If Y generates X then $Y' := Y \cap H$ also generates X, for every general $H \in |\mathcal{O}_X(1)|$.*

Proof. Let Z be an arbitrary hypersurface. By Proposition (4.3) it will be sufficient to show that $Y' \cap Z \neq \emptyset$. Since Y generates X, by Proposition (4.3), $Y \cap Z \neq \emptyset$, and since $\dim(Y) \geq 2$, $\dim(Y \cap Z) \geq 1$ (because the ambient space is smooth). This implies that $H \cap (Y \cap Z) \neq \emptyset$ (since H is a hyperplane section), whence $Y' \cap Z \neq \emptyset$. □

Theorem (4.5). (i) (Chow [Ch2]) *Let X be a projective homogeneous space over k, and let Y be a closed irreducible subvariety of X. Then the natural map $K(X) \to K(X_{/Y})$ is an algebraic field extension if and only if Y generates X in the sense of Definition (4.1).*

(ii) *In the hypotheses of (i), assume that Y generates X. Let $\zeta \in K(X_{/Y})$ be a formal-rational function which is algebraic of degree d over $K(X)$. Then there exists a finite étale morphism $f : X' \to X$ of degree d from a smooth irreducible projective variety X' such that: $\tilde{f}^*(\zeta) \in K(X')$, the irreducible components of $f^{-1}(Y)$ are mutually disjoint, numerically equivalent to each other as cycles on X' and are mapped by f isomorphically onto Y.*

Proof. The implication "Y generates X" $\Longrightarrow$ "the field extension $K(X) \to K(X_{/Y})$ is algebraic" of (i) is the difficult part of Theorem (4.5). Chow's proof given in [Ch2] follows the idea of the proof of the analytic counterpart of it (see [Ch1]) plus arguments involving non-archimedean function theory. He hope to come back in a future paper with a scheme-theoretic proof of this implication.

Note that the reverse implication of (i) is easy. Indeed, assume that $K(X) \to K(X_{/Y})$ is an algebraic field extension. Then Y must generate X. Assume not; since $X = G/P$ we can take as the reference point p just $e \mod P$. Let $Q := G_Y$ be the (closed) subgroup of G generated by $G_{Y,p}$. Since $Y \subseteq Q/P$, Y is mapped to a point by the canonical proper surjective morphism $f : X = G/P \to G/Q$. Since

$\dim(G/Q) \geq 1$, it follows from a general remark (see [HM], remark (2.10)) that the field extension $K(X) \to K(X_{/Y})$ is transcendental, a contradiction.

We proceed to the proof of (ii). We follow a beautifully simple idea of Chow [Ch1]. By Theorem (2.5), (iv) there exists a finite surjective morphism $f : X' \to X$, with X' normal and irreducible, a closed subvariety Y' of X' with the following properties: $f(Y') \subseteq Y$, the restriction $f|Y' : Y' \to Y$ is an isomorphism, f is étale in a neighbourhood of Y', and $\tilde{f}^*(\zeta) \in K(X')$. We will show that f is everywhere étale.

Let $\Delta' \subset X'$ be the ramification locus of f, and let $\Delta := f(\Delta')$ be the branch locus of f. Since f is étale along Y', $Y' \cap \Delta' = \emptyset$. Let u be the generic point of the group G, and let v be the generic point of Y. If $s : G \times X \to X$ is the morphism arising from the action of G on X, then by transitivity of this action we get that $s(u, v) = u \cdot v$ is the generic point of X. It follows that $u \cdot Y = s(\{u\} \times Y)$ is dense in X, whence $f^{-1}(u \cdot Y)$ is also dense in X', and in particular, $f^{-1}(u \cdot Y)$ is an irreducible scheme over $k(u) = K(G)$ (because X' is irreducible). Consider the incidence variety $Z := \{(gy, g) \mid g \in G, y \in Y\} \subset X \times G$, and let $\psi : Z \to X$ and $\pi : Z \to G$ be the morphisms defined by $\psi(gy, g) = gy$ (ψ is surjective because G acts transitively on X) and $\pi(gy, g) = g$. Clearly $\psi|\pi^{-1}(u)$ defines a $k(u)$-isomorphism between $\pi^{-1}(u)$ and $u \cdot Y$. It follows that the $k(u)$-scheme of finite type $f^{-1}(u \cdot Y)$ is irreducible. If $\overline{k(u)}$ is an algebraic closure of $k(u)$ we infer that the irreducible components of the $\overline{k(u)}$-scheme $(\pi \circ \varphi)^{-1}(u) \otimes_{k(u)} \overline{k(u)} \cong f^{-1}(u \cdot Y) \otimes_{k(u)} \overline{k(u)}$ are all conjugate over $k(u)$ under the natural action of the Galois group $\mathrm{Gal}(\overline{k(u)}|k(u))$. Thus we get the commutative diagram with cartesian squares

$$\begin{array}{ccccc}
f^{-1}(u \cdot Y) \otimes_{k(u)} \overline{k(u)} & \longrightarrow & (u \cdot Y) \otimes_{k(u)} \overline{k(u)} & \longrightarrow & \mathrm{Spec}(\overline{k(u)}) \\
\downarrow & & \downarrow & & \downarrow \\
Z' = X' \times_X Z & \xrightarrow{\varphi} & Z & \xrightarrow{\pi} & G \\
\rho \downarrow & & \downarrow \psi & & \\
X' & \xrightarrow{f} & X & &
\end{array}$$

Clearly the ramification locus of φ is $\Delta_1 := \rho^{-1}(\Delta')$. On the other hand, by the proof of Theorem (2.7) we know that Y' is an isolated connected component of $f^{-1}(Y)$, which is isomorphic to Y. It follows that $(\pi \circ \varphi)^{-1}(e)$ has an isolated irreducible component (corresponding to Y') which is contained in the étale locus of φ. Since this component is a specialization of an isolated irreducible component T of the geometric generic fiber $f^{-1}(u \cdot Y) \otimes_{k(u)} \overline{k(u)}$, it follows that T does not intersect the ramification locus of the morphism $f^{-1}(u \cdot Y) \otimes_{k(u)} \overline{k(u)} \to (u \cdot Y) \otimes_{k(u)} \overline{k(u)}$. Since the irreducible components of $f^{-1}(u \cdot Y) \otimes_{k(u)} \overline{k(u)}$ are conjugate under the action of $\mathrm{Gal}(\overline{k(u)}|k(u))$, we infer that the morphism

$$f^{-1}(u \cdot Y) \otimes_{k(u)} \overline{k(u)} \longrightarrow (u \cdot Y) \otimes_{k(u)} \overline{k(u)}$$

is étale and every irreducible component of $f^{-1}(u \cdot Y) \otimes_{k(u)} \overline{k(u)}$ is isolated and maps isomorphically onto $(u \cdot Y) \otimes_{k(u)} \overline{k(u)}$. In particular, $(\pi \circ \varphi)^{-1}(u) \cap \Delta_1 = \emptyset$. It follows that $(\pi \circ \varphi)^{-1}(u) \cap \Delta_1 = \emptyset$, which implies that $u \notin \pi(\varphi(\Delta_1))$. Set $U := G \setminus \pi(\varphi(\Delta_1))$. Since the morphisms π and φ are proper, $\pi(\varphi(\Delta_1))$ is a proper closed subset of G, whence U is a non-empty open subset of G. It follows that for all $g \in U$, $f^{-1}(g \cdot Y) \cap \Delta' = \emptyset$. In other words, $(g \cdot Y) \cap \Delta = \emptyset$ for all $g \in U$.

Now, since Y generates X, $g \cdot Y$ also generates X, and since $(g \cdot Y) \cap \Delta = \emptyset$ for all $g \in U$, $\mathrm{codim}_X(\Delta) \geq 2$, by Proposition (4.3). Finally, by the purity of the branch locus (X is smooth), $\mathrm{codim}_X(\Delta) \geq 2$ implies $\Delta = \emptyset$, i.e. f is everywhere étale. In particular, since X is smooth and f étale, X' is also smooth.

Actually we got even more: the irreducible components of $f^{-1}(g \cdot Y)$ are mutually disjoint, all isomorphic to $g \cdot Y$, and numerically equivalent to each other for for all $g \in G$. □

As an application of Theorem (4.5) we recover the following result of Hironaka and Matsumura [HM], with a completely different proof:

Corollary (4.6) (Hironaka–Matsumura [HM]). *Let Y be an irreducible subvariety of an abelian variety X, let $i : Y \to \mathrm{Alb}(Y)$ be the canonical morphism into the strict Albanese variety $\mathrm{Alb}(Y)$ of Y in the sense of [HM], and let $f : \mathrm{Alb}(Y) \to X$ be the morphism induced by the inclusion $Y \to X$ arising from the universal property of strict Albanese varieties (f is a homomorphism of abelian varieties if we assume that Y passes through the origin 0_X of X). Then:*

(i) *Y generates X if and only if f is surjective.*

(ii) *The morphism i is a closed embedding and $i(Y)$ is $G3$ in $\mathrm{Alb}(Y)$.*

(iii) *If Y generates X then Y is $G2$ in X. Moreover, Y is $G3$ in X if and only if Y generates X and the canonical morphism f has connected fibers.*

Proof. Part (i) is trivial, as well as the fact that i is a closed embedding (because $f \circ i$ coincides with the inclusion $Y \subset X$). Since $i(Y)$ generates $A := \mathrm{Alb}(Y)$, by Theorem (4.5), the map $K(A) \to K(A_{/i(Y)})$ is an algebraic field extension. Assume that there is a formal-rational function $\zeta \in K(A_{/i(Y)}) \setminus K(A)$. Then by part (ii) of Theorem (4.5) there is a finite étale morphism $g : A' \to A$ such that $\tilde{g}^*(\zeta) \in K(A')$ and such that the embedding $i : Y \to A$ lifts to an embedding $i' : Y \to A'$. Then by a theorem of Serre–Lang (see [M], chapter IV, §18) there is a unique structure of abelian variety on A' such that g becomes an isogeny of abelian varieties. Then by the universal property of $\mathrm{Alb}(Y)$, there is a unique morphism $h : A \to A'$ such that $h \circ i = i'$. It follows that $g \circ h = \mathrm{id}_A$, which would disconnect A' if $\deg(g) > 1$. Thus $\deg(g) = 1$. Recalling that $\deg(g) = \deg_{K(A)}(\zeta)$ it follows that $\zeta \in K(A)$, i.e. $i(Y)$ is $G3$ in A. So, part (ii) is proved.

Part (iii) follows now from (ii). Indeed, look at the map $\tilde{f}^* : K(X_{/Y}) \to K(A_{/i(Y)})$. By (ii), $K(A_{/i(Y)}) = K(A)$, whence $\tilde{f}^*(K(X_{/Y})) \subseteq K(A)$. But the extension $f^* : K(X) \to K(A)$ is finitely generated, so that the subextension $K(X) \to K(X_{/Y})$ is also finitely generated. Being also algebraic, it follows that it is finite, i.e. Y is $G2$ in X. To prove the last statement of (iii), by Corollary (2.4) the extension $K(X) \subseteq K(X_{/Y})$ is finite and separable, whence there exists a formal-rational function $\zeta \in K(X_{/Y})$ such that $K(X_{/Y}) = K(X)(\zeta)$. Therefore by Theorem (4.5), (ii) there exists a finite étale morphism $u : X' \to X$ with the following properties: X' is a smooth irreducible projective variety, the inclusion $Y \subset X$ lifts to an inclusion $Y \subset X'$ and $\tilde{u}^*(\zeta) \in K(X)$. This implies that the image of Y in X' is $G3$ in X'. Since X is an abelian variety, X' is also an abelian variety and u is a isogeny of abelian varieties (by the theorem of Serre–Lang already used above). By the universal property of $\text{Alb}(Y)$ the morphism f factors as $\text{Alb}(Y) \to X' \to X$, and this yields immediately the conclusion. □

Corollary (4.7). *Let X be a projective homogeneous space over k, and let Y be a closed irreducible subvariety of X that generates X. Assume furthermore that one of the following two conditions is satisfied:*

(i) *X is a rational variety, or*

(ii) *$Y \cdot Y \neq 0$ (in the Chow ring of cycles modulo numerical equivalence).*

Then the natural map $K(X) \to K(X_{/Y})$ is an isomorphism, i.e. Y is $G3$ in X.

Proof. Since Y generates X, by Theorem (4.5), (i) the extension $K(X) \subseteq K(X_{/Y})$ is algebraic. Let $\zeta \in K(X_{/Y})$ be an arbitrary formal-rational function. Since ζ is algebraic over $K(X)$, by Theorem (4.5), (ii) we find a finite étale morphism $f : X' \to X$ of degree equal to $\deg_{K(X)}(\zeta)$, with X' normal, such that a closed subvariety Y' of X' with the following properties: $f(Y') \subseteq Y$, the restriction $f|Y' : Y' \to Y$ is an isomorphism, and $\tilde{f}^*(\zeta) \in K(X')$. Under hypothesis (i), X is algebraically simply connected (as any rational variety, see e.g. [SGA1]), hence f is an isomorphism. This implies that $\deg_{K(X)}(\zeta) = 1$, i.e. $\zeta \in K(X)$. Thus the map $\alpha : K(X) \to K(X_{/Y})$ is an isomorphism in case (i).

Assume now that we are in the hypothesis (ii), and assume that f is not an isomorphism. Since f is a finite étale morphism, it follows that $d := \deg(f) \geq 2$. Then $f^{-1}(Y)$ consists of n irreducible components $Y_1 = Y', Y_2, \ldots, Y_d$. By Theorem (4.5), (ii), Y_i is numerically equivalent to $Y_1 = Y'$, $i = 2, \ldots, d$, and these components do not meet each other. It follows that

$$Y_1 \cdot f^*(Y) = Y_1 \cdot (Y_1 + Y_2 + \cdots + Y_d) = dY_1 \cdot Y_2 = 0.$$

Then, by the projection formula, we get $0 = f_*(Y_1 \cdot f^*(Y)) = f_*(Y_1) \cdot Y = Y \cdot Y$, contradicting the hypothesis (ii). This completes the proof of Corollary (4.7). □

Corollary (4.7) is an algebraic analogue of an analytic result of Chow [Ch1]. To illustrate it let us recall the following problem raised by Hartshorne [H1], page 208, problem 4.8): let Y be a smooth closed subvariety of a smooth projective variety of dimension n such that $\dim(Y) \geq \frac{n}{2}$ and the normal bundle $N_{Y|X}$ is ample. Is then Y G3 in X?

The following proposition answers affirmatively this problem in the case when X is a projective homogeneous space.

Proposition (4.8). *Let Y be a smooth closed subvariety of a projective homogeneous space X of dimension n such that $\dim(Y) \geq \frac{n}{2}$ and the normal bundle $N_{Y|X}$ is ample. Then Y is G3 in X.*

Proof. Since the normal bundle $N_{Y|X}$ is ample, by a result of Hartshorne (see [Ha2], theorem (6.7)), Y is G2 in X. (Alternatively, the ampleness of $N_{Y|X}$ implies that Y generates X, so that, instead of using Hartshorne's result, one could use Chow's Theorem (4.5), (i) above to get that the field extension $K(X_{/Y})|K(X)$ is algebraic.) So, to prove G3, in view of Corollary (4.7), (ii) it will be sufficient to check that $Y \cdot Y \neq 0$. To do this, we first apply a result of Kleiman [K] (valid if $\mathrm{char}(k) = 0$) which says that for general $g \in G$ (with G the connected algebraic group acting transitively on X) either $Y \cap gY = \varnothing$, or Y and gY have a non-empty proper intersection. Clearly, we have to rule out the first possibility. Since $N_{Y|X}$ is ample and $\dim(Y) + \dim(gY) = 2\dim(Y) \geq n$ (by hypothesis), we can conclude that $Y \cap gY \neq \varnothing$ by applying a result of Lübke (see [Lü], or also [FL2] for a slightly more general result, with a different proof). □

For further consequences of Theorem (4.5) we need to recall the following:

Definition (4.9) (Ran-Debarre, see [R], and [De2]). Let Y be a d-dimensional irreducible subvariety ($d > 0$) of an n-dimensional abelian variety X, and let $s \geq 0$ be a non-negative integer. Then Y is said to be *s-geometrically nondegenerate* in X if for every abelian subvariety K of X the image under the canonical morphism $X \to X/K$ is either X/K, or has dimension $\geq d - s$. If Y is 0-geometrically nondegenerate in X one simply says that Y is geometrically nondegenerate ([R]).

An irreducible subvariety Y is s-geometrically nondegenerate precisely if for every closed subvariety Z of X such that $\dim(Y) + \dim(Z) \geq \dim(X) + s$ one has $Y \cap Z \neq \varnothing$ (see [De2], (1.5)). Examples of s-geometrically nondegenerate subvarieties are smooth irreducible (positive dimensional) subvarieties Y of an abelian variety X whose normal bundle $N_{Y|X}$ is s-ample in the sense of Sommese [So], see [De2], (1.3). Every irreducible (positive-dimensional) subvariety Y of a simple abelian variety X is geometrically nondegenerate.

Corollary (4.10). *Let Y be an irreducible d-dimensional subvariety of an n-dimensional abelian variety X, and assume that $d \geq \frac{n}{2}$ and that Y is $(d - [\frac{n}{2}])$-geometrically nondegenerate in X. Then Y is G3 in X. In particular, every closed irreducible subvariety of dimension $d \geq \frac{n}{2}$ of a simple n-dimensional abelian variety X is G3 in X.*

Proof. Since $d - [\frac{n}{2}] \leq d - 1$, the hypothesis implies that for every proper abelian subvariety K of X (i.e. $K \neq X$), the image of Y in X/K is at least one dimensional. It follows that Y cannot be contained in any proper abelian subvariety of X, hence Y generates X. Thus, by Corollary (4.6) (iii) of Hironaka–Matsumura, Y is $G2$ in X.

To get the result, it will be sufficient to check that condition (ii) of Corollary (4.7) is fulfilled. By [K], for general $x \in X$ the intersection of Y with the translate $t_x(Y) = Y + x$ is either empty, or smooth of dimension $2d - n$. In particular, Y and $t_x(Y)$ are in general position for $x \in X$ general. To show that $Y \cdot Y \neq 0$, it remains to show that for every $x \in X$, $Y \cap t_x(Y) \neq \emptyset$. This follows from the above mentioned intersection properties of s-geometrically nondegenerate subvarieties [De2], (1.5). $\square$

Note that the last part of Corollary (4.10) was first proved by Hironaka–Matsumura [HM], and in the analytic context, by Barth [Ba].

Corollary (4.11) (Sommese [So]). *Let Y be a positive dimensional connected subvariety of an abelian variety X. Assume that $\dim(Y) - \mathrm{codim}(Y) \geq m$, where m is the maximum dimension of a proper abelian subvariety of X. Then Y is $G3$ in X.*

Proof. The inequality $\dim(Y) - \mathrm{codim}(Y) \geq m$ implies that Y cannot be contained in any proper subvariety of X, whence Y generates X, and so, by Corollary (4.6) (iii), Y is $G2$ in X. As in the proof of Corollary (4.10), it will then be sufficient to show that $Y \cap t_x(Y) \neq \emptyset$, for all $x \in X$. Let K be an arbitrary proper abelian subvariety of X, and denote by $\varphi : X \to X/K$ the canonical morphism. Clearly, $\dim(\varphi(Y)) \geq \dim(Y) - \dim(K) \geq \dim(Y) - m$. It follows that Y is m-geometrically nondegenerate, and since $\dim(Y) + \dim(t_x(Y)) = 2\dim(Y) \geq \dim(X) + m$ (by hypothesis), by [De2], (1.5) it follows that $Y \cap t_x(Y) \neq \emptyset$ for all $x \in X$, as claimed. $\square$

Corollary (4.12) (Babakarian–Hironaka). *Let Y be an irreducible closed subvariety of positive dimension of a Grassmann variety X. Then Y is $G3$ in X.*

Proof. Corollary (4.12) is a result due to Babakarian–Hironaka [BH] proved in arbitrary characteristic. However, in characteristic zero it is a consequence of Chow's result (Theorem (4.5)). Indeed, if X is a Grassmann variety, then X is a homogeneous space $X = G/P$, with P a maximal parabolic subgroup of G, whence for every closed subvariety Y of positive dimension we have $G_Y = G$, i.e. Y generates X. Then the conclusion follows from Corollary (4.7), (i). $\square$

4.13. Before stating the last application of Theorem (4.5), let us recall Goldstein's notion of coampleness of a projective homogeneous space X, (see [Go]). For such a homogeneous space X the tangent bundle T_X is generated by its global sections. Consider the projective bundle $\mathbb{P}(T_X)$ associated to T_X. It follows that the tautological line bundle $\mathcal{O}_{\mathbb{P}(T_X)}(1)$ is also generated by its global sections. Consider then the morphism $\varphi : \mathbb{P}(T_X) \to \mathbb{P}^N$ associated to the complete linear system $|\mathcal{O}_{\mathbb{P}(T_X)}(1)|$. By definition the *ampleness* of X, denoted $\mathrm{amp}(X)$, is the maximum fiber dimension of the morphism φ. By definition the *coampleness* of X, denoted $\mathrm{ca}(X)$, is given by the formula $\mathrm{ca}(X) := \dim(X) - \mathrm{amp}(X)$. Goldstein studied the coampleness of a rational projective homogeneous space $X = G/P$ over $\mathbb{C}$, showing that for G a semi-simple complex Lie group (which we can always assume) $\mathrm{ca}(X) \geq r$, where r is the minimum of ranks of the simple factors of G. In particular, for a complex rational projective homogeneous space X, $\mathrm{ca}(X) \geq 1$. More precisely, Goldstein computes $\mathrm{ca}(X)$ explicitly in each of the cases corresponding to the Dynkin diagrams (see [Go]). For example, in the case $G = \mathrm{SL}_{n+1}(\mathbb{C})$ one has $\mathrm{ca}(X) = n$ for every homogeneous space of the form $X = G/P$, with P an arbitrary parabolic subgroup of G.

Next we will show that the closed subvarieties of homogeneous spaces with small codimension generate the ambient variety.

Lemma (4.14). *Let X be a projective homogeneous space and let Y be a closed irreducible subvariety of X. If $\mathrm{codim}_X(Y) < \mathrm{ca}(X)$ then Y generates X.*

Proof. The argument is completely formal. Let $X = G/P$. Indeed, assume that Y does not generate X. Taking $p = e \mod P$, let $Q := G_Y$ be the subgroup of G generated by $G_{Y,p}$. Q is a closed subgroup of G containing P. Moreover, Y is a closed subvariety of Q/P. Consider the canonical fibration $X = G/P \to G/Q$ with fiber Q/P. It follows that the normal bundle $N_{F|X}$ of $F := Q/P$ in $X = G/P$ is trivial, and since it is a quotient of $T_X|F$, we get

$$\mathrm{amp}(X) = \mathrm{amp}(T_X) \geq \mathrm{amp}(T_X|F) \geq \mathrm{amp}(N_{F|X}) = \dim(F).$$

It follows that $\mathrm{codim}_X(F) \geq \mathrm{ca}(X)$. Since Y is a subvariety of F, $\mathrm{codim}_X(Y) \geq \mathrm{codim}_X(F)$, whence $\mathrm{codim}_X(Y) \geq \mathrm{ca}(X)$, contradicting the hypothesis. $\square$

Corollary (4.15) (Faltings [F1]). *Let Y be a closed irreducible subvariety of a rational projective homogeneous space $X = G/P$ over $\mathbb{C}$. If $\mathrm{codim}_X(Y) < \mathrm{ca}(X)$ then Y is $G3$ in X.*

Proof. Combine Lemma (4.14) with Corollary (4.7), (i). $\square$

Following [BS] we shall give an independent proof of Corollary (4.15) (in the case when Y is smooth) as an application of the main result proved in loc. cit.

Proof. Since the normal bundle $N_{Y|X}$ is a quotient of $T_X|Y$, and since T_X is $(\dim(X) - \mathrm{ca}(X))$-ample, it follows that $N_{Y|X}$ is $(\dim(X) - \mathrm{ca}(X))$-ample. The inequality assumed in Corollary (4.15) is equivalent to $\dim(Y) - 1 \geq \dim(X) - \mathrm{ca}(X)$. It follows in particular that $N_{Y|X}$ is $(d-1)$-ample, with $d := \dim(Y)$. Therefore Y is $G2$ in X by theorem 1 in [BS]. This conclusion will avoid using the difficult part (i) of Chow's Theorem (4.5). To get that Y is $G3$ in X use the fact that Y is $G2$ and Corollary (4.7), (i). $\square$

Another application of Corollary (4.7) is the following:

Theorem (4.16) ([BS]). *If X is a homogeneous space over $\mathbb{C}$ as above, then the diagonal Δ_X of $X \times X$ is $G3$ in $X \times X$.*

Proof. The normal bundle of Δ_X in $X \times X$ is just the tangent bundle T_X of X. According to Goldstein [Go], $\mathrm{ca}(X) > 1$. In particular, T_X is $(d-1)$-ample, with $d := \dim(X)$. Therefore we can apply theorem 2 of [BS] to deduce that Δ_X is $G2$ in $X \times X$.

To get $G3$ from $G2$ in this case one proceeds exactly as in the proof of Corollary (4.15) (using Corollary (4.7), (i)), since $X \times X$ is a rational projective homogeneous space. □

Note that the argument to deduce $G3$ from $G2$ in Corollary (4.15) and Theorem (4.16), based on Corollary (4.7), (i), is a lot easier than the argument given in [BS] (which was based on Faltings' connectivity theorem for rational homogeneous spaces, see [F1]).

References

[A]   A. Andreotti, Théorèmes de dépendence algébrique sur les espaces pseudoconcaves, Bull. Soc. Math. France 91 (1963), 1–38.

[BH]   A. Babakarian and H. Hironaka, Formal functions over Grassmannians, Illinois J. Math. 26 (1982), 201–211.

[B]   L. Bădescu, Algebraic Barth-Lefschetz theorems, Nagoya Math. J. 142 (1996), 17–38.

[BS]   L. Bădescu and M. Schneider, A criterion for extending meromorphic functions, Math. Annalen 305 (1996), 393–402.

[BSt]   C. Bănică and O. Stănășilă, Algebraic Methods in the Global Theory of Complex Spaces, John Wiley, New York 1976.

[Ba]   W. Barth, Fortsetzung meromorpher Funktionen in Tori und komplex-projektiven Räumen, Invent. Math. 5 (1968), 42–62.

[Bo]   A. Borel, Linear algebraic groups, W. A. Benjamin, New York–Amsterdam 1969.

[Ch1]   W. L. Chow, On meromorphic maps of algebraic varieties, Ann. of Math. 89 (1969), 391–403.

[Ch2]   W. L. Chow, Formal functions on homogeneous spaces, Invent. Math. 86 (1986), 115–130.

[De1]   O. Debarre, Sur un théorème de connexité de Mumford pour les espaces homogènes, Manuscripta Math. 89 (1996), 407–425.

[De2]   O. Debarre, Fulton-Hansen and Barth-Lefschetz theorems for subvarieties of abelian varieties, J. Reine Angew. Math. 467 (1995), 187–197.

[F1]   G. Faltings, Formale Geometrie und homogene Räume, Invent. Math. 64 (1981), 123–165.

[F2]   G. Faltings, A contribution to the theory of formal meromorphic functions, Nagoya Math. J. 77 (1980), 99–106.

[FH]   W. Fulton and J. Hansen, A connectedness theorem for projective varieties with applications to intersections and singularities of mappings, Ann. of Math. 110 (1979), 159–166.

[FL1]   W. Fulton and R. Lazarsfeld, Connectivity and its applications in algebraic geometry, in: Lecture Notes in Math. 862, Springer-Verlag, Berlin–Heidelberg–New York 1981, 26–92.

[FL2]   W. Fulton and R., Lazarsfeld, Positivity and excess intersection, in: Enumerative Geometry and Classical Algebraic Geometry, Progr. Math. 24, Birkhäuser, Boston–Basel–Stuttgart 1982, 97–105.

[Gi]   D. Gieseker, On two theorems of Griffiths about embeddings with ample normal bundle, Amer. J. Math. 99 (1977), 1137–1150.

[Go]   N. Goldstein, Ampleness and connectedness in complex G/P, Trans. Amer. Math. Soc. 274 (1982), 361–373.

[Gr]   Ph. A. Griffiths, The extension problem in complex analysis II. Embeddings with positive normal bundle, Amer. J. Math. 88 (1965), 366–446.

[EGAI] A. Grothendieck and J. Dieudonné, Eléments de Géométrie Algébrique I, Springer-Verlag, Berlin–Heidelberg–New York 1971.

[EGAIII] A. Grothendieck and J. Dieudonné, Eléments de Géométrie Algébrique III, Inst. Hautes Études Sci. Publ. Math. 11 (1961).

[EGAIV] A. Grothendieck and J. Dieudonné, Eléments de Géométrie Algébrique IV (2), Inst. Hautes Études Sci. Publ. Math. 24 (1965).

[SGA1] A. Grothendieck, Revêtements étales et groupe fondamental, Lecture Notes in Math. 224, Springer-Verlag, Berlin 1971.

[SGA2] A. Grothendieck, Cohomologie locale des faisceaux cohérents et théorèmes de Lefschetz locaux et globaux, North Holland, Amsterdam 1968.

[Ha1]   R. Hartshorne, Ample Subvarieties of Algebraic Varieties, Lecture Notes in Math. 156, Springer-Verlag, Berlin 1970.

[Ha2]   R. Hartshorne, Cohomological dimension of algebraic varieties, Ann. of Math. 88 (1968), 403–450.

[H]   H. Hironaka, On some formal embeddings, Illinois J. Math. 12 (1968), 587–602.

[HM]   H. Hironaka and H. Matsumura, Formal functions and formal embeddings, J. Math. Soc. Japan 20 (1968), 52–82.

[J]   J.-P. Jouanolou, Théorèmes de Bertini et Applications, Birkhäuser, Boston–Basel–Stuttgart 1983.

[KaSa] J. Kajiwara and E. Sakai, Generalization of a theorem of Oka-Levi on meromorphic functions, Nagoya Math. J. 29 (1967), 75–84.

[K]   S. L. Kleiman, The transversality of a general translate, Compositio Math. 28 (1974), 287–297.

[Lü]   M. Lübke, Beweis einer Vermutung von Hartshorne für den Fall homogener Mannigfaltigkeiten, J. Reine Angew. Math. 316 (1980), 215–220.

[M]   D. Mumford, Abelian Varieties, Tata Inst. Fund. Res. Lect. Math. Bombay 1968.

[R]   Z. Ran, On subvarieties of abelian varieties, Invent. Math. 62 (1981), 459–479.

[Ro]   H. Rossi, Continuation of subvarieties of projective varieties, Amer. J. Math. 91 (1969), 565–575.

[RVdV]   R. Remmert and A. Van de Ven, Zur Funktionentheorie homogener komplexer Mannigfaltigkeiten, Topology 2 (1963), 137–157.

[GAGA]   J.-P. Serre, Géométrie algébrique et géométrie analytique, Ann. Inst. Fourier 6 (1955), 1–42.

[Se]   F. Severi, Alcune proprietà fondamentali dell'insieme dei punti singolari di una funzione analitica di più variabili, Mem. Accad. Ital. 3 (1932).

[So]   A. J. Sommese, Submanifolds of abelian varieties, Math. Ann. 233 (1978), 229–256.

[Sp]   R. Speiser, Cohomological dimension and abelian varieties, Amer. J. Math. 95 (1973), 1–34.

[Z]   O. Zariski, Theory and applications of holomorphic functions on algebraic varieties over arbitrary ground fields, Mem. Amer. Math. Soc. 5, Amer. Math. Soc., Providence, RI, 1951.

L. Bădescu, Institute of Mathematics of the Romanian Academy and Department of Mathematics, University of Bucharest, P.O. Box 1-764, 70700 Bucharest Romania
E-mail: lbadescu@stoilow.imar.ro

On algebraic 1-motives related to Hodge cycles

Luca Barbieri-Viale

Abstract. The goal of this paper is to introduce *Hodge 1-motives* of algebraic varieties and to state a corresponding cohomological Grothendieck–Hodge conjecture, generalizing the classical Hodge conjecture to arbitrarily singular proper schemes.

We also construct generalized cycle class maps from the (Quillen) $\mathcal{K}$-cohomology groups $H^{p+i}(\mathcal{K}_p)$ to the sub-quotiens $W_{2p}H^{2p+i}/W_{2p-2}$ given by the weight filtration. However, in general, the image of this cycle map (as well as the image of the canonical map from motivic cohomology) is strictly smaller than the rational part of the Hodge filtration F^p on H^{2p+i}.

2000 Mathematics Subject Classification: 14F42, 14C30, 32S35, 19E15

0.	Introduction	26
	0.1. A short survey of the subject	27
	0.2. An outline of the conjectural picture	27
	0.3. Towards Hodge mixed motives	30
1.	Filtrations on Chow groups	31
	1.1. Regular homomorphisms	31
	1.2. Extensions	32
2.	Hodge 1-motives	33
	2.1. Algebraic construction	34
	2.2. Analytic construction	35
	2.3. Hodge conjecture for singular varieties	38
3.	Local Hodge theory	40
	3.1. Infinite dimensional mixed Hodge structures	40
	3.2. Zariski sheaves of mixed Hodge structures	41
	3.3. Hodge structures and Zariski cohomology	43
	3.4. Local-to-global properties	45
4.	Edge maps	46
	4.1. Bloch–Ogus theory	46
	4.2. Coniveau filtration	49

Algebraic Geometry. A Volume in Memory of Paolo Francia
M. C. Beltrametti, F. Catanese, C. Ciliberto, A. Lanteri, C. Pedrini (Eds.) © Walter de Gruyter 2002

4.3. Exotic (1, 1)-classes   51
4.4. $\mathcal{K}$-cohomology and motivic cohomology   53
5. Examples   55
5.1. Bloch's example   55
5.2. Srinivas' example   56

0. Introduction

Let X be an algebraic $\mathbb{C}$-scheme. The singular cohomology groups $H^*(X, \mathbb{Z}(\cdot))$ carry a mixed Hodge structure, see [11, III]. Deligne theory of 1-motives (see [11, III]) is an algebraic framework in order to deal with *some* mixed Hodge structures extracted from $H^*(X, \mathbb{Z}(\cdot))$, i.e., those having non-zero Hodge numbers in the set $\{(0, 0), (0, -1), (-1, 0), (-1, -1)\}$. Therefore, these cohomological invariants of algebraic varieties would be algebraically defined as 1-motives over arbitrary base fields or schemes. Note that a general theory of mixed motives can be regarded as an algebraic framework in order to deal with *all* mixed Hodge structures $H^*(X, \mathbb{Z}(\cdot))$.

A 1-motive M over a scheme S is given by an S-homomorphism of group schemes

$$M = [L \xrightarrow{u} G]$$

where G is an extension of an abelian scheme A by a torus T over S, and the group scheme L is, locally for the étale topology on S, isomorphic to a given finitely-generated free abelian group. There are Hodge, De Rham and ℓ-adic realizations (see [11, III] and [12]).

If X is a smooth proper $\mathbb{C}$-scheme then $H^i(X, \mathbb{Z}(j))$ is pure of weight $i - 2j$. If $i = 2p$ is even a natural 1-motive would be given by the lattice of Hodge cycles in $H^{2p}(X, \mathbb{Z}(p))$, i.e., of those integral cohomology classes (modulo torsion) which are of type $(0, 0)$. Classical Hodge conjecture claims that (over $\mathbb{Q}$) such a 1-motive would be obtained from classes of algebraic cycles on X only. For $i = 2p + 1$ odd the 1-motive corresponding to $H^{2p+1}(X, \mathbb{Z}(p+1))$ is given by the abelian variety associated to the largest sub-Hodge structure whose types are $(-1, 0)$ or $(0, -1)$. Grothendieck–Hodge conjecture characterize (over $\mathbb{Q}$) this sub-Hodge structure as the coniveau $\geq p$ subspace, i.e., the abelian variety as the algebraic part of the intermediate jacobian.

Grothendieck–Hodge conjectures are concerned with the quest of an algebraic definition for the named 1-motives. In fact, the usual Hodge conjecture can be reformulated by saying that the Hodge realization of the algebraically defined $\mathbb{Q}$-vector space of codimension p algebraic cycles modulo numerical (or homological) equivalence is the 1-motivic part of $H^{2p}(X, \mathbb{Q}(p))$. Moreover, the 1-motivic part of $H^{2p+1}(X, \mathbb{Q}(p+1))$ would be the Hodge realization of the isogeny class of the universal regular quotient.

The main task of this paper is to define *Hodge 1-motives* of singular varieties and to state a corresponding cohomological Grothendieck–Hodge conjecture by dealing with their Hodge realizations.

0.1. A short survey of the subject

The classical Hodge conjecture along with a tantalizing overview can be found in [13]. Recall that Grothendieck corrected the general Hodge conjecture in [19]. The book of Lewis [27] is a very good compendium of methods and results.

Recall that Jannsen [23] formulated an homological version of the Hodge conjecture for singular varieties. Moreover, Bloch in a letter to Jannsen (see the Appendix A in [23] cf. Section 5), gave a counterexample to a naive cohomological Hodge conjecture for curves on a singular 3-fold. However, in the same letter, Bloch was guessing that the Hodge conjecture for divisors, i.e., $F^1 \cap H^2(X, \mathbb{Z})$ is generated by c_1 of line bundles on X, holds true in the singular setting "because one has the exponential". Anyways, jointly with V. Srinivas, we gave a counterexample to this claim and questioned a reformulation of the Hodge conjecture for divisors in [3] by restricting to Zariski locally trivial cohomology classes, i.e., let $L^p H^*(X, \mathbb{Z})$ be the filtration induced by the Leray spectral sequence along the canonical continuous map $X_{\text{an}} \to X_{\text{Zar}}$, is $F^1 \cap L^1 H^2(X, \mathbb{Z})$ given by c_1 of line bundles on X? Still, this reformulation does not hold in general, e.g., see [7] where it is also proved for X normal.

From the work of Carlson (see [9] and [10]) and the theory of Albanese and Picard 1-motives [4] it now appears that the theory of 1-motives is a natural setting for a formulation of a cohomological version of the Hodge conjectures for singular varieties. For example, $F^1 \cap H^2(X, \mathbb{Z})$ is simply given by c_1 of simplicial line bundles on a smooth proper hypercovering $\pi : X_\bullet \to X$ via universal cohomological descent $\pi^* : H^2(X, \mathbb{Z}) \cong H^2(X_\bullet, \mathbb{Z})$. This Néron–Severi group NS $(X_\bullet) \cong F^1 \cap H^2(X, \mathbb{Z})$ is actually independent of the choice of the smooth simplicial scheme. Furthermore, NS $(X_\bullet)$ admits an algebraic definition as the quotient of the simplicial Picard group scheme $\mathbf{Pic}_{X_\bullet}$ by its connected component of the identity (cf. [4]). However, the 1-motivic part of $H^2(X, \mathbb{Z})$ is still larger than NS $(X_\bullet)$. Therefore the largest algebraic part of $H^2(X, \mathbb{Z})$ will be detected from a honest 1-motive only (see [2]).

Note that the rank of the usual NS (X) (= the image of Pic (X) in $H^2(X, \mathbb{Z})$) is actually smaller than NS $(X_\bullet)$, in general. Moreover $F^1 \cap W_0 H^2(X, \mathbb{Q}) = 0$ thus NS $(X_\bullet)_\mathbb{Q}$ is naturally a subspace of $H^2(X, \mathbb{Q})/W_0$.

Considering the Leray filtration $L^p H^{2p}(X, \mathbb{Q})$ a natural question formulated in [7] is if $F^p \cap L^p H^{2p}(X, \mathbb{Q})$ will be given by higher Chern classes. However, this would not be true without some extra hypothesis on X and does not tell enough about the algebraic part of all $H^{2p}(X, \mathbb{Q})$.

0.2. An outline of the conjectural picture

Let X be a proper integral $\mathbb{C}$-scheme. Let $H \stackrel{\text{def}}{=} H^{2p+i}(X)$ be our mixed Hodge structure on $H^{2p+i}(X, \mathbb{Z})/(\text{torsion})$ for a fixed pair of integers $p \geq 0$ and $i \in \mathbb{Z}$. First remark that we always have an extension

$$0 \to \operatorname{gr}^W_{2p-1} H \to W_{2p} H / W_{2p-2} H \to \operatorname{gr}^W_{2p} H \to 0.$$

An extension always defines an extension class map

$$e^p : H_{\mathbb{Z}}^{p,p} \to J^p(H)$$

which is not, in general, a 1-motive. In fact, $J^p(H)$ is a complex torus which is not an abelian variety, in general. Recall that Carlson [9] studied abstract extensions of Hodge structures showing their geometric content for low-dimensional varieties. For higher dimensional schemes consider the largest abelian subvariety $A^p(H)$ of the torus $J^p(H)$. Denote $H^p(H)$ the group of Hodge cycles, that is the preimage in $H_{\mathbb{Z}}^{p,p}$ of $A^p(H)$ under the extension class map. Note that an example due to Srinivas shows that the group $H^p(H)$ of Hodge cycles in this sense can be strictly smaller than $H_{\mathbb{Z}}^{p,p}$ (see Section 5.2).

Define the *Hodge 1-motive* of the mixed Hodge structure H the so obtained 1-motive

$$e^p : H^p(H) \to A^p(H).$$

Conversely, Deligne's theory of 1-motives [11] grant us of a mixed Hodge structure H^h corresponding to this 1-motive (see Section 2.2 for details). According to Deligne's philosophy of 1-motives there should be an algebraically defined 1-motive whose Hodge realization is H^h. The algebraic definition (see Section 2.1) is predictable *via* Grothendieck–Hodge conjectures and Bloch–Beilinson motivic world as follows.

Assume X smooth. Consider the filtration F_a^* on the Chow group $CH^p(X)$ given by $F_a^0 = CH^p(X)$, $F_a^1 = CH^p(X)_{\mathrm{alg}}$ the subgroup of cycles algebraically equivalent to zero and $F_a^2 =$ the kernel of the Abel–Jacobi map. Thus the graded pieces are $\mathrm{gr}^0_{F_a} = NS^p(X)$, the Néron–Severi group, and $\mathrm{gr}^1_{F_a} = J_a^p(X) =$ the group of $\mathbb{C}$-points of an abelian subvariety of the intermediate jacobian. We then get an extension

$$0 \to J_a^p(X) \to CH^p(X)/F_a^2 \to NS^p(X) \to 0.$$

Note that Grothendieck–Hodge conjecture claims that $J_a^p(X) = A^p(H^{2p-1}(X))$ up to isogeny, *i.e.*, $J_a^p(X)$ is the largest abelian subvariety of the intermediate jacobian.

If X is not smooth then let $X_\bullet$ be a smooth proper simplicial scheme along with $\pi : X_\bullet \to X$, a universal cohomological descent morphism (*cf.* [18]). In zero characteristic, such $X_\bullet$ was firstly provided by the construction of hypercoverings in [11], then by that of cubical hyperresolutions in [16] where the dimensions of the components are bounded or by the method of hyperenvelopes given in [15]. In [14] such a simplicial scheme is provided in positive characteristics.

The above extension, given by the filtration F_a^* on the Chow groups of each component of the so obtained simplicial scheme $X_\bullet$, yields a short exact sequence of complexes. Let $(NS^p)^\bullet$ and $(J_a^p)^\bullet$ denote such complexes. By taking homology groups we then get boundary maps

$$\lambda_a^i : H^i((NS^p)^\bullet) \to H^{i+1}((J_a^p)^\bullet).$$

We conjecture that the boundary map λ_a^i behave well with respect to the extension class map e^p yielding a motivic cycle class map, *i.e.*, the following diagram

$$\begin{array}{ccc} H^i((\mathrm{NS}^p)^\bullet) & \xrightarrow{\lambda_a^i} & H^{i+1}((J_a^p)^\bullet) \\ \downarrow & & \downarrow \\ H^{2p+i}(X)^{p,p} & \xrightarrow{e^p} & J^p(H^{2p+i}(X)) \end{array}$$

commutes. Note that all maps in the square are canonically defined. We guess that the image 1-motive (up to isogeny!) is the above Hodge 1-motive of the mixed Hodge structure (see Conjecture 2.3.4 for a full statement). In fact, we may expect J_a^p would be obtained as the universal regular quotient of $\mathrm{CH}^p(X)_{\mathrm{alg}}$ and that the filtration F_a^* would be induced by the motivic filtration conjectured by Bloch, Murre and Beilinson. Accordingly we can sketch an algebraic definition of such Hodge 1-motives (see Section 2.1).

If X is singular one is then puzzled by the role of

$$\mathrm{Hom}_{\mathrm{MHS}}(\mathbb{Z}(-p), H^{2p+i}(X))$$

where the Hom is taken in the abelian category of mixed Hodge structures. That is the integral part of F^p (the Hodge filtration on $H^{2p+i}(X, \mathbb{C})$) which is contained in the kernel of the extension class map e^p above. Note that $F^p \cap W_{2p-2} H^{2p+i}(X, \mathbb{Q}) = 0$ here. In the smooth case, such a target is usually reached by algebraic cycles. In order to obtain cycle class maps we may use local higher Chern classes and edge maps in coniveau spectral sequences (see [8] and [1]).

In the singular case, we show that such edge maps can be recovered by weight arguments. In order to do this we define *Zariski sheaves* of mixed Hodge structures, obtaining *infinite dimensional* mixed Hodge structures on their cohomology (see Section 3). The main example is given by the Zariski sheaf $\mathcal{H}_X^*$ associated to the presheaf $U \subset X \mapsto H^*(U)$ of mixed Hodge structures. On the smooth simplicial scheme $X_\bullet$ we also have a simplicial sheaf $\mathcal{H}_{X_\bullet}^*$ of mixed Hodge structures. Since $\pi : X_\bullet \to X$ yields $H^*(X) \cong \mathbb{H}^*(X_\bullet)$ we then obtain a local-to-global spectral sequence

$$L_2^{p,q} = \mathbb{H}^p(X_\bullet, \mathcal{H}_{X_\bullet}^q) \Rightarrow H^{p+q}(X)$$

in the category of infinite dimensional mixed Hodge structures. The sheaf $\mathcal{H}_{X_\bullet}^q$ has weights $\leq 2q$ and the same holds for its cohomology. There is an edge map (see Section 4)

$$s\ell^{p+i} : \mathbb{H}^{p+i}(X_\bullet, \mathcal{H}_{X_\bullet}^p)/W_{2p-2} \to W_{2p} H^{2p+i}(X)/W_{2p-2}.$$

We expect that the image of $s\ell^{p+i}$ is the mixed Hodge structure $H^{2p+i}(X)^h$ corresponding to the Hodge 1-motive.

Moreover, consider $\mathcal{K}$-cohomology groups $\mathbb{H}^*(X_\bullet, \mathcal{K}_p)$ where $\mathcal{K}_p$ are the simplicial sheaves associated to Quillen's higher K-theory. Recall that there are local higher Chern classes

$$c_p : \mathcal{K}_p \to \mathcal{H}^p_{X_\bullet}(p)$$

for each $p \geq 0$. We thus obtain a generalized cycle class map

$$c\ell^{p+i} : \mathbb{H}^{p+i}(X_\bullet, \mathcal{K}_p)_\mathbb{Q} \to W_{2p}H^{2p+i}(X, \mathbb{Q})/W_{2p-2}.$$

However the image of $c\ell^{p+i}$ is not $F^p \cap H^{2p+i}(X, \mathbb{Q})$, i.e., the rational part of the Hodge filtration can be larger (see Section 5.1 where Bloch's counterexample is explained). The same applies to the canonical map

$$H^{2p+i}_m(X, \mathbb{Q}(p)) \to H^{2p+i}(X, \mathbb{Q}(p))$$

from motivic cohomology.

0.3. Towards Hodge mixed motives

Any reasonable theory of mixed motives would include the theory of 1-motives, i.e., it would be a fully faithful functor from the $\mathbb{Q}$-linear category of 1-motives to that of mixed motives. This is the case of the triangulated category of geometrical motives introduced by Voevodsky (see [39, 3.4], cf. [26] and [29]). Hanamura's construction (see [22] and [21]) does not apparently provide such a property as yet.

As remarked by Grothendieck [19, §2] and Deligne [13, §5] the Hodge conjecture yields nice properties of the Hodge realization of pure motives, i.e., the usual Hodge conjecture means that the Hodge realization functor is fully faithful. It would be interesting to investigate such a property in the mixed case, e.g., if this formulation of the Hodge conjecture provide such a property of mixed motives.

We remark that M. Saito recently observed (see [36, 2.5 (ii)] and [37]) that the canonical functor from arithmetic Hodge structures to mixed Hodge structures is not full. Even if the Hodge realization factors through arithmetic Hodge structures, this non-fullness does not imply the non-fullness of the Hodge realization of mixed motives (as noticed by M. Saito as well).

However, the first natural attempt to go further with Hodge mixed motives is to provide an intrinsic definition of such objects internally. In fact, since 1-motives provide mixed motives we may claim that such Hodge mixed motives exist and would be naturally defined over any field or base scheme.

Acknowledgements. Papers are written because many people co-operate with one another. Often the author is just one of a whole group of people who pitch in, and that was the case here. Gratitude is due therefore, firstly, to V. Srinivas: he has been exceedingly generous to me with his time, shared enthusiasm over the years and told quite a few tricks. A huge intellectual debt is due to P. Deligne who also provided me with advice that helped on several portions of the manuscript.

I am grateful also to S. Bloch, O. Gabber, H. Gillet, M. Hanamura, U. Jannsen, M. Levine, J. D. Lewis, J. Murre, A. Rosenschon, M. Saito, C. Soulé and V. Voevodsky for discussions on some matters treated herein.

Finally, I would like to thank Paolo Francia for his helpful insight and invaluable guidance in the vast field of algebraic geometry by dedicating this paper to his memory.

Note that this research was carried out with smooth efficiency thanks to several foundations. I would like to thank the Tata Institute of Fundamental Research and the Institut Henri Poincaré for their support and hospitality.

1. Filtrations on Chow groups

Following the general framework of mixed motives (*e.g.*, see [12], [26] and [24] for a full overview) we may expect the following picture for non-singular algebraic varieties over a field k (algebraically closed of characteristic zero for simplicity).

Let X be a smooth proper k-scheme. Bloch, Beilinson and Murre (*cf.* [24], [25] and [34]) conjectured the existence of a finite filtration F_m^* on Chow groups $\mathrm{CH}^p(X) \stackrel{\mathrm{def}}{=} \mathcal{Z}^p(X)/\equiv_{\mathrm{rat}}$ of codimension p cycles modulo rational equivalence such that

- $F_m^0 \mathrm{CH}^p(X) = \mathrm{CH}^p(X)$,

- $F_m^1 \mathrm{CH}^p(X)$ is given by $\mathrm{CH}^p(X)_{\mathrm{hom}}$, *i.e.*, by the subgroup of those codimension p cycles which are homologically equivalent to zero for some Weil cohomology theory,

- $F_m^* \mathrm{CH}^p(X)$ should be functorial and compatible with the intersection pairing, and

- this filtration should be motivic, *e.g.*, $\mathrm{gr}^l_{F_m} \mathrm{CH}^p(X)_{\mathbb{Q}}$ depends only on the Grothendieck motive $h^{2p-i}(X)$.

1.1. Regular homomorphisms

Consider the subgroup $\mathrm{CH}^p(X)_{\mathrm{alg}}$ of those cycles in $\mathrm{CH}^p(X)$ which are algebraically equivalent to zero, *i.e.*, $\mathrm{CH}^p(X)_{\mathrm{alg}} \stackrel{\mathrm{def}}{=} \ker(\mathrm{CH}^p(X) \to \mathrm{NS}^p(X))$. Denote $\mathrm{CH}^p(X)_{\mathrm{ab}}$ the subgroup of $\mathrm{CH}^p(X)_{\mathrm{alg}}$ of those cycles which are abelian equivalent to zero, *i.e.*, $\mathrm{CH}^p(X)_{\mathrm{ab}}$ is the intersection of all kernels of regular homomorphisms from $\mathrm{CH}^p(X)_{\mathrm{alg}}$ to abelian varieties.

Assume the existence of a universal regular homomorphism $\rho^p : \mathrm{CH}^p(X)_{\mathrm{alg}} \to A_{X/k}^p(k)$ to (the group of k-points) of an abelian variety $A_{X/k}^p$ defined over the base field k (cf. [30]). This is actually proved for $p = 1, 2, \dim(X)$ (see [32]). Note that $\mathrm{CH}^p(X)_{\mathrm{ab}}$ is then a divisible group (see [6]).

Thus $\mathrm{CH}^p(X)_{\mathrm{alg}} \subseteq \mathrm{CH}^p(X)_{\mathrm{hom}}$ and there would be an induced functorial filtration F_a^* on $\mathrm{CH}^p(X)$ such that

- $F_a^0 \mathrm{CH}^p(X) = \mathrm{CH}^p(X)$,

- $F_a^1 \mathrm{CH}^p(X) = \mathrm{CH}^p(X)_{\mathrm{alg}}$ is the subgroup of cycles algebraically equivalent to zero, and

- $F_a^2 \mathrm{CH}^p(X) = \mathrm{CH}^p(X)_{\mathrm{ab}}$, i.e., is the kernel of the universal regular homomorphism ρ^p defined above.

Note that the existence of the abelian variety $A_{X/k}^p$ is not explicitly mentioned in the context of mixed motives but is a rather natural property after the case $k = \mathbb{C}$.

For X smooth and proper over $\mathbb{C}$ one obtains that the motivic filtration is such that (i) $F_m^1 \mathrm{CH}^p(X) = \mathrm{CH}^p(X)_{\mathrm{hom}}$ is the subgroup of cycles whose cycle class in $H^{2p}(X, \mathbb{Z}(p))$ is zero, and (ii) $F_m^2 \mathrm{CH}^p(X)$ is contained in the kernel of the Abel–Jacobi map $\mathrm{CH}^p(X)_{\mathrm{hom}} \to J^p(X)$ and $F_m^2 \mathrm{CH}^p(X) \cap \mathrm{CH}^p(X)_{\mathrm{alg}}$ is the kernel of the Abel–Jacobi map $\mathrm{CH}^p(X)_{\mathrm{alg}} \to J^p(X)$. In this case, $\mathrm{CH}^p(X)_{\mathrm{ab}}$ will be the kernel of the Abel–Jacobi map $\mathrm{CH}^p(X)_{\mathrm{alg}} \to J^p(X)$, i.e., (iii) $A_{X/\mathbb{C}}^p$ is the algebraic part of the intermediate jacobian. Is well known that the image $J_a^p(X)$ of $\mathrm{CH}^p(X)_{\mathrm{alg}}$ into $J^p(X)$ yields a sub-torus of $J^p(X)$ which is an abelian variety: moreover, is known to be universal for $p = 1, 2, \dim(X)$ (see [32] and [33]).

In the following, for the sake of simplicity, the reader can indeed assume that $k = \mathbb{C}$ and (i)–(iii) are satisfied by the first two steps of the filtration F_m^i. In fact, for $k = \mathbb{C}$, S. Saito has obtained (up to torsion!) such a result (see [38, Prop. 2.1], cf. [28] and [25]). Moreover, in the following, the reader could also avoid reference to the motivic filtration by dealing with the first two steps of the "algebraic" filtration F_a^i defined above.

1.2. Extensions

Let X be smooth and proper over k. For our purposes just consider the following extension

$$0 \to \mathrm{gr}_{F_m}^1 \mathrm{CH}^p(X) \to \mathrm{CH}^p(X)/F_m^2 \to \mathrm{gr}_{F_m}^0 \mathrm{CH}^p(X) \to 0 \tag{1}$$

Note that $A_{X/k}^p(k)$ is contained in $\mathrm{gr}_{F_m}^1 \mathrm{CH}^p(X)$ (since $F_a^2 = F_m^2 \cap \mathrm{CH}^p(X)_{\mathrm{alg}}$) and $\mathrm{gr}_{F_m}^0 \mathrm{CH}^p(X)$ has finite rank.

For $p = 1$ this extension is the usual extension associated to the connected component of the identity of the Picard functor, i.e., $A_{X/k}^1 = \mathrm{Pic}_{X/k}^0$ and $\mathrm{gr}_{F_m}^0$ is the Néron–Severi of X. If $p = \dim X$ then F_m^1 will be the kernel of the degree map and F_m^2 is the Albanese kernel, i.e., $A_{X/k}^{\dim X}$ is the Albanese variety and $\mathrm{gr}_{F_m}^0 = \mathbb{Z}^{\oplus c}$ where c is the number of components of X.

However, if $1 < p < \dim X$ then $\mathrm{CH}^p(X)_{\mathrm{alg}} \neq \mathrm{CH}^p(X)_{\mathrm{hom}}$ in general. Let $\mathrm{Grif}^p(X)$ denote the quotient group, i.e., the Griffiths group of X. As $\mathrm{gr}^1_{F_a} \mathrm{CH}^p(X) = A^p_{X/k}(k)$ and $\mathrm{gr}^0_{F_a} \mathrm{CH}^p(X) = \mathrm{NS}^p(X)$ we then have a diagram with exact rows and columns

$$
\begin{array}{ccccccc}
& & & & 0 & & 0 \\
& & & & \downarrow & & \downarrow \\
& & 0 & \to & F_m^2/F_a^2 & \to & \mathrm{Grif}^p(X) \\
& & \downarrow & & \downarrow & & \downarrow \\
0 \to & A^p_{X/k}(k) & \to & \mathrm{CH}^p(X)/F_a^2 & \to & \mathrm{NS}^p(X) & \to 0 \quad (2)\\
& \downarrow & & \downarrow & & \downarrow & \\
0 \to & \mathrm{gr}^1_{F_m} \mathrm{CH}^p(X) & \to & \mathrm{CH}^p(X)/F_m^2 & \to & \mathrm{gr}^0_{F_m} \mathrm{CH}^p(X) & \to 0 \\
& & & & \downarrow & & \downarrow \\
& & & & 0 & & 0
\end{array}
$$

Note that these extensions still fail to be of the same kind of the Pic extension. However, considering the extension

$$0 \to \mathrm{CH}^p(X)_{\mathrm{alg}} \to \mathrm{CH}^p(X) \to \mathrm{NS}^p(X) \to 0$$

we may hope for a natural regular homomorphism to k-points of an abstract extension in the category of group schemes (locally of finite type over k)

$$0 \to A \to G \to N \to 0$$

where G is a commutative group scheme which is an extension of a discrete group of finite rank N, associated to the abelian group of codimension p cycles modulo numerical equivalence, by an abelian variety A, isogenous to the universal regular quotient (cf. [17]). If $k = \mathbb{C}$ it is easy to see that such extension $G(\mathbb{C})$ exists trascendentally.

2. Hodge 1-motives

Let k be a field, for simplicity, algebraically closed of characteristic zero. Consider the $\mathbb{Q}$-linear abelian category $1 - \mathrm{Mot}_k$ of 1-motives over k with rational coefficients (see [11] and [4]). Denote $M_\mathbb{Q}$ the isogeny class of a 1-motive $M = [L \to G]$. The category $1 - \mathrm{Mot}_k$ contains (as fully faithful abelian sub-categories) the tensor category of finite dimensional $\mathbb{Q}$-vector spaces as well as the semi-simple abelian category of isogeny classes of abelian varieties. The Hodge realization (see [11] and [4]) is a fully faithful functor

$$T_{\mathrm{Hodge}} : 1 - \mathrm{Mot}_\mathbb{C} \hookrightarrow \mathrm{MHS}, \qquad M_\mathbb{Q} \mapsto T_{\mathrm{Hodge}}(M_\mathbb{Q})$$

defining an equivalence of categories between $1 - \mathrm{Mot}_\mathbb{C}$ and the abelian sub-category of mixed $\mathbb{Q}$-Hodge structures of type $\{(0,0), (0,-1), (-1,0), (-1,-1)\}$ such that gr^W_{-1} is polarizable. Under this equivalence a $\mathbb{Q}$-vector space corresponds to a

$\mathbb{Q}$-Hodge structure purely of type $(0, 0)$ and an isogeny class of an abelian variety corresponds to a polarizable $\mathbb{Q}$-Hodge structure purely of type $\{(0, -1), (-1, 0)\}$.

2.1. Algebraic construction

Let X be a proper scheme over k. We perform such a construction for simplicial schemes $X_\bullet$ coming from universal cohomological descent morphisms $\pi : X_\bullet \to X$ (cf. [11], [16], [15], [22] and [14]).

Let $X_\bullet$ be such a proper smooth simplicial scheme over the base field k. By functoriality, the filtration $F_m^j \mathrm{CH}^p$ on each component X_i of $X_\bullet$ yields a complex

$$(F_m^j \mathrm{CH}^p)^\bullet : \cdots \to F_m^j \mathrm{CH}^p(X_{i-1}) \xrightarrow{\delta_{i-1}^*} F_m^j \mathrm{CH}^p(X_i) \xrightarrow{\delta_i^*} F_m^j \mathrm{CH}^p(X_{i+1}) \to \cdots$$

where δ_i^* is the alternating sum of the pullback along the face maps $\partial_i^k : X_{i+1} \to X_i$ for $0 \le k \le i+1$.

The complex of Chow groups $(\mathrm{CH}^p)^\bullet$, induced from the simplicial structure as above, is filtered by sub-complexes:

$$0 \subseteq (F_m^p \mathrm{CH}^p)^\bullet \subseteq \cdots \subseteq (F_m^1 \mathrm{CH}^p)^\bullet \subseteq (F_m^0 \mathrm{CH}^p)^\bullet = (\mathrm{CH}^p)^\bullet.$$

Define $F_m^*(\mathrm{CH}^p)^\bullet \stackrel{\text{def}}{=} (F_m^* \mathrm{CH}^p)^\bullet$.

The extension (1) given by the filtration $F_m^* \mathrm{CH}^p(X_i)$ on each component X_i of the simplicial scheme $X_\bullet$, for a fixed $p \ge 0$, yields the following short exact sequence of complexes

$$0 \to \mathrm{gr}^1_{F_m}(\mathrm{CH}^p)^\bullet \to (\mathrm{CH}^p)^\bullet / F_m^2 \to \mathrm{gr}^0_{F_m}(\mathrm{CH}^p)^\bullet \to 0 \qquad (3)$$

Note that $\mathrm{gr}^1_{F_m}(\mathrm{CH}^p)^i$ contains the group of k-points of the abelian variety $A^p_{X_i/k}$ and, moreover $\mathrm{gr}^0_{F_m}(\mathrm{CH}^p)^i_{\mathbb{Q}}$ is the finite dimensional vector space of codimension p cycles on X_i modulo homological (or numerical) equivalence.

From (3) we then get a long exact sequence of homology groups and, in particular, we obtain boundary maps

$$\lambda_m^i : H^i(\mathrm{gr}^0_{F_m}(\mathrm{CH}^p)^\bullet) \to H^{i+1}(\mathrm{gr}^1_{F_m}(\mathrm{CH}^p)^\bullet).$$

Denote $A^p_{X_\bullet/k}$ the complex of abelian varieties $A^p_{X_i/k}$. Since $A^p_{X_\bullet/k}(k)$ is a subcomplex of $\mathrm{gr}^1_{F_m}(\mathrm{CH}^p)^\bullet$ we then get induced (functorial) maps on homology groups

$$\theta^i : H^{i+1}(A^p_{X_\bullet/k}(k)) \to H^{i+1}(\mathrm{gr}^1_{F_m}(\mathrm{CH}^p)^\bullet).$$

Note that (3) is involved in a functorial diagram (2). The corresponding complex of Néron–Severi groups $(\mathrm{NS}^p)^\bullet$ yield boundary maps

$$\lambda_a^i : H^i((\mathrm{NS}^p)^\bullet) \to H^{i+1}(A^p_{X_\bullet/k}(k)). \qquad (4)$$

These maps fit into the following commutative square

$$\begin{array}{ccc} H^i(\mathrm{gr}^0_{F_m}(\mathrm{CH}^p)^\bullet) & \xrightarrow{\lambda^i_m} & H^{i+1}(\mathrm{gr}^1_{F_m}(\mathrm{CH}^p)^\bullet) \\ \uparrow & & \uparrow \theta^i \\ H^i((\mathrm{NS}^p)^\bullet) & \xrightarrow{\lambda^i_a} & H^{i+1}(A^p_{X_\bullet/k}(k)). \end{array}$$

Moreover, the kernel of θ^i is clearly equal to the image of the boundary map

$$\tau^i : H^i(\mathrm{gr}^1_{F_m}(\mathrm{CH}^p)^\bullet / A^p_{X_\bullet/k}(k)) \to H^{i+1}(A^p_{X_\bullet/k}(k)).$$

Problem 2.1.1. *Is the image of the connected component of the identity of $H^{i+1}(A^p_{X_\bullet/k})$ under θ^i an abelian variety, e.g., is $\tau^i = 0$ up to a finite group?*

This is clearly the case if $p = 1$ (see below for the case $k = \mathbb{C}$). For $k = \mathbb{C}$ this question is related to Griffiths Problem E in [17] asking for a description of the "invertible points" of the intermediate jacobians (also *cf.* Mumford–Griffiths Problem F in [17]).

Assume that the above question has a positive answer and denote $H^{i+1}(A^p_{X_\bullet/k})^\dagger$ the so obtained abelian variety. We then obtain an algebraically defined 1-motive as follows.

Let $H^i(\mathrm{gr}^0_{F_m}(\mathrm{CH}^p)^\bullet)^\dagger$ be the subgroup of those elements in $H^i(\mathrm{gr}^0_{F_m}(\mathrm{CH}^p)^\bullet)$ which are mapped to $H^{i+1}(A^p_{X_\bullet/k})^\dagger$ under the boundary map λ^i_m above.

Definition 2.1.2. Let $X_\bullet$ be such a smooth proper simplicial scheme over k. Denote

$$\Xi^{i,p} \stackrel{\mathrm{def}}{=} [H^i(\mathrm{gr}^0_{F_m}(\mathrm{CH}^p)^\bullet)^\dagger \xrightarrow{\lambda^i_m} H^{i+1}(A^p_{X_\bullet/k})^\dagger]_\mathbb{Q}$$

the isogeny 1-motive obtained from the construction above. Call $\Xi^{i,p}$ the *Hodge 1-motive* of the simplicial scheme.

We expect that $\Xi^{i,p}$ is independent of the choice of $\pi : X_\bullet \to X$. The main motivation for questioning the existence of such a purely algebraic construction is given by the following analytic counterpart.

2.2. Analytic construction

Let MHS be the abelian category of usual Deligne's mixed Hodge structures [11]. An object H of MHS is defined as a triple $H = (H_\mathbb{Z}, W, F)$ where $H_\mathbb{Z}$ is a finitely generated $\mathbb{Z}$-module, W is a finite increasing filtration on $H_\mathbb{Z} \otimes \mathbb{Q}$ and F is a finite decreasing filtration on $H_\mathbb{Z} \otimes \mathbb{C}$ such that W, F and $\bar{F}$ is a system of opposed filtrations.

Let $H \in \mathrm{MHS}$ be a torsion free mixed Hodge structure with positive weights. Let $W_* H$ denote the sub-structures defined by the intersections of the weight filtration and $H_\mathbb{Z}$. Let p be a fixed positive integer and assume that $\mathrm{gr}^W_{2p-1} H$ is polarizable.

Consider the following extension in the abelian category MHS

$$0 \to \operatorname{gr}^W_{2p-1} H \to \frac{W_{2p} H}{W_{2p-2} H} \to \operatorname{gr}^W_{2p} H \to 0 \tag{5}$$

Taking $\operatorname{Hom}(\mathbb{Z}(-p), -)$ we get the extension class map

$$e^p : \operatorname{Hom}(\mathbb{Z}(-p), \operatorname{gr}^W_{2p} H) \to \operatorname{Ext}(\mathbb{Z}(-p), \operatorname{gr}^W_{2p-1} H)$$

where $\operatorname{Hom}(\mathbb{Z}(-p), \operatorname{gr}^W_{2p} H) = H^{p,p}_\mathbb{Z}$ is the sub-structure of (p, p)-classes in $\operatorname{gr}^W_{2p} H$ and

$$\operatorname{Ext}(\mathbb{Z}(-p), \operatorname{gr}^W_{2p-1} H) \cong J^p(H) \stackrel{\text{def}}{=} \frac{\operatorname{gr}^W_{2p-1} H_\mathbb{C}}{F^p + \operatorname{gr}^W_{2p-1} H_\mathbb{Z}}$$

is a compact complex torus. Note that (cf. [9])

$$\operatorname{Ext}(H^{p,p}_\mathbb{Z}, \operatorname{gr}^W_{2p-1} H) \cong \operatorname{Hom}(H^{p,p}_\mathbb{Z}, J^p(H)).$$

Thus $e^p \in \operatorname{Hom}(H^{p,p}_\mathbb{Z}, J^p(H))$ corresponds to a unique extension class

$$0 \to \operatorname{gr}^W_{2p-1} H \to H^e \to H^{p,p}_\mathbb{Z} \to 0 \tag{6}$$

which is the pull-back extension associated to $H^{p,p}_\mathbb{Z} \hookrightarrow \operatorname{gr}^W_{2p} H$ and (5). Moreover, since we always have $\operatorname{gr}^W_{2p-1} H \cap F^p = 0$ then

$$F^p \cap H^e_\mathbb{C} = \ker(H^{p,p}_\mathbb{Z} \stackrel{e^p}{\to} J^p(H)).$$

Now, if $\operatorname{gr}^W_{2p-1} H$ is (polarizable) of level 1 then the torus $J^p(H)$ is an abelian variety and H^e is the Hodge realization of the 1-motive over $\mathbb{C}$ defined by the extension class map e^p above.

In general, let H' be the largest sub-structure of $W_{2p-1} H$ which is polarizable and purely of type $\{(p-1, p), (p, p-1)\}$ modulo $W_{2p-2} H$, i.e., if

$$H^{2p-1}_a \stackrel{\text{def}}{=} (H^{p-1,p} + H^{p,p-1})_\mathbb{Z}$$

is the polarizable sub-structure of $\operatorname{gr}^W_{2p-1} H$ of those elements which are purely of the above type, then H' is defined by the following pull-back extension

$$0 \to W_{2p-2} H \to H' \to H^{2p-1}_a \to 0,$$

along the canonical projection $W_{2p-1} H \twoheadrightarrow \operatorname{gr}^W_{2p-1} H$.

Let $H'' \subseteq W_{2p} H$ be defined by the following pull-back extension

$$0 \to W_{2p-1} H \to H'' \to H^{p,p}_\mathbb{Z} \to 0,$$

along the canonical projection $W_{2p} H \twoheadrightarrow \operatorname{gr}^W_{2p} H$. Thus, the extension (6) can be regarded as the push-out of such extension involving H'' along $W_{2p-1} H \twoheadrightarrow \operatorname{gr}^W_{2p-1} H$.

Namely, we obtain that
$$\frac{H''}{W_{2p-2}H} = H^e$$
fitting in the extension
$$0 \to \frac{H'}{W_{2p-2}H} \to H^e \to \frac{H''}{H'} \to 0. \tag{7}$$

Let
$$h^p : \mathrm{Hom}\left(\mathbb{Z}(-p), \frac{H''}{H'}\right) \to \mathrm{Ext}\left(\mathbb{Z}(-p), \frac{H'}{W_{2p-2}H}\right)$$
be the extension class map obtained from (7).

Proposition 2.2.1. *The map h^p above yields a 1-motive over $\mathbb{C}$ which is just the restriction of $e^p : H_{\mathbb{Z}}^{p,p} \to J^p(H)$ to the largest abelian subvariety in $J^p(H)$. In particular:* $\ker(h^p) = \ker(e^p)$.

Proof. Since we have that $H'/W_{2p-2}H = H_a^{2p-1}$ by construction we also get that $\mathrm{Ext}(\mathbb{Z}(-p), H_a^{2p-1})$ is the largest abelian sub-variety of $\mathrm{Ext}(\mathbb{Z}(-p), \mathrm{gr}_{2p-1}^W H) = J^p(H)$. Moreover, note that we also have induced extensions
$$0 \to \frac{H'}{W_{2p-2}H} \to \mathrm{gr}_{2p-1}^W H \to \frac{W_{2p-1}H}{H'} \to 0$$
and
$$0 \to \frac{W_{2p-1}H}{H'} \to \frac{H''}{H'} \to H_{\mathbb{Z}}^{p,p} \to 0$$
yielding, together with (6) and (7), the following commutative diagram with exact rows

$$\begin{array}{ccccccc}
\mathrm{Hom}\left(\mathbb{Z}(-p), \frac{W_{2p-1}H}{H'}\right) & \to & \mathrm{Hom}\left(\mathbb{Z}(-p), \frac{H''}{H'}\right) & \to & \mathrm{Hom}\left(\mathbb{Z}(-p), H_{\mathbb{Z}}^{p,p}\right) & \to & \cdots \\
\| & & \downarrow h^p & & \downarrow e^p & & \| \\
\mathrm{Hom}\left(\mathbb{Z}(-p), \frac{W_{2p-1}H}{H'}\right) & \to & \mathrm{Ext}\left(\mathbb{Z}(-p), \frac{H'}{W_{2p-2}H}\right) & \to & \mathrm{Ext}\left(\mathbb{Z}(-p), \mathrm{gr}_{2p-1}^W H\right) & \to & \cdots
\end{array}$$

where the last group on the right part of the diagram is $\mathrm{Ext}\left(\mathbb{Z}(-p), \frac{W_{2p-1}H}{H'}\right)$. Since $\mathrm{Hom}\left(\mathbb{Z}(-p), \frac{W_{2p-1}H}{H'}\right) = 0 = \mathrm{Hom}(\mathbb{Z}(-p), H_a^{2p-1})$ everything follows from a diagram chase. □

Definition 2.2.2. Let $A^p(H) \stackrel{\mathrm{def}}{=} \mathrm{Ext}(\mathbb{Z}(-p), H_a^{2p-1})$ denote the abelian part of the compact complex torus $J^p(H)$. Denote $H^p(H) \stackrel{\mathrm{def}}{=} \mathrm{Hom}(\mathbb{Z}(-p), H''/H')$ the group of Hodge cycles, i.e., the subgroup of $H_{\mathbb{Z}}^{p,p}$ mapped to $A^p(H)$ under the extension class map e^p. Define
$$H^h \stackrel{\mathrm{def}}{=} T_{\mathrm{Hodge}}([H^p(H) \stackrel{h^p}{\to} A^p(H)])$$

the mixed Hodge structure corresponding to the 1-motive defined from (7) above. Call this 1-motive the *Hodge 1-motive* of H.

We remark that the mixed Hodge structure H^h clearly corresponds to the following extension

$$0 \to H_a^{2p-1} \to H^h \to H^p(H) \to 0,$$

obtained by pulling back $H^p(H) = F^p \cap (H''/H')_{\mathbb{Z}}$ along the induced projection $H^e \twoheadrightarrow H''/H'$ in (7). In particular $H^h \subseteq H^e$ and $F^p \cap H_{\mathbb{Z}} \subseteq F^p \cap H_{\mathbb{Z}}^h = F^p \cap H_{\mathbb{Z}}^e$.

2.3. Hodge conjecture for singular varieties

Let X be a proper smooth $\mathbb{C}$-scheme. The coniveau or arithmetic filtration (*cf.* [19])

$$N^i H^j(X) \stackrel{\text{def}}{=} \ker(H^j(X) \to \varinjlim_{\text{codim}_X Z \geq i} H^j(X - Z))$$

yields a filtration by Hodge sub-structures of $H^j(X)$. We have that $N^i H^j(X)$ is of level $j - 2i$ and

$$N^i H^j(X)_{\mathbb{Q}} \subseteq H^j(X, \mathbb{Q}) \cap F^i H^j(X). \tag{8}$$

Conjecture 2.3.1 (Grothendieck–Hodge conjecture [19])**.** *The left hand side of* (8) *is the largest subspace of the right hand side, generating a subspace of $H^j(X, \mathbb{C})$ which is a sub-Hodge structure.*

Let X be a proper (integral) $\mathbb{C}$-scheme. Recall that the weight filtration on $H^*(X, \mathbb{Q})$ is given by the canonical spectral sequence of mixed $\mathbb{Q}$-Hodge structures

$$E_1^{s,t} = H^t(X_s) \Rightarrow \mathbb{H}^{s+t}(X_\bullet)$$

for any smooth and proper hypercovering $\pi : X_\bullet \to X$ and universal cohomological descent $H^*(X) \cong \mathbb{H}^*(X_\bullet)$ (see [11]). In fact, the spectral sequence degenerates at E_2. Denote $(H^t)^\bullet$ the complexes $E_1^{\cdot,t}$ of E_1-terms. We clearly have

$$H^i((H^t)^\bullet) = \mathrm{gr}_t^W H^{t+i}(X).$$

Consider the complexes $(N^l H^t)^\bullet$ induced from the coniveau filtration $N^l H^t(X_i)$ on the components X_i of the simplicial scheme $X_\bullet$. We then have a natural map of Hodge structures

$$\nu^{i,l} : H^i((N^l H^t)^\bullet) \to \mathrm{gr}_t^W H^{t+i}(X).$$

Note that the image of $\nu^{i,l}$ is contained in the subspace $\mathrm{gr}_t^W H^{t+i}(X, \mathbb{Q}) \cap F^l$.

Conjecture 2.3.2. *The image of $\nu^{i,l}$ is the largest subspace of $\mathrm{gr}_t^W H^{t+i}(X, \mathbb{Q}) \cap F^l$ which is a sub-Hodge structure of $\mathrm{gr}_t^W H^{t+i}(X)$.*

It is reasonable to expect that such a statement will follow from the original Grothendieck–Hodge conjecture and abstract Hodge theory.

Grothendieck–Hodge conjecture (for coniveau p and degrees $2p, 2p+1$) can be reformulated as follows (*cf.* Grothendieck's remark on motives in [19]). Consider $\mathrm{gr}^0_{F_m} \mathrm{CH}^p(X)$ and $A^{p+1}_{X/k} \subseteq \mathrm{gr}^1_{F_m} \mathrm{CH}^{p+1}(X)$ (for $k = \mathbb{C}$ this is the algebraic part of $J^{p+1}(X)$) as 1-motives with rational coefficients. The Hodge realization of these algebraically defined 1-motives is $N^p H^{2p}(X)$ and $N^p H^{2p+1}(X)$ respectively.

Conjecture 2.3.3. *Let X be smooth and proper over $\mathbb{C}$. Then*
$$T_{\mathrm{Hodge}}([\mathrm{gr}^0_{F_m} \mathrm{CH}^p(X) \to 0]_{\mathbb{Q}}) = H^{p,p}_{\mathbb{Q}}$$
and
$$T_{\mathrm{Hodge}}([0 \to A^{p+1}_{X/k}]_{\mathbb{Q}}) = (H^{p,p+1} + H^{p+1,p})_{\mathbb{Q}}.$$

Note that $\mathrm{gr}^0_{F_m} \mathrm{CH}^p(X)$ would be better defined as $Z^p(X)/\equiv_{\mathrm{num}}$, up to torsion.

Now apply to the mixed $\mathbb{Q}$-Hodge structure $H = H^{2p+i}(X)$ the construction performed in the previous section. For a fixed pair (i, p) of integers recall that (5) is an extension of $\mathrm{gr}^W_{2p} H^{2p+i}(X)$ by $\mathrm{gr}^W_{2p-1} H^{2p+i}(X)$, where:

$$H^{i+1}((H^{2p-1})^\bullet) = \mathrm{gr}^W_{2p-1} H^{2p+i}(X) = \frac{\ker(H^{2p-1}(X_{i+1}) \to H^{2p-1}(X_{i+2}))}{\mathrm{im}\,(H^{2p-1}(X_i) \to H^{2p-1}(X_{i+1}))}$$

and

$$H^i((H^{2p})^\bullet) = \mathrm{gr}^W_{2p} H^{2p+i}(X) = \frac{\ker(H^{2p}(X_i) \to H^{2p}(X_{i+1}))}{\mathrm{im}\,(H^{2p}(X_{i-1}) \to H^{2p}(X_i))}.$$

We then have that $J^p(H) = J^p(H^{i+1}((H^{2p-1})^\bullet)))$ is isogenous to $H^{i+1}((J^p)^\bullet))$ where $(J^p)^\bullet$ is the complex of jacobians $J^p(X_i)$ of the components X_i.

Consider the complex $A^p_{X_\bullet/\mathbb{C}}$ of abelian sub-varieties given by the algebraic part of intermediate jacobians. The complex $A^p_{X_\bullet/\mathbb{C}}(\mathbb{C})$ is a sub-complex of the complex of compact tori $(J^p)^\bullet$. Therefore, the induced maps

$$H^{i+1}(A^p_{X_\bullet/\mathbb{C}}(\mathbb{C})) \to J^p(H)$$

are holomorphic mappings (which factor through θ^i) and whose image is isogenous to an abelian sub-variety of the maximal abelian sub-variety $A^p(H)$ of $J^p(H)$.

Moreover, the homology of the complex of (p, p)-classes is mapped to $H^{p,p}_{\mathbb{Q}} = F^p \cap H^i((H^{2p})^\bullet)$. Thus, there are canonical maps

$$H^i((\mathrm{NS}^p)^\bullet) \to H^{p,p}_{\mathbb{Q}}$$

which factor through $H^i(\mathrm{gr}^0_{F_m}(\mathrm{CH}^p)^\bullet)$.

We expect that the Hodge 1-motive of the simplicial scheme $X_\bullet$ would be canonically isomorphic to the Hodge 1-motive of $H^{2p+i}(X)$.

Conjecture 2.3.4. *Let X be a proper $\mathbb{C}$-scheme and let $\pi : X_\bullet \to X$ be a smooth and proper hypercovering. Let $H^{2p+i}(X)$ denote Deligne's mixed $\mathbb{Q}$-Hodge structure on $H^{2p+i}(X,\mathbb{Q})$, i.e., obtained by the universal cohomological descent isomorphism $H^{2p+i}(X,\mathbb{Q}) \cong \mathbb{H}^{2p+i}(X_\bullet,\mathbb{Q})$.*

1. *The following square*

$$\begin{array}{ccc} H^i((\mathrm{NS}^p)^\bullet) & \stackrel{\lambda_a^i}{\to} & H^{i+1}(A^p_{X_\bullet/\mathbb{C}}(\mathbb{C})) \\ \downarrow & & \downarrow \\ H^{2p+i}(X)^{p,p} & \stackrel{e^p}{\to} & J^p(H^{2p+i}(X)) \end{array}$$

commutes, yielding a motivic "cycle class map".

2. *The image of the motivic "cycle class map" is the Hodge 1-motive of the $\mathbb{Q}$-Hodge structure $H^{2p+i}(X)$.*

3. *We have that*

$$T_{\mathrm{Hodge}}(\Xi^{i,p}) \cong H^{2p+i}(X)^h.$$

Remark 2.3.5. Note that if X is smooth and proper then $H^{2p+i}(X)$ is pure and $H^{2p+i}(X)^h \neq 0$ if and only if $i = 0, -1$ (p fixed). In this case, the above conjecture follows from the reformulation of Grothendieck–Hodge conjecture for $H^{2p}(X)$ and $H^{2p-1}(X)$.

3. Local Hodge theory

See [11] for notations, definitions and properties of mixed Hodge structures.

3.1. Infinite dimensional mixed Hodge structures

Let MHS denote the abelian category of usual Deligne's A-mixed Hodge structures [11], i.e., for A a noetherian subring of $\mathbb{R}$ such that $A \otimes \mathbb{Q}$ is a field, an object H of MHS is defined as a triple $H = (H_A, W, F)$ where H_A is a finitely generated A-module, W is a finite increasing filtration on $H_A \otimes \mathbb{Q}$ and F is a finite decreasing filtration on $H_A \otimes \mathbb{C}$ such that W, F and $\overline{F}$ is a system of opposed filtrations.

Definition 3.1.1. An ∞-mixed Hodge structure H is a triple (H_A, W, F) where H_A is any A-module, W is a finite increasing filtration on $H_A \otimes \mathbb{Q}$ and F is a finite decreasing filtration on $H_A \otimes \mathbb{C}$ such that W, F and $\overline{F}$ is a system of opposed filtrations.

Denote MHS$^\infty$ the category of ∞-mixed Hodge structures or "infinite dimensional" mixed Hodge structures, i.e., where the morphisms are those which are compatible with the filtrations.

The category MHS$^\infty$ is abelian and MHS is a fully faithful abelian subcategory of MHS$^\infty$. Note that similar categories of infinite dimensional mixed Hodge structures already appeared in the literature, see Hain [20] and Morgan [31]. For example the category of limit mixed Hodge structures MHS$^{\lim}$, i.e., whose objects and morphisms are obtained by formally add to MHS (small) filtered colimits of objects in MHS with colimit morphisms.

Consider the case $A = \mathbb{Q}$. In this case, in the category MHS$^\infty$ we have infinite products of those families of objects $\{H_i\}_{i \in I}$ such that the induced families of filtrations $\{W_i\}_{i \in I}$ and $\{F_i\}_{i \in I}$ are finite. Moreover such a (small) product of epimorphisms is an epimorphism.

For the sake of exposition we often call mixed Hodge structures the objects of MHS as well as those of MHS$^\infty$ (or MHS$^{\lim}$).

3.2. Zariski sheaves of mixed Hodge structures

Let X denote the (big or small) Zariski site on an algebraic $\mathbb{C}$-scheme. However, most of the results in this section are available for any topological space or Grothendieck site.

Denote $X_\bullet$ a simplicial object of the category of algebraic $\mathbb{C}$-schemes over X: recall that (see [11, 5.1.8]) simplicial sheaves on $X_\bullet$ can be regarded as objects of a Grothendieck topos with enough points.

Consider presheaves of mixed Hodge structures. Note that a presheaf of usual Deligne's A-mixed Hodge structures will have its stalks in MHS$^{\lim}$.

Consider those presheaves (resp. simplicial presheaves) of $\mathbb{Q}$-mixed Hodge structures on X (resp. on $X_\bullet$) such that the filtrations are finite as filtrations of subpresheaves on X. These presheaves can be sheafified to sheaves having finite filtrations and preserving the above conditions on the stalks.

Make the following working definition of sheaves (or simplicial sheaves) of mixed Hodge structures. Let A, $\mathbb{Q}$ and $\mathbb{C}$ denote as well the constant sheaves on X (or $X_\bullet$) associated to the ring A, the rationals and the complex numbers.

Definition 3.2.1. A (simplicial) sheaf $\mathcal{H}$ of A-mixed Hodge structures, or "A-mixed sheaf" for short, is given by the following set of datas:

i) a (simplicial) sheaf $\mathcal{H}_A$ of A-modules,

ii) a finite (exhaustive) increasing filtration $\mathcal{W}$ by $\mathbb{Q}$-subsheaves of $\mathcal{H}_\mathbb{Q} \stackrel{\text{def}}{=} \mathcal{H}_A \otimes \mathbb{Q}$,

iii) a finite (exhaustive) decreasing filtration $\mathcal{F}$ by $\mathbb{C}$-subsheaves of $\mathcal{H}_\mathbb{C} \stackrel{\text{def}}{=} \mathcal{H}_A \otimes \mathbb{C}$;

satisfying the condition that $\mathcal{W}$, $\mathcal{F}$ and $\overline{\mathcal{F}}$ is a system of opposed filtrations, i.e., we have that

$$\mathrm{gr}\,{}^p_{\mathcal{F}}\,\mathrm{gr}\,{}^q_{\overline{\mathcal{F}}}\,\mathrm{gr}\,{}^W_n(\mathcal{H}) = 0$$

for $p + q \neq n$.

There is a canonical decomposition
$$\mathrm{gr}_n^W(\mathcal{H}) = \bigoplus_{p+q=n} \mathcal{A}^{p,q}$$
where $\mathcal{A}^{p,q} \stackrel{\text{def}}{=} \mathcal{F}^p \cap \overline{\mathcal{F}}^q$ and conversely.

In the case of a simplicial sheaf assume that the filtrations are given by simplicial subsheaves, i.e., the simplicial structure should be compatible with the filtrations on the components. A simplicial A-mixed sheaf $\mathcal{H}_{X_\bullet}$ on the simplicial space $X_\bullet$ can be regarded (cf. [11, 5.1.6]) as a family of A-mixed sheaves $\mathcal{H}_{X_i}$ (on the components X_i) such that the simplicial structure is also compatible with the filtrations $\mathcal{W}_{X_i}$ and $\mathcal{F}_{X_i}$ of $\mathcal{H}_{X_i}$.

A morphism of A-mixed sheaves is a morphism of (simplicial) sheaves of A-modules which is compatible with the filtrations. Denote $\mathcal{MHS}_X$ and $\mathcal{MHS}_{X_\bullet}$ the corresponding categories.

In order to show the following Lemma one can just use Deligne's Theorem [11, 1.2.10].

Lemma 3.2.2. *The categories $\mathcal{MHS}_X$ and $\mathcal{MHS}_{X_\bullet}$ of $\mathbb{Q}$-mixed sheaves are abelian categories. The kernel (resp. the cokernel) of a morphism $\varphi : \mathcal{H} \to \mathcal{H}'$ has underlying $\mathbb{Q}$ and $\mathbb{C}$-vector spaces the kernels (resp. the cokernels) of $\varphi_\mathbb{Q}$ and $\varphi_\mathbb{C}$ with induced filtrations. Any morphism is strictly compatible with the filtrations. The functors gr_W and $\mathrm{gr}_\mathcal{F}$ are exacts.*

Note that if $X = \{\infty\}$ is the singleton then $\mathcal{MHS}_\infty$ is equal to MHS^∞. Examples of $\mathbb{Q}$-mixed sheaves are clearly given by constant sheaves associated to $\mathbb{Q}$-mixed Hodge structures, yielding a canonical fully faithful functor
$$\mathrm{MHS}^\infty \to \mathcal{MHS}_X.$$

Stalks of a $\mathbb{Q}$-mixed sheaf $\mathcal{H}$ are in MHS^∞, the filtrations being induced stalkwise. In fact, the condition on the filtrations given with any $\mathbb{Q}$-mixed sheaf is local, at any point of X. Skyscraper sheaves $x_*(H)$ associated to an object $H \in \mathrm{MHS}^\infty$ and a point x of X are in $\mathcal{MHS}_X$. There is a natural isomorphism
$$\mathrm{Hom}_{\mathrm{MHS}^\infty}(\mathcal{H}_x, H) \cong \mathrm{Hom}_{\mathcal{MHS}_X}(\mathcal{H}, x_*(H)).$$

More generally, a presheaf in MHS^∞, with finite filtrations presheaves, can be sheafified to an A-mixed sheaf, in a canonical way, by applying the usual sheafification process to the filtrations together with the presheaf.

Definition 3.2.3. Say that an A-mixed sheaf $\mathcal{H}$ is flasque if $\mathcal{H}_A$ is a flasque sheaf.

For a given $\mathbb{Q}$-mixed sheaf $\mathcal{H}$ we then dispose of a canonical flasque $\mathbb{Q}$-mixed sheaf
$$\prod_{x \in X} x_*(\mathcal{H}_x)$$
where the product is taken over a set of points of X.

3.3. Hodge structures and Zariski cohomology

We show that, if $\mathcal{H}$ is a $\mathbb{Q}$-mixed sheaf then there is a unique $\mathbb{Q}$-mixed Hodge structure on the sections such that $\Gamma(-, \text{gr}(\dagger)) = \text{gr}\,\Gamma(-, \dagger)$. In fact, the mixed Hodge structure on the $\mathbb{Q}$-vector space of (global) sections is such that the following

$$\Gamma(X, \mathcal{H}_{\mathbb{Q}}) \subset \prod_{x \in X} \mathcal{H}_x$$

is strictly compatible with the filtrations; in the same way, for a simplicial sheaf, the following

$$\Gamma(X_\bullet, \mathcal{H}_{\mathbb{Q},\bullet}) \subset \ker \prod_{x \in X_0} \mathcal{H}_x \to \prod_{x \in X_1} \mathcal{H}_x$$

is strictly compatible with the filtrations.

Proposition 3.3.1. *Let $\mathcal{H}_X \in \mathcal{MHS}_X$ and $\mathcal{H}_{X_\bullet} \in \mathcal{MHS}_{X_\bullet}$ as above. There are left exact functors*

$$\mathcal{H}_X \mapsto \Gamma(X, \mathcal{H}_X) \qquad \mathcal{MHS}_X \to \text{MHS}^\infty$$

and

$$\mathcal{H}_{X_\bullet} \mapsto \Gamma(X_\bullet, \mathcal{H}_{X_\bullet}) \qquad \mathcal{MHS}_{X_\bullet} \to \text{MHS}^\infty$$

These functors yield $\mathbb{Q}$-mixed Hodge structures on the usual cohomology, which we denote by $H^*(X, \mathcal{H}_X)$ and $H^*(X_\bullet, \mathcal{H}_{X_\bullet})$ respectively, such that if

$$0 \to \mathcal{H}' \to \mathcal{H} \to \mathcal{H}'' \to 0$$

is exact in $\mathcal{MHS}_X$, respectively in $\mathcal{MHS}_{X_\bullet}$, then

$$\cdots \to H^i(X, \mathcal{H}_X) \to H^i(X, \mathcal{H}''_X) \to H^{i+1}(X, \mathcal{H}'_X) \to \cdots$$

is exact in MHS^∞, and respectively for the cohomology of $X_\bullet$: moreover, in the latter case we have a spectral sequence

$$E_1^{p,q} = H^q(X_p, \mathcal{H}_{X_p}) \Rightarrow \mathbb{H}^{p+q}(X_\bullet, \mathcal{H}_{X_\bullet})$$

in the category MHS^∞.

Proof. In fact, there is an extension in $\mathcal{MHS}$

$$0 \to \mathcal{H} \to \prod_{x \in X} x_*(\mathcal{H}_x) \to \mathcal{Z}^1 \to 0$$

where $\mathcal{Z}^1$ has the quotient $\mathbb{Q}$-mixed structure; as usual, we then get another extension

$$0 \to \mathcal{Z}^1 \to \prod_{x \in X} x_*(\mathcal{Z}_x^1) \to \mathcal{Z}^2 \to 0$$

and so on. We therefore get a flasque resolution

$$\prod_{x \in X} x_*(\mathcal{H}_x) \to \prod_{x \in X} x_*(\mathbb{Z}_x^1) \to \prod_{x \in X} x_*(\mathbb{Z}_x^2) \to \cdots$$

in $\mathcal{MHS}$: the canonical $\mathbb{Q}$-mixed flasque resolution.

If $\mathcal{H}_X$ is flasque then $\Gamma(X, \mathcal{H}_X)$ has a canonical $\mathbb{Q}$-mixed Hodge structure as claimed above; note that the filtrations would be given by flasque sub-sheaves.

In general, by construction, the cohomology is the homology of the complex of sections in MHS$^\infty$. Thus $H^*(X, \mathcal{H}_X)$ has a canonical $\mathbb{Q}$-mixed Hodge structure. The same argument applies to the total complex of the double complex of flasque $\mathbb{Q}$-mixed sheaves given by the canonical resolutions on each component of a simplicial sheaf.

Refer to [SGA4] and [26, Chapter IV] for a construction of canonical Godement resolutions available on any site and compare [11, 1.4.11] for the existence of bifiltered resolutions. □

In particular, the mixed Hodge structure $H^*(X, \mathcal{H}_X)$ is such that $H^*(X, \mathcal{H}_X)_\mathbb{Q} = H^*(X, \mathcal{H}_\mathbb{Q})$, $WH^*(X, \mathcal{H}_X)_\mathbb{Q} = H^*(X, \mathcal{W}\mathcal{H}_\mathbb{Q})$ and $FH^*(X, \mathcal{H}_X)_\mathbb{C} = H^*(X, \mathcal{F}\mathcal{H}_\mathbb{C})$. There is a decomposition

$$\mathrm{gr}_n^W H^*(X, \mathcal{H}_X) = \bigoplus_{p+q=n} H^*(X, \mathcal{A}_X^{p,q}).$$

Remark 3.3.2. Note that any (non-canonical) $\mathbb{Q}$-mixed flasque resolution in $\mathcal{MHS}$ yields a bifiltered complex and a bifiltered quasi-isomorphism with the canonical resolution. Therefore, the so obtained ∞-mixed Hodge structure on the cohomology is unique up to isomorphism.

Considering complexes in $\mathcal{MHS}_X$ and $\mathcal{MHS}_{X_\bullet}$ we construct the derived categories of $\mathbb{Q}$-mixed sheaves $\mathcal{D}^*(\mathcal{MHS}_X)$ and $\mathcal{D}^*(\mathcal{MHS}_{X_\bullet})$ as usual, as well as $\mathcal{D}^*(\mathrm{MHS}^\infty)$. We have a total derived functor

$$R\Gamma(X, -) : \mathcal{D}^*(\mathcal{MHS}_X) \to \mathcal{D}^*(\mathrm{MHS}^\infty)$$

sending a complex of $\mathbb{Q}$-mixed sheaves to the total complex of sections of its canonical resolution. Moreover, if $f : X \to Y$ is a continuous map, we have a higher direct image $\mathbb{Q}$-mixed sheaf $R^q f_*(\mathcal{H}_X)$ on Y, which corresponds as well to an exact functor

$$Rf_* : \mathcal{D}^*(\mathcal{MHS}_X) \to \mathcal{D}^*(\mathcal{MHS}_Y).$$

There is an inverse image exact functor $f^* : \mathcal{D}^*(\mathcal{MHS}_Y) \to \mathcal{D}^*(\mathcal{MHS}_X)$. Moreover, Grothendieck's six standard operations can be obtained in the derived category of $\mathbb{Q}$-mixed sheaves.

3.4. Local-to-global properties

Let X be an algebraic $\mathbb{C}$-scheme and let X_{an} be the associated analytic space. For any Zariski open subset $U \subseteq X$ the corresponding integral cohomology $H^r(U_{\text{an}}, \mathbb{Z}(t))$ carries a mixed Hodge structure (see [11, 8.2]) such that the restriction maps $H^r(U_{\text{an}}, \mathbb{Z}(t)) \to H^r(V_{\text{an}}, \mathbb{Z}(t))$ for $V \subseteq U$ are maps of mixed Hodge structures. Thus the presheaf of mixed Hodge structures

$$U \mapsto H^r(U_{\text{an}}, \mathbb{Q}(t)) \qquad (9)$$

can be sheafified to a Zariski $\mathbb{Q}$-mixed sheaf. In fact, for a fixed r, the resulting non-zero Hodge numbers of $H^r(U_{\text{an}}, \mathbb{Q})$, for any U, are in the finite set $[0, r] \times [0, r]$ (see [11, 8.2.4]).

Definition 3.4.1. Denote $\mathcal{H}^r_X(\mathbb{Q}(t))$ the $\mathbb{Q}$-mixed sheaf obtained hereabove. For $X_\bullet$ a simplicial $\mathbb{C}$-scheme denote $\mathcal{H}^r_{X_\bullet}$ the simplicial $\mathbb{Q}$-mixed sheaf given by $\mathcal{H}^r_{X_p}$ on the component X_p.

If X has algebraic dimension n then all its Zariski open affines U do have dimension $\leq n$, thus $\mathcal{H}^r_X = 0$ for $r > n$.

Scholium 3.4.2. *The Zariski cohomology groups $H^*(X, \mathcal{H}^r_X)$ carry ∞-mixed Hodge structures. Possibly non-zero Hodge numbers of $H^*(X, \mathcal{H}^r_X)$ are in the finite set $[0, r] \times [0, r]$. The Zariski cohomology $\mathbb{H}^*(X_\bullet, \mathcal{H}^r_{X_\bullet})$ carries ∞-mixed Hodge structure and the canonical spectral sequence*

$$E_1^{p,q} = H^q(X_p, \mathcal{H}^r_{X_p}) \Rightarrow \mathbb{H}^{p+q}(X_\bullet, \mathcal{H}^r_{X_\bullet})$$

is in the category MHS^∞.

Let $\omega : X_{\text{an}} \to X_{\text{Zar}}$ be the continuous map of sites induced by the identity mapping. We then have a Leray spectral sequence

$$L_2^{q,r} = H^q(X_{\text{Zar}}, R^r\omega_*(\mathbb{Z})) \Rightarrow H^{q+r}(X_{\text{an}}, \mathbb{Z})$$

of abelian groups. Since $R^r\omega_*(\mathbb{Q}) \cong \mathcal{H}^r_X$, these sheaves can be regarded as $\mathbb{Q}$-mixed sheaves and their Zariski cohomology carry ∞-mixed Hodge structures as above.

For $X_\bullet$ a simplicial scheme we thus have $\omega_\bullet : (X_\bullet)_{\text{an}} \to (X_\bullet)_{\text{Zar}}$ as above and a Leray spectral sequence

$$L_2^{q,r} = \mathbb{H}^q(X_\bullet, R^r(\omega_\bullet)_*(\mathbb{Z})) \Rightarrow \mathbb{H}^{q+r}(X_\bullet, \mathbb{Z})$$

where $R^r(\omega_\bullet)_*(\mathbb{Z}) \cong \mathcal{H}^r_{X_\bullet}$.

Claim 3.4.3 (Local-to-global). *There are spectral sequences*

$$L_2^{q,r} = H^q(X, \mathcal{H}^r_X) \Rightarrow H^{q+r}(X_{\text{an}}, \mathbb{Q})$$

and
$$L_2^{q,r} = \mathbb{H}^q(X_\bullet, \mathcal{H}_{X_\bullet}^r) \Rightarrow \mathbb{H}^{q+r}(X_\bullet, \mathbb{Q})$$

in the category of ∞-mixed Hodge structures.

The proof of this compatibility result will appear elsewhere; however, for smooth $\mathbb{C}$-schemes and using (12) below, the compatibility follows from [35, Corollary 4.4].

4. Edge maps

Recall that the classical cycle class maps can be obtained *via* edge homomorphisms in the coniveau spectral sequence. This is a consequences of Bloch's formula [8]. Working simplicially we then construct certain cycle class maps for singular varieties *via* edge maps in the local-to-global spectral sequence. We first show that the results of [8] hold in the category of ∞-mixed Hodge structures.

4.1. Bloch–Ogus theory

From Deligne [11, 8.2.2 and 8.3.8] the cohomology groups $H_Z^*(X) (= H^*(X \bmod X - Z, \mathbb{Z})$ in Deligne's notation) carry a mixed Hodge structure fitting into long exact sequences

$$\cdots \to H_Z^j(X) \to H_T^j(X) \to H_{T-Z}^j(X - Z) \to H_Z^{j+1}(X) \to \cdots \quad (10)$$

for any pair $Z \subset T$ of closed subschemes of X.

Since classical Poincaré duality is compatible with the mixed Hodge structures involved, then the functors

$$Z \subseteq X \mapsto (H_Z^*(X), H_*(Z))$$

yield a Poincaré duality theory with supports (see [8] and [23]) in the abelian tensor category of mixed Hodge structures. Furthermore we have that the above theory is appropriate for algebraic cycles in the sense of [1].

Let X^p be the set of codimension p points in X. For $x \in X^p$ let

$$H_x^*(X) \stackrel{\text{def}}{=} \varinjlim_{U \subset X} H_{\overline{\{x\}} \cap U}^*(U).$$

Taking direct limits of (10) over pairs $Z \subset T$ filtered by codimension and applying the exact couple method to the resulting long exact sequence we obtain the coniveau spectral sequence

$$C_1^{p,q} = \coprod_{x \in X^p} H_x^{q+p}(X) \Rightarrow H^{p+q}(X)$$

in the abelian category MHS$^\infty$ (*cf.* [1]).

Consider X smooth over $\mathbb{C}$. By local purity, we have that $H_x^{q+p}(X, \mathbb{Z}(r)) \cong H^{q-p}(x, \mathbb{Z}(r-p))$ if x is a codimension p point in X, i.e., here we have set

$$H^*(x) \stackrel{\text{def}}{=} \varinjlim_{V \text{ open } \subset \overline{\{x\}}} H^*(V).$$

Sheafifying the (limit) sequences (10), we obtain the following exact sequences of $\mathbb{Q}$-mixed sheaves on X:

$$0 \to \mathcal{H}_{Z^p}^r \to \coprod_{x \in X^p} x_*(H^{r-2p}(x)) \to \mathcal{H}_{Z^{p+1}}^{r+1} \to 0 \qquad (11)$$

where $\mathcal{H}_{Z^p}^r$ is the $\mathbb{Q}$-mixed sheaf associated to the presheaf

$$U \mapsto \varinjlim_{\text{codim}_X Z \geq p} H_{Z \cap U}^r(U).$$

In fact, the claimed short exact sequences (11) are obtained *via* the "locally homologically effaceable" property (see [8, Claim p. 191]), i.e., the following map of sheaves on X

$$\mathcal{H}_{Z^{p+1}}^* \xrightarrow{\text{zero}} \mathcal{H}_{Z^p}^*$$

vanishes for all $p \geq 0$.

Proposition 4.1.1 (Arithmetic resolution). *Let $\mathcal{H}_X^q(\mathbb{Q}(t))$ be the $\mathbb{Q}$-mixed sheaf defined in (9). Assuming X smooth over $\mathbb{C}$ then*

$$0 \to \mathcal{H}^q(\mathbb{Q}(t)) \to \coprod_{x \in X^0} x_*(H^q(x)(t)) \to \coprod_{x \in X^1} x_*(H^{q-1}(x)(t-1)) \to \cdots \qquad (12)$$

$$\cdots \to \coprod_{x \subset X^{q-1}} x_*(H^1(x)(t-q+1)) \to \coprod_{x \in X^q} x_*(\mathbb{Q}(t-q)) \to 0$$

is a flasque resolution in the category $\mathcal{MHS}_X$. Therefore, the coniveau spectral sequence

$$C_2^{p,q} = H^p(X, \mathcal{H}^q(\mathbb{Q}(t))) \Rightarrow H^{p+q}(X, \mathbb{Q}(t)) \qquad (13)$$

is in the category MHS^∞.

Proof. Follows from construction as sketched above. In fact, all axioms stated in [8, Section 1] are verified in MHS and the results in [8, Sections 3–4] can be obtained in MHS^∞. □

In particular, consider the presheaf of vector spaces

$$U \mapsto F^i H^*(U) \qquad (\text{resp. } U \mapsto W_i H^*(U))$$

and the associated Zariski sheaves $\mathcal{F}^i \mathcal{H}^*$ (resp. $\mathcal{W}_i \mathcal{H}^*$) on X filtering the sheaves $\mathcal{H}^*(\mathbb{C})$ (resp. $\mathcal{H}^*(\mathbb{Q})$). These filtrations are defining the sheaf of mixed Hodge structures $\mathcal{H}_X^*$ above according to (9). From Lemma 3.2.2 (cf. [11, Theorems 1.2.10 and 2.3.5]) the functors F^i, W_i, gr_i^W and gr_F^i (any $i \in \mathbb{Z}$) from the category of $\mathbb{Q}$-mixed sheaves to that of ordinary sheaves are exact. Applying these functors to the arithmetic resolution (12) we obtain resolutions of $\mathcal{F}^i \mathcal{H}^*$ (resp. $\mathcal{W}_i \mathcal{H}^*$) as follows.

Scholium 4.1.2. *The arithmetic resolution* (12) *yields a bifiltered quasi-isomorphism*

$$(\mathcal{H}^*, \mathcal{F}^\dagger, \mathcal{W}_\sharp) \xrightarrow{\sim} (\coprod_{x \in X^\odot} x_* H^{*-\odot}(x), \coprod_{x \in X^\odot} x_* F^{\dagger - \odot}, \coprod_{x \in X^\odot} x_* W_{\sharp - 2\odot}),$$

i.e., there are flasque resolutions

$$0 \to \mathcal{F}^i \mathcal{H}^q \to \coprod_{x \in X^0} x_*(F^i H^q(x)) \to \coprod_{x \in X^1} x_*(F^{i-1} H^{q-1}(x)) \to \cdots$$

and

$$0 \to \mathcal{W}_j \mathcal{H}^q \to \coprod_{x \in X^0} x_*(W_j H^q(x)) \to \coprod_{x \in X^1} x_*(W_{j-2} H^{q-1}(x)) \to \cdots$$

as well as

$$0 \to \mathrm{gr}_F^i \mathrm{gr}_j^W \mathcal{H}^q(\mathbb{C}) \to \coprod_{x \in X^0} x_*(\mathrm{gr}_F^i \mathrm{gr}_j^W H^q(x)) \to \cdots$$

$$\cdots \to \coprod_{x \in X^q} x_*(\mathrm{gr}_F^{i-q} \mathrm{gr}_{j-2q}^W H^0(x)) \to 0.$$

Consider the twisted Poincaré duality theory $(F^n H^*, F^{-m} H_*)$ where the integers n and m play the role of twisting, *i.e.*, we have

$$F^{d-n} H_Z^{2d-k}(X) \cong F^{-n} H_k(Z)$$

for X smooth of dimension d. From the arithmetic resolution of $\mathcal{F}^i \mathcal{H}^q$ (see the Scholium 4.1.2) we obtain the following:

Scholium 4.1.3. *Assume X smooth and let i be a fixed integer. We then have a coniveau spectral sequence*

$$F^i C_2^{p,q} = H^p(X, \mathcal{F}^i \mathcal{H}^q) \Rightarrow F^i H^{p+q}(X) \tag{14}$$

where $H^p(X, \mathcal{F}^i \mathcal{H}^q) = 0$ if $q < \min(i, p)$.

Concerning the Zariski sheaves $\mathrm{gr}_F^i \mathcal{H}^q$ and $\overline{\mathcal{F}}^i$ we indeed obtain corresponding coniveau spectral sequences as above.

Note that the spectral sequence $F^i C_2$ can be obtained applying F^i to the coniveau spectral sequence (13).

Remark 4.1.4. i) Note that applying F^i (resp. W_i) to the long exact sequences (10), taking direct limits over pairs $Z \subset T$ filtered by codimension and sheafifying, we do obtain the claimed flasque resolutions of $\mathcal{F}^i \mathcal{H}^*$ and $\mathcal{W}_i \mathcal{H}^*$ without reference to the category of $\mathbb{Q}$-mixed sheaves.

ii) Observe that for X (equidimensional) of dimension d, the fundamental class η_X belongs to $W_{-2d} H_{2d}(X) \cap F^{-d} H_{2d}(X)$ so that "local purity" yields the shift by two for the weight filtration and the shift by one for the Hodge filtration. Therefore one has to keep care of Tate twists when dealing with arithmetic resolutions.

iii) For $x \in X^0$ we have $H^q(x) = \mathcal{H}^q_x$, and there is a natural projection in $\mathcal{MHS}_X$

$$\prod_{x \in X} x_*(\mathcal{H}^q_x) \to \coprod_{x \in X^0} x_*(H^q(x))$$

which is the identity on $\mathcal{H}^q$. The mixed Hodge structure induced by the arithmetic resolution on $H^*(X, \mathcal{H}^q)$ is not the canonical one (which is the one induced by the canonical $\mathbb{Q}$-mixed flasque resolution) but yields a mixed Hodge structure which is naturally isomorphic to the canonical one (being induced by a natural isomorphism in the derived category $\mathcal{D}^*(\mathcal{MHS}_X)$).

4.2. Coniveau filtration

Let X be a smooth $\mathbb{C}$-scheme. The coniveau filtration (cf. [19]) $N^i H^j(X)$ is a filtration by (mixed) sub-structures of $H^j(X)$. This filtration is clearly induced from the coniveau spectral sequence (13) via (10). Remark that from the coniveau spectral sequence (13)

$$\operatorname{gr}^{i-1}_N H^j(X) = C^{i-1,j-i+1}_\infty$$

which is a substructure of $H^{i-1}(X, \mathcal{H}^{j-i+1})$ for $i \leq 2$. In fact, from the arithmetic resolution we have that $C^{p,q}_2 = H^p(X, \mathcal{H}^q(t)) = 0$ for $p > q$.

Case $i = 1$. Let X be a proper smooth $\mathbb{C}$-scheme. Note that we have $N^1 H^j(X) = \ker(H^j(X) \to H^0(X, \mathcal{H}^j)) = \{$ Zariski locally trivial classes in $H^j(X) \}$. Thus

$$\frac{H^j(X, \mathbb{Q}) \cap F^1 H^j(X)}{N^1 H^j(X)} = \operatorname{gr}^0_N H^j(X) \cap F^1 \subseteq H^0(X, \mathcal{H}^j) \cap F^1.$$

We remark that $\mathcal{H}^j / \mathcal{F}^1$ is the constant sheaf associated to $H^j(X, \mathcal{O}_X)$. Thus

$$F^1 \cap H^0(X, \mathcal{H}^j) \cong \ker(H^0(X, \mathcal{H}^j) \to H^j(X, \mathcal{O}_X)).$$

If $j = 1$ then $H^1(X) = H^0(X, \mathcal{H}^1)$ from (13) and (8) is trivially an equality. If $j = 2$ then $F^1 \cap H^0(X, \mathcal{H}^2) = 0$ from the exponential sequence. But for $j = 3$ and X the threefold product of an elliptic curve with itself Grothendieck's argument in [19] yields a non-trivial element in $F^1 \cap H^0(X, \mathcal{H}^3)$.

Case $i = p$ and $j = 2p$. Let X be a smooth $\mathbb{C}$-scheme. If $j = 2p$ we then have $N^i H^{2p}(X)(t) = 0$ for $i > p$ and $N^p H^{2p}(X)(t) = C_\infty^{p,p}$. Moreover, from (13) there is an induced edge map

$$s\ell_0^p : H^p(X, \mathcal{H}_X^p(p)) \to H^{2p}(X)(p)$$

which is a map of ∞-mixed Hodge structures and whose image is $N^p H^{2p}(X)(p)$. This equal the image of the classical cycle class map $c\ell^p : \mathrm{CH}^p(X) \to H^{2p}(X)(p)$. In fact, by [8, 7.6], the cohomology group

$$H^p(X, \mathcal{H}_X^p(p)) \cong \mathrm{coker}(\coprod_{x \in X^{p-1}} H^1(x) \to \coprod_{x \in X^p} \mathbb{Z})$$

coincide with NS$^p(X)$, the group of algebraic cycles of codimension p in X modulo algebraic equivalence. Thus $c\ell^p$ factors through $s\ell_0^p$ and the canonical projection (see [1]).

Recall that $F^i H^p(X, \mathcal{H}^q) \stackrel{\mathrm{def}}{=} H^p(X, F^i \mathcal{H}^q) \hookrightarrow H^p(X, \mathcal{H}^q(\mathbb{C}))$ is injective and

$$H^p(X, F^p \mathcal{H}^p) \cong \mathrm{coker}(\coprod_{x \in X^{p-1}} F^1 H^1(x) \to \coprod_{x \in X^p} \mathbb{C})$$

whence the canonical map $H^p(X, F^p \mathcal{H}^p) \to \mathrm{NS}^p(X) \otimes \mathbb{C}$ is also surjective. As an immediate consequence of this fact, e.g., from the coniveau spectral sequence (14), we get the following.

Scholium 4.2.1. *Let X be a proper smooth $\mathbb{C}$-scheme. Then*

$$F^0 H^p(X, \mathcal{H}^p(p)) \stackrel{\mathrm{def}}{=} H^p(X, F^0 \mathcal{H}^p(p)) \cong \mathrm{NS}^p(X) \otimes \mathbb{C}$$

and the image of the cycle map is in $H^{2p}(X, \mathbb{Q}(p)) \cap F^0 H^{2p}(X, \mathbb{C}(p))$.

Now $\mathrm{im} c\ell_\mathbb{Q}^p = N^p H^{2p}(X, \mathbb{Q}(p))$ and $H^{2p}(X, \mathbb{Q}(p)) \cap F^0 H^{2p}(X, \mathbb{C}(p))$ is equal to $H_\mathbb{Q}^{p,p}$, i.e., the sub-structure of rational (p, p)-classes in $H^{2p}(X)$. Note that $H_\mathbb{Q}^{p,p}$ corresponds to the 1-motivic part of $H^{2p}(X)(p)$. For X a smooth proper $\mathbb{C}$-scheme, the Hodge conjecture then claims that $\mathrm{im}\, c\ell_\mathbb{Q}^p = H_\mathbb{Q}^{p,p}$.

In this case $\mathrm{gr}_N^{p-1} H^{2p}(X)(p) = C_\infty^{p-1,p+1}$ is a quotient of $H^{p-1}(X, \mathcal{H}_X^{p+1}(p))$. For example: $\mathrm{gr}_N^1 H^4(X)(2) = H^1(X, \mathcal{H}_X^3(2))$.

Problem 4.2.2. *Is $F^2 \cap H^1(X, \mathcal{H}_X^3) = 0$?*

Case $i = p$ and $j = 2p + 1$. Let X be a smooth $\mathbb{C}$-scheme. If $j = 2p + 1$ then $N^i H^{2p+1}(X) = 0$ for $i > p$ and $N^p H^{2p+1}(X)(t) = C_\infty^{p,p+1}$ which is a quotient of $H^p(X, \mathcal{H}_X^{p+1})$, i.e., there is an edge map

$$s\ell_{-1}^{p+1} : H^p(X, \mathcal{H}_X^{p+1}(p+1)) \to H^{2p+1}(X)(p+1)$$

with image $N^p H^{2p+1}(X)(p+1)$. In this case the Grothendieck–Hodge conjecture characterize $N^p H^{2p+1}(X)$ as the largest sub-Hodge structure of type $\{(p, p+1), (p+1, p)\}$. This is the same as the 1-motivic part of $H^{2p+1}(X)(p+1)$.

This 1-motivic part yields an abelian variety which is the maximal abelian subvariety of the intermediate jacobian $J^{p+1}(X)$. On the other hand, it is easy to see that $N^p H^{2p+1}(X)(p+1)$ yields the algebraic part of $J^{p+1}(X)$, i.e., defined by the images of codimension $p+1$ cycles on X which are algebraically equivalent to zero modulo rational equivalence (cf. [19] and [33]).

4.3. Exotic (1, 1)-classes

Consider X singular. We briefly explain the Conjecture 2.3.4 for $p = 1$. Moreover we show that there are edge maps generalizing the cycle class maps constructed in the previous section.

For X a proper irreducible $\mathbb{C}$-scheme, consider the mixed Hodge structure on $H^{2+i}(X, \mathbb{Z})$ modulo torsion. The extension (5) is the following

$$0 \to H^{1+i}((H^1)^\bullet) \to W_2 H^{2+i}(X)/W_0 \to H^i((H^2)^\bullet) \to 0. \qquad (15)$$

Since the complex $(H^1)^\bullet$ is made of level 1 mixed Hodge structures then

$$H^{2+i}(X)^h = H^{2+i}(X)^e$$

in our notation.

If X is nonsingular then $H^{2+i}(X)$ is pure and there are only two cases where this extension is non-trivial. In the case $i = -1$ the above conjecture corresponds to the well known fact that $H_1(\text{Pic}^0(X)) = H^1(X, \mathbb{Z})$. The case $i = 0$ corresponds to the celebrated theorem by Lefschetz showing that the subgroup $H_\mathbb{Z}^{1,1}$ of $H^2(X, \mathbb{Z})$ of cohomology classes of type $(1, 1)$ is generated by c_1 of line bundles on X. Since homological and algebraic equivalences coincide for divisors, the Néron–Severi group $\text{NS}^1(X)$ coincide with $H_\mathbb{Z}^{1,1}$. For such a nonsingular variety X we then have

$$\text{NS}^1(X) = F^1 \cap H^2(X, \mathbb{Z}) = H_\mathbb{Z}^{1,1} = H^1(X, \mathcal{H}_X^1) = N^1 H^2(X, \mathbb{Z}).$$

For $i = -1$ and X possibly singular, the conjecture corresponds to the fact (proved in [4]) that the abelian variety corresponding to $\text{gr}_1^W H^1$ is

$$\ker^0(\text{Pic}^0(X_0) \to \text{Pic}^0(X_1)).$$

For $i = 0$ the Conjecture 2.3.4 is quite easily verified by checking the claimed compatibility of the extension class map. Such a statement then corresponds to a Lefschetz $(1, 1)$-theorem for complete varieties with arbitrary singularities.

For $i \geq 1$ we may get *exotic* $(1, 1)$-*classes* in the higher cohomology groups $H^{2+i}(X)$ of an higher dimensional singular variety X. We ignore the geometrical meaning of these exotic $(1, 1)$-classes. It will be interesting to produce concrete examples. The conjectural picture is as follows.

Let $\pi : X_\bullet \to X$ be an hypercovering. Let $(H^q(\mathcal{H}^1))^\bullet$ be the complex of $E_1^{\bullet,q}$-terms of the spectral sequence in Corollary 3.4.2 for $r = 1$. Now $E_1^{i,q} = H^q(X_i, \mathcal{H}^1_{X_i}) = 0$ for $q \geq 2$ (where X_i are the smooth components of the hypercovering $X_\bullet$ of X) and all non-zero terms are pure Hodge structures: therefore the spectral sequence degenerates at E_2. Thus, from Corollary 3.4.2, we get an extension

$$0 \to H^{1+i}((H^0(\mathcal{H}^1))^\bullet) \to \mathbb{H}^{1+i}(X_\bullet, \mathcal{H}^1_{X_\bullet}) \to H^i((H^1(\mathcal{H}^1))^\bullet) \to 0 \quad (16)$$

in the category of mixed $\mathbb{Q}$-Hodge structures. We have

$$H^{i+1}((H^0(\mathcal{H}^1))^\bullet) = H^{i+1}((H^1)^\bullet) = \operatorname{gr}_1^W H^{2+i}$$

and

$$H^i((H^1(\mathcal{H}^1))^\bullet) = H^i((\mathrm{NS})^\bullet) = H^i((N^1 H^2)^\bullet).$$

Moreover, from the local-to-global spectral sequence in Claim 3.4.3 and cohomological descent we get the following edge map

$$s\ell^{1+i} : \mathbb{H}^{1+i}(X_\bullet, \mathcal{H}^1_{X_\bullet}) \to W_2 H^{2+i}(X)/W_0.$$

In fact, first observe that $W_0 H^{2+i}(X) = \mathbb{H}^{2+i}(X_\bullet, \mathcal{H}^0_{X_\bullet})$. From (16) above we then see that $W_0 \mathbb{H}^{1+i}(X_\bullet, \mathcal{H}^1_{X_\bullet}) = 0$. The map $s\ell^{1+i}$ is then easily obtained as an edge homomorphism of the cited local-to-global spectral sequence and weight arguments. This cycle map will fit in a diagram

$$\begin{array}{ccccccccc}
0 \to & H^{1+i}((H^1)^\bullet) & \to & W_2 H^{2+i}(X)/W_0 & \to & H^i((H^2)^\bullet) & \to 0 \\
& \uparrow \| & & \uparrow s\ell^{1+i} & & \uparrow & \\
0 \to & H^{1+i}((H^0(\mathcal{H}^1))^\bullet) & \to & \mathbb{H}^{1+i}(X_\bullet, \mathcal{H}^1_{X_\bullet}) & \to & H^i((H^1(\mathcal{H}^1))^\bullet) & \to 0
\end{array}$$

mapping the extension (16) to (15).

Scholium 4.3.1. *The image of the map $s\ell^{1+i}$ is $H^{2+i}(X)^e$.*

Following [4] consider the simplicial sheaf $\mathcal{O}^*_{X_\bullet}$ and the corresponding Zariski cohomology groups $\mathbb{H}^{1+i}(X_\bullet, \mathcal{O}^*_{X_\bullet})$. Since the components of $X_\bullet$ are smooth, the canonical spectral sequence

$$E_1^{p,q} = H^q(X_p, \mathcal{O}^*_{X_p}) \Rightarrow \mathbb{H}^{p+q}(X_\bullet, \mathcal{O}^*_{X_\bullet})$$

yields a long exact sequence

$$H^{1+i}((H^0(\mathcal{O}^*))^\bullet) \to \mathbb{H}^{1+i}(X_\bullet, \mathcal{O}^*_{X_\bullet}) \to H^i((\mathrm{Pic}\,)^\bullet) \xrightarrow{d^i} H^{2+i}((H^0(\mathcal{O}^*))^\bullet) \to \cdots$$

According to [4] (see the construction in [2]) we may regard $\mathbb{H}^{1+i}(X_\bullet, \mathcal{O}^*_{X_\bullet})$ as the group of k-points of a group scheme whose connected component of the identity yields a semi-abelian variety

$$0 \to H^{1+i}((H^0(\mathcal{O}^*))^\bullet)/\sigma \to \mathbb{H}^{1+i}(X_\bullet, \mathcal{O}^*_{X_\bullet})^0 \to H^i((\mathrm{Pic}^0))^\bullet)^0 \to 0$$

where σ is a finite group. The Hodge realization of the so obtained isogeny 1-motive is

$$T_{\mathrm{Hodge}}([0 \to \mathbb{H}^{1+i}(X_\bullet, \mathcal{O}^*_{X_\bullet})^0]_\mathbb{Q}) = W_1 H^{1+i}(X, \mathbb{Q})(1).$$

This last claim is clearly related to Deligne's conjecture [11, 10.4.1]. For $i = -1, 0$ this is actually proven in [4] and for all i in [2].

Recall the existence of a canonical map of sheaves $c_1 : \mathcal{O}^*_{X_\bullet} \to \mathcal{H}^1_{X_\bullet}$ yielding a map

$$c_1 : \mathbb{H}^{1+i}(X_\bullet, \mathcal{O}^*_{X_\bullet}) \to \mathbb{H}^{1+i}(X_\bullet, \mathcal{H}^1_{X_\bullet}).$$

By composing $s\ell^{1+i}$ and c_1 we then obtain a cycle map

$$\mathbb{H}^{1+i}(X_\bullet, \mathcal{O}^*_{X_\bullet}) \to W_2 H^{2+i}(X)/W_0.$$

We may regard the image of this cycle map as the discrete part of $\mathbb{H}^{1+i}(X_\bullet, \mathcal{O}^*_{X_\bullet})$. Over $\mathbb{Q}$, it is clearly equal to $F^1 \cap H^{2+i}(X, \mathbb{Q})$. The reader can easily check that this is the case, e.g., $\mathbb{H}^{2+i}(X_\bullet, \mathcal{O}^*_{X_\bullet}[-1])$ coincides with Deligne–Beilinson cohomology (see [5, 5.4]). In general, we may expect the following picture for cycle maps.

4.4. $\mathcal{K}$-cohomology and motivic cohomology

Let $X_\bullet$ be a smooth simplicial scheme. Consider the local-to-global spectral sequence in Claim 3.4.3. For a fixed p we then obtain a spectral sequence

$$W_{2p} \mathbb{H}^q(X_\bullet, \mathcal{H}^r_{X_\bullet})/W_{2p-2} \Rightarrow W_{2p} \mathbb{H}^{q+r}(X_\bullet, \mathbb{Q})/W_{2p-2}.$$

The sheaf $\mathcal{H}^r_{X_\bullet}$ has weights $\leq 2r$ and so the mixed Hodge structure on its cohomology has weights $\leq 2r$. Thus $W_{2p}\mathbb{H}^q(X_\bullet, \mathcal{H}^p_{X_\bullet}) = \mathbb{H}^q(X_\bullet, \mathcal{H}^p_{X_\bullet})$ and $W_{2p}\mathbb{H}^q(X_\bullet, \mathcal{H}^r_{X_\bullet})/W_{2p-2} = 0$ if $r < p$. Thus, there is an edge map

$$s\ell^{p+i} : \mathbb{H}^{p+i}(X_\bullet, \mathcal{H}^p_{X_\bullet})/W_{2p-2} \to W_{2p}\mathbb{H}^{2p+i}(X_\bullet)/W_{2p-2}. \tag{17}$$

Note that if $X_\bullet = X$ is constant then $s\ell^{p+0} = s\ell^p_0$ and $s\ell^{p-1} = s\ell^p_{-1}$ in the notation of Section 4.2, modulo W_{2p-2}.

Scholium 4.4.1. *The image of the edge map $s\ell^{p+i}$ is $H^{2p+i}(X)^h$.*

Consider Quillen's higher K-theory. Consider Zariski sheaves associated to Quillen's K-functors. The $\mathcal{K}$-cohomology groups are $\mathbb{H}^*(X_\bullet, \mathcal{K}_p)$ (as usual we consider Zariski simplicial sheaves $\mathcal{K}_p$). Local higher Chern classes give us maps of simplicial sheaves $c_p : \mathcal{K}_p \to \mathcal{H}_{X_\bullet}^p(p)$ for each $p \geq 0$ (cf. [1]). We thus obtain a map

$$c_p : \mathbb{H}^{p+i}(X_\bullet, \mathcal{K}_p) \to \mathbb{H}^{p+i}(X_\bullet, \mathcal{H}_{X_\bullet}^p(p)). \tag{18}$$

Note that in the canonical spectral sequence

$$E_1^{s,t} = H^t(X_s, \mathcal{K}_p) \Rightarrow \mathbb{H}^{s+t}(X_\bullet, \mathcal{K}_p)$$

we have $H^t(X_s, \mathcal{K}_p) = 0$ if $t > p$. The same hold for the sheaf $\mathcal{H}_{X_\bullet}^p$. We then have a commutative square

$$\begin{array}{ccc} \mathbb{H}^{p+i}(X_\bullet, \mathcal{K}_p) & \xrightarrow{c_p} & \mathbb{H}^{p+i}(X_\bullet, \mathcal{H}_{X_\bullet}^p(p)) \\ \downarrow & & \downarrow \\ H^i((CH^p)^\bullet) & \longrightarrow & H^i((NS^p)^\bullet). \end{array}$$

Thus the image of c_p in $H^i((NS^p)^\bullet)$ is clearly contained in the kernel of the map λ_a^i defined in (4). We also have the following commutative square

$$\begin{array}{ccc} \mathbb{H}^{p+i}(X_\bullet, \mathcal{H}_{X_\bullet}^p)/W_{2p-2} & \xrightarrow{s\ell^{p+i}} & W_{2p}\mathbb{H}^{2p+i}(X_\bullet)/W_{2p-2} \\ \downarrow & & \downarrow \\ H^i((NS^p)^\bullet) & \longrightarrow & \mathrm{gr}_{2p}\mathbb{H}^{2p+i}(X_\bullet). \end{array}$$

Composing $s\ell^{p+i}$ and c_p above we then obtain a simplicial cycle map

$$c\ell^{p+i} : \mathbb{H}^{p+i}(X_\bullet, \mathcal{K}_p) \to W_{2p}H^{2p+i}(X_\bullet)/W_{2p-2}. \tag{19}$$

Let X be a proper $\mathbb{C}$-scheme and let $X_\bullet \to X$ be a universal cohomological descent morphism. By descent, $\mathbb{H}^*(X_\bullet) \cong H^*(X)$ as mixed Hodge structures. Let H denote the mixed Hodge structure on $H^{2p+i}(X, \mathbb{Z})/(\text{torsion})$. Let F^p denote the Hodge filtration. Note that

$$F^p \cap H_\mathbb{Z} = \mathrm{Hom}_{\mathrm{MHS}}(\mathbb{Z}(-p), H) = \mathrm{Hom}_{\mathrm{MHS}}(\mathbb{Z}(-p), W_{2p}H).$$

Moreover

$$\mathrm{Hom}_{\mathrm{MHS}}(\mathbb{Z}(-p), W_{2p}H) \subseteq \mathrm{Hom}_{\mathrm{MHS}}(\mathbb{Z}(-p), W_{2p}H/W_{2p-2}H) = F^p \cap H_\mathbb{Z}^e.$$

Thus

$$F^p \cap H_\mathbb{Z} \subseteq F^p \cap H_\mathbb{Z}^e = F^p \cap H_\mathbb{Z}^h = \ker(H_\mathbb{Z}^{p,p} \xrightarrow{e^p} J^p(H)).$$

Therefore we have the following natural question.

Problem 4.4.2. *Let X be a proper $\mathbb{C}$-scheme and let $\pi : X_\bullet \to X$ be a proper smooth hypercovering. Is $F^p \cap H^{2p+i}(X, \mathbb{Q})$ the image of the cycle class map $c\ell^{p+i}$ in (19)?*

Bloch's counterexample answer this question in the negative (see Section 5.1 below). Note that here we actually deal with the simplicial scheme $X_\bullet$ (not just X) as, e.g., in [3] it is shown that $F^1 \cap H^2(X, \mathbb{Z})$ can be larger than the image of Pic (X), if X is singular. However, $F^1 \cap H^2(X, \mathbb{Z})$ is the image of the Pic of any hypercovering of X. However, $F^2 \cap H^4(X, \mathbb{Q})$ is larger than the image of $\mathbb{H}^2(X_\bullet, \mathcal{K}_2)$ if X is the singular 3-fold in Section 5.1 below.

Let's then consider the case $p = 2$ in the above. In this case we have that $H^q(X_i, \mathcal{H}^2_{X_i})$ is purely of weight $q + 2$ by the coniveau spectral sequence (14). Thus the canonical spectral sequence in Corollary 3.4.2 degenerates yielding the following extension

$$0 \to H^{1+i}((H^1(\mathcal{H}^2))^\bullet) \to \mathbb{H}^{2+i}(X_\bullet, \mathcal{H}^2_{X_\bullet})/W_2 \to H^i((\mathrm{NS}^2)^\bullet) \to 0. \qquad (20)$$

Note that $H^1(X_i, \mathcal{H}^2_{X_i}) = N^1 H^3(X_i)$ and $W_2 \mathbb{H}^j(X_\bullet, \mathcal{H}^2_{X_\bullet}) = H^j((H^0(\mathcal{H}^2))^\bullet)$. The map in (17) is mapping the extension (20) to the following canonical extension

$$0 \to H^{1+i}((H^3)^\bullet) \to W_4 H^{4+i}(X)/W_2 \to H^i((H^4)^\bullet) \to 0.$$

According to Conjectures 2.3.2–2.3.4 we may expect that the image of the mixed Hodge structure $\mathbb{H}^{2+i}(X_\bullet, \mathcal{H}^2_{X_\bullet})/W_2$ under this map is $H^{4+i}(X)^h$.

Finally, making use of the triangulated category of motives (see [39] and [26]) let $H_m^*(X, \mathbb{Q}(\cdot))$ denote the motivic cohomology of the proper $\mathbb{C}$-scheme X. Since motivic cohomology is universal we may get a canonical map $H_m^{2p+i}(X, \mathbb{Q}(p)) \to H^{2p+i}(X, \mathbb{Q}(p))$ compatibly with the weight filtrations. This map will factor through Beilinson's absolute Hodge cohomology [5]. However, in general, its image will not be larger than $c\ell^{p+i}$ in (19), i.e., smaller than the rational part of $F^p H^{2p+i}(X, \mathbb{C})$. In fact, we can see that only Beilinson's absolute Hodge cohomology (or Deligne–Beilinson cohomology) would have image equal to $F^2 \cap H^4(X, \mathbb{Q})$ if X is the singular 3-fold in Bloch's counterexample below.

5. Examples

We finally discuss a couple of examples where one can test the conjectures.

5.1. Bloch's example

We now consider Bloch's example explained in a letter to U. Jannsen, reproduced in the Appendix A of [23] (see also Appendix A.I in [27]). This example, originally requested by Mumford, is a counterexample to a naive extension of the cohomological Hodge conjecture to the singular case. Moreover (as indicated by Bloch's Remark 1 in [23, Appendix A]) it shows that no cohomological invariants of algebraic varieties, that agree with Chow groups of non-singular varieties, can provide all Hodge cycles for singular varieties.

Let P be the blow-up of $\mathbb{P}^3$ at a point x in $S_0 \subset \mathbb{P}^3$ a smooth hypersurface of degree ≥ 4 over $\overline{\mathbb{Q}}$. The point x is assumed $\overline{\mathbb{Q}}$-generic. Let S be the blow-up of S_0 at x over $\mathbb{C}$. Thus $S \subset P$ and $H^3(S, \mathbb{Q}) = 0$.

Let X be the gluing of two copies of P along S, i.e., the singular projective variety defined as the pushout

$$\begin{array}{ccc} S \coprod S & \xrightarrow{i \coprod i} & P \coprod P \\ c \downarrow & & \downarrow f \\ S & \xrightarrow{j} & X \end{array}$$

Such a Mayer–Vietoris diagram always defines a cohomological descent morphism $X_\bullet \to X$ (e.g., a distinguished (semi)simplicial resolution in the sense of Carlson [9, §3 and §13]).

Thus we obtain a short exact sequence

$$0 \to H^4(X, \mathbb{Q}(2)) \to H^4(P, \mathbb{Q}(2))^{\oplus 2} \to H^4(S, \mathbb{Q}(2)) \to 0$$

where $CH^2(P)_\mathbb{Q} \cong H^4(P, \mathbb{Q}(2))$. Thus $H^4(X, \mathbb{Z})$ has rank 3, is purely of type $(2, 2)$ and the Hodge 1-motive is $[H^4(X, \mathbb{Q}(2)) \to 0]$. From (20) we obtain

$$\mathbb{H}^2(X_\bullet, \mathcal{H}^2_{X_\bullet})/W_2 = H^0((NS^2)^\bullet)) = H^4(X, \mathbb{Q}(2)).$$

Since the Albanese of S vanishes, the algebraically defined Hodge 1-motive is given by $H^0((NS^2)^\bullet)) = \ker(NS^2(P)^{\oplus 2} \to NS^2(S))$ and we clearly have that

$$[H^0((NS^2)^\bullet)) \to 0] \cong [H^4(X, \mathbb{Q}(2)) \to 0]$$

as predicted by Conjecture 2.3.4. However $H^0((CH^2)^\bullet)) = \ker(CH^2(P)^{\oplus 2} \to CH^2(S))$ has rank 2, and it is strictly smaller than $H^4(X, \mathbb{Q}(2))$, as Bloch observed. Moreover, from the above we may regard $\mathbb{H}^2(X_\bullet, \mathcal{K}_2)_\mathbb{Q}$ mapping to both $H^0((CH^2)^\bullet))$ and $H^4(X, \mathbb{Q}(2))$. Then $c\ell^2$ in (19) is not surjective as $H^0((CH^2)^\bullet)) \neq H^4(X, \mathbb{Q}(2))$. The same argument applies to motivic cohomology $H^4_m(X, \mathbb{Q}(2))$. In fact, if Y is a smooth variety $H^4_m(Y, \mathbb{Q}(2)) \cong CH^2(Y)_\mathbb{Q}$. However, Beilinson's absolute Hodge cohomology is $H^4_\mathcal{D}(X, \mathbb{Z}(2)) \otimes \mathbb{Q} \cong H^4(X, \mathbb{Q}(2))$. Thus, Beilinson's formulation of the Hodge conjecture in [5, §6] does not hold in the singular case.

Note that in this example, all Hodge classes are involved, as the Hodge structure is pure.

5.2. Srinivas' example

The following example has been produced by Srinivas upon author's request. It is similar to Bloch's example however, in this example, the space of "Hodge cycles" is strictly smaller than $H^{2,2}_\mathbb{Q}$.

Let Y be a smooth projective complex 4-fold with $H^1(Y, \mathbb{Z}) = H^3(Y, \mathbb{Z}) = 0$, and with an algebraic cycle $\alpha \in \mathrm{CH}^2(Y)$ whose singular cohomology class $\bar{\alpha} \in H^4(Y, \mathbb{Q})$ is a non-zero primitive class. For example, Y could be a smooth quadric hypersurface in $\mathbb{P}^5$, and $\alpha \in \mathrm{CH}^2(Y)$ the difference of the classes of two planes, taken from the two distinct connected families of planes in Y. Let Z be a general hypersurface section of Y of any fixed degree d such that $H^{3,0}(Z) \neq 0$ (this holds for any large enough degree d; for example, if Y is a quadric then we may take $Z = Y \cap H$ to be the intersection with a general hypersurface H of any degree ≥ 3).

Then Z is a smooth projective 3-fold, and by the theorem of Griffiths, if $i : Z \to Y$ is the inclusion, then $i^*\alpha \in \mathrm{CH}^2(Z)$ is homologically trivial, but no non-zero multiple of $i^*\alpha$ is algebraically equivalent to 0. In fact, $H^3(Z, \mathbb{Q})$ has no proper Hodge substructures, and so (because $H^{3,0}(Z) \neq 0$) the Abel–Jacobi map vanishes on the group $\mathrm{CH}^2(Z)_{\mathrm{alg}}$ of cycle classes algebraically equivalent to 0; on the other hand, the Abel–Jacobi image of $i^*\alpha$ is non-torsion.

Now let X be the singular projective variety defined as a push-out

$$\begin{array}{ccc} Z \coprod Z & \xrightarrow{i \coprod i} & Y \coprod Y \\ c \downarrow & & \downarrow f \\ Z & \xrightarrow{j} & X \end{array}$$

so that X is obtained by gluing two copies of Y along Z.

Consider the simplicial scheme $X_\bullet$ obtained as above (e.g., the Čech hypercovering of X, with $X_0 \to X$ taken to be the quotient map $f : Y \coprod Y \to X$). Then $H^*(X_\bullet, \mathbb{Q}) \cong H^*(X, \mathbb{Q})$ as mixed Hodge structures, and we have an exact sequence of mixed Hodge structures (of which all terms except $H^4(X, \mathbb{Z})$ are in fact pure)

$$0 \to H^3(Z, \mathbb{Z}) \to H^4(X, \mathbb{Z}) \to H^4(Y, \mathbb{Z})^{\oplus 2} \xrightarrow{s} H^4(Z, \mathbb{Z})$$

where $s(a, b) = i^*a - i^*b$. Then $(\bar{\alpha}, 0)$ and $(0, \bar{\alpha})$ are linearly independent elements of $\ker s$, since $i^*\alpha = 0$ in $H^4(Z, \mathbb{Q})$ (this is essentially the definition of $\bar{\alpha}$ being a primitive cohomology class).

In this situation, the group of Hodge classes in $H^4(X, \mathbb{Q})/W_3$ is non-trivial, but since $H^3(Z, \mathbb{Q})$ has no non-trivial sub-Hodge structures, the intermediate Jacobian $J^2(Z)$ has no non-trivial abelian subvariety. The extension of Hodge structures determined by the Hodge classes is not split; for example the extension class of the pullback of

$$0 \to H^3(Z, \mathbb{Z}) \to H^4(X, \mathbb{Z}) \to \ker s \to 0$$

under $\mathbb{Z}(-2) \to \ker s$ determined by $(\bar{\alpha}, 0)$ is (up to sign) the Abel–Jacobi image of $i^*\alpha$, which is non-torsion. Here rank $(\ker s) = 3$, so we get an extension class map $\mathbb{Z}^3 \to J^2(Z)$; one checks that the image has rank 1, generated by the image of $(\bar{\alpha}, 0)$ (or equivalently by the image of $(0, \bar{\alpha})$).

So the lattice for the corresponding Hodge 1-motive is, by definition

$$\ker(\mathbb{Z}^3 \to J^2(Z)) = F^2 \cap H^4(X, \mathbb{Z}),$$

which is strictly smaller than the lattice of all Hodge classes in $H^4(X, \mathbb{Z})$. Moreover, since $H^1(Z, \mathcal{H}^2) = N^1 H^3(Z) = 0$, from the extension (20) we obtain

$$\mathbb{H}^2(X_\bullet, \mathcal{H}^2_{X_\bullet})/W_2 \cong H^0((\mathrm{NS}^2)^\bullet).$$

Finally, the cycle $\alpha \in \mathrm{CH}^2(Y)$ projects to a cycle in $\mathrm{NS}^2(Y)$ which restricts to a non-zero class $i^*\alpha \in \mathrm{NS}^2(Z)_\mathbb{Q}$ by construction. Since $\mathrm{CH}^2(Z)_{\mathrm{ab}} = \mathrm{CH}^2(Z)_{\mathrm{alg}}$ then the algebraically defined 1-motive is given by the image of $H^0((\mathrm{NS}^2)^\bullet) = \ker(\mathrm{NS}^2(Y)^{\oplus 2} \to \mathrm{NS}^2(Z))$ in $H^4(X, \mathbb{Z})$, providing generators for $\ker(\mathbb{Z}^3 \to J^2(Z))$ as claimed in Conjecture 2.3.4.

References

[1] L. Barbieri-Viale, $\mathcal{H}$-cohomologies versus algebraic cycles, Math. Nachr. 184 (1997), 5–57.

[2] L. Barbieri-Viale, A. Rosenschon and M. Saito, Deligne's conjecture on 1-motives, preprint math.AG/0102150, 2001, to appear in Ann. of Math.

[3] L. Barbieri-Viale and V. Srinivas, The Néron-Severi group and the mixed Hodge structure on H^2, J. Reine Angew. Math. 450 (1994), 37–42.

[4] L. Barbieri-Viale and V. Srinivas, Albanese and Picard 1-motives, Mém. Soc. Math. Fr. (N.S.) 87, Paris 2001 (Preliminary note: C. R. Acad. Sci. Paris Sér. I Math. 326 (1998), 1397–1401.)

[5] A. Beilinson, Notes on absolute Hodge cohomology, in: Applications of Algebraic K-Theory to Algebraic Geometry and Number Theory (Spencer J. Bloch et al., eds.), Part 1, Contemp. Math. 55, Amer. Math. Soc., Providence, RI, 1986, 35–68.

[6] M. Beltrametti and P. Francia, A property of regular morphisms, Indag. Math. 46 (1984), 361–368.

[7] J. Biswas and V. Srinivas, A Lefschetz (1,1)-Theorem for normal projective varieties, Duke Math. J. 101 (2000), 427–458.

[8] S. Bloch and A. Ogus, Gersten's conjecture and the homology of schemes, Ann. Sci. Ecole Norm. Sup. 7 (1974), 181–202.

[9] J. A. Carlson, The obstruction to splitting a mixed Hodge structure over the integers, I, University of Utah, Salt Lake City 1979.

[10] J. A. Carlson, The one-motif of an algebraic surface, Compositio Math. 56 (1985), 271–314.

[11] P. Deligne, Théorie de Hodge II, III, Inst. Hautes Études Sci. Publ. Math. 40 (1972), 5–57, Inst. Hautes Études Sci. Publ. Math. 44 (1974), 5–78.

[12] P. Deligne, A quoi servent les motifs?, in: Motives (Uwe Jannsen et. al., eds.), Part 1, Proc. Sympos. Pure Math 55, Amer. Math. Soc., Providence, RI, 1994, 143–161.

[13] P. Deligne, The Hodge conjecture, Millennium Prize Problems, available browsing from http://www.claymath.org/

[14] A. J. de Jong, Smoothness, semistability and alterations, Inst. Hautes Études Sci. Publ. Math. 83 (1996), 51–93.

[15] H. Gillet and C. Soulé, Descent, motives and K-theory, J. Reine Angew. Math. 478 (1996), 127–176.

[16] F. Guillen and V. Navarro-Aznar, Un critère d'extension d'un foncteur défini sur les schémas lisses, preprint, 1995.

[17] P. A. Griffiths, Some transcendental methods in the study of algebraic cycles, Lecture Notes in Math. 185, Springer-Verlag, Heidelberg 1971, 1–46.

[18] A. Grothendieck, Technique de descente et théorèmes d'existence en géométrie algébrique I–VI, Fondements de la Géométrie Algébrique, Extraits du Séminaire Bourbaki 1957/62.

[19] A. Grothendieck, Hodge's general conjecture is false for trivial reasons, Topology 8 (1969), 299–303.

[20] R. Hain, The de Rham homotopy theory of complex algebraic varieties I, II, K-theory 1 (1987), 271–324, 481–497.

[21] M. Hanamura, Mixed motives and algebraic cycles, I, II, III, preprints.

[22] M. Hanamura, The mixed motive of a projective variety, in: The arithmetic and geometry of algebraic cycles (B. Brent Gordon et al., eds.), CRM Proc. Lecture Notes 24, Amer. Math, Soc., Providence, RI, 2000, 183–193.

[23] U. Jannsen, Mixed motives and algebraic K-theory, Lecture Notes in Math. 1400, Springer-Verlag, Berlin 1990.

[24] U. Jannsen, Motivic sheaves and filtrations on Chow groups, in: Motives (U. Jannsen et. al., eds.), Part 1, Proc. Sympos. Pure Math 55, Amer. Math. Soc., Providence, RI, 1994, 245–302.

[25] U. Jannsen, Equivalence relations on algebraic cycles, in: The arithmetic and geometry of algebraic cycles (B. Brent Gordon et al., eds.) NATO ASI Ser., Ser. C, Math. Phys. Sci. 548, Kluwer Academic Publishers, Dordrecht, 2000, 225–260.

[26] M. Levine, Mixed Motives, Math. Surveys Monogr. 57, Amer. Math. Soc., Providence, RI, 1998.

[27] J. D. Lewis, A survey of the Hodge conjecture, 2nd ed. (with an appendix by B. Brent Gordon), CRM Monogr. Ser. 10, Amer. Math. Soc., Providence, RI, 1999.

[28] J. D. Lewis, A filtration on the Chow groups of a complex projective variety, to appear in Compositio Math.

[29] S. Lichtenbaum, Suslin homology and Deligne 1-motives, in: Algebraic K-theory and algebraic topology (P. G. Goerss et al., eds.), NATO Sci. Ser. C Math. Phys. Sci. 407, Kluwer Academic Publishers, Dordrecht 1993, 189–197.

[30] D. Lieberman, Intermediate jacobians, in: Algebraic Geometry, Oslo, Wolters-Nordhoff, 1972, 125–139.

[31] D. Morgan, The algebraic topology of smooth algebraic varieties, Inst. Hautes Études Sci. Publ. Math. 48 (1978), 137–204 (Errata Corrige in Inst. Hautes Études Sci. Publ. Math. 64).

[32] J. Murre, Applications of algebraic K-theory to the theory of algebraic cycles, in: Proc. Algebraic Geometry, Sitges, Lecture Notes in Math. 1124, Springer-Verlag, Berlin 1985, 216–261.

[33] J. Murre, Abel-Jacobi equivalence versus incidence equivalence for algebraic cycles of codimension two, Topology 24 (1985), 361–367.

[34] J. Murre, On a conjectural filtration on the Chow groups of an algebraic variety, I, II, Indag. Math. New Ser. 4 (1993), 177–188, 189–201.

[35] K. H. Paranjape, Some spectral sequences for filtered complexes and applications, J. Algebra 186 (1996), 793–806.

[36] M. Saito, Arithmetic mixed sheaves, preprint `math.AG/9907189`.

[37] M. Saito, Refined cycle maps, preprint `math.AG/0103116`.

[38] S. Saito, Motives, algebraic cycles and Hodge theory, in: The arithmetic and geometry of algebraic cycles (B. Brent Gordon et al., eds.), CRM Proc. Lecture Notes 24, 2000, 235–253.

[39] V. Voevodsky, Triangulated categories of motives over a field, in: Cycles, Transfers, and Motivic Homology Theories, Ann. of Math. Stud. 143, Princeton University Press, Princeton, NJ, 2000, 188–238.

L. Barbieri-Viale, Dipartimento di Metodi e Modelli Matematici, Università degli Studi di Roma "La Sapienza", Via A. Scarpa, 16, 00161 Roma, Italy

E-mail: `barbieri@dmmm.uniroma1.it`

The Szpiro inequality for higher genus fibrations

Arnaud Beauville

Abstract. Let $f : S \to B$ be a non-trivial family of semi-stable curves of genus g, N the number of critical points of f and s the number of singular fibres. We prove the inequality $N < (4g + 2)(s + 2g(B) - 2)$.

2000 Mathematics Subject Classification: 14H10

1. Introduction

The aim of this note is to prove the following result:

Proposition. *Let $f : S \to B$ be a non-trivial semi-stable fibration of genus $g \geq 2$, N the number of critical points of f and s the number of singular fibres. Then*

$$N < (4g + 2)(s + 2g(B) - 2).$$

Recall that a semi-stable fibration of genus g is a surjective holomorphic map of a smooth projective surface S onto a smooth curve B, whose generic fibre is a smooth curve of genus g and whose singular fibres are allowed only ordinary double points; moreover we impose that each smooth rational curve contained in a fibre meets the rest of the fibre in at least 2 points (otherwise by blowing up non-critical points of f in a singular fibre we could arbitrarily increase N keeping s fixed).

The corresponding inequality $N \leq 6(s + 2g(B) - 2)$ in the case $g = 1$ has been observed by Szpiro; it was motivated by the case of curves over a number field, where an analogous inequality would have far-reaching consequences [S]. The higher genus case is considered in the recent preprint [BKP], where the authors prove the slightly weaker inequality $N \leq (4g + 2)s$ for hyperelliptic fibrations over $\mathbf{P}^1$. Their method is topological, and in fact the result applies in the much wider context of symplectic Lefschetz fibrations. We will show that in the more restricted algebraic-geometric set-up, the proposition is a direct consequence of two classical inequalities in surface theory. It would be interesting to know whether the proof of [BKP] can be extended to non-hyperelliptic fibrations.

2. Proof of the proposition

The main numerical invariants of a surface S are the square K_S^2 of the canonical bundle, the Euler–Poincaré characteristic $\chi(\mathcal{O}_S)$ and the topological Euler–Poincaré characteristic $e(S)$; they are linked by the Noether formula $12\,\chi(\mathcal{O}_S) = K_S^2 + e(S)$. For a semi-stable fibration $f : S \to B$ it has become customary to modify these invariants as follows. Let b be the genus of B, and $K_f = K_X \otimes f^* K_B^{-1}$ the relative canonical bundle of X over B; then we consider:

$$K_f^2 = K_X^2 - 8(b-1)(g-1)$$
$$\chi_f := \deg f_*(K_f) = \chi(\mathcal{O}_X) - (b-1)(g-1)$$
$$e_f := N = e(X) - 4(b-1)(g-1).$$

Observe that we have again $12\,\chi_f = K_f^2 + e_f$. We will use the Xiao inequality ([X], Theorem 2)

$$K_f^2 \geq \left(4 - \frac{4}{g}\right) \chi_f$$

and the "strict canonical class inequality" ([T], lemma 3.1)

$$K_f^2 < 2(g-1)(s + 2b - 2) \quad \text{for } s > 0.$$

Let us prove the proposition. If $s = 0$, we have $N = 0$ and $g(B) \geq 2$ (otherwise the fibration would be trivial), so the inequality of the proposition holds. Assume $s > 0$; the Xiao inequality gives

$$3g\,K_f^2 \geq 12(g-1)\,\chi_f = (g-1)(K_f^2 + e_f),$$

hence, using the strict canonical class inequality,

$$N = e_f \leq \frac{2g+1}{g-1} K_f^2 < (4g+2)(s + 2b - 2). \qquad \square$$

Example. We constructed in [B] a semi-stable genus 3 fibration over $\mathbf{P}^1$ with 5 singular fibres; each of these has 8 double points. Therefore

$$N = 40 \quad \text{and} \quad (4g+2)(s-2) = 42.$$

Remark. M. Kim pointed out to me that one gets a finer inequality by taking into account the dimension g_0 of the fixed part of the Jacobian fibration associated to f. Indeed one can deduce (with some work) from Arakelov's seminal paper [A] the inequality $\chi_f \leq \frac{1}{2}(g - g_0)(s + 2b - 2)$ (see also [D] for a Hodge-theoretical proof); combined with Xiao's inequality this gives

$$N \leq (4g+2)(s + 2b - 2)\left(1 - \frac{g_0}{g}\right)$$

(note however that the inequality is no longer strict).

References

[A]   S. Arakelov, Families of algebraic curves with fixed degeneracies, Math. U.S.S.R. Izv. 5 (1971), 1277–1302.

[B]   A. Beauville, Le nombre minimum de fibres singulières d'une courbe stable sur $\mathbf{P}^1$, Astérisque 86 (1981), 97–108.

[BKP] F. Bogomolov, L. Katzarkov, T. Pantev, Hyperelliptic Szpiro inequality, Preprint math.GT/0106212.

[D]   P. Deligne, Un théorème de finitude pour la monodromie, Discrete groups in geometry and analysis (New Haven, 1984), Progr. Math. 67, Birkhäuser, Boston (1987), 1–19.

[S]   L. Szpiro, Discriminant et conducteur des courbes elliptiques, Séminaire sur les Pinceaux de Courbes Elliptiques (Paris, 1988), Astérisque 183 (1990), 7–18.

[T]   S.-L. Tan, The minimal number of singular fibers of a semistable curve over $\mathbf{P}^1$, J. Algebraic Geom. 4 (1995), 591–596.

[X]   G. Xiao, Fibered algebraic surfaces with low slope, Math. Ann. 276 (1987), 449–466.

A. Beauville, Institut Universitaire de France et Laboratoire J.-A. Dieudonné, UMR 6621 du CNRS, Université de Nice, Parc Valrose, 06108 Nice Cedex 02, France
E-mail: beauville@math.unice.fr

On regular surfaces of general type with $p_g = 2$ and non-birational bicanonical map

Giuseppe Borrelli

Abstract. In the present note we classify regular surfaces of general type with $p_g = 2$ and non-birational bicanonical map under the assumption that the canonical system has no fixed part.

1. Introduction

The behaviour of pluricanonical maps for surfaces of general type is a topic of the theory of surfaces classically studied by several authors (Enriques, Franchetta, Zariski, Kodaira, Shafarevich). Pluricanonical maps are the rational maps

$$\Phi_n := \Phi_{|nK|} : S \to \mathbf{P}(H^0(S, \mathcal{O}_S(nK))^\vee) =: \mathbb{P}_{P_n-1}, \quad n \geq 1,$$

associated to the linear systems $|nK|$, where K is a canonical divisor of the surface S.

In the epochal paper [B] Bombieri proved, among other things, that if $K^2 \geq 10$, $p_g(S) = h^0(S, \mathcal{O}_S(K)) \geq 6$ and the bicanonical map Φ_2 fails to be birational, there exists a rational map $\varphi : S \to B$ onto a curve B such that the general fibre F is a smooth irreducible curve of genus 2.

On the other hand if there is such a φ, then the bicanonical map Φ_2 of S is not birational.

We shall refer to the above exception to the birationality of the bicanonical map as to the *standard case*.

More recently I. Reider has proved in [R] that the hypothesis $K^2 \geq 10$ is sufficient in order to ensure that if Φ_2 is not birational, then we have the standard case.

From these results it follows that there is only a finite number of families of minimal surfaces of general type not presenting the standard case for the non-birationality of the bicanonical map (briefly the *non-standard cases*).

The classification of these surfaces has become an object of study for many authors.

In fact, Du Val considered this problem for regular surfaces, i.e. with $q(S) := h^1(S, \mathcal{O}_S) = 0$. In his article [DV], under further hypotheses implicitly assumed, he obtained a detailed classification of these surfaces.

Recently the non-standard cases with $p_g(S) \geq 4$ have been classified by C. Ciliberto, P. Francia and M. Mendes Lopes, in [CFM], whose results essentially confirm the list proposed by Du Val. In particular, every non-standard case with $p_g(S) \geq 4$ is regular.

Concerning the surfaces presenting the non-standard case with $p_g(S) = 3$: C. Ciliberto, F. Catanese and M. Mendes Lopes in the article [CCM] proved that if $q(S) > 0$, then $q(S) = p_g(S) = 3$, $K^2 = 6$ and S is the symmetric product of a smooth irreducible curve of genus 3; whereas C. Ciliberto and M. Mendes Lopes, in [CM,1], proved that if $q(S) = 0$ then S is one of the surfaces appearing in Du Val's list, or one of their specializations.

Recently C. Ciliberto and M. Mendes Lopes in [CM,2], [CM,3] have worked out also the case $p_g = 2$ and $q > 0$, showing that if S does not present the standard case, then it is birationally equivalent to a double cover of a principally polarized abelian surface (A, Θ) branched along a divisor $B \in |2\Theta|$ with at most non essential singularities.

In particular in these papers the hypotheses implicitly assumed by Du Val, that is that the surface is regular and the general canonical curve is smooth and irreducible, have been removed.

Another contribution to the study of this problem is Xiao Gang's paper [X]. He mainly took the point of view of the projective study of the image of the bicanonical map determining a list of numerical possibilities for the invariants of the cases which might occur.

In his article Du Val suggested a way, that we shall describe soon, to produce and classify the surfaces under consideration also for $p_g < 3$. This is the object of my thesis [Bo] where I proved:

Theorem. *Let S be a smooth minimal surface of general type with $p_g(S) = 2$, $q(S) = 0$ and non-birational bicanonical map. Assume that $|K_S|$ has no fixed part. Then either we are in the standard case or S is a Du Val surface and $4 \leq K_S^2 \leq 8$.*

In this note we describe the strategy of the proof; for a full treatment we refer the reader to [Bo].

Acknowledgements. I would like to thank Professor Ciro Ciliberto, who suggested the problem, for his advice and encouragement.

My deep gratitude goes also to the memory of Professor Paolo Francia for his encouragement and especially for very interesting conversations.

I would like to thank the referee for the very careful reading of the paper and several useful remarks.

2. Notations and conventions

We will denote by S a projective algebraic surface over the complex field. Usually S will be smooth.

We will say that S presents the *standard case* for the non birationality of the bicanonical map, or simply that S presents the *standard case*, if S is a surface of general type with non birational map and such that there exists a map $\varphi : S \to B$ onto a curve B, whose general fibre is a smooth irreducible curve of genus two. This is what one calls a pencil of curves of genus 2.

We denote by K_S a canonical divisor, by $p_g(S) = h^0(S, \mathcal{O}_S(K_S))$ the geometric genus and by $q(S) = h^1(S, \mathcal{O}_S)$ the *irregularity of S*.

By a *curve* on S we mean an effective non zero divisor on S. We denote by $C.D$ the intersection number of the divisors C, D, by C^2 the selfintersection of a divisor C and by $p_a(C) = \frac{1}{2}C.(C + K_S) + 1$ the arithmetic genus of a curve C on S.

Recall that S is minimal of general type if and only if $K_S.C \geq 0$ for every curve C of S (i.e. K_S is *nef*) and $K_S^2 > 0$.

We denote by $\operatorname{mult}_P(C)$ the multiplicity of the curve C at a point P and by $\operatorname{Sing}(C)$ the set of singular points of C.

We denote by $\sim$ the linear equivalence for divisor on S. $|D|$ will be the complete linear system of the effective divisors $D' \sim D$ and

$$\varphi_{|D|} : S \to \mathbf{P}(H^0(S, \mathcal{O}_S(D))^\vee) = |D|^\vee$$

the natural rational map defined by $|D|$.

By abuse of terminology, we will say that an infinitely near point to a curve *lies on the curve*.

We will say that a curve C on S has a singularity of *type* $[k, k]$ at a point $P \in C$ (or that P is a point of *type* $[k, k]$ or more simply a $[k, k]$-point) if P is a k-tuple point which has another k-tuple point infinitely near to it.

We will say that C has an *essential singularity* at P if $\operatorname{mult}_P(C) \geq 4$ or if P is a $[3, 3]$-point.

We will say that a smooth reduced curve $C \subset S$ is a $(-n)$-*curve* if $C \cong \mathbb{P}_1$ and $C^2 = -n$.

3. Surfaces with non birational bicanonical map

As mentioned before, Du Val is probably the first author who dealt in a systematic way with the problem under consideration. He supposed that the bicanonical map is not birational for the surface S and implicitly made the following assumptions:

(a) the surface is regular, i.e. $q(S) = 0$;

(b) the general canonical curve $C \in |K_S|$ is smooth and irreducible.

Notice that if (a) and (b) hold, then C is hyperelliptic; conversely if the general canonical curve is smooth and hyperelliptic then the bicanonical map cannot be birational.

We state Du Val's result [DV]:

Theorem 1. *Let S be a smooth minimal surface of general type. Assume that $p_g(S) \geq 3$, $q(S) = 0$, the general canonical curve $C \in |K_S|$ is smooth and irreducible and the bicanonical map of S is not birational. Then either we are in the standard case, i.e. S has a pencil of curves of genus 2, or S is one of the so called Du Val surfaces described below. Their invariants are shown in the following table:*

$$
\begin{array}{c|cccccccc}
p_g & & & & K_S^2 & & & & \\
6 & 8^* & & 9^* & & & & & \\
  & \downarrow & \swarrow & \downarrow & & & & & \\
5 & 7 & & 8^* & & & & & \\
  & \downarrow & \swarrow & \downarrow & & & & & \\
4 & 6 & & 7 & & 8^* & & & \\
  & \downarrow & \swarrow & \downarrow & \swarrow & \downarrow & & & \\
3 & 5 & & 6 & 6 & & 7 & 8^* & 2^*
\end{array}
\qquad (1)
$$

For each pair of invariants (p_g, K_S^2) in the above table, the corresponding surfaces fill up an irreducible family of double covers of which we will describe the general member below. The explanation for the arrows in the table is the following:

$\downarrow$ means that one imposes a $[3, 3]$-point to the branch curve of the double cover. As it is well known, this drops the geometric genus and K_S^2 both by 1;

$\swarrow$ means that one imposes a 4-tuple or 5-tuple ordinary point to the branch curve of the double cover, which drops the geometric genus by 1 and K_S^2 by 2.

Observe that by imposing the aforementioned singularities to the branch curve of the double covers marked with the asterisk (the Du Val *ancestors*), one obtains the others. Hence it is sufficient to describe only the marked double covers.

First of all we give the following:

Definition. We will say that a smooth surface S is a Du Val surface if it is birationally equivalent to one of the following:

D_1) a degree two cover $X_1 \to \mathbb{F}_2$ branched along a curve $B_1 = B_1' + C_0$, where C_0 is the (-2)-section and, denoting by Γ a fibre of $\pi : \mathbb{F}_2 \to \mathbb{P}_1$, $B_1' \sim 7(C_0 + 2\Gamma)$; the singularities of B_1 are at most 4-tuple or 5-tuple ordinary points or $[3, 3]$-points (these come from the ancestor with $p_g = 6$, $K^2 = 9$);

D_2) a degree two cover $X_2 \to \mathbb{P}_2$ with branch curve of the form $G = G' + a_1 + \cdots + a_n$, $n \in \{0, 1, 2, 3, 4\}$, where $a_1, \ldots, a_n$ are distinct lines meeting at a point γ and G' is a curve of degree $10 + n$ whose singularities are: a $(n + 2)$-tuple point at γ, four $[4, 4]$-points $\alpha_i \in a_i$, different from each other and from γ, where the infinitely near 4-tuple point to α_i lies on the line a_i, $i = 1, 2, 3, 4$ and perhaps other 4-tuple or 5-tuple ordinary points or [3, 3]-points (these come from the ancestor with $p_g = 6 - n$, $K^2 = 8$);

D3) a degree two cover $X_3 \to \mathbb{P}_2$ with a smooth branch curve B_3 of degree 8 ($p_g = 3$, $K^2 = 2$).

More details about Du Val surfaces can be found in [C], [CFM].

We will consider a smooth minimal surface S of general type such that:
(*, 1) $p_g(S) = 2$ and $q(S) = 0$;
(*, 2) the bicanonical map Φ_2 is not birational and
(*, 3) $|K_S|$ has no fixed part.

The canonical system is a rational pencil with $d := K_S^2$ base points $p_1, \ldots, p_d$, whose general curve $M \in |K_S|$ is irreducible, otherwise M would be composed of $k \geq 2$ curves of a rational pencil and therefore

$$p_g(S) = h^0(S, \mathcal{O}_S(K_S)) = k + 1 \geq 3.$$

We begin with a general result:

Lemma 2 (cf. [CM], Propositions 2.1 and 2.2). *Let M be the general curve in $|K_S|$, then*

(i) *M is smooth and hyperelliptic;*

(ii) *the base points $p_1, \ldots, p_d$ are distinct Weierstrass points on M.*

The hyperelliptic involution of general curves in $|K_S|$ induces an involution σ of S which is a morphism since S is minimal. We remark that the bicanonical map of S is composed with σ.

The fixed locus of σ is the union of a smooth curve R and of k isolated points $q_1, \ldots, q_k$. We denote by $\rho : S \to X := S/\sigma$ the projection onto the quotient, by B_X the image of R and by Q_i the image of q_i, $i = 1, \ldots, k$. The surface X is normal and $Q_1, \ldots, Q_k$ are ordinary double points, which are the only singularities of X. In particular, the singularities of X are canonical and by adjunction formula $K_S \sim \rho^*(K_X) + R$.

Let $\pi : S' \to S$ be the blow-up of S at $q_1, \ldots, q_k$ and $E_1, \ldots, E_k$ be the exceptional curves of π. It is easily seen that σ induces an involution σ' of S' whose fixed locus is the smooth curve union of $R' := \pi^{-1}(R)$ and $E_1, \ldots, E_k$. We recall that a canonical divisor for S is

$$K_{S'} \sim \pi^*(K_S) + E_1 + \cdots + E_k.$$

Let $\rho' : S' \to \Sigma := S'/\sigma'$ be the canonical projection onto the quotient and set $B' := \rho'(R')$, $C_i' := \rho'(E_i)$, $i = 1, \ldots, k$.

The surface Σ is smooth, ρ' is a double cover branched along the smooth curve $B' + C_1' + \cdots + C_k'$ and the C_i' are disjoint (-2)-curves.

Let $\overline{M} := \pi^*(M) - \sum_{i=1}^{k} \mathrm{mult}_{q_i}(M) E_i$ be the strict transform of the general curve $M \in |K_S|$ and $\sigma'_{\overline{M}}$ be the hyperelliptic involution on it. Then the image $\Gamma' := \overline{M}/\sigma'_{\overline{M}} = \rho'(\overline{M})$ is a rational curve on Σ. Moreover $h^0(\Sigma, \mathcal{O}_\Sigma(\Gamma')) = h^0(S, \mathcal{O}_S(M)) = 2$. Hence the linear system $|\Gamma'|$ is a rational pencil of rational curves and therefore Σ is a rational surface.

We denote by $\eta : \Sigma \to X$ the morphism induced by π. The map η is the minimal resolution of the singularities of X and there is the commutative diagram:

$$\begin{array}{ccc} S' & \xrightarrow{\pi} & S \\ \rho' \downarrow & & \downarrow \omega \\ \Sigma & \xrightarrow{\eta} & X \end{array}$$

Our strategy is to study the cover $\rho' : S' \to \Sigma$. The first basic step is given by the following lemma.

Lemma 3. *In the above situation:*

(i) $p_1, \ldots, p_d$ *are isolated fixed points of σ;*

(ii) $k = K_S^2$.

Hence the only isolated fixed points of σ are the base points of $|K_S|$.

Proof. We omit the easy proof of (i).

As for (ii), by the theory of double covers, there exists some $L \in \mathrm{Pic}(\Sigma)$ such that $2L \sim B' + C'_1 + \cdots + C'_k$ and $\rho'_* \mathcal{O}_{S'} = \mathcal{O}_\Sigma \oplus L^{-1}$. The adjunction formula gives

$$2K_{S'} \sim \rho'^*(2K_\Sigma + B' + C'_1 + \cdots + C'_k)$$

and by the projection formula for double covers:

$$H^0(S', 2K_{S'}) = H^0(\Sigma, 2K_\Sigma + L) \oplus H^0(\Sigma, 2K_\Sigma + B' + C'_1 + \cdots + C'_k).$$

Since the bicanonical map is composed with σ, it is easy to see that $h^0(\Sigma, 2K_\Sigma + L) = 0$. We remark that $2K_X + B_X$ is *nef* and *big*, since $2K_S \sim \omega^*(2K_X + B_X)$ is *nef* and *big*. We have the equality of $\mathbb{Q}$-divisors:

$$K_\Sigma + L = \frac{1}{2}(2K_\Sigma + B') + \frac{1}{2}(C'_1 + \cdots + C'_k).$$

The divisor $\frac{1}{2}(2K_\Sigma + B') = \frac{1}{2}\eta^*(2K_X + B_X)$ is *nef* and *big* and the divisor $\frac{1}{2}(C'_1 + \cdots + C'_k)$ is effective with normal crossings support and zero integral part. Thus $h^i(\Sigma, 2K_\Sigma + L) = 0$ for $i > 0$ by Kawamata–Viehweg vanishing and so:

$$0 = h^0(\Sigma, 2K_\Sigma + L) = \chi(2K_\Sigma + L) = 1 + K_\Sigma^2 + \frac{3}{2}L.K_\Sigma + \frac{1}{2}L^2.$$

Using again the projection formula for double covers, we get

$$3 = \chi(\mathcal{O}_{S'}) = \chi(\mathcal{O}_\Sigma) + \chi(K_\Sigma + L) = 1 + \chi(K_\Sigma + L),$$

which by Riemann–Roch is equivalent to
$$L^2 + L.K_\Sigma = 2.$$
Finally we have
$$k = K_S^2 - K_{S'}^2 = K_S^2 - 2(L + K_\Sigma)^2.$$
Using equalities above, this equation can be written as:
$$k = K_S^2 - 2h^0(\Sigma, 2K_\Sigma + L) = K_S^2. \qquad \square$$

Hence we have $\overline{M} = \pi^*(M) - E_1 - \cdots - E_d$ and by the theory of double covers, $(\Gamma')^2 = \frac{1}{2}(\overline{M})^2 = 0$. Therefore the map $\varphi_{|\Gamma'|} : \Sigma \to \mathbb{P}_1$ is a morphism and Σ is a rational *ruled* surface. Note that the curves C_i' are sections of $\varphi_{|\Gamma'|}$.

Let $\mathbb{F}_2$ denote the Hirzebruch surface $\mathbf{P}(\mathcal{O}_{\mathbb{P}_1} \oplus \mathcal{O}_{\mathbb{P}_1}(-2))$. If Σ is not minimal, then there is a birational morphism $\epsilon : \Sigma \to \mathbb{F}_2$ such that $\Gamma := \epsilon(\Gamma')$ and $C_0 := \epsilon(C_1')$ are respectively a fibre and the only (-2)-section of the morphism $f : \mathbb{F}_2 \to \mathbb{P}_1$ induced by $\varphi_{|\Gamma'|}$ (notice that since $d = K_S^2 > 0$, then there exists C_1'; in particular Γ and C_0 generate $\mathrm{Pic}(\mathbb{F}_2)$).

Remark that ϵ is obtained by contracting all (-1)-cycles in fibres of $\varphi_{|\Gamma'|}$ which do not meet C_1' and that it is the blow up of the surface $\mathbb{F}_2$ at $n > 0$ points, which we will denote by $P_1, \ldots, P_n$.

A priori some of these points could be infinitely near to the surface $\mathbb{F}_2$, for semplicity we will make the following assumption:

(†) *the points $P_1, \ldots, P_n$ are n distinct points on $\mathbb{F}_2$ (i.e. there are no infinitely near points).*

The discussion of the other case is only technically more complicated, but no new idea really comes into play.

So we have a generically finite morphism of degree two
$$\rho := \epsilon \circ \rho' : S' \to \mathbb{F}_2,$$
branched along the curve $B := \epsilon_*(B' + \sum_{i=1}^d C_i')$.

Remark that since ρ' is finite, then a curve D' on Σ is contracted by ϵ if and only if there exists a curve D on S' such that $\rho'(D) = D'$ and ρ contracts D.

Since no curve meeting C_1' is contracted, one has:
$$(B - C_0) \cap C_0 = \emptyset,$$
because $B' + \sum_{i=1}^d C_i'$ is smooth.

On the other hand the morphism $\rho\mid_{\overline{M}} : \overline{M} \to \Gamma \cong \mathbb{P}_1$ is a double cover for each general $M \in |K_S|$, so by Hurwitz formula:
$$B \sim (2K_S^2 + 4)C_0 + (4K_S^2 + 6)\Gamma.$$

A similar argument applies to the curves $C_i := \epsilon(C_i') = \rho(E_i)$, $i = 2, \ldots, d$, to show that they are distinct sections of f such that $C_i \cap C_0 = \emptyset$ and $C_i \sim C_0 + 2\Gamma$, $i = 2, \ldots, d$.

In particular $C_i^2 = 2 = (C_i')^2 + 4$. Hence by our assumption (†), it is easily seen that for every $i = 2, \ldots, d$, there exist exactly four curves on S' intersecting E_i, each contracted by ρ to a point on C_i.

Let z be a singular point of $B = \rho_*(R' + E_1 + \cdots + E_d)$. Since the curve $R' + E_1 + \cdots + E_d$ is smooth, there is some irreducible curve C on S' contracted by ρ to z, in particular we have that C is contained in some reducible curve of $|\overline{M}|$ and $C \neq E_i$, $i = 1, \ldots, d$.

As it is well known, the contraction of the curve C produces a non essential singularity at z if and only if C is a (-2)-curve, thus if and only if

$$0 = C.K_{S'} = C.(\overline{M} + 2\sum_{i=1}^{d} E_i) = 2\sum_{i=1}^{d} C.E_i.$$

Hence z is a non essential singular point of B if and only if $C \cap E_i = \emptyset$, $i = 1, \ldots, d$, and then if and only if $z \notin C_i$, $i = 1, \ldots, d$.

So we can write $B = \widehat{B} + C_0 + \Psi$ where $\Psi := \sum_{i=2}^{d} C_i$ and $\widehat{B}$ is effective. Summarizing we have

Proposition 4. *Let B and Ψ be as above. Then* $\mathrm{Sing}(B) \subseteq \{P_1, \ldots, P_n\}$ *and*

(i) *the essential singular points of the branch curve B all lie on the curve $\Psi = \sum_{i=2}^{d} C_i$;*

(ii) *every singular point of B lying on Ψ is an essential singular point of B;*

(ii) *on each C_i, $i = 2, \ldots, d$, there are exactly four singular points of B.* □

Note that if $d = 1$ then $|K_S|$ is a pencil of curves of genus

$$p_a(M) = \frac{1}{2}M.(M + K_S) + 1 = d + 1 = 2$$

and S presents the *standard* case. Hence we can assume $\Psi \neq \emptyset$ and by the above proposition, there exist at least four essential singular points of B. Without loss of generality we can assume that $\mathrm{Sing}(B) = \{P_1, \ldots, P_{n'}\}$, $4 \leq n' \leq n$.

Since B is the branch curve of a morphism of degree two, then there is some $F \in \mathrm{Pic}(\mathbb{F}_2)$ such that $B \in |2F|$ and we can consider the double cover $f_0 : S_0 \to \mathbb{F}_2$ branched along B. Since B is singular, S_0 is also singular and we introduce the *canonical resolution* S^* (cf. [HQ]).

In particular S^* is the double cover of a non singular surface W which in our hypotheses is the blow-up of $\mathbb{F}_2$ at $n' + n''$ points $P_1, \ldots, P_{n'}, x_1, \ldots, x_{n''}$,

$$\begin{array}{ccc} S^* & \to & S_0 \\ \downarrow & & \downarrow \\ W & \stackrel{a}{\to} & \mathbb{F}_2 \end{array}$$

Remark that $x \in \{x_1, \ldots, x_{n''}\}$ if and only if there exists an $i \in \{1, \ldots, n'\}$ such that $\text{mult}_{P_i}(B)$ is odd and x is an infinitely near point to P_i which lies on B (cf. [HQ]).

Now we have introduced all the elements we need and the announced classification may be obtained in 3 steps.

Step I. $S' \cong S^*$.

Proposition 5. *Under the above assumptions:*

(i) $S^* \cong S'$;

(ii) $\text{Sing}(B) = \{P_1, \ldots, P_n\}$, $\{x_1, \ldots, x_{n''}\} = \emptyset$ and

(iii) $\text{mult}_{P_i}(B)$ is even, $i = 1, \ldots, n$.

Proof. (i) A trivial verification shows that the morphism $\rho : S' \to \mathbb{F}_2$ is relatively minimal, i.e. there are no (-1)-curves contracted by ρ. Then S' is the *minimal resolution* of singularities of S_0 (cf. [H,III, Lemma 2.1]), and there is a birational morphism $\vartheta : S^* \to S'$.

The exceptional curves of the morphism ϑ are disjoint (-1)-curves contained in the ramification curve of the double cover $S^* \to W$ (cf. [CaFe], Theorem 10.1 and Lemma 10.2); if ϑ were not an isomorphism, there would be at least one exceptional curve Θ contracted by ϑ to a point $x \in S'$.

Since the ramification curve of the double cover $S^* \to W$ is smooth, it follows that x would be an isolated fixed point of the involution σ', which is impossible.

(ii) is a consequence of (i. Indeed we see that $W \cong \Sigma$, hence $W \to \mathbb{F}_2$ is the blow up of $\mathbb{F}_2$ at the points $P_1, \ldots, P_n$. In particular $\{x_1, \ldots, x_{n''}\} = \emptyset$ and (iii) follows by the above remark. $\square$

Definition. We will say that a surface X is a double plane if it is birationally equivalent to a degree two cover $Y \to \mathbb{P}_2$.

Step II. S' is a double plane.

The following theorem yields information about the singularities of B:

Proposition 6. *Let P_i be a singular point of B. Then*

$$\text{mult}_{P_i}(B) = 2\,\text{mult}_{P_i}(\Psi) + 2.$$

Proof. Consider the commutative diagram:

$$\begin{array}{ccc} S' \cong S^* & \xrightarrow{\pi} & S \\ \rho' \downarrow & \searrow \rho & \\ \Sigma \cong W & \xrightarrow{\epsilon} & \mathbb{F}_2 \end{array}$$

By the theory of double covers, the branch curve of ρ' is
$$B_* \sim \rho'_*(K_{S'} - \rho'^*(K_W)).$$

Let $\mathcal{E}_1, \ldots, \mathcal{E}_n$ be the exceptional curves of ϵ, which by our assumption (†) are disjoint (-1)-curves, so that
$$K_W \sim \epsilon^*(K_{\mathbb{F}_2}) + \sum_{i=1}^n \mathcal{E}_i \sim \epsilon^*(-2C_0 - 4\Gamma) + \sum_{i=1}^n \mathcal{E}_i.$$

By Lemma 3, $K_{S'} \sim \overline{M} + 2\sum_{i=1}^d E_i$, and by the commutativity of the diagram above it is easy to check that
$$\rho'_*(\overline{M}) \sim 2\epsilon^*(\Gamma),$$
$$\rho'_*(E_1) \sim \epsilon^*(C_0),$$
$$\rho'_*\left(\sum_{i=2}^d E_i\right) \sim \epsilon^*(\Psi) - \sum_{i=1}^n \mathrm{mult}_{P_i}(\Psi)\mathcal{E}_i$$
$$\sim (d-1)\epsilon^*(C_0 - 2\Gamma) - \sum_{i=1}^n \mathrm{mult}_{P_i}(\Psi)\mathcal{E}_i;$$

so we get
$$B_* \sim (2d+4)\epsilon^*(C_0) + (4d+6)\epsilon^*(\Gamma) - \sum_{i=1}^n (2\,\mathrm{mult}_{P_i}(\Psi) + 2)\mathcal{E}_i.$$

On the other hand since S^* is the *canonical resolution*,
$$B_* \sim \epsilon^*(B) - \sum_{i=1}^n \mu_i \mathcal{E}_i \sim \epsilon^*((2K_S^2 + 4)C_0 + (4K_S^2 + 6)\Gamma) - \sum_{i=1}^n \mu_i \mathcal{E}_i,$$

where $\mu_i := 2\left[\frac{\mathrm{mult}_{P_i}(B)}{2}\right]$, (cf. [HQ]).

Combining these we conclude that $\mu_i = 2\,\mathrm{mult}_{P_i}(\Psi) + 2$ and by part (iii) of Proposition 5, it follows that $\mathrm{mult}_{P_i}(B) = 2\,\mathrm{mult}_{P_i}(\Psi) + 2\ i = 1, \ldots, n$. □

Since we can assume $\Psi \neq \emptyset$, by Proposition 4 there is at least one point $x \in \Psi$ which is an essential singular point of B. Now we consider the "projection" from the point x of $\mathbb{F}_2$ onto the plane $\mathbb{P}_2$
$$\varphi' : \mathbb{F}_2 \to \mathbb{P}_2,$$
that is the rational map defined by the linear system $|C_0 + 2\Gamma - x|$. Let
$$\varphi := \varphi' \circ \rho : S' \to \mathbb{P}_2$$

be the composition, which is a generically finite rational map of degree two branched along the curve:
$$\beta := \varphi_*(R' + E_1 + \cdots + E_d).$$

Since x is an essential singular point of B and $S' \cong S^*$, one checks that φ is a morphism (indeed blowing up x is a step in the canonical resolution) and that S^* is the *canonical resolution* of the double cover of $\mathbb{P}_2$ branched along β.

Setting $\delta := \mathrm{mult}_x(\Psi)$ we state the fundamental

Proposition 7. *The branch curve β of the morphism φ is such that:*

(i) $\beta \sim (4K_S^2 + 4 - 2\delta)l$, $1 \leq \delta \leq d - 1$ ($\mathcal{O}_{\mathbb{P}_2}(l) \cong \mathcal{O}_{\mathbb{P}_2}(1)$);

(ii) $\beta = \beta' + \Psi'$, *where* $\Psi' := \varphi(E_2 + \cdots + E_d)$ *and* $\beta' > 0$;

(iii) $\Psi' = \sum_{i=2}^{\delta+1} l_i + \sum_{j=\delta+2}^{d} c_j$, *where the* $l_i = \varphi(E_i)$, $i = 2, \ldots, \delta + 1$ *are δ distinct lines and the* $c_j = \varphi(E_i)$, $j = \delta + 2, \ldots, d$ *are $d - \delta - 1$ distinct smooth conics.*

Moreover, denoting by $m_y := \mathrm{mult}_y(\beta)$ *and by* $n_y := \mathrm{mult}_y(\Psi')$:

(iv) $q = \varphi(E_1)$ *is a point and β has a singularity of type* $\left[2K_S^2 + 1 - 2\delta, 2K_S^2 + 1 - 2\delta\right]$ *at q*;

(v) $q \notin l_i$, $i = 2, \ldots, \delta + 1$, *and* $q \in c_j$, $j = \delta + 2, \ldots, d$;

(vi) *if $y \neq q$ is an essential singular point of β, then $y \in \Psi'$ and $m_y = 2n_y + 2$.*

Proof. It follows by proposition 6 and by observing that each irreducible curve $C \in |C_0 + 2\Gamma|$ is mapped by φ' either to a line on $\mathbb{P}_2$ if $x \in C$, or to a smooth conic if $x \notin C$. □

Since the essential singularities of β all lie on $\Psi' = \varphi(E_2 + \cdots + E_d)$ and the multiplicity of Ψ' at a singular point determines the multiplicity of β, it is important to study the curve Ψ'. First of all we observe that since the curves $E_2, \ldots, E_d$ on S' are (-1)-curves, there are exactly three distinct or infinitely near points lying on every l_i, each corresponding to a curve on S' contracted by φ to a singular point of β, which (similarly to what we saw before in an analogous situation) is essential. Analogously there are exactly six such singular points on every c_j.

Clearly each singular point of $\Psi' \subset \beta$ is singular also for β. Without loss of generality we can choose x such that $\delta = \mathrm{mult}_x(\Psi)$ is maximum.

The following result is merely combinatorial:

Proposition 8. *The curve $\Psi' = \sum_{i=2}^{\delta+1} l_i + \sum_{j=\delta+2}^{d} c_j$ is the union of $1 \leq \delta \leq d - 1$ distinct lines and of $d - \delta - 1$ distinct smooth conics such that:*

(i) *each line l_i meets the other curves of Ψ' in at most three points, including the infinitely near ones;*

(ii) *the conics c_j, $j = \delta + 2, \ldots, d$, are all tangent at the point q, which does not lie on any line l_i, $i = 2, \ldots, \delta + 1$;*

(iii) *each conic c_j meets the other curves of Ψ' in at most four points different by q, including the infinitely near ones;*

(iv) *for every point $y \in \Psi'$, one has $\mathrm{mult}_y(\Psi') \leq \delta$.*

According to (i), ..., (iv) the possible configurations of Ψ' are the following:

either $\delta = K_S^2 - 1$ and all the lines pass through the same point,

or $4 \leq K_S^2 \leq 8$ and:

(a$_\mathrm{I}$) $K_S^2 = 4$, $\delta = 3$: *the three lines form a triangle;*

(a$_\mathrm{II}$) $K_S^2 = 4$, $\delta = 2$: *the two lines meet at a point x and the conic meets the lines at two distinct pairs of points none of which contains x;*

(b$_\mathrm{I}$) $K_S^2 = 5$, $\delta = 4$: *the four lines are the edges of a quadrilateral;*

(b$_\mathrm{II}$) $K_S^2 = 5$, $\delta = 4$: *three of the lines meet at one point which does not belong to the fourth line;*

(c$_\mathrm{I}$) $K_S^2 = 5$, $\delta = 3$: *the three lines form a triangle and the conic passes at least through two of the three vertices of the triangle;*

(c$_\mathrm{II}$) $K_S^2 = 6$, $\delta = 5$: *four of the lines are the edges of a quadrilateral and the fifth line joins two opposite vertices;*

(d$_\mathrm{I}$) $K_S^2 = 6$, $\delta = 4$: *the four lines are the edges of a quadrilateral and the conic passes through four of the vertices;*

(d$_\mathrm{II}$) $K_S^2 = 7$, $\delta = 6$: *the six lines are the sides of a complete quadrangle;*

(e) $K_S^2 = 7$, $\delta = 5$: *four of the lines are the edges of a quadrilateral, the fifth line joins two opposite vertices, say x_1 and x_2, and the conic passes through four of the vertices such that: two of them are x_1 and x_2, and no three of them are collinear.*

(a$_\mathrm{I}$) $K_S^2 = 8$, $\delta = 7$: *the six lines are the sides of a complete quadrangle and the conic passes through the four triple vertices of quadrangle.*

We will denote by X_{a_I}, $X_{a_\mathrm{II}}, \ldots, X_{d_\mathrm{II}}$, X_e the double planes corresponding respectively to the configurations (a$_\mathrm{I}$), (a$_\mathrm{II}$), ..., (d$_\mathrm{II}$), (e) of Ψ' and by X_* one of them.

Step III. Conclusion.

Now we give a sketch of the proof of

Theorem 9. *Let S be a smooth minimal surface of general type with $p_g(S) = 2$, $q(S) = 0$ and non-birational bicanonical map. Assume that $|K_S|$ has no fixed part. Then either we are in the standard case or S is a Du Val surface and $4 \leq K_S^2 \leq 8$.*

Proof. As we have seen above S is birationally equivalent to a surface S' which is a double plane whose branch curve $\beta = \beta' + \Psi'$ is described in propositions 7 and 8. It is easy to see that if Ψ' is composed by $\delta = d - 1$ lines passing through the same point x, then S' has a genus two pencil, that is the pull-back of the pencil of lines through x. Hence if S does not present the *standard* case, the surface S' is birationally equivalent to a double plane X_*.

So for the proof it is sufficient to show that each double plane X_* is birationally equivalent to a Du Val surface.

A case by case analysis of the ten situations described above shows that for each one of them there is a quadratic transformation of the plane $Q_* : \mathbb{P}_2 \dashrightarrow \mathbb{P}_2$ such that, denoting by β_* the transform of β via Q_*, the double plane branched along β_* is a Du Val surface.

For example we discuss the case (e). The branch curve of the morphism $\varphi : S' \to \mathbb{P}_2$ is $\beta = \beta' + \Psi'$ where β' is a curve of degree 24 and Ψ' is the union of 6 lines $l_2, \ldots, l_7$, which are the sides of a complete quadrangle Ξ, and a conic c_8 which passes through the four triple vertices x_1, x_2, x_3, x_4 of Ξ. Let y_1, y_2, y_3 be the double vertices of Ξ. By Proposition 7 the only singularities of β are: four ordinary 8-tuple points at x_i, $i = 1, \ldots, 4$, three ordinary 6-tuple points at y_j, $j = 1, 2, 3$ and a $[5, 5]$-point at $q \in c_8$.

Let us make the quadratic transformation of the plane based at three of the points x_i.

The transformation contracts to three distinct points $\alpha_1, \alpha_2, \alpha_3$ the three lines on which two of the three base points lie. The other three lines of Ξ are mapped to different lines a_1, a_2, a_3 such that $\alpha_i \in a_j$ if and only if $i = j$. The conic c_8 is mapped to a line a_4.

The four lines a_i all meet at a point γ and the curve β' is in turn mapped to a curve β_*' of degree 14 with an ordinary 6-tuple point at γ and four $[4, 4]$-tuple points at α_i where the infinitely near 4-tuple point to α_i lies on the line a_i, $i = 1, 2, 3, 4$. Note that a_4 is the image via Q_* of q.

Hence S' is a Du Val surface (cf. Definition D_2). A similar analysis works for the other cases. We leave the details to the reader. □

References

[B] E. Bombieri, Canonical models of surfaces of general type, Inst. Hautes Études Sci. Publ. Math. 42 (1973), 447–495.

[Bo] G. Borrelli, Sulle superfici minimali di tipo generale con $p_g = 2, q = 0$ e mappa bicanonica non birazionale, tesi di dottorato, Università di Genova, 2002.

[CaFe] A. Calabri , R. Ferraro, Explicit resolution of double point singularities of surfaces, Collect. Math. 55 (2002), 99–131.

[CCM] F. Catanese, C. Ciliberto, M. Mendes Lopes, On the classification of irregular surfaces of general type with non birational bicanonical map, Trans. Amer. Math. Soc. 350 (1998), 275–308.

[C] C. Ciliberto, The bicanonical map for surfaces of general type, Proc. Sympos. Pure Math. 62 (1997), 57–84.

[CFM] C. Ciliberto, P. Francia, M. Mendes Lopes, Remarks on the bicanonical map for surfaces of general type, Math. Z. 224 (1997), 137–166.

[CM,1] C. Ciliberto, M. Mendes Lopes, On regular surfaces of general type with $p_g(S) = 3$ and non birational bicanonical map, J. Math. Kyoto Univ. 40 (2000), 79–117.

[CM,2] C. Ciliberto, M. Mendes Lopes, On surfaces of general type with $p_g(S) = q = 2$ and non birational bicanonical map, to appear in Advances in Geom.

[CM,3] C. Ciliberto, M. Mendes Lopes, On surfaces with $p_g = 2, q = 1$ and non-birational bicanonical map, in: Algebraic Geometry (M. C. Beltrametti, F. Catanese, C. Ciliberto, A. Lanteri and C. Pedrini, eds.), Walter de Gruyter, Berlin 2002, 117–126..

[DV] P. Du Val, On surfaces whose canonical system is hyperelliptic, Canad. J. Math. 4 (1952), 204–221.

[HQ] E. Horikawa, On deformations of quintic surfaces, Invent. Math. 31 (1975), 43–85.

[HIII] E. Horikawa, Algebraic surfaces of general type with small c_1^2, III , Invent. Math. 47 (1978), 209–248.

[R] I. Reider, Vector bundles of rank 2 and linear systems on algebraic surfaces, Ann. of Math. 127 (1988), 309–316.

[X] G. Xiao, Hyperelliptic surfaces of general type, Amer. J. Math. 112 (1990), 309–316.

G. Borrelli, Via Volci 49, 00052 Cerveteri (Roma), Italy
E-mail: borrelli@mat.uniroma3.it

Canonical projections of irregular algebraic surfaces

Fabrizio Catanese and Frank-Olaf Schreyer*

We dedicate this article to the memory of Paolo Francia

1. Introduction

In Chapter VIII of his book "Le superficie algebriche" F. Enriques raised the question to describe concretely the canonical surfaces with $p_g = 4$ and discussed possible constructions of the regular ones of low degree.

A satisfactory answer to the existence question for degree ≤ 10 was given by Ciliberto [1981], and for the case of regular surfaces a satisfactory structure theorem for the equations of the image surface and its singular locus was achieved by the first author [1984b].

The main purpose of this paper is to extend those results to the case of irregular surfaces.

Irregular varieties are often easy to construct via transcendental methods, as is the case for elliptic curves or Abelian varieties. But the problem of explicitly describing the equations of their projective models has always been a challenge for algebraic geometry (cf. Enriques [1949], Mumford [1966–67]).

From an algebraic point of view, we might say that irregularity is responsible for the failure of the Cohen–Macaulay condition for the canonical ring of a variety of general type.

In this context therefore the method of Hilbert resolutions must be replaced by another method, and we show in this paper that Beilinson's theorem [1978] allows to give a suitable generalization of the structure theorem of the first author [1984b].

The first two sections are devoted to this extension, and the situation that we consider is the following: φ is a morphism to $\mathbb{P}^3$ given by four independent sections of the canonical bundle K_S of a minimal surface of general type, and we assume that the degree of φ is at most two. The following is the main theorem, concerning the case where $\deg \varphi = 1$.

---

*The present research started in 1993 while both authors were visiting the Max Planck Institut für Mathematik in Bonn. The full results presented here were later obtained in 1994, when also the second author visited the University of Pisa. The research continued in the framework of the Schwerpunkt "Globale Methode in der komplexen Geometrie", and of EAGER.

Theorem 2.9. *The datum $\varphi : S \to Y \subset \mathbb{P}^3 = \mathbb{P}(V)$ of a good birational canonical projection determines a homomorphism*

$$(\mathcal{O}_\mathbb{P} \oplus E)^*(-5) \xrightarrow{\alpha} (\mathcal{O}_\mathbb{P} \oplus E),$$

where $E = (K^2 - q + p_g - 9)\mathcal{O}_\mathbb{P}(-2) \oplus q\Omega^1_\mathbb{P}(-1) \oplus (p_g - 4)\Omega^2_\mathbb{P}$, such that

   (i) *α is symmetric,*

   (ii) *$\det \alpha$ is an irreducible polynomial (defining Y),*

   (iii) *α satisfies the ring condition (cf. 2.7), and,*

   - *defining $\mathcal{F}$ as the sheaf of $\mathcal{O}_Y$-algebras given by the module $\mathrm{coker}(\alpha)$ provided with the ring structure determined by α as in 2.7(1),*

   (iv) *$\mathrm{Spec}\,\mathcal{F}$ is a surface with at most rational double points as singularities.*

Conversely, given α satisfying (i), (ii), (iii) and (iv), $X = \mathrm{Spec}\,\mathcal{F}$ is the canonical model of a minimal surface S and

$$\varphi : S \to X \to Y \subset \mathbb{P}^3 = \mathbb{P}(V)$$

is a good birational canonical projection.

We recall (cf. 2.7) that the ring condition, or rank condition, on a matrix α is the condition that the ideal sheaf generated by the top minors of the matrix α' obtained by deleting the first row of α equals the ideal sheaf generated by the minors of α of the same size.

The following sections are more in the spirit of Enriques, and we discuss explicitly the irregular canonical surfaces with $p_g = 4$ of lowest degree, for instance $K^2 = 12$ in the case of irregularity $q = 1$.

We classify completely the above surfaces with $p_g = 4, q = 1$: they can be described as a genus three non hyperelliptic fibration over an elliptic base curve.

This classification shows that the corresponding moduli space has only one irreducible component (cf. Theorem 5.10), and we dwell over the geometry of a dense open set of the moduli space, corresponding to surfaces which we name "of the main stream".

Examples with $p_g = 4, K^2 = 12, q = 3$ are the polarizations of type $(1, 1, 2)$ on an Abelian threefold: a remarkable subfamily of surfaces for which the canonical map becomes a degree two covering of a canonical surface with $K^2 = 6$ is given by the "special" surfaces which are the pull backs, under a degree two isogeny, of the theta divisor of a principal polarization.

A further example, with $p_g = 4, q = 2$ and $K^2 = 18$ is provided by certain Abelian covers with Galois group $(\mathbb{Z}/2)^2$ of a principally polarized Abelian surface.

The results presented in this paper were announced in [Catanese, 1997] and very recently A. Canonaco [2002] was able to extend the method for canonical projections to a weighted 3-dimensional projective space.

There are still many questions which this paper leaves open:

- A precise description of the structure theorem in the general degree two case, without the assumption that Y be normal
- The extension of the structure theorem to the case of higher dimensional varieties (cf. however [Catanese 1985] in the "pluriregular case")
- The complete classification of canonical surfaces of low degree (this is still open also in the regular case, as soon as $K^2 \geq 8$, see [Catanese 1984b] for the case $K^2 = 7$)
- The construction of new examples without transcendental methods but via computer algebra.
- Decide when is the corresponding moduli space unirational and, in this case, give an explicit rationally parametrized family

Notation

$S :=$ the minimal model of a surface of general type;
$\mathcal{R} = \mathcal{R}(S) :=$ the canonical ring of S;
$X = \mathrm{Proj}(\mathcal{R})$ is the canonical model of S;
$\pi : S \to X$ is the canonical morphism.

K is a canonical divisor on X or on S (note: $\pi^*(K_X) = K_S$);
$\varphi : S \to Y \subset \mathbb{P}^3$ is a good canonical projection, i.e., φ is given by a base point free 4-dimensional subspace V^* of $H^0(\mathcal{O}_S(K))$ (here V^* denotes the dual vector space of V, and $\mathbb{P}^3 = \mathbb{P}(V)$ is the projective space of 1-dimensional vector subspaces of V).

φ is said to be almost generic if it is good and yields a morphism which is either birational to Y or of degree 2 onto Y (in [Catanese 1984b] φ was said to be quasi-generic if moreover Y is normal in case $\deg(\varphi) = 2$).

$\varphi : S \to Y$ factors through π and a finite morphism $\psi : X \to Y$;
$\{y_0, y_1, y_2, y_3\}$ is a basis of $V^* \subset H^0(\mathcal{O}_S(K)) \cong H^0(\mathcal{O}_X(K))$;
$V^* \oplus W^* \cong H^0(\mathcal{O}_S(K)) \cong H^0(\mathcal{O}_X(K))$ is a fixed splitting;
$\mathcal{A}$ is the graded polynomial ring $\mathbb{C}[y_0, y_1, y_2, y_3] = \mathrm{Sym}(V^*)$.

$M^\sim$, for a graded $\mathcal{A}$-module M, denotes the associated sheaf on $\mathbb{P}^3$; therefore we shall consider

$$\mathcal{R}^\sim = \varphi_* \mathcal{O}_S = \psi_* \mathcal{O}_X = (\psi_* \omega_X)(-1).$$

$\mathcal{T}_Z$ will denote the tangent sheaf of a quasi-projective scheme Z.

Given Cartier divisors D, D', $D \equiv D'$ means that D is linearly equivalent to D', while $D \sim D'$ means that D is algebraically equivalent to D'.

2. Determinantal presentations of canonical projections

By Beilinson's theorem [1978], for any coherent sheaf $\mathcal{F}$ on a projective space $\mathbb{P}$ there is a complex

$$\cdots \to \mathcal{C}^i(\mathcal{F}) = \bigoplus_{q-p=i} H^q(\mathcal{F}(-p)) \otimes_{\mathbb{C}} \Omega_{\mathbb{P}}^p(p) \to \mathcal{C}^{i+1}(\mathcal{F}) \to \cdots$$

whose cohomology is concentrated in degree 0 and yields $\mathcal{F}$. For a new proof we also refer to the paper of Eisenbud, Fløystad and Schreyer [2001]. In particular, there is a spectral sequence with $E_1^{-p,q}$ equal to $H^q(\mathcal{F}(-p)) \otimes_{\mathbb{C}} \Omega_{\mathbb{P}}^p(p)$, and with $d_1^{-p,q} : E_1^{-p,q} \to E_1^{-p+1,q}$ given by the identity tensor

$$\text{id} = (\Sigma_j y_j \otimes y_j^*) \in V^* \otimes_{\mathbb{C}} V = H^0(\mathcal{O}_{\mathbb{P}}(1)) \otimes_{\mathbb{C}} \text{Hom}(\Omega_{\mathbb{P}}^1(1), \mathcal{O}_{\mathbb{P}})$$

which acts according to the tensor rule

$$(x \otimes e)(s \otimes \omega) = (xs) \otimes (\omega \neg e),$$

$\neg$ denoting contraction of a contravariant tensor with a vector.

From now on we let $\mathcal{F}(m)$ be $= \psi_* \mathcal{O}_X(m)$, with $m = 0, 2$ or 3.
Later on, we shall simply denote $\mathcal{F}(0)$ by $\mathcal{F}$.
The Beilinson table in the particular case $m = 3$ reads out as:

0	0	0	0
$H^2(\mathcal{O}_X)$	$H^2(\mathcal{O}_X(K))$	0	0
$H^1(\mathcal{O}_X)$	$H^1(\mathcal{O}_X(K))$	0	0
$H^0(\mathcal{O}_X)$	$H^0(\mathcal{O}_X(K))$	$H^0(\mathcal{O}_X(2K))$	$H^0(\mathcal{O}_X(3K))$

since $H^3(\psi_* \mathcal{O}_X(i)) = 0$ for all i,

$$H^2(\psi_* \mathcal{O}_X(i)) = H^2(\mathcal{O}_X(iK)) = H^0(\mathcal{O}_X(-(i-1)K))^* = 0 \quad \text{for } i \geq 2,$$

and

$$H^1(\psi_* \mathcal{O}_X(i)) = H^1(\mathcal{O}_X(iK)) = H^1(\mathcal{O}_X(-(i-1)K))^* = 0 \quad \text{for } i \neq 0, 1.$$

From Beilinson's theorem we obtain with a simpler proof a stronger version of a result by Ciliberto:

Theorem 2.1 (Ciliberto, Thm. 2.4 (iii) 1983). *If $p_g \geq 4$ and $|K|$ has no base points, then $\mathcal{R}$ is generated in degrees ≤ 3 as an $\mathcal{A}$-module.*

Proof. The d_1-differential on the top row of the Beilinson spectral sequence for $\mathcal{F}(3) = \psi_* \mathcal{O}_X(3)$ has the form

$$H^2(\mathcal{O}_X) \otimes_{\mathbb{C}} \Omega_{\mathbb{P}}^3(3) = (W \oplus V) \otimes_{\mathbb{C}} \Omega_{\mathbb{P}}^3(3) \to H^2(\mathcal{O}_X(K)) \otimes_{\mathbb{C}} \Omega_{\mathbb{P}}^2(2) = \Omega_{\mathbb{P}}^2(2)$$

where the first summand maps to 0, and the second maps according to the twisted Serre dual of the Euler sequence: therefore it is surjective with kernel $\mathcal{K}$ isomorphic to a direct sum $(W \otimes_{\mathbb{C}} \mathcal{O}_{\mathbb{P}}(-1)) \oplus \mathcal{O}_{\mathbb{P}}(-2) \cong [(p_g - 4)]\mathcal{O}_{\mathbb{P}}(-1) \oplus \mathcal{O}_{\mathbb{P}}(-2)$.

Hence not only $\mathcal{F}(3)$ is a quotient of $H^0(\mathcal{O}_X(3K)) \otimes_{\mathbb{C}} \mathcal{O}_{\mathbb{P}}$, but $\mathcal{F}(3)$ has a locally free resolution by sheaves which are direct sums of sheaves $\mathcal{G}$ isomorphic to either $\mathcal{O}_{\mathbb{P}}(-2)$, or $\Omega_{\mathbb{P}}^p(p)$. Such sheaves $\mathcal{G}$ have the property that $H^j(\mathcal{G}(m)) = 0$ for $j > 0$ and $m \geq 0$.

Therefore, if we tensor this resolution by $\mathcal{O}_{\mathbb{P}}(m)$, $m \geq 0$, it remains exact on global sections, in particular we have that

$$H^0(\mathcal{O}_X(3K)) \otimes_{\mathbb{C}} H^0(\mathcal{O}_{\mathbb{P}}(m-3)) \to H^0(\mathcal{O}_X(mK))$$

is surjective for $m \geq 3$. $\square$

In the case $m = 2$ the Beilinson table reads out as

0	0	0	0
$H^2(\mathcal{O}_X(-K))$	$H^2(\mathcal{O}_X)$	$H^2(\mathcal{O}_X(K))$	0
0	$H^1(\mathcal{O}_X)$	$H^1(\mathcal{O}_X(K))$	0
0	$H^0(\mathcal{O}_X)$	$H^0(\mathcal{O}_X(K))$	$H^0(\mathcal{O}_X(2K))$

and the symmetry with respect of the middle point of the second row from the bottom takes each vector space to its Serre dual and each d_1-differential to its Serre dual map.

Sublemma 2.2. *The d_1-differential from*

$$H^0(\mathcal{O}_X) \otimes \Omega_{\mathbb{P}}^2(2) \to H^0(\mathcal{O}_X(K)) \otimes \Omega_{\mathbb{P}}^1(1)$$

is an isomorphism onto a subbundle.

Proof. d_1 factors through the natural map of $\Omega_{\mathbb{P}}^2(2) \to V^* \otimes \Omega_{\mathbb{P}}^1(1)$ and the subbundle inclusion of $V^* \otimes \Omega_{\mathbb{P}}^1(1)$ inside $H^0(\mathcal{O}_X(K)) \otimes \Omega_{\mathbb{P}}^1(1)$, hence it suffices to show that the first map is a subbundle inclusion. But this follows from the Beilinson complex for $\mathcal{O}_{\mathbb{P}}(2)$ which yields the exact sequence

$$0 \to \Omega_{\mathbb{P}}^2(2) \to V^* \otimes \Omega_{\mathbb{P}}^1(1) \to \text{Sym}^2(V^*) \otimes \mathcal{O}_{\mathbb{P}} \to \mathcal{O}_{\mathbb{P}}(2) \to 0. \quad (2.1)$$

$\square$

ψ is either birational onto Y or $2 : 1$ by assumption. In the second case one can treat separately the case where Y is a quadric surface (this leads to $K_S^2 = 4, q = 0$, cf. Enriques [1949] pages 270-271, i.e., to a double cover of $\mathbb{P}^1 \times \mathbb{P}^1$ branched on a curve of type $(3, 3)$).

Therefore we assume from now on that

Assumption 2.3. *Y is not a quadric.*

In algebraic terms, this means that $H^0(\mathcal{O}_{\mathbb{P}}(2)) = \text{Sym}^2(V^*)$ is a direct summand of $H^0(\mathcal{O}_X(2K))$, so we can choose a splitting

$$H^0(\mathcal{O}_X(2K)) \cong \text{Sym}^2(V^*) \oplus U^*. \quad (2.2)$$

Moreover, by (2.1), we can replace the Beilinson complex by a homotopic one, which gives a new diagram

$$
\begin{array}{cccc}
0 & 0 & 0 & 0 \\
\mathcal{O}(-3) \oplus (U \otimes \mathcal{O}(-1)) & W \otimes \Omega^2(2) & 0 & 0 \\
0 & H^1(\mathcal{O}_X) \otimes \Omega^2(2) & H^1(\mathcal{O}_X(K)) \otimes \Omega^1(1) & 0 \\
0 & 0 & W^* \otimes \Omega^1(1) & (U^* \otimes \mathcal{O}) \oplus \mathcal{O}(2)
\end{array}
$$

Again, here, there is a symmetry taking vector spaces and linear maps to their Serre duals.

Let us denote by E the vector bundle on $\mathbb{P}^3$ defined by

$$E(2) = (U^* \otimes \mathcal{O}_{\mathbb{P}}) \oplus (H^1(\mathcal{O}_X(K)) \otimes \Omega^1_{\mathbb{P}}(1)) \oplus (W \otimes \Omega^2_{\mathbb{P}}(2)). \tag{2.3}$$

We have therefore concluded that $\mathcal{F} = \psi_* \mathcal{O}_X$ admits a locally free resolution of length 1 of the form

$$0 \to (\mathcal{O}_{\mathbb{P}} \oplus E)^*(-5) \to \mathcal{O}_{\mathbb{P}} \oplus E \to \mathcal{F} = \psi_* \mathcal{O}_X \to 0. \tag{2.4}$$

Remark that again here, for each twist $m \geq 2$, (2.4) is exact on global sections. Now, the two locally free terms of the resolution are dual to each other up to a twist, and we shall see that one can indeed achieve that the resolution itself is given by a symmetric map

$$\alpha : (\mathcal{O}_{\mathbb{P}} \oplus E)^*(-5) \to \mathcal{O}_{\mathbb{P}} \oplus E \quad \text{(that is, } \alpha = \alpha^*(-5)\text{)}.$$

For simplicity of notation we denote by α^t the map $\alpha^*(-5)$, and by F the vector bundle $\mathcal{O}_{\mathbb{P}} \oplus E$.

As a first step we state

Lemma 2.4. *Let $\mathcal{F}'$ be the cokernel of another homomorphism $\beta : F^*(-5) \to F$. Then every homomorphism of $\mathcal{F}$ to $\mathcal{F}'$ has a lift to a homomorphism of complexes. Furthermore, any lift of an isomorphism is an isomorphism of complexes.*

Proof. By the exact sequence

$$\mathrm{Hom}(F, F) \to \mathrm{Hom}(F, \mathcal{F}') \to \mathrm{Ext}^1(F, F^*(-5))$$

to get a lifting it suffices to have that

$$\mathrm{Ext}^1(F, F^*(-5)) = \mathrm{Ext}^1(F(2), F^*(-3)) = 0$$

This holds true since the summands for $F(2)$, $F^*(-3)$ are either $\Omega^j(j)$'s or of rank one (and of degree 0 or 2 for $F(2)$): moreover, by Bott's formula $H^1(\Omega^j(m)) = 0$, unless $j = 1$ and $m = 0$, and by Lemma 2 of Beilinson [1978], $\mathrm{Ext}^p(\Omega^j(j), \Omega^i(i)) = 0$ for $p \geq 1$. So every homomorphism of $\mathcal{F}$ to $\mathcal{F}'$ has a lift to a homomorphism $f : F \to F$.

Moreover f restricted to $F^*(-5)$ factors through $g : F^*(-5) \to F^*(-5)$. This proves the first statement.

Since also every homomorphism of $\mathcal{F}'$ to $\mathcal{F}$ has a lift by the same argument, it suffices to prove the second statement for $\mathcal{F}' = \mathcal{F}$ and for the identity homomorphism of $\mathcal{F}$. One lift (f, g) is therefore the identity on the complex and this is an isomorphism. Every other lift (f_1, g_1) of the same automorphism differs by a homotopy, i.e. $f_1 = f + \alpha \circ h$ and $g_1 = g + h \circ \alpha$ for some homorphism $h \in \text{Hom}(F, F^*(-5))$.

Since $H^0(\Omega_\mathbb{P}^i(m)) = 0$ for $i \geq 1, m \leq i$, and again by Beilinson's lemma, it follows that

$$\text{Hom}(F, F^*(-5)) = \text{Hom}(F(2), F^*(-3))$$
$$= \left[\text{Hom}(W, H^1(\mathcal{O}_X)) \otimes \text{End}(\Omega_\mathbb{P}^2(2))\right]$$
$$\oplus \left[\text{Hom}(W, W^*) \otimes \text{Hom}(\Omega_\mathbb{P}^2(2), \Omega_\mathbb{P}^1(1))\right]$$
$$\oplus \left[\text{Hom}(H^1(\mathcal{O}_X(K)), W^*) \otimes \text{End}(\Omega_\mathbb{P}^1(1))\right]$$

On the other hand, if we look at the summands of α involving $\text{Hom}(\Omega_\mathbb{P}^i(i), \Omega_\mathbb{P}^j(j))$, with $i, j = 1$ or 2, the only non zero one is the term in

$$\text{Hom}(H^1(\mathcal{O}_X) \otimes \Omega_\mathbb{P}^2(2), H^1(\mathcal{O}_X(K)) \otimes \Omega_\mathbb{P}^1(1)).$$

Therefore the compositions $\alpha \circ h$ and $h \circ \alpha$ are nilpotent and f_1, g_1 are isomorphisms, since f, g are identity matrices. $\square$

To obtain a symmetric resolution we apply $\mathcal{H}om_{\mathcal{O}_\mathbb{P}}(\ , \omega_\mathbb{P}(-1))$ to the sequence (2.4) and get the exact sequence

$$0 \to F^*(-5) \to F \to \mathcal{E}xt^1_{\mathcal{O}_\mathbb{P}}(\mathcal{F}, \omega_\mathbb{P}(-1)) \to 0 \qquad (2.5)$$

But $\mathcal{F}(1) = \psi_*\omega_X$, thus by relative duality for ψ, the last term is isomorphic to $\psi_*\mathcal{O}_X = \mathcal{F}$. Pick an isomorphism $\varepsilon : \mathcal{F} \to \mathcal{E}xt^1_{\mathcal{O}_\mathbb{P}}(\mathcal{F}, \omega_\mathbb{P}(-1))$. By Lemma 2.4 there is a lift

$$\begin{array}{ccccccccc}
0 & \to & F^*(-5) & \xrightarrow{\alpha} & F & \to & \mathcal{F} & \to & 0 \\
& & \downarrow g & & \downarrow f & & \downarrow \varepsilon & & \\
0 & \to & F^*(-5) & \xrightarrow{\alpha^t} & F & \to & \mathcal{E}xt^1_{\mathcal{O}_\mathbb{P}}(\mathcal{F}, \omega_\mathbb{P}(-1)) & \to & 0
\end{array}$$

Applying once more $\mathcal{H}om_{\mathcal{O}_\mathbb{P}}(\ , \omega_\mathbb{P}(-1))$ to the above diagram we get

$$\begin{array}{ccccccccc}
0 & \to & F^*(-5) & \xrightarrow{\alpha} & F & \to & \mathcal{E}xt^1_{\mathcal{O}_\mathbb{P}}(\mathcal{E}xt^1_{\mathcal{O}_\mathbb{P}}(\mathcal{F}, \omega_\mathbb{P}(-1)), \omega_\mathbb{P}(-1)) & \to & 0 \\
& & \downarrow f^t & & \downarrow g^t & & \downarrow \varepsilon' & & \\
0 & \to & F^*(-5) & \xrightarrow{\alpha^t} & F & \to & \mathcal{E}xt^1_{\mathcal{O}_\mathbb{P}}(\mathcal{F}, \omega_\mathbb{P}(-1)) & \to & 0
\end{array}$$

and we obtain a canonical isomorphism

$$\mathcal{E}xt^1_{\mathcal{O}_\mathbb{P}}(\mathcal{E}xt^1_{\mathcal{O}_\mathbb{P}}(\mathcal{F}, \omega_\mathbb{P}(-1)), \omega_\mathbb{P}(-1)) \cong \mathcal{F} \tag{2.6}$$

induced by the identity of F, since $(\alpha^t)^t = \alpha$. Under this identification $\varepsilon' = \mathcal{E}xt^1_{\mathcal{O}_\mathbb{P}}(\varepsilon, \omega_\mathbb{P}(-1))$ is the isomorphism $\mathcal{F} \to \mathcal{E}xt^1_{\mathcal{O}_\mathbb{P}}(\mathcal{F}, \omega_\mathbb{P}(-1))$ induced by g^t. Let's assume now that φ is birational: then $\text{Hom}_{\mathcal{O}_\mathbb{P}}(\mathcal{F}, \mathcal{F}) = \mathbb{C}$ and we have $\varepsilon' = \lambda \varepsilon$ for some $\lambda \in \mathbb{C}^*$. Moreover by (2.6)

$$\varepsilon = \mathcal{E}xt^1_{\mathcal{O}_\mathbb{P}}(\varepsilon', \omega_\mathbb{P}(-1)), \quad \text{since both are induced by } f,$$
$$= \mathcal{E}xt^1_{\mathcal{O}_\mathbb{P}}(\lambda \varepsilon, \omega_\mathbb{P}(-1)) = \lambda \, \mathcal{E}xt^1_{\mathcal{O}_\mathbb{P}}(\varepsilon, \omega_\mathbb{P}(-1))$$
$$= \lambda \, \varepsilon' = \lambda^2 \varepsilon,$$

i.e. $\lambda = \pm 1$. (We will see in the end that actually $\lambda = 1$.) Thus f and λg^t cover the same isomorphism ε, and so does $(f + \lambda g^t)/2$. Furthermore $(f + \lambda g^t)/2$ is an isomorphism by Lemma 2.4. We claim that $\beta = ((f + \lambda g^t)/2) \circ \alpha$ is the desired symmetric matrix. Indeed β and α have isomorphic cokernels, hence both resolve $\mathcal{F} = \psi_* \mathcal{O}_X$ and

$$\beta^t = \alpha^t \circ ((f + \lambda g^t)/2)^t = (\alpha^t \circ f^t + \lambda \alpha^t \circ g)/2 = (g^t \circ \alpha + \lambda f \circ \alpha)/2 = \lambda \beta$$

is either symmetric or skew-symmetric depending on the value of λ. Since φ is birational $\det \beta$ gives the equation of Y. In particular $\det \beta$ is not a square. Hence β cannot be skew-symmetric, i.e. $\lambda = 1$.

Proposition 2.5. *If $\varphi \colon S \to Y$ is birational then there is a resolution*

$$0 \to (\mathcal{O}_\mathbb{P} \oplus E)^*(-5) \xrightarrow{\alpha} \mathcal{O}_\mathbb{P} \oplus E \to \mathcal{F} = \psi_* \mathcal{O}_X \to 0$$

given by a symmetric map α (that is, $\alpha = \alpha^*(-5)$).

Proof. Take for α the matrix β as above. □

Remark 2.6. The matrix α is a block matrix with entries as indicated below:

Summands	$\mathcal{O}(-3)$	$\oplus\ U \otimes \mathcal{O}(-1)$	$\oplus\ H^{0,1} \otimes \Omega^2(2)$	$\oplus\ W^* \otimes \Omega^1(1)$
$\mathcal{O}(2)$	$S_5 V^*$	$S_3 V^*$	$H^0(\Lambda^2 \mathcal{T}_\mathbb{P})$	$H^0(\mathcal{T}_\mathbb{P})$
$\oplus$				
$U^* \otimes \mathcal{O}$	$S_3 V^*$	$V^* \cong \Lambda^3 V$	$\Lambda^2 V$	V
$\oplus$				
$H^{2,1} \otimes \Omega^1(1)$	$H^0(\Lambda^2 \mathcal{T}_\mathbb{P})$	$\Lambda^2 V$	V	0
$\oplus$				
$W \otimes \Omega^2(2)$	$H^0(\mathcal{T}_\mathbb{P})$	V	0	0

Note that the $h^{2,1} \times h^{0,1}$ block is actually skew-symmetric, since it is induced from wedge-product

$$H^1(\mathcal{O}_X) \times H^1(\mathcal{O}_X) \to H^2(\mathcal{O}_X) = H^0(\mathcal{O}_X(K))^*$$

composed with the projection

$$H^0(\mathcal{O}_X(K))^* \to H^0(\mathcal{O}_{\mathbb{P}}(1))^* = V$$

However each element of

$$V \cong \mathrm{Hom}(\Omega_{\mathbb{P}}^2(2), \Omega_{\mathbb{P}}^1(1) \cong \mathrm{Hom}((\Omega_{\mathbb{P}}^1(1))^*, (\Omega_{\mathbb{P}}^2(2))^*)$$

gives a skew-symmetric morphism of bundles. So the resulting morphism

$$H^{0,1} \otimes \Omega_{\mathbb{P}}^2(2) \to H^{2,1} \otimes \Omega_{\mathbb{P}}^1(1)$$

is symmetric again. For general sign patterns in Beilinson monads of symmetric or skew-symmetric sheaves see Eisenbud and Schreyer [2001].

For a morphism $\alpha : G \to F$ of vector bundles on a scheme Z we denote by $I_r(\alpha)$ the ideal sheaf of $r \times r$ minors of α, i.e. the image of $\wedge^r G \otimes (\wedge^r F)^* \to \mathcal{O} := \mathcal{O}_Z$ under the natural map induced by $\wedge^r(\alpha)$. So $I_r(\alpha)$ is the $(\mathrm{rank}(F) - r)^{\mathrm{th}}$ Fitting ideal of $\mathrm{coker}(\alpha)$.

Theorem 2.7 (Catanese [1984b], de Jong and van Straten [1990]). *Let $\alpha = (\alpha_1, \alpha')$: $G \to \mathcal{O} \oplus E$ be a morphism of vector bundles with $r = \mathrm{rank}\, E = \mathrm{rank}\, G - 1$. Suppose $\det(\alpha)$ is a non zero-divisor and $((\det(\alpha)) = I_{r+1}(\alpha))\, \mathrm{depth}(I_r(\alpha'), \mathcal{O}) \geq 2$. Let $Y \subset Z$ denote the subscheme defined by $\det(\alpha)$. Then the following are equivalent:*

(1) $\mathcal{F} = \mathrm{coker}(\alpha)$ *carries the structure of a sheaf of commutative $\mathcal{O}_Y$-algebras with $1 \in \mathcal{F}$ given by the image of $1 \in \Gamma(Z, \mathcal{O}) \subset \Gamma(Z, \mathcal{O} \oplus E)$,*

(R.C.) $I_r(\alpha) = I_r(\alpha')$.

Proof. Since $\det(\alpha)$ is a non-zero divisor,

$$0 \to G \xrightarrow{\alpha} F \to \mathcal{F} \to 0$$

with $F = \mathcal{O} \oplus E$ is exact. As an $\mathcal{O}_Y$-module $\mathcal{F}$ has an infinite periodic resolution

$$\cdots \to \mathcal{B}^2 \otimes F_Y \xrightarrow{\beta_Y} \mathcal{B} \otimes G_Y \xrightarrow{\alpha_Y} \mathcal{B} \otimes F_Y \xrightarrow{\beta_Y} G_Y \xrightarrow{\alpha_Y} F_Y \to \mathcal{F} \to 0$$

where $\mathcal{B} = \wedge^{r+1} G \otimes (\wedge^{r+1} F)^*$, $-_Y = - \otimes \mathcal{O}_Y$ denotes restriction to Y, and the map

$$\beta : \wedge^{r+1} G \otimes \wedge^{r+1} F^* \otimes F \to G$$

is induced by $\wedge^r(\alpha)$, i.e. β is given by the matrix of cofactors of α. Exactness follows, since $\alpha \cdot \beta = \det(\alpha)\, \mathrm{id}_F$ and $\beta \cdot \alpha = \det(\alpha)\, \mathrm{id}_G$ and $\det(\alpha)$ is a non-zerodivisor, by Eisenbud [1980].

The above resolution is obtained by truncating the infinite exact periodic complex

$$\cdots \to \mathcal{B}^2 \otimes F_Y \xrightarrow{\beta_Y} \mathcal{B} \otimes G_Y \xrightarrow{\alpha_Y} \mathcal{B} \otimes F_Y \xrightarrow{\beta_Y} G_Y$$

$$\xrightarrow{\alpha_Y} F_Y \xrightarrow{\beta_Y} \mathcal{B}^{-1} \otimes G_Y \xrightarrow{\alpha_Y} \mathcal{B}^{-1} \otimes F_Y \to \cdots$$

Similarly we have an exact infinite periodic complex

$$\cdots \longrightarrow \mathcal{B} \otimes F_Y^* \xrightarrow{\alpha_Y^t} \mathcal{B} \otimes G_Y^* \xrightarrow{\beta_Y^t} F_Y^* \xrightarrow{\alpha_Y^t} G_Y^* \xrightarrow{\beta_Y^t} \mathcal{B}^{-1} F_Y^* \longrightarrow \cdots$$

It follows then that $\mathcal{C} := \mathcal{H}om_Y(\mathcal{F}, \mathcal{O}_Y)$ satisfies $\mathcal{F} = \mathcal{H}om_Y(\mathcal{C}, \mathcal{O}_Y)$.

Recall that $F = \mathcal{O} \oplus E$ whence $\mathcal{O}_Y \subset \mathcal{F}$ by Cramer's rule. By the assumption $\text{depth}(I_r(\alpha'), \mathcal{O}) \geq 2$ the quotient $\mathcal{F}/\mathcal{O}_Y$, which is annihilated by $I_r(\alpha')$, satisfies $\mathcal{H}om_Y(\mathcal{F}/\mathcal{O}_Y, \mathcal{O}_Y) = 0$. Therefore

$$\mathcal{O}_Y = \mathcal{H}om_Y(\mathcal{O}_Y, \mathcal{O}_Y) \supset \mathcal{C} = \mathcal{H}om_Y(\mathcal{F}, \mathcal{O}_Y)$$
$$\cong \ker(F_Y^* \to G_Y^*) \cong \text{im}(\mathcal{B} \otimes G_Y^* \to F_Y^*),$$

so $\mathcal{C} \subset \mathcal{O}_Y$ is the ideal sheaf

$$\text{im}(\mathcal{B} \otimes G_Y^* \to \mathcal{O}_Y) = I_r(\alpha')/(\det(\alpha)).$$

Suppose $\mathcal{F}$ is a ring. Then $\mathcal{C} \subset \mathcal{O}_Y$ is called the conductor of $\mathcal{O}_Y \subset \mathcal{F}$ and $\mathcal{F} \subset \mathcal{H}om_Y(\mathcal{C}, \mathcal{C})$, i.e. the image of $\mathcal{C} \times \mathcal{F}$ in $\mathcal{O}_Y$ is contained in $\mathcal{C} \subset \mathcal{O}_Y$: in fact for each (c, m) in $\mathcal{C} \times \mathcal{F}$ the image $c(m) \in \mathcal{O}_Y$ carries $\mathcal{F}$ into $\mathcal{O}_Y$ since $c(m)\mathcal{F} = c(m\mathcal{F}) \subset \mathcal{O}_Y$. Thus

$$\mathcal{F} \subset \mathcal{H}om_Y(\mathcal{C}, \mathcal{C}) \subset \mathcal{H}om_Y(\mathcal{C}, \mathcal{O}_Y) = \mathcal{F}.$$

In particular $\mathcal{H}om_Y(\mathcal{C}, \mathcal{C}) = \mathcal{H}om_Y(\mathcal{C}, \mathcal{O}_Y)$. Now the last equality means that every $\varphi \in \mathcal{H}om_Y(\mathcal{C}, \mathcal{O}_Y) = \mathcal{F} = \text{coker}(G_Y \to F_Y)$ has image in $\mathcal{C} = I_r(\alpha')/(\det(\alpha))$. That is the pairing induced by β_Y

$$F_Y \times (\mathcal{B} \otimes G_Y^*) \to \mathcal{O}_Y$$

has image in $I_r(\alpha')\mathcal{O}_Y$. Since β is the matrix of cofactors of α, $\mathcal{H}om_Y(\mathcal{C}, \mathcal{C}) = \mathcal{H}om_Y(\mathcal{C}, \mathcal{O}_Y)$ is equivalent to $I_r(\alpha) = I_r(\alpha')$.

Conversely, if (R.C.) holds, we have $\mathcal{F} = \mathcal{H}om_Y(\mathcal{C}, \mathcal{O}_Y) = \mathcal{H}om_Y(\mathcal{C}, \mathcal{C})$ is an $\mathcal{O}_Y$-algebra under composition. The structure is commutative, since $\mathcal{C}$ as $\mathcal{O}_Y$-module is invertible on $Y - V(I_r(\alpha'))$ and $\text{depth}(I_r(\alpha'), \mathcal{F}) \geq 1$ by assumption. $\square$

Remark 2.8. We call the condition (R.C.) $I_r(\alpha) = I_r(\alpha')$ the Ring Condition (or Rank Condition). For a symmetric matrix

$$\alpha = \begin{pmatrix} \alpha_{11} & \alpha_{12} \\ \alpha_{12}^t & \alpha'' \end{pmatrix} : (\mathcal{O} \oplus E)^*(-5) \to \mathcal{O} \oplus E$$

the ring condition implies the further rank condition

$$I_{r-1}(\alpha') = I_{r-1}(\alpha'')$$

for lower minors of $(\alpha')^t = (\alpha_{12}, \alpha'') : E^*(-5) \to \mathcal{O} \oplus E$ if $\mathrm{depth}(I_{r-1}(\alpha''), \mathcal{O}) = 3$ has the expected maximal value.

Cf. Prop. 4.1 of Mond and Pellikaan [1987], and Prop. 5.8 and 5.10 of Catanese [1984b].

Theorem 2.9. *The datum $\varphi : S \to Y \subset \mathbb{P}^3 = \mathbb{P}(V)$ of a good birational canonical projection determines a morphism*

$$(\mathcal{O}_\mathbb{P} \oplus E)^*(-5) \xrightarrow{\alpha} (\mathcal{O}_\mathbb{P} \oplus E),$$

where $E = (K^2 - q + p_g - 9)\mathcal{O}_\mathbb{P}(-2) \oplus q\Omega_\mathbb{P}^1(-1) \oplus (p_g - 4)\Omega_\mathbb{P}^2$, such that

(i) *α is symmetric,*

(ii) *$\det \alpha$ is an irreducible polynomial (defining Y),*

(iii) *α satisfies the ring condition, and*

- *defining $\mathcal{F}$ as the sheaf of $\mathcal{O}_Y$-algebras given by the module $\mathrm{coker}(\alpha)$ provided with the ring structure determined by α as in 2.7(1),*

(iv) *$\mathrm{Spec}\,\mathcal{F}$ is a surface with at most rational double points as singularities.*

Conversely, given α satisfying (i), (ii), (iii) *and* (iv), *$X = \mathrm{Spec}\,\mathcal{F}$ is the canonical model of a minimal surface S and*

$$\varphi : S \to X \to Y \subset \mathbb{P}^3 = \mathbb{P}(V)$$

is a good birational canonical projection.

Proof. The first statement follows by combining 2.7 and 2.5. Notice that, since $\det \alpha$ is irreducible (in one direction, this is a consequence of the birationality of φ, in the other direction, it holds by assumption), the ring condition $I_r(\alpha) = I_r(\alpha')$ implies $\mathrm{depth}(I_r(\alpha'), \mathcal{O}_\mathbb{P}) \geq 2$, since otherwise all $r \times r$ minors of α would have a common irreducible factor, whose square would divide $\det(\alpha)$. So the assumptions of 2.7 are satisfied.

For the converse, we note that duality applied to

$$\psi : X = \mathrm{Spec}\,\mathcal{F} \to Y \subset \mathbb{P}^3$$

gives

$$\mathcal{F}(1) = \psi_* \omega_X,$$

since α is symmetric. So $\psi^* \mathcal{O}_\mathbb{P}(1) \cong \omega_X$, and, because X has only rational double points as singularities, $\varphi^* \mathcal{O}_\mathbb{P}(1) \cong \omega_S$ holds on the desingularization S of X. So $\varphi : S \to Y \subset \mathbb{P}^3$ is a quasi generic birational canonical projection. □

Remark 2.10. Given a symmetric matrix $\alpha : (\mathcal{O} \oplus E)^*(-5) \to \mathcal{O} \oplus E$ as in the theorem, we denote by $\alpha' : (\mathcal{O} \oplus E)^*(-5) \to E$ and $\alpha'' : E^*(-5) \to E$ the distinguished submatrices. Then $\det(\alpha'')$ defines the adjoint surface F of degree $K^2 - 5$. F intersects Y precisely in the non-normal locus Γ of Y, which is defined by $I_r(\alpha')$. F is singular at the points of the subscheme T defined by $I_{r-1}(\alpha'')$, Γ has embedding dimension 3 at T, and the points of T are at least triple for Y.

Typically the points of T correspond to triple points of Y, and F has ordinary quadratic singularities in T. In terms of the invariants $d = K^2$, q and p_g of S we have

$$\deg \Gamma = 1/2\, d^2 - 5/2\, d + 1$$

and

$$\deg T = 1/6\, d^3 - 5/2\, d^2 + 37/3\, d - 4(1 - q + p_g).$$

Proposition 2.11. *Suppose $\varphi : S \to Y \subset \mathbb{P}^3 = \mathbb{P}(V)$ is a good canonical projection with Y not a quadric. If the Albanese image of S is a curve then*

$$K^2 \geq 4q + 4 + 2(p_g - 4) = 2p_g + 4q - 4.$$

If equality holds then the map is not birational.

Proof. Since $Y \subset \mathbb{P}^3$ is a surface we have the presentation 2.4 of $\phi_*\mathcal{O}_S = \psi_*\mathcal{O}_X$ given by a matrix α with entries as indicated in Remark 2.6 (with α perhaps not symmetric). If the Albanese dimension is one then

$$H^0(\Omega_S^1) \times H^0(\Omega_S^1) \to H^0(\Omega_S^2)$$

is the zero map, and so is $H^1(\mathcal{O}_S) \times H^1(\mathcal{O}_S) \to H^2(\mathcal{O}_S)$ by Hodge symmetry. Thus we have a large block of zeroes. On the other hand the determinant of α equals the equation of Y to the power $\deg(\varphi)$, in particular $\det \alpha \neq 0$. Thus the $(q+p_g-4) \times (q+p_g-4)$ block cannot be too big. More precisely,

$$1 + \dim U \geq 3(q + p_g - 4).$$

Since $1 + \dim U = 1 + K^2 + 1 - q + p_g - \dim S_2 V = K^2 - q + p_g - 8$ the desired inequality follows. Moreover in case of equality we have

$$\det \alpha = \lambda[\det(\mathcal{O}(-3) \oplus (U \otimes \mathcal{O}(-1)) \to (H^{2,1} \otimes \Omega^1(1)) \oplus (W \otimes \Omega^2(2))]^2$$

for some scalar $\lambda \in \mathbb{C}$. Thus φ cannot be birational. $\square$

Remark 2.12. The same argument, but without the assumption that the Albanese image of S be a curve, gives in general the inequality $K^2 \geq 2p_g - 2q - 4$ which is however weaker than Noether's inequality $K^2 \geq 2p_g - 4$. For $q \geq 1$ we may get $K^2 \geq 2p_g - 2q + 2$, which is still however weaker than the inequality given by Debarre [1982] for irregular surfaces, namely, $K^2 \geq 2p_g$.

3. The case of double covers

Suppose the good canonical projection $\psi : S \to Y \subset \mathbb{P}^3$ is $2:1$ and that Y is not a quadric. Then $\mathcal{F} = \psi_*\mathcal{O}_X$ still has a locally free resolution of length 1 of the form (2.4):

$$0 \to (\mathcal{O}_\mathbb{P} \oplus E)^*(-5) \to \mathcal{O}_\mathbb{P} \oplus E \to \mathcal{F} = \psi_*\mathcal{O}_X \to 0.$$

with E the vector bundle on $\mathbb{P}^3$ defined by

$$E(2) = (U^* \otimes \mathcal{O}_\mathbb{P}) \oplus (H^1(\mathcal{O}_X(K)) \otimes \Omega^1_\mathbb{P}(1)) \oplus (W \otimes \Omega^2_\mathbb{P}(2)).$$

However the proof that the resolution can be chosen symmetric needs further arguments. Let $\rho : Z \to Y$ denote the normalization of Y. Then $\psi : X \to Y$ factors over $\varepsilon : X \to Z$, where ε is $2:1$. The covering involution $\Phi : X \to X$ induces a decomposition of $\varepsilon_*\mathcal{O}_X$ into invariant and anti-invariants parts:

$$\varepsilon_*\mathcal{O}_X = \mathcal{O}_Z \oplus \mathcal{H}. \tag{3.1}$$

On Y this induces the decomposition

$$\mathcal{F} = \psi_*\mathcal{O}_X = \rho_*\mathcal{O}_Z \oplus \rho_*\mathcal{H}. \tag{3.2}$$

This in turn decomposes the Beilinson cohomology groups of $\mathcal{F}$ and (assuming that Y is not a quadric), this gives a decomposition of (2.4). There are two cases how the isomorphism $\mathcal{F} \to \mathcal{E}xt^1_{\mathcal{O}_\mathbb{P}}(\mathcal{F}, \omega_\mathbb{P}(-1))$ can respect the summands (which are generically of rank 1 on Y). Either

(a) $\rho_*\mathcal{O}_Z \cong \mathcal{E}xt^1_{\mathcal{O}_\mathbb{P}}(\rho_*\mathcal{O}_Z, \omega_\mathbb{P}(-1))$

or

(b) $\rho_*\mathcal{O}_Z \cong \mathcal{E}xt^1_{\mathcal{O}_\mathbb{P}}(\rho_*\mathcal{H}, \omega_\mathbb{P}(-1))$.

Case (a) occurs when y_0, y_1, y_2, y_3 pullback to Φ invariant sections of $H^0(\mathcal{O}_X(K))$ i.e. $V^* \subset H^0(\mathcal{O}_X(K))^+$.
  Case (b) occurs when $V^* \subset H^0(\mathcal{O}_X(K))^-$.

The Φ-invariant case (a). Since $V^* \subset H^0(\mathcal{O}_X(K))^+$, the isomorphism $\psi_*\omega_X$ with $\psi_*\mathcal{O}_X(1)$ respects the invariant and anti-invariant summands. Therefore,

$$(\psi_*\omega_X)^+ = \rho_*\omega_Z = \mathcal{E}xt^1_{\mathcal{O}_\mathbb{P}}(\rho_*\mathcal{O}_Z, \omega_\mathbb{P})$$

is isomorphic to $\rho_*\mathcal{O}_Z(1)$ as asserted. Since moreover ρ is birational, $\rho^*\rho_*\mathcal{O}_Z(1) = \mathcal{O}_Z(1)$ and by the projection formula we infer that $\mathcal{O}_Z(1) \cong \omega_Z$, in particular, that Z is Gorenstein, whence the canonical model of a surface of general type.
  Similarly, we see that $(\psi_*\omega_X)^- = \mathcal{E}xt^1_{\mathcal{O}_\mathbb{P}}(\rho_*\mathcal{H}, \omega_\mathbb{P}) = \rho_*\mathcal{H}(1)$.
  Decomposing then

$$E = E_+ \oplus E_- \tag{3.3}$$

into parts coming from $\rho_*\mathcal{O}_Z \oplus \rho_*\mathcal{H}$ we obtain in this case a decomposition of (2.4) into

$$0 \to (\mathcal{O}_\mathbb{P} \oplus E_+)^*(-5) \xrightarrow{\alpha_+} \mathcal{O}_\mathbb{P} \oplus E_+ \to \rho_*\mathcal{O}_Z \to 0 \tag{3.4}$$

and

$$0 \to (E_-)^*(-5) \xrightarrow{\alpha_-} E_- \to \rho_*\mathcal{H} \to 0. \tag{3.5}$$

The argument of 2.5 shows that both α_+ and α_- can be chosen to be symmetric matrices. We can also apply Theorem 2.9 to Z, since as we observed $Z = X/\Phi$ is a canonical model of a surface. (3.4 is the determinantal description of the good birational canonical projection $\rho : Z \to Y \subset \mathbb{P}^3 = \mathbb{P}(V)$.

We leave aside for the time being the task of describing the $\mathcal{O}_Z$ module structure on $\rho_*\mathcal{H}$: we simply observe that the bilinear map

$$(E_-) \times (E_-) \to \mathcal{O}_\mathbb{P}$$

is given as in [Catanese 1981] by the adjoint matrix of α_-, which solves the problem in the very particular case where Y is normal.

The Φ-anti-invariant case (b). Here (2.4) decomposes into

$$0 \to (E_-)^*(-5) \xrightarrow{\alpha_+} \mathcal{O}_\mathbb{P} \oplus E_+ \to \rho_*\mathcal{O}_Z \to 0 \tag{3.6}$$

and

$$0 \to (\mathcal{O}_\mathbb{P} \oplus E_+)^*(-5) \xrightarrow{\alpha_-} E_- \to \rho_*\mathcal{H} \to 0 \tag{3.7}$$

and we may choose $\alpha_- = (\alpha_+)^t$. Notice that α_+ satisfies the ring condition (R.C.).

To recover X from (3.4) and (3.5) or from (3.6) we need in addition to describe the map

$$S_2(\rho_*\mathcal{H}) \to \rho_*\mathcal{O}_Z. \tag{3.8}$$

4. Generalities on irregular surfaces with $p_g = 4$

As explained in the introduction, one of the main purposes of our investigation is to understand the equations of the projections of irregular canonical surfaces in $\mathbb{P}^3$.

For this reason the most natural case to consider is the case where $p_g = 4$, and there is no choice whatsoever to make for the projection.

We recall, for the reader's benefit, some important inequalities for irregular surfaces

- Castelnuovo's Theorem [1905]: $\chi(S) \geq 1$ if the surface S is not ruled.

From Castelnuovo's theorem follows that if $p_g = 4$, then the irregularity $q(S)$ is ≤ 4. We have also the

- Inequality of Castelnuovo–Beauville (Beauville [1982]): $p_g \geq 2q - 4$, equality holding if and only if S is a product of a curve of genus 2 with a curve of genus $g \geq 2$.

Whence follows that, if $p_g = q = 4$, then S is the product of two curves of genus 2 and its canonical map is a $(\mathbb{Z}/2)^2$-Galois covering of a smooth quadric.

Debarre has moreover shown [1982] that for an irregular surface one has the following

- Debarre's inequality: if S is irregular, then $K^2 \geq 2p_g$.

Therefore, in our case, the above inequality yields more than the more general inequalities by Castelnuovo [1891] and by Horikawa [1976b]], Reid [1979], Beauville [1979] and Debarre [1982]: $K^2 \geq 3p_g + q - 7$ if the canonical map is birational.

Our inequality $K^2 \geq 2p_g + 4q - 4$ if the Albanese image is a curve severely restricts the numerical possibilities if $p_g = 4$, $q = 3$, since by the Bogomolov-Miyaoka-Yau inequality we always have $K^2 \leq 9\chi$, thus $16 \leq K^2 \leq 18$ if the Albanese map is a pencil.

This case is completely solved by using the following inequalities for surfaces fibred over curves:

- Arakelov's inequality: let $f : S \to B$ be a fibration onto a curve B of genus b, with fibres curves of genus $g \geq 2$. Then $K_S^2 \geq 8(b-1)(g-1)$, equality holding only if the fibration has constant moduli.

- Beauville's inequality: let $f : S \to B$ be a fibration onto a curve B of genus b, with fibres curves of genus $g \geq 2$. Then $\chi(S) \geq (b-1)(g-1)$, equality holding if and only if the fibration is an etale bundle (there is an etale cover $B' \to B$ such that the pull back is a product $B' \times F$).

By Beauville's inequality follows that if $q = 3$ and the Albanese image is a curve, then (take as B the genus 3 curve which is the Albanese image, and f the Albanese map) the Albanese map is an etale bundle with fibre F of genus 2.

In particular, we have $K_S^2 = 16$ and all our surfaces are obtained as follows.

Let G be a finite group acting faithfully on a curve F of genus 2 in such a way that $F/G \cong \mathbb{P}^1$, and take an etale G-cover $B' \to B$ of a curve B of genus 3: then our surfaces S with $p_g = 4, q = 3, K_S^2 = 16$ are exactly the quotients $(F \times B')/G$ of the product $(F \times B')$ by the diagonal action of G.

The groups G as above were classified by Bolza [1888] (cf. also [Zucconi 1994]).

If instead the Albanese image is a curve of genus $q = 2$, then, since we assume $p_g = 4$, then $\chi(S) = 3 \geq (g - 1)$, and the genus of the Albanese fibres is at most 4.

The case $g = 4$ gives again rise to an etale bundle with fibre F of genus 4, so that our surface S is a quotient $S = F \times B'/G$ where $F/G \cong \mathbb{P}^1$ and $B = B'/G$ is a curve of genus $b = 2$ (one must impose the condition that G operate freely on the product).

In particular, we have $K_S^2 = 24$. A concrete example is furnished by $G = \mathbb{Z}/2$, which operates freely on the genus 3 curve B'. Then $H^0(K_S) = H^0(K_{F \times B'})^+ =$

$H^0(K_F)^- \otimes H^0(K_{B'})^-$, and since $H^0(K_{B'})^-$ is 1-dimensional, the canonical system is a pencil and the canonical image of S is the canonical image of F, namely, a twisted cubic curve in $\mathbb{P}^3$.

It would be also interesting to determine what is the minimal value of K^2 for an irregular surface with $p_g = 4$ such that the canonical map is birational.

In the next section we shall see that the answer to this question is: $K^2 = 12$ in the case $q = 1$, and in this case we shall give a complete classification of the surfaces for which the canonical map is birational.

On the other hand, the problem remains for $q = 2, 3$. In the forthcoming sections, for $q = 3$ we provide examples where $K^2 = 12$ (of course the Albanese image is a surface, as we already observed), while for $q = 2$ we show that there are examples with $K^2 = 18$ (this is not the maximum allowed by the B–M–Y inequality, yielding $K^2 \le 9\chi = 27$, but still not so low).

For the case where the degree of the canonical map is 2, we recall

Theorem 4.1 (Debarre's theorem 4.8 [1982]). *Assume that $1 \le q \le 3$ and that the canonical map has degree 2. Then $K^2 \ge 2p_g + q - 1$.*

Finally, we recall the following inequality for fibred surfaces (cf. Xiao [1987], Konno [1993])

- Xiao–Konno inequality: if $f : S \to B$ is a morphism to a curve B of genus b, with fibres of genus g, and without constant moduli, then the slope $\lambda(f) = (K_S - f^*(K_B))^2 / \deg(f_*\omega_{S|B}) = \frac{K_S^2 - 8(b-1)(g-1)}{\chi(S) - (b-1)(g-1)}$ satisfies $4(g-1)/g \le \lambda(f) \le 12$, the first inequality being an inequality iff all the fibres are hyperelliptic.

It follows that $K_S^2 \le 12\chi(S) - 4(b-1)(g-1)$, and this implies that if the Albanese image is a curve of genus $q \ge 2$, then $K_S^2 \le 12p_g - 12(q-1) - 4(q-1)(g-1) = 12p_g - 4(q-1)(g+2)$.

Therefore, if $p_g = 4$ and if the Albanese image is a curve of genus $q \ge 2$, we obtain indeed $16 \le K_S^2 \le 48 - 4(q-1)(g+2)$. For $q = 3$ this confirms that we must have $K^2 = 16$ and $g = 2$, a case that we have already illustrated. While, if $q = 2$ we have $12 \le K_S^2 \le 48 - 4(g+2)$: whence, $g \le 5$ which is weaker than the inequality $g \le 4$ we have already obtained.

5. Irregular surfaces with $p_g = 4, q = 1$

Horikawa [1981] proved that for an irregular surface with $K^2 < 3\chi$ ($3\chi = 12$ here), the Albanese map has a curve as image, and the fibres have either genus 2 or they have genus 3 and are hyperelliptic: moreover only the first possibility occurs if $K^2 < 8/3\chi$.

In this section we shall restrict our attention to the case $p_g = 4, q = 1$, thus $\chi = 4$ and $K^2 = 8$ is the smallest value for K^2, while $K^2 = 12$ is the smallest value for which we can have a birational canonical map (in view of the quoted result by Horikawa).

Let $a : S \to A$ be the Albanese map of S. By the above inequality for the slope, if $K^2 = 12$ then $g \leq 4$, and if $g = 4$ all the fibres of a are hyperelliptic. But if the Albanese fibres are hyperelliptic, then the canonical map ϕ factors through the hyperelliptic involution i. We make therefore the following

Assumption 5.1. $p_g = 4, q = 1$, and the general Albanese fibre is non hyperelliptic of genus $g = 3$.

Under the above assumption, let $\mathcal{V}$ be the vector bundle on A defined by

$$\mathcal{V} = a_* \omega_S, \tag{5.1}$$

which enjoys the base change property. $\mathcal{V}$ has rank $g = 3$, and $h^0(\mathcal{V}) = p_g = 4$. More generally, we can consider the vector bundles $\mathcal{V}_i = a_*(\omega_S^{\otimes i})$, which have zero H^1-cohomology groups and have degree $\deg(\mathcal{V}_i) = \chi + \frac{i(i-1)}{2} K^2 = 4 + \frac{i(i-1)}{2} K^2$.

Under the assumption 5.1, we have an exact sequence

$$0 \to \mathrm{Sym}^2(\mathcal{V}) \to \mathcal{V}_2 \to \mathcal{C} \to 0, \tag{5.2}$$

where $\mathcal{C}$ is a torsion sheaf.

Since the degree of $\mathrm{Sym}^2(\mathcal{V})$ equals 16, we obtain

Proposition 5.2. *Under the assumption 5.1 we have $K^2 \geq 12$, equality holding iff there is no hyperelliptic fibre, that is, $\mathrm{Sym}^2(\mathcal{V}) \cong \mathcal{V}_2$.*

By the theorem of Fujita [1978] $\mathcal{V}$ is semipositive, moreover by a corollary of a theorem of Simpson [1993] observed for instance in 2.1.7 of Zucconi [1994], there is in general a splitting

$$\mathcal{V} = \left(\bigoplus_{\tau \in \mathrm{Pic}(A)_{\mathrm{tors}} - \{0\}} L_\tau \right) \oplus \left(\bigoplus_i W_i \right) \tag{5.3}$$

where, as indicated, L_τ is a non trivial torsion line bundle, and instead W_i is an indecomposable bundle of strictly positive degree.

We are interested mostly in the case where the canonical map ϕ is birational; since $\mathcal{V}$ has rank 3, the hypothesis that ϕ be birational implies that $\mathcal{V}$ must be generically generated by global sections. Thus we make the following

Assumption 5.3. *$\mathcal{V}$ is generically generated by global sections.*

In particular, there are no summands of type L_τ in (5.3).

Moreover, by Atiyah [1957], Lemma 15, page 430, setting $\deg(W_i) = d_i$ and $\mathrm{rank}(W_i) = r_i$, we have $h^0(W_i) = \deg(W_i) = d_i$, and therefore $d_i \geq r_i$. Conversely, by loc. cit. Theorem 6, page 433) and by induction follows

Proposition 5.4. *If W is an indecomposable vector bundle on an elliptic curve of degree $d \geq r = \mathrm{rank}(W)$, then W is generically generated by global sections.*

Therefore, since $\Sigma_i d_i = 4$, $\Sigma_i r_i = 3$, and the r_i's, d_i's are > 0, we have only the following possibilities for the pairs (r_i, d_i) which are ordered by the slope d/r:

(i) (3,4)

(ii) (2,3) (1,1)

(iii) (1,2) (2,2)

(iv) (1,2) (1,1) (1,1).

The structure of these bundles is then clear by the quoted results of Atiyah: for each line bundle $\mathcal{O}_A(p)$ of degree one, where $p \in A = \text{Pic}^1(A)$, there are a point $u \in A$ and line bundles $\mathcal{L}, \mathcal{L}' \in \text{Pic}^0(A)$ such that $\mathcal{V}' = \mathcal{V} \otimes \mathcal{O}_A(-p)$ is respectively equal to

$$
\begin{aligned}
&\text{(i)} \quad E_3(u) \\
&\text{(ii)} \quad E_2(u) \oplus \mathcal{L} \\
&\text{(iii)} \quad \mathcal{O}_A(u) \oplus (\mathcal{L} \otimes F_2) \\
&\text{(iv)} \quad \mathcal{O}_A(u) \oplus \mathcal{L} \oplus \mathcal{L}',
\end{aligned}
\tag{5.4}
$$

where $E_1(u) := \mathcal{O}_A(u)$ and $E_i(u)$ is defined inductively as the unique non trivial extension

$$0 \to \mathcal{O}_A \to E_i(u) \to E_{i-1}(u) \to 0,$$

while $F_1 := \mathcal{O}_A$ and F_i is defined inductively as the unique non trivial extension

$$0 \to \mathcal{O}_A \to F_i \to F_{i-1} \to 0.$$

Lemma 5.5. *Assume that X is the canonical model of a surface with $p_g = 4$, $q = 1$, $K^2 = 12$, and non hyperelliptic Albanese fibres.*

Then the relative canonical map $\omega : X \to \text{Proj}(\mathcal{V}) = \mathbb{P}$ is an embedding. Moreover, there is a point $p \in A$ such that, setting $\mathcal{V}' = \mathcal{V} \otimes \mathcal{O}_A(-p)$, $(\det \mathcal{V}') \cong \mathcal{O}_A(p)$ and X is a divisor in the linear system $|4D|$, D being the tautological divisor of $\text{Proj}(\mathcal{V}')$. (Notice that p is defined only up to 4-torsion).

Conversely, if $\mathcal{V}$ is as in (5.4), any divisor X in $|4D|$ with at most R.D.P.'s as singularities is the canonical model of a surface with $p_g = 4$, $q = 1$, $K^2 = 12$, and non hyperelliptic Albanese fibres.

Proof. By 5.2, the relative canonical map is an embedding of the canonical model X of S if and only if the relative bicanonical map is an embedding. Let therefore F be a fibre of the Albanese map $a : X \to A$. Since $\mathcal{V}_2$ enjoys the base change property, we are just asking whether $\mathcal{O}_F(2K_X)$ is very ample on F.

By Catanese and Franciosi [1996] or Catanese, Franciosi, Hulek and Reid [1999] we get that very ampleness holds provided that F is 2-connected, i.e., there is no decomposition $F = F_1 + F_2$ with F_1, F_2 effective and with $F_1 F_2 \leq 1$.

If such a decomposition would exist, we claim that we may then assume $F_1 F_2 = 1$.

Otherwise $F_1 F_2 \leq 0$ and, since also $F_i^2 \leq 0$, it follows by Zariski's Lemma that $F = 2F_1$. Since F_1 has genus 2 the exact sequence

$$\mathcal{O}_{F_1}(K_X - F_1) \to \mathcal{O}_F(K_X) \to \mathcal{O}_{F_1}(K_X) \to 0$$

shows easily that the hyperelliptic involution on F_1 extends to a hyperelliptic involution on F, a contradiction.

Since the genus of F equals 3, $K_X F = 4 = K_X F_1 + K_X F_2$. Since moreover $K_X F_i \equiv F_i^2 \pmod{2} \equiv -F_1 F_2 \pmod{2}$, we may assume w.l.o.g. that $K_X F_1 = 1$. So F_1 is an elliptic tail, while F_2 has genus 2.

More precisely, we have $h^0(\mathcal{O}_{F_1}(K_X)) = 1$, $h^0(\mathcal{O}_{F_1}(2K_X)) = 2$, contradicting the fact that $\text{Sym}^2(H^0(K_F)) = (H^0(2K_F))$, since in fact $h^1(\mathcal{O}_{F_2}(2K_X - F_1)) = 0$.

We proved now that X is embedded in $\mathbb{P}$, whence it follows that the surjection $\text{Sym}^4(\mathcal{V}) \to \mathcal{V}_4$ has as kernel an invertible sheaf L on the elliptic curve A. The exact sequence

$$0 \to L \to \text{Sym}^4(\mathcal{V}) \to \mathcal{V}_4 \to 0,$$

and the easy calculation: $\deg(\text{Sym}^4(\mathcal{V})) = 80$, $\deg(\mathcal{V}_4) = 76$ shows that $\deg(L) = 4$, so there is point $p \in A$ such that $L = \mathcal{O}_A(4p)$, and we have the following linear equivalence in $\mathbb{P}$: $X \equiv 4H - 4F$, where H is the tautological hyperplane divisor, and F is the fibre of a over p.

If we write $\mathbb{P}$ as $\text{Proj}(\mathcal{V}') = \text{Proj}(\mathcal{V}(-p))$, and let D be the corresponding hyperplane divisor, then X is a divisor in the linear system $|4D|$.

Conversely, since $K_\mathbb{P} \equiv -3H + \omega^*(\det \mathcal{V})$, if we choose a divisor $X \equiv 4H - \omega^*(\det \mathcal{V})$ we get that $K_X \equiv H$, so that $\mathcal{V}$ is the direct image of the canonical sheaf $\mathcal{O}_X(K_X)$. It is then clear that $L = \omega^*(\det \mathcal{V})$, and since we chose p such that $X \equiv 4H - 4F$, $K_X \equiv D + \omega^*(\det \mathcal{V}') \equiv D + F$, we have proven that $(\det \mathcal{V}') \cong \mathcal{O}_A(p)$.

Moreover, we have that $p_g(X) = h^0(\mathcal{V}) = 4$, $q = 1 + h^1(\mathcal{V}) = 1 + 0 = 1$, while $K_X^2 = (D+F)^2(4D)$. By the Leray–Hirsch formula we get $D^2 = FD$, moreover $D^2 F = 1$, whence $K_X^2 = 4(1+2) = 12$. $\square$

Remark 5.6. Observe that $\det(\mathcal{V}) \cong \mathcal{O}_A(4p)$, whence in the notation of the previous lemma we have $\det(\mathcal{V}') \cong \mathcal{O}_A(p)$. Thus $p \equiv u$ in case (i), while $p \equiv u + \mathcal{L}$, $p \equiv u + 2\mathcal{L}$, $p \equiv u + \mathcal{L} + \mathcal{L}'$, in the respective cases (ii), (iii), (iv).

It follows that the pair $(A, \mathcal{V})$ has 1 modulus in case (i), 2 moduli in case (ii), while we are going to show next that the pair has 1 modulus in case (iii), and 1 or 2 moduli in the last case (iv).

There remains as a first problem the question about the existence of the surfaces under consideration, that is, whether the general element in the linear system $|4D|$ has only Rational Double Points as singularities. The result is a consequence of techniques developed in Catanese and Ciliberto [1993].

Proposition 5.7. *Let $\mathcal{V}$ be as in (5.4) a rank 3 bundle over an elliptic curve, and X as in Lemma 5.5 a general divisor in the linear system $|4D|$ on $\mathbb{P} = \text{Proj}(\mathcal{V})$. Then X*

is smooth in cases (i) *and* (ii). *Instead, in case* (iii), *the general element X has only Rational Double Points as singularities if and only if* $\mathcal{L}^4 \cong \mathcal{O}_A$.

Finally, in case (iv), *the general element X has Rational Double Points as singularities if and only if one of the bundles* $\mathcal{L}^k \otimes \mathcal{L}'^{4-k}$ *is trivial.*

Proof. In case (i), the linear system $|4D|$ is very ample on $\mathbb{P}$ by Theorem 1.21 of Catanese and Ciliberto [1993], so a general X is smooth by Bertini's Theorem.

Notice that to apply Bertini's theorem it suffices to show that the general element of the linear system $|4D|$ is smooth along the base locus of $|4D|$. To show that $|4D|$ is base point free is in turn sufficient the condition that the vector bundle $\text{Sym}^4(\mathcal{V}')$ be generated by global sections.

In case (ii), we observe that $\text{Sym}^4(\mathcal{V}')$ is a direct sum

$$\bigoplus_{k=1,..4} \text{Sym}^k(E_2(v_k)) \bigoplus \mathcal{L}^4,$$

where the v_k's are suitable points on A. The symmetric powers with $k \geq 2$ are generated by global sections by virtue of Theorem 1.18 of Catanese and Ciliberto [1993].

Whereas a bundle of the form $E_2(v_k)$ has only one section, which is nowhere vanishing. Therefore, the base locus of $|4D|$ is contained in the section Δ of $\mathbb{P}$ dual to the subbundle $E_2(u)$, and there, if x, y are local equations for Δ, then the Taylor development of the equation of a divisor in $|4D|$ only fails at most to have a term of type x. Therefore in this case the general element X is smooth.

Let us then consider case (iv): it is immediate to remark that the linear system $|4D|$ has as fixed part the projective $\mathbb{P}^1$-subbundle $\mathbb{P}'$ annihilated by $\mathcal{O}_A(u)$ in the case where no line bundle $\mathcal{L}^k \otimes \mathcal{L}'^{4-k}$ is trivial.

This cannot occur, so assume that one of such line bundles is trivial.

Case (iv, I''): $\mathcal{L}'^4$ *and* $\mathcal{L}^4$ *are trivial.* This is the easy case where $|4D|$ has no base points, whence Bertini's theorem applies.

Case (iv, I'): $\mathcal{L}'^4$ *is not trivial but* $\mathcal{L}^4$ *is trivial.* This is the case where the base locus of $|4D|$ is the section Δ annihilated by $\mathcal{O}_A(u) \oplus \mathcal{L}$.

At each point of Δ exists then a term of order 1 in the Taylor expansion of the equation f of X, except at the point $t' = 0$, where $t' = 0$ lies over the unique point $v \in A$ such that $v \equiv u + 3\mathcal{L}'$.

In this point we have then local coordinates x, z, t' with $\Delta = \{z = x = 0\}$ and we surely get monomials z^2, zt', zx (zx corresponds to the fact that the unique section of $\mathcal{O}_A(u) \otimes \mathcal{L} \otimes \mathcal{L}'^2$ does not vanish in our point $t' = 0$, otherwise $\mathcal{L} \cong \mathcal{L}'$, contradicting our assumption (iv, I')). Since moreover we get the monomial x^4, we certainly obtain for general X at worst a Rational Double Point of type A_3.

Case (iv, II): $\mathcal{L}'^4$ *and* $\mathcal{L}^4$ *are not trivial but* $\mathcal{L}^2 \otimes \mathcal{L}'^2$ *is trivial.* In this case the base locus of $|4D|$ is given by the two sections Δ, Δ', where Δ' is annihilated by $\mathcal{O}_A(u) \oplus \mathcal{L}'$.

In this case, by symmetry, let us study the singularity of a general X along Δ.

We get, as in the previous case, a section zt', and sections z^2, x^2, thus a singularity of type A_1 at worst.

Case (iv, III): $\mathcal{L}'^4$, $\mathcal{L}^2 \otimes \mathcal{L}'^2$ and $\mathcal{L}^4$ are not trivial but $\mathcal{L}^3 \otimes \mathcal{L}'$ is trivial. In this case again the base locus of $|4D|$ is given by the two sections Δ, Δ'. For Δ we get monomials of type x^3, z^2, zt', thus a singularity of type A_2 at worst, while for Δ' we get monomials of type x, z^2, zt', thus no singularity at all for general X along Δ'.

We can finally analyse case (iii). To this purpose, recall

Theorem 5.8 (Atiyah's Theorem 9 in [57]). *Let F_2 be the indecomposable bundle on an elliptic curve with trivial determinant and of rank 2: then $\mathrm{Sym}^k(F_2) \cong F_{k+1}$.*

We observe then that the tensor product of F_r with a line bundle $\mathcal{M}$ is generated by global sections if $\deg(\mathcal{M}) \geq 2$, whereas, if $d := \deg(\mathcal{M}) \leq 1$, $F_r \otimes \mathcal{M}$ is generated by global sections outside a unique point where all sections vanish if $d = 1$, whereas for $d = 0$, $F_r \otimes \mathcal{M}$ has no global sections unless the line bundle $\mathcal{M}$ is trivial. In this last case, there is only one non zero section, which vanishes nowhere.

After these remarks, it is clear that, in case (iii), if $\mathcal{L}^4$ is non trivial, then the fixed part of $|4D|$ contains the $\mathbb{P}^1$-bundle $\mathbb{P}'$ annihilated by $\mathcal{O}_A(u)$. Otherwise, the base locus consists of a section Δ and a fibre F_v of $\mathbb{P}'$.

At the intersection point of these two curves, we have local coordinates z, x, t such that $z = 0$ defines $\mathbb{P}'$, $z = x = 0$ defines Δ, $z = t = 0$ defines F_v.

In the Taylor expansion of the equation of X we get x^4, zt, z^2, therefore, by an argument we already used, in this point we get at worst a singularity of type A_3, while the other points of the base locus are smooth for general X. □

We derive from the previous result some preliminary information on the moduli space of the above surfaces.

We need however to slightly simplify our previous presentation. Observe therefore that we can exchange the roles of $\mathcal{L}$, $\mathcal{L}'$, and we can tensor $\mathcal{V}'$ by a line bundle of 4-torsion.

Therefore, in the last case (iv), we may simplify our treatment to consider only the subcases

- (iv, I) $\mathcal{L} \cong \mathcal{O}_A$

- (iv, II) $\mathcal{L}' \cong \mathcal{L}^{-1}$, $\mathcal{L}$ not of 4-torsion

- (iv, III) $\mathcal{L}' \cong \mathcal{L}^{-3}$, $\mathcal{L}$ not of 4-torsion

(the reader may in fact observe that cases (iv, I') and (iv, I'') are just special subcases of (iv, I)).

Corollary 5.9. *Consider the open set $\mathcal{M}$ of the moduli space of the surfaces with $p_g = 4$, $q = 1$, $K^2 = 12$ such that assumptions 5.1 and 5.3 are verified.*

Then $\mathcal{M}$ consists of the following 10 locally closed subsets:

- $\mathcal{M}(\text{i})$, *of dimension 20, corresponding to case* (i)

- $\mathcal{M}(\text{ii}, 0)$ *corresponding to the case* $\mathcal{L}^4 \cong \mathcal{O}_A$, *and* $\mathcal{M}(\text{ii}, 1)$ *corresponding to the case* $\mathcal{L}^4 \not\cong \mathcal{O}_A$, *both of dimension 19*

- $\mathcal{M}(\text{iii})$ *corresponding to case* (iii), *of dimension 18*

- $\mathcal{M}(\text{iv}, \text{I})$ *of dimension 18, corresponding to the case where* $\mathcal{L} \cong \mathcal{O}_A$, *but* $\mathcal{L}'$ *is neither of 3-torsion nor of 4-torsion*

- $\mathcal{M}(\text{iv}, \text{II})$ *of dimension 19, corresponding to the case* (iv, II) ($\mathcal{L}$ *not of 4-torsion*)

- $\mathcal{M}(\text{iv}, \text{III})$ *of dimension 19, corresponding to the case* (iv, III) ($\mathcal{L}$ *not of 4-torsion*)

- $\mathcal{M}(\text{iv}, \text{I}, 1/4)$ *of dimension 18, corresponding to case* (iv, I), *where we may assume* $\mathcal{L} \cong \mathcal{O}_A$ *and* $\mathcal{L}'$ *of 4-torsion but not of 2-torsion*

- $\mathcal{M}(\text{iv}, \text{I}, 1/2)$ *of dimension 19, corresponding to case* (iv, I), *where we may assume* $\mathcal{L} \cong \mathcal{O}_A$ *and* $\mathcal{L}'$ *of 2-torsion, but non trivial*

- $\mathcal{M}(\text{iv}, \text{I}, 1/3)$ *of dimension 18, corresponding to case* (iv, I), *where we may assume* $\mathcal{L} \cong \mathcal{O}_A$ *and* $\mathcal{L}'$ *of 3-torsion and non trivial*

- $\mathcal{M}(\text{iv}, \text{I}, 1)$ *of dimension 19, corresponding to case* (iv, I), *where we may assume* $\mathcal{L} \cong \mathcal{O}_A$ *and also* $\mathcal{L}' \cong \mathcal{O}_A$.

Proof. Observe that the moduli space of elliptic curves together with a torsion sheaf of torsion precisely n is irreducible of dimension 1.

Observe moreover that the hypersurface X (the canonical model of S) moves in a linear system $|4D|$ in $\mathbb{P}$, whose dimension is given by $h^0(\text{Sym}^4(\mathcal{V}')) - 1$. By Riemann–Roch, since $\mathcal{V}'$ has rank $= 3$ and degree $= 1$, we have that $h^0(\text{Sym}^4(\mathcal{V}')) = 20 + h^1(\text{Sym}^4(\mathcal{V}'))$.

Moreover we have that the dimension of each stratum, since each surface of general type has a finite automorphism group, equals $1 + h^0(\text{Sym}^4(\mathcal{V}')) - h^0(\text{End}(\mathcal{V}'))$. This justifies the assertion about the dimensions.

Furthermore, we may observe that the conditions that a vector bundle be indecomposable is an open one, while the condition that a line bundle be of n-torsion is a closed one. $\square$

The previous corollary allows us to conclude that our moduli space is irreducible: we use for this purpose a lower bound for the dimension of the moduli space which is a consequence of a general principle stated by Ziv Ran [1995], and turned into a precise theorem by Herb Clemens [2000].

Theorem 5.10. *The open set of the moduli space of surfaces with* $p_g = 4$, $q = 1$, $K^2 = 12$, *with non hyperelliptic Albanese fibres, and with* $\mathcal{V}$ *as in* (5.4), *i.e., generically generated by global sections, is irreducible of dimension 20.*

Proof. In view of the previous corollary, our moduli space has a stratification by locally closed sets, of which one only, $\mathcal{M}(i)$, has dimension 20, while the others have strictly smaller dimension. Since $\mathcal{M}(i)$ is clearly irreducible, it suffices to show that the dimension of the moduli space is at least 20 in each point. Equivalently, since the germ of the moduli space at the point corresponding to the surface S is the quotient of the base of the Kuranishi family of S by the finite group of automorphisms of S, it suffices to show that the dimension of the Kuranishi family is at least 20.

Now, in any case, the dimension of the Kuranishi family is always at least

$$h^1(S, T_S) - \dim(\mathrm{Obs}(S)),$$

but in this case the obstruction space $\mathrm{Obs}(S)$ is not the full cohomology group $H^2(S, T_S)$. Because we have a natural Hodge bilinear map

$$\gamma : H^0(S, \Omega^1_S) \times H^0(S, \Omega^2_S) \to H^0(S, \Omega^1_S \otimes \Omega^2_S),$$

and the natural subspace $H := \mathrm{Im}(H^0(S, \Omega^1_S) \otimes H^0(S, \Omega^2_S)) \subset H^0(S, \Omega^1_S \otimes \Omega^2_S)$ determines by Serre duality a quotient map $\gamma^\vee : H^2(S, T_S) \to H^\vee$. By Theorem 10.1 of Clemens [2000] we have that $\gamma^\vee(\mathrm{Obs}(S)) = 0$.

Since in this case it is obvious that $\dim(H) = 4$, γ being non degenerate, it follows that the base of the Kuranishi family has dimension $\geq -\chi(S, T_S) + 4 = 10\chi(S) - 2K_S^2 + 4 = 20$. $\square$

Theorem 5.11. *Assume that X is the canonical model of a surface with $p_g = 4, q = 1$, $K^2 = 12$, and non hyperelliptic Albanese fibres, and $\mathcal{V}$ is as in (5.4), i.e., generically generated by global sections. Then in cases (i), (ii) the canonical map ϕ is always a birational morphism, whereas in the other cases ϕ is birational for a general choice of X in the given linear system. The case $\deg(\varphi) = 3$ never occurs.*

Proof. Since $\mathcal{V}$ is generically generated by global sections, and the general fibre F_a is a non hyperelliptic curve of genus 3, it follows that F_a maps isomorphically to a plane quartic curve Γ_a. Let H_a be the plane containing Γ_a: since $K_X \equiv D + F$, then the pull-back divisor of H_a splits as $F_a + D_{-a}$, where $D_{-a} \sim D$.

Since $12 \geq \deg \phi \cdot \deg \Sigma$, and there are plane sections H_a which intersect Σ in a curve containing Γ_a, $\deg \Sigma \geq 4$, and the only possibility to exclude is that $\deg \phi = 2$ or 3.

In the case where $\deg \phi = 2$ let $\iota : X \to X$ be the corresponding biregular involution.

ι acts also on the Albanese variety A, and in a non trivial way, since a general fibre F_a is embedded by the canonical map ϕ, and let us then denote $a' := \iota(a)$.

If ι had no fixpoints on A, then $X \to X/\iota := Y$ would be unramified, so that $K_Y^2 = 6$, $\chi(\mathcal{O}_Y) = 2$, $q(Y) = 1$, whence $p_g(Y) = 2$, contradicting the fact that ϕ factors through Y. We may therefore assume that $a' = -a$ for a suitable choice of the origin in A.

Therefore, the inverse image of H_a contains $F_a + F_{a'}$, and we can write a linear equivalence $K_X \equiv F_a + F_{a'} + C_a$, where C_a is effective and $C_a = C_{-a}$.

Observe that $K_X \cdot F = 4$, $K_X \cdot C_a = 4$, $C_a \cdot F = 4$ whence C_a is not vertical for the Albanese map. Moreover, $12 = K_X^2 = (2F + C_a)^2 = 16 + C_a^2$, whence $C_a^2 = -4$.

In particular the algebraic system C_a has a fixed part.

We obtain a contradiction as follows.

First of all, since $|K_X - F_a - F_{a'}| \neq \emptyset$, we get $H^0(A, \mathcal{V}(-a-a')) \neq 0$, and this leaves out only the cases (5.4) (iii) and (iv), and moreover with $a + a' \equiv u + p$ on A.

We saw that $A/\iota \cong \mathbb{P}^1$, so that all the curves C_a are linearly equivalent. Indeed, a closer look reveals that all the curves C_a are the intersection of X with a fixed $\mathbb{P}^1$-subbundle of $\mathbb{P}$, thus we may consider the curve $C = C_a$ for all $a \in A$.

The curve C maps to a line L under the two dimensional linear system corresponding to $H^0(A, W)$, where we write $\mathcal{V} = \mathcal{O}_A(u + p) \oplus W$.

Before we further investigate the geometry of the situation, remark that ι acts equivariantly on X and A, therefore $\mathcal{V}$ is isomorphic to $\iota^*(\mathcal{V})$ and indeed we have an action of ι on $\mathcal{V}$.

This however implies that $\mathcal{L}$ is of 2-torsion in case (iii), while in case (iv) $\mathcal{L} \cong -\mathcal{L}'$. Once these conditions are satisfied, it is clear that we have an involution ι on $\mathbb{P}$ and that the system $|\mathcal{O}_\mathbb{P}(1)|$ is invariant, but it remains to be seen whether the hypersurface X is also ι-invariant (notice that the involution is completely determined by the four fixed points O such that $2O \equiv u + p$).

It is easy to verify that for a general choice of X in $|4D|$, this does not hold.

Claim. $\deg(\phi) = 3$ *never occurs.*

Consider in fact the possibility that $\deg(\phi) = 3$: then Γ_a is a full hyperplane section of Σ, and K_X is base-point free (in general, if $|K_X| = |M| + \Psi$, with Ψ a non trivial fixed part, then $M^2 = K_X^2 - K_X \cdot \Psi - M \cdot \Psi < K_X^2$, if then $|M|$ has base points, then $M^2 > \deg(\phi) \cdot \deg(\Sigma)$: while here $K_X^2 = 12 = \deg(\phi) \cdot \deg(\Sigma)$).

Observe that the surface Σ is normal, since it has a smooth hyperplane section.

Let $\pi : \tilde{\Sigma} \to \Sigma$ be a minimal resolution of Σ and denote by $\tilde{X}$ a minimal resolution of the fibre product $\tilde{\Sigma} \times_\Sigma X$: since $\tilde{X}$ is birational to X, $R^1 p_* \mathcal{O}_{\tilde{X}} = 0$, $p : \tilde{X} \to \Sigma$ being the composite morphism. Whence it follows that $R^1 \pi_* \mathcal{O}_{\tilde{\Sigma}} = 0$, i.e., Σ has only rational double points as singularities, and $\tilde{\Sigma}$ is a smooth $K3$-surface.

We will now consider the ramification formula for ϕ. Let B be the reduced branch divisor of ϕ, set $\phi^*(B) = R + R'$, R being the ramification divisor, and observe that $R' \geq 1/2 R$. The fact that Σ is a $K3$ with R.D.P.'s implies that $R \equiv K_X$, i.e., there is a hyperplane divisor H with $R = \phi^*(H)$.

On the other hand, since $\deg(\phi) = 3$ it follows that $\phi_*(R_{\text{red}}) = B$, whence B is the reduced divisor of the plane section which pulls back to H: this is a contradiction since then $R = \phi^*(H) \geq \phi^*(B) = R + R'$, while $R' > 0$ (this follows since $R > 0$, and $R' \geq 1/2 R$). $\square$

Remark 5.12. We saw that we have several strata of the above irreducible moduli space. The stratum of maximal dimension, such that the moduli space is just its closure will be called the 'Main Stratum', and we shall say that the surfaces which belong to this Main Stratum are of the "Main Stream".

It is certainly, as we shall see, the one which is most interesting and related to the geometry of elliptic space curves of degree 4.

We should also remark that a detailed and more general study of surfaces with irregularity $q = 1$ and with $K^2 = 3\chi$ was undertaken in the 1996 Thesis of T. Takahashi. However his results are weaker than ours in the case where $p_g = 4$, so we could not use this reference.

We finally come to a discussion of the geometry of the surfaces of the "Main Stream" (case (i)).

Let A be an elliptic curve of degree 4 in $\mathbb{P}^3$. Then, as it is well known, A is the complete intersection of 2 quadric surfaces Q, Q'.

We may indeed without loss of generality assume that the pencil of quadrics be Heisenberg invariant, in other words that:

$$A = \{(x) \mid x_0^2 + x_2^2 - \lambda^2 x_1 x_3 = 0, x_1^2 + x_3^2 - \lambda^2 x_0 x_2 = 0\}.$$

It is also well known (Atiyah [1957], Catanese and Ciliberto [1988]) that in case (i) the projective bundle $\mathbb{P}$ is nothing else than the triple symmetric product of the elliptic curve A, $\mathbb{P} = A^{(3)}$.

In this context the canonical mapping of X is induced by a morphism $\phi : \mathbb{P} \to (\mathbb{P}^3)^\vee$ which can be explained without formulae as follows: consider a point P of $\mathbb{P}$, i.e., P is a divisor of degree 3 on A. Then there is a unique plane $\phi(P)$ containing this divisor.

This geometric explanation shows that the degree of ϕ is 4 (as it had to be, since F being a fibre of the Albanese map a, $(D + F)^3 = 4$ by the Leray–Hirsch formula which we already mentioned).

In fact the reason of the above is that

- the projection onto the elliptic curve (the Albanese map) associates to a divisor $P = P_1 + P_2 + P_3$ the sum of the three points P_1, P_2, P_3 in the elliptic curve A,

- the tautological divisor D on $\mathbb{P}$ consists of the divisors where P_1 is fixed (whence, $D^3 = 1$).

- Let $I = \text{Proj}(\mathcal{T}_{\mathbb{P}^3})$ be the incidence correspondence, $I \subset \mathbb{P}^3 \times (\mathbb{P}^3)^\vee$: then we claim that $I \cap (A \times (\mathbb{P}^3)^\vee) = A^{(3)}$.

  Proof. The isomorphism is given by $(x, h) \to \text{div}_A(h) - x$. □

Observe only that the first projection does not correspond precisely to the Albanese map, but only to the composition of the Albanese map with an involution of A, since to a divisor $P = P_1 + P_2 + P_3$ corresponds the point x such that $x + P_1 + P_2 + P_3$ is linearly equivalent to the hyperplane divisor of A.

- We claim that the second projection is given by the linear system $|D + F|$.

  Proof. Any hyperplane in $(\mathbb{P}^3)^\vee$ is the hyperplane H_x dual to a point $x \in \mathbb{P}^3$. Let $x \in A$: then the inverse image of H_x is given by the divisors P' of degree 3 on A such that P'_1, P'_2, P'_3 span a plane containing x. Thus, we have two possibilities: either we take the divisors P such that $P + x$ is linearly equivalent to the hyperplane divisor on A, and thus we get a fibre F, or we simply take the divisors P' of degree 3 for which $P' \geq x$, i.e., we get a divisor of type D.

Set for convenience $W := \mathbb{P} = A^{(3)}$ and observe that the pull back H_2 of the hyperplane divisor in $(\mathbb{P}^3)^\vee$ is thus linearly equivalent to $D + F$. Observe also that the pull back of the hyperplane divisor in $(\mathbb{P}^3)$ is linearly equivalent to $4F$. Therefore, the desired canonical model $X \subset W$ is in the linear system $|4D| = |4H_2 - H_1|$.

We can perhaps summarize these observations as follows:

Proposition 5.13. *The canonical model of a surface with $p_g = 4$, $q = 1$, $K^2 = 12$ of the Main Stream, i.e., of type (i), is a divisor of bidegree $(-1, 4)$ on the variety W given by the intersection of the incidence variety $I \subset \mathbb{P}^3 \times (\mathbb{P}^3)^\vee$ (itself a divisor of bidegree $(1, 1)$) with the pull back of the elliptic curve A under the first projection.*

Thus W is a complete intersection of type $(1, 1)$, $(2, 0)(2, 0)$, but X is not a complete intersection. The canonical divisor on X is induced by the divisor of bidegree $(0, 1)$ on $\mathbb{P}^3 \times (\mathbb{P}^3)^\vee$.

From this it is easy to produce equations of explicit examples of these surfaces via computer algebra. The method is based on the following

Remark 5.14. Since W is a complete intersection in $M := \mathbb{P}^3 \times (\mathbb{P}^3)^\vee$ it follows easily that the restriction homomorphism $H^0(\mathcal{O}_M(n, 4)) \to H^0(\mathcal{O}_W(n, 4))$ is surjective as soon as $n \geq 2$.

Fix therefore a divisor $\bar{B}$ in $|\mathcal{O}_M(3, 0)|$, for instance the pull back of the three planes $\{x \mid x_0 x_1 x_2 = 0\}$. Then the linear system $|4D|$ on W is the residual system $|H^0(\mathcal{O}_W(3, 4)(-\bar{B}))|$.

6. Surfaces with $p_g = 4$, $q = 3$ and canonical map of degree 1 or 2

In this section we shall consider surfaces with $p_g = 4$, $q = 3$, $K^2 = 12$, contained in an Abelian 3-fold as a polarization of type $(1, 1, 2)$: we will first show that this family is stable by small deformation.

Later, we will show that for a general such surface the canonical map is a birational morphism onto a surface of degree twelve in $\mathbb{P}^3$, whereas, for all the surfaces which are the pull back of a theta divisor on a principally polarized Abelian 3-fold, then the canonical map is of degree 2 onto an interesting sextic surface.

More precisely, our situation will be as follows: we let J be a principally polarized Abelian variety of dimension 3, which is the Jacobian of a curve C of genus 3, and we let Θ be its principal polarization. We let $\pi : A \to J$ be an isogeny of degree 2 and S a smooth divisor in the complete linear system $|\pi^*\Theta|$ associated to the pull back $\pi^*\Theta$.

Since at some step we will also need theta functions, we represent the Jacobian variety J as $J = \mathbb{C}^3/\mathbb{Z}^3 + \Omega\mathbb{Z}^3 := \mathbb{C}^3/\Lambda(\Omega)$, with Ω in the Siegel upper half-space (we have thus already represented the theta divisor as a symmetric divisor with respect to the origin in J).

We set then $A = \mathbb{C}^3/\Lambda$ with $\Lambda \subset \Lambda(\Omega)$ of index 2 dual under the symplectic pairing to $\Lambda(\Omega) + \mathbb{Z}b$, with $b \in 1/2\mathbb{Z}^3$ (e.g., $b = 1/2e_3$).

Proposition 6.1. *A basis of* $H^0(A, \mathcal{O}_A(\pi^*\Theta))$ *is given by even functions.*

Proof. Let $c \in \{0, b\}$, and consider the basis given by the following two elements:
$$\theta[0, c](z, \Omega) := \sum_{n \in \mathbb{Z}^3} \exp(2\pi i\ (1/2\,{}^t n\Omega n +{}^t n(z + c))).$$

An elementary calculation shows that
$$\theta[0, c](-z, \Omega) = \sum_{n \in \mathbb{Z}^3} \exp(2\pi i\ (1/2\,{}^t n\Omega n +{}^t n(-z + c)))$$
$$= \sum_{m=-n \in \mathbb{Z}^3} \exp(2\pi i\ (1/2\,{}^t m\Omega m +{}^t m(z - c)))$$
$$= \theta[0, c](z, \Omega)$$
since, for all $m \in \mathbb{Z}^3$, $\exp(2\pi i({}^t m(-2c))) = 1$. $\square$

Proposition 6.2. *Let S be a smooth surface in a polarization of type $(1, 1, 2)$ in an Abelian 3-fold A. Then the invariants of S are $p_g = 4$, $q = 3$, $K^2 = 12$.*

Proof. Let us consider the exact sequence
$$0 \to \mathcal{O}_A \to \mathcal{O}_A(S) \to \omega_S \to 0$$
and observe that $H^i(\mathcal{O}_A(S)) = 0$ for $i = 1, 2$. Whence, $p_g = h^0(\omega_S) = 4$, and $q := h^1(\mathcal{O}_S)(= h^1(\omega_S)$ by Serre duality$) = 3$.
Moreover, we have $K_S^2 = S^3 = 12$. $\square$

Proposition 6.3. *Let S be a smooth surface in a polarization of type $(1, 1, 2)$ in an Abelian 3-fold A. Then any small deformation is a surface of the same kind.*

Proof. Since the canonical divisor of A is trivial, the normal bundle of S in A is $N_S = \omega_S$, whence its cohomology groups have respective dimensions $h^0(N_S) = 4$, $h^1(N_S) = 3$, $h^2(N_S) = 1$.

The tangent sheaf sequence reads out as follows:

$$0 \to \mathcal{T}_S \to \mathcal{T}_A \otimes \mathcal{O}_S \cong \mathcal{O}_S^3 \to N_S \to 0,$$

whose exact cohomology sequence is:

$$0 \to H^0(N_S)/H^0(\mathcal{O}_S^3) \cong \mathbb{C} \to H^1(\mathcal{T}_S) \to H^1(\mathcal{T}_A \otimes \mathcal{O}_S) \to H^1(N_S) \to H^2(\mathcal{T}_S) \to \cdots.$$

We get a smooth 7-dimensional family by varying A in its 6-dimensional local moduli space (Siegel's upper half space), and S in the corresponding 1-dimensional linear system.

This family will be shown to coincide with the Kuranishi family once we prove that $H^1(\mathcal{T}_S)$ has dimension 7, or, equivalently, (since $H^1(\mathcal{T}_A \otimes \mathcal{O}_S) \cong H^1(\mathcal{O}_S^3) \cong \mathbb{C}^9$) we show the surjectivity of $H^1(\mathcal{T}_A \otimes \mathcal{O}_S) \to H^1(N_S)$.

To understand this map, consider an element $\sum_{i=1,2,3} \xi_i \otimes \psi_i \in H^1(\mathcal{T}_A \otimes \mathcal{O}_S)$, where ξ_1, ξ_2, ξ_3 yield a basis of $H^0(\mathcal{T}_A)$, $\psi_i \in H^1(\mathcal{O}_S) \cong H^1(\mathcal{O}_A)$.

Let $\{U_\alpha\}$ be an open cover of A such that $S \cap U_\alpha = \mathrm{div}(f_\alpha)$, and let $f_\alpha = g_{\alpha,\beta} f_\beta$ in $U_\alpha \cap U_\beta$: then the image of $\sum_{i=1,2,3} \xi_i \otimes \psi_i$ is given by $\sum_{i=1,2,3} \xi_i(f_\alpha) \otimes \psi_i$.

We use moreover the isomorphism $H^1(N_S) \cong H^2(\mathcal{O}_A)$: since for a vector field ξ we have $\xi(f_\alpha) = g_{\alpha,\beta} \xi(f_\beta) \pmod{f_\beta}$, the image of $\sum_{i=1,2,3} \xi_i \otimes \psi_i$ into $H^2(\mathcal{O}_A)$ is the cohomology class $\sum_{i=1,2,3} \xi_i(g_{\alpha,\beta}) \cup \psi_i$.

We are quickly done, since

- the map $\xi \in H^0(\mathcal{T}_A) \to \xi(g_{\alpha,\beta}) \in H^1(\mathcal{O}_A)$ is an isomorphism, being the tangent map at the origin of the isogeny $\tau : A \to \mathrm{Pic}(A)$ such that $\tau(x) = S - (S+x)$

- $H^1(\mathcal{O}_A) \cup H^1(\mathcal{O}_A) \to H^2(\mathcal{O}_A)$ is onto. $\square$

Theorem 6.4. *Let S be a smooth divisor yielding a polarization of type $(1, 1, 2)$ on an Abelian 3-fold: then the canonical map of S is in general a birational morphism onto a surface Σ of degree 12.*

In the special case where S is the inverse image of the theta divisor in a principally polarized Abelian 3-fold, the canonical map is a degree two morphism onto a sextic surface Σ in $\mathbb{P}^3$. In this case, the singularities of Σ are in general: a plane cubic Γ which is a double curve of nodal type for Σ and, moreover, a strictly even set of 32 nodes for Σ. Also, in this special case, the normalization of Σ is in fact the quotient of S by an involution ι on A having only isolated fixed points (on A), of which exactly 32 lie on S.

Proof. Observe that the natural map $H^0(\Omega_A^2) \to H^0(\Omega_S^2)$ is injective because S is not a subabelian variety, moreover we get in this way a linear subsystem of $|K_S|$ which is base point free, since S embeds into A.

Canonical projections of irregular algebraic surfaces 107

It is easy to observe that each translation, and also each involution ι with fixed points on A (multiplication by -1 for a suitable choice of an origin) acts trivially on the vector space $H^0(\Omega_A^2)$.

On the other hand, considering the exact sequence in Prop. 6.2,

$$0 \to \omega_A \to \omega_A(S) \to \omega_S \to 0$$

we see that the 3-dimensional system generated by $H^0(\Omega_A^2)$ maps isomorphically to $H^1(\omega_A)$, whereas $H^0(\mathcal{O}_A(S)) \cong H^0(\omega_A(S))$ maps to $H^0(\omega_S)$ under the following explicit map

$$f(z) \to f(z)(dz_1 \wedge dz_2 \wedge dz_3)/d\theta(z),$$

where $S = \mathrm{div}(\theta(z))$, $f(z)$, $\theta(z) \in H^0(\mathcal{O}_A(S))$ are expressed by even functions, and $(dz_1 \wedge dz_2 \wedge dz_3)/d\theta(z)$ stands for the Poincaré Residuum $\eta := \eta_i := (dz_1 \wedge dz_2 \wedge dz_3) \neg (\partial/\partial z_i)(\partial\theta(z)/\partial z_i)^{-1}$ ($\neg$ is the contraction operator).

Whence follows that the involution $z \to -z$ acts on the image of $H^0(\mathcal{O}_A(S))$ in $H^0(\mathcal{O}_S(K_S))$ as multiplication by -1.

Let us now choose in particular a surface S which is the inverse image of a theta divisor Θ on J: then the subspace V_{++} coming from $H^0(\Omega_A^2)$ is the pull back of $H^0(\Omega_\Theta^2)$, so it consists of the sections in $H^0(\Omega_S^2) = H^0(\mathcal{O}_S(K_S))$ which are invariant under the fixed point free covering involution $z \to z + \eta$ for the double cover $\pi : S \to \Theta$.

On our particular surface S acts the group $(\mathbb{Z}/2)^2$ generated by $z \to z + \eta$ and by $z \to -z$ for our choice of the origin (c.f. Prop. 6.1), and we see that, if we define V_{--} as the one dimensional space coming from $H^0(\omega_A(S))$, then V_{--} is an eigenspace with eigenvalue -1 for both the involutions above.

In particular, it follows that the involution ι defined by $\iota(z) = -z + \eta$ acts trivially on the space $H^0(\mathcal{O}_S(K_S))$. Therefore the canonical map of such a special S factors through the involution ι.

Geometry of the situation for special surfaces

Let $Z := S/\iota$.

Lemma 6.5. *The involution ι has exactly 32 isolated fixed points on S.*

Proof. Let us find the fixed points of ι recalling that $\iota(z) = -z + \eta$. Then z yields a fixed point on A iff $2z \equiv \eta \pmod{\Lambda}$. The fixed points moreover lie on S if and only if they project in $J = \mathbb{C}^3/\Lambda(\Omega)$ to a (2-torsion) point which lies on Θ, i.e., to an odd thetacharacteristic.

Set $\Lambda' := \Lambda(\Omega)$, thus $\eta \in \Lambda'$, whence for such a fixed point $2z \in \Lambda'$ and its image in $\Lambda'/\Lambda \cong \mathbb{Z}/2$ is non trivial.

Therefore the number of the odd thetacharacteristics which are image of such a fixed points are in bijection with the set $N \subset ((\mathbb{Z}/2)^3)^2$ defined by the following equations:

$$N = \{(x, y) \mid {}^t xy = 1, x_1 = 1\}.$$

Whence, card$(N) = 16$ and there are exactly 32 fixed points on S. □

Remark 6.6. Since the double cover $S \to Z$ is ramified exactly on the 32 corresponding nodes of Z, these form an even set according to the definition of Catanese [1981].

Note moreover that ι acts as multiplication by -1 on the space of global 1-forms, therefore the quotient surface Z has $q(Z) = 0$, $p_g(Z) = 4$, $K_Z^2 = 6$.

Then the canonical map of S, for S special, factors through Z. In turn, since V_{++} is base point free, this means that there is a point $O \in \mathbb{P}^3 - \Sigma$ so that the projection with centre O to $\mathbb{P}^2$ yields the composition of the projection onto Θ with the canonical map of Θ.

On the other hand, as well known, Θ is the symmetric product $C^{(2)}$ of a curve C of genus 3, which, since Θ is smooth, is a smooth plane quartic curve $C = C_4 \subset \mathbb{P}^2$.

Claim. *The canonical map of $C^{(2)}$ sends the divisor $P + Q$ to the line generated by P and Q.*

Proof of the claim. If $\omega_1, \omega_2, \omega_3$ are a basis of $H^0(\Omega_C^1)$, then a basis of the canonical system of $C^{(2)}$ is given, on the Cartesian product C^2, by $\omega_i(P) \wedge \omega_j(Q) + \omega_i(Q) \wedge \omega_j(P)$, but this vector is the wedge product of the two vectors $\omega_i(P)$ and $\omega_j(Q)$.

That this is a morphism follows e.g. since its base locus on C^2 is just the diagonal Δ, but $\epsilon^* |K_{C^{(2)}}| = |K_{C^2} - \Delta|$, whence $|K_{C^{(2)}}|$ is free from base points. □

We let now Y be the quotient of Θ by the multiplication by -1, whence $Y = S/(\mathbb{Z}/2)^2$: Y has $K_Y^2 = 3$, $q(Y) = 0$, $p_g(Y) = 3$ and its canonical map is a triple cover of $\mathbb{P}^2$, branched on the dual curve $C^\vee$ of C. In fact, multiplication by -1 on Θ corresponds to residuation with respect to K_C on $C^{(2)}$.

Y has 28 nodes, corresponding to the odd thetacharacteristics of C. The covering $Z \to Y$ is etale, except over 12 of the nodes of Y: as we saw, Z has exactly 32 nodes lying above the remaining 16 nodes of Y, over these 12 nodes lie instead 12 smooth points of Z.

Remark 6.7. 1) The bicanonical system of $C^{(2)}$ (cf. Catanese, Ciliberto and Mendes Lopes [1998]) factors through the bicanonical system of Y, which embeds Y in $\mathbb{P}^6$, since it is induced by the sections of $H^0(J, \mathcal{O}_J(2\Theta))$.

2) The monodromy of $\Theta \to \mathbb{P}^2$ is the full symmetric group $\mathcal{S}_4$. The monodromy of the canonical map of Z is instead the symmetric group $\mathcal{S}_3$.

Lemma 6.8. *The image Σ of Z is a surface of degree 6 (hence, birational to Z).*

Proof. Consider the morphism $f : Z \to \mathbb{P}^2$, obtained as the composition of the canonical map ϕ of Z with the projection p with centre O of Σ to $\mathbb{P}^2$.

It cannot be that $\deg(\Sigma) = 2$, otherwise p would be branched on a plane conic, whereas the branch curve of f is the irreducible curve $C^\vee$.

If instead $\deg(\Sigma) = 3$, then there would be a covering involution i for ϕ. Since there is already a covering involution j for f, gotten from the double cover $Z \to Y$, we let G be the group of covering involutions for f. Since the canonical map of S does not factor through the one of Θ, it follows that $i \neq j$.

Then G is a group of order $h \geq 4$ with h dividing 6, thus $h = 6$ and f should be Galois.

This is however a contradiction, since the inverse image of the branch curve $C^\vee$ has components of multiplicity both 1 and 2; this holds because $Z \to Y$ is etale in codimension 1, while $Y \to \mathbb{P}^2$ has simple branching on the curve $C^\vee$, and the general tangent to $C^\vee$ is not a bitangent. □

With the result of the previous lemma in our hands, we can finish the proof of the theorem. Assume that the canonical map of S were always not birational.

Since for special S the degree equals 2, we would have that the canonical map always factors through the involution ι. But, since S always admits the involution $z \to -z$, then S would be stable under the involution $z \to z + \eta$, i.e., would be a pull back of a theta divisor. Contradicting that the Kuranishi family has dimension 7 and not 6.

Finally, in the special case, the surface Z is a canonical model with $K^2 = 6$, $p_g = 4$, $q = 0$ and with birational canonical map. Therefore, the double curve of Σ is a plane cubic, cf. Catanese [1984b]. □

In the special case, the equations of Σ can be written explicitly. In fact giving an unramified double covering of a non hyperelliptic curve C of genus 3 is equivalent (cf. e.g. Catanese [1981]) to writing the equation of its canonical model as the determinant of a 2×2 symmetric matrix of quadratic forms.

We have, more precisely, coordinates x_0, x_1, x_2 in $\mathbb{P}^2$ and quadratic forms $Q_{33}(x)$, $Q_{34}(x)$, $Q_{44}(x)$ such that

$$C = \{(x_0, x_1, x_2) \mid Q_{33}(x)Q_{44}(x) - Q_{34}(x)^2 = 0\};$$

moreover, the double unramified covering C' of C is the genus 5 curve whose canonical model in $\mathbb{P}^5$ is defined as the following intersection of three quadrics:

$$C' = \{(x_0, x_1, x_2, y_3, y_4) \mid y_3^2 = Q_{33}(x), y_4^2 = Q_{44}(x), y_3 y_4 = Q_{34}(x)\}.$$

Now, there is a natural surjection of $(C')^2$ onto S. In fact, Θ is the symmetric square of C, and thus dominated by C^2, and S is the quotient of $(C')^2$ under the $(\mathbb{Z}/2)^2$ action permuting the coordinates and acting with the diagonal action of the involution $\iota : C' \to C'$.

We can then read the canonical map of S as the map corresponding to the $(\mathbb{Z}/2)^2$-invariant sections of $K_{(C')^2}$.

Recall that on the first curve of the product $(C')^2$ a basis of $H^0(K_{C'})^+$ is given by x_0, x_1, x_2, and a basis for $H^0(K_{C'})^-$ is given by y_3, y_4. Similarly we have a basis w_0, w_1, w_2, z_3, z_4 for the second curve.

We find therefore that a basis for the $(\mathbb{Z}/2)^2$-invariant sections of $K_{(C')^2}$ is provided by $u_0 := x_1 w_2 - x_2 w_1$, $u_1 := x_0 w_2 - x_2 w_0$, $u_2 := x_0 w_1 - x_1 w_0$, $v := y_3 z_4 - y_4 z_3$ (these are just ι-invariant Plücker coordinates of the line spanned by the two points of C').

Let a, b, c be the symmetric 3×3 matrices yielding the respective quadratic forms $Q_{33}(x)$, $Q_{34}(x)$, $Q_{44}(x)$: then the entries of the matrix α are polynomial functions in the respective entries of a, b, c and in the coordinates (u_0, u_1, u_2, v) on $\mathbb{P}^3$.

The shape of α^+ is

$$\begin{pmatrix} v^5 + Av^3 + Bv & C \\ C & v \end{pmatrix}$$

where for instance $A =^t u(-2\Lambda^2 b + \Lambda^2(a+c) - \Lambda^2 a - \Lambda^2 b)u$, and

$$C = \det \begin{pmatrix} ^txax & ^twax & ^twaw \\ ^txbx & ^twbx & ^twbw \\ ^txcx & ^twcx & ^twcw \end{pmatrix}.$$

We have not yet found a compact expression for B, the one we have is too long to be reproduced anywhere.

7. Irregular surfaces with $p_g = 4, q = 2$

This section will be devoted to the description of another interesting example, of surfaces with the following invariants: $p_g = 4, q = 2, K^2 = 18$ and birational canonical morphism onto its image.

The surfaces are obtained as $(\mathbb{Z}/2\mathbb{Z})^2$-Galois covers of a principally polarized Abelian surface A, with branch locus consisting of 3 divisors D_1, D_2, D_3 which are algebraic equivalent to the theta divisor Θ. We shall follow the notation of Catanese [1984a].

We choose then L_1, L_2, L_3 divisors which are also algebraically equivalent to Θ, and such that

$$2L_i \equiv D_j + D_k \quad \text{for all } i \neq j \neq k \neq i.$$

We take the corresponding $(\mathbb{Z}/2\mathbb{Z})^2$-Galois cover $\pi : S \to A$ such that

$$\pi_* \mathcal{O}_S = \mathcal{O}_A \bigoplus \left(\oplus_{i=1,2,3} \mathcal{O}_A(-L_i) \right), \pi_* \omega_S = \mathcal{O}_A \bigoplus (\oplus_{i=1,2,3} \mathcal{O}_A(L_i)).$$

It follows immediately that the constructed surfaces have the numerical invariants as desired: for instance, since K_S is the ramification divisor R, and $2R \equiv \pi^*(D)$, where $D = D_1 + D_2 + D_3$, we have $K_S^2 = R^2 = D^2 = 9\Theta^2 = 18$.

We recall the standard notation, by which $D_i = \mathrm{div}(x_i)$, $R_i = \mathrm{div}(z_i)$ so that S is defined by the equations

$$w_i^2 = x_j x_k, \quad w_i x_i = w_j w_k$$

in the rank 3 bundle $(\oplus_{i=1,2,3} \mathcal{O}_A(L_i))$, and we have $z_i^2 = x_i$, $w_i = z_j z_k$.

We also let ϕ_i be the unique section of $\mathcal{O}_A(L_i)$, and $C_i := \mathrm{div}(\phi_i)$.

With this notation, there are 4 sections of the canonical sheaf ω_S, namely: $\omega := z_1 z_2 z_3$, and, for all $i = 1, 2, 3$, $\omega_i := \omega \phi_i / w_i = z_i \phi_i$.

We obtain immediately that the base locus of the canonical system projects down in A to the set $D \cap (\cap_{i=1,2,3}(D_i + C_i))$.

Remark 7.1. The surface S has base point free canonical system provided the 6 curves D_i, C_i have no point common to three of them. Since any three of the six curves can be chosen as arbitrary translates of the theta divisor, it follows easily that for a general choice there are no base points of K_S.

We assume henceforth the canonical system to be base-point free, so that we have the canonical morphism

$$\Phi : S \to \Sigma \subset \mathbb{P}^3$$

and we use the characters of the Galois group in order to study the geometry of the map Φ and more generally the canonical ring of S.

We have here $\mathcal{R}(S) = \bigoplus_{m=0}^{\infty} H^0(\mathcal{O}_S(mK_S))$ where

$$H^0(\mathcal{O}_S(2mK_S)) = H^0(\mathcal{O}_A(mD)) \bigoplus H^0(\oplus_i \mathcal{O}_A(-L_i + mD))$$

$$H^0(\mathcal{O}_S((2m+1)K_S)) = \omega \, H^0(\mathcal{O}_A(mD)) \bigoplus (\oplus_i z_i \, H^0(\mathcal{O}_A(+L_i + mD))).$$

We have 4 generators for $\mathcal{R}(S)$ in degree 1, namely, ω, ω_1, ω_2, ω_3, moreover we observe the following dimensions for the four respective eigenspaces in degree 2: $\dim(\mathcal{R}(S)_0) = 9$, $\dim(\mathcal{R}(S)_i) = 4$.

Lemma 7.2. $\Phi(S)$ *is not a quadric (if the canonical system is base point free).*

Proof. It suffices to show the linear independence of the 10 monomials ω^2, ω_i^2 and $\omega \omega_i$, $\omega_j \omega_k$. Any linear relation is a sum of linear relations in each eigenspace, and clearly, if i, j, k are distinct indices, then $\omega \omega_i$, $\omega_j \omega_k$ are independent since their divisors are $2R_i + R_j + R_k + C_i'$, $R_j + R_k + C_j' + C_k'$ respectively, C_i' being the inverse image of the divisor C_i.

Moreover, a linear relation of the form $\Sigma_{i=0,1,2,3} c_i \omega_i^2 = 0$ would translate into a relation $c_0 x_1 x_2 x_3 + \Sigma_{i=1,2,3} c_i x_i \phi_i^2 = 0$ and since w.l.o.g. we may assume that $c_3 = 1$, we obtain that ω_3 vanishes at the points where $x_1 = x_2 = 0$, contradicting that the canonical system is base point free. $\square$

Theorem 7.3. *For general choice of the three divisors D_i the canonical map Φ is birational onto its image.*

Proof. Consider the 8 points $P \in C'_1 \cap C'_2 = \{\phi_1 = \phi_2 = 0\}$. They map to $(z_1 z_2 z_3(P), 0, 0, z_3 \phi_3(P))$. Moreover, the inverse image of these points is contained in $\omega_1 = \omega_2 = 0$ which consists of these 8 points, plus points in R_1 or in R_2 which therefore map to points where the first coordinate equals 0.

Thus, the inverse image of the punctured line $y_1 = y_2 = 0$, $y_0 \neq 0$ consists of these 8 points, which form two $(\mathbb{Z}/2)^2$ orbits. For each point $(a, 0, 0, b)$ in the image, since by generality we may assume $a \neq b \neq 0 \neq a$, the inverse image consists therefore of either 2 or 4 points. However, $\deg(\Phi) \deg(\Sigma) = 18$, whence the only possibility that Φ may not be birational is that $\deg(\Phi) = 2$.

Assume this to be the case: then, since $(\mathbb{Z}/2)^2$ acts Φ-equivariantly on S and on $\mathbb{P}^3$, then we would have an involution i on A which would lift to S, and actually in such a way to centralize the Galois action. This however implies that i leaves the three branching divisors D_i invariant.

Consider then the curve D_1, which has genus 2. It possesses then only the hyperelliptic involution, or at most a finite number of involutions whose quotient is an elliptic curve. Since however we may choose D_2 to cut D_1 in any assigned pair of points of D_1, we easily get the desired contradiction. □

References

[A-O]   V. Ancona, G. Ottaviani, An introduction to the derived categories and the theorem of Beilinson, Atti Accad. Peloritana Pericolanti Cl. Sci. Fis. Mat. Natur. LXVII (1989), 99–110.

[A-S]   E. Arbarello, E. Sernesi, The equation of a plane curve, Duke Math. J. 46 (1979), 469–485.

[Ash]   T. Ashikaga, A remark on the geometry of surfaces with birational canonical morphisms, Math. Ann. 290 (1991), 63–76.

[At]   M. F. Atiyah, Vector bundles over an elliptic curve Proc. Lond. Math. Soc. III. Ser. 7, (1957), 414–452.

[Bar]   W. Barth, Counting singularities of quadratic forms on vector bundles, in: Vector Bundles and Differential equations, Proc. Nice 1979, Progr. Math. 7, Birkhäuser, 1980, 1–29.

[B-P-V]   W. Barth, C. Peters, A. van de Ven, Compact complex surfaces, Ergeb. Math. Grenzgeb. (3) 4, Springer-Verlag, Berlin–Heidelberg 1984.

[Bau]   I. C. Bauer, Surfaces with $K^2 = 7$ and $p_g = 4$, Mem. Amer. Math. Soc. 721 (2001).

[Bea78]   A. Beauville, Surfaces algebriques complexes, Astérisque 54, Soc. Math. France, Paris (1978).

[Bea79]   A. Beauville, L'application canonique pour les surfaces de type general, Invent. Math. 55 (1979), 121–140.

[Bea82]   A. Beauville, L'inegalité $p_g \geq 2q - 4$ pour les surfaces de type general, Bull. Soc. Math. France 110 (1982), 344–346.

[Bei]   A. Beilinson, Coherent sheaves on $\mathbb{P}^n$ and problems of linear algebra, Funct. Anal. Appl. 12 (1978), 214–216 (translated from Funktsional. Anal. i Prilozhen. 12 (3) (1978), 68–69).

[B-G-G]   I.N. Bernshtein, I. M. Gel'fand, S. I. Gel'fand, Algebraic bundles on $\mathbb{P}^n$ and problems of linear algebra, Funct. Anal. Appl. 12 (1978), 212–214 (translated from Funktsional. Anal. i Prilozhen. 12 (3) (1978), 66–68).

[Bol]   O. Bolza, On binary sextics with linear transformations into themselves, Amer. J. Math. 10 (1888), 47–70.

[Bom]   E. Bombieri, Canonical models of surfaces of general type, Inst. Hautes Études Sci. Publ. Math. 42 (1973), 171–219.

[B-E]   D. A. Buchsbaum, D. Eisenbud, What annihilates a module?, J. Algebra 47 (1977), 231–243.

[Can00]   A. Canonaco, A Beilinson-type theorem for coherent sheaves on weighted projective spaces, J. Algebra 225 (2000), 28–46.

[Can02]   A. Canonaco, Thesis, Scuola Normale Superiore Pisa (2002)

[C-E]   G. Casnati, T. Ekedahl, Covers of algebraic varieties I. General structure theorem, covers of degree 3,4 and Enriques surfaces, J. Algebraic Geom. 5 (1996), 439–460.

[Cas91]   G. Castelnuovo, Osservazioni intorno alla geometria sopra una superficie, I, II, Istit. Lombardo Accad. Sci. Lett. Rend. A (II) 24 (1891); also in 'Memorie scelte', Zanichelli (1937), Bologna 1937, 245–265.

[Cas05]   G. Castelnuovo, Sulle superficie aventi il genere aritmetico negativo, Rend. Circ. Mat. Palermo 20 (1905), 55–60.

[Cat81]   F. Catanese, Babbage's conjecture, contact of surfaces, symmetric determinantal varieties and applications, Invent. Math. 63 (1981), 433–465.

[Cat84a]   F. Catanese, On the moduli space of surfaces of general type, J. Differential Geom. 19 (1984), 483–515.

[Cat84b]   F. Catanese, Commutative algebra methods and equations of regular surfaces, in: Algebraic Geometry – Bucharest 1982, Lecture Notes in Math. 1056, Springer-Verlag, Berlin 1984, 68–111.

[Cat85]   F. Catanese, Equations of pluriregular varieties of general type, in: Geometry today – Roma 1984, Progr. Math. 60, Birkhäuser, Basel 1985, 47–67.

[Cat97]   F. Catanese, Homological algebra and algebraic surfaces, in: Algebraic geometry (J. Kollár et al., eds.), Proceedings of the Summer Research Institute, Santa Cruz, CA, USA, July 9–29, 1995, Proc. Sympos. Pure Math. 62, Amer. Math. Soc., Providence, RI, 1997, 3–56.

[C-F]   F. Catanese, M. Franciosi, Divisors of small genus on algebraic surfaces and projective embeddings, in: Proceedings of the Hirzebruch 65 conference on algebraic geometry (Teicher, Mina, ed.), Bar-Ilan University, Ramat Gan, Israel, May 2–7, 1993. Israel Math. Conf. Proc. 9 (1996), Ramat-Gan 1996, 109–140.

[C-F-H-R] F. Catanese, M. Franciosi, K. Hulek, M. Reid, Embeddings of curves and surfaces, Nagoya Math. J. 154 (1999), 185–220.

[C-C91]   F. Catanese, C. Ciliberto, Surfaces with $p_g = q = 1$, in: Problems in the theory of surfaces and their classification (F. Catanese et al., eds.), Papers from the meeting held at the Scuola Normale Superiore, Cortona, Italy, October 10–15, 1988, Symp. Math. 32, Academic Press, London 1991, 49–79.

[C-C93]   F. Catanese, C. Ciliberto, Symmetric products of elliptic curves and surfaces of general type with $p_g = q = 1$, J. Algebraic Geom. 2 (1993), 389–411.

[C-C-ML] F. Catanese, C. Ciliberto, M. Mendes Lopes, On the classification of irregular surfaces of general type with nonbirational bicanonical map, Trans. Amer. Math. Soc. 350 (1998), 275–308.

[Cil81]   C. Ciliberto, Canonical surfaces with $p_g = p_a = 4$ and $K^2 = 5, \ldots, 10$, Duke Math. J. 48 (1981), 121–157.

[Cil83]   C. Ciliberto, Sul grado dei generatori dell' anello canonico di una superficie di tipo generale, Rend. Sem. Mat. Torino 41 (3) (1983), 83–111.

[Clem00] H. Clemens, Cohomology and obstructions I: on the geometry of formal Kuranishi theory, math.AG/9901084 v2.

[De]   O. Debarre, Inegalites numeriques pour les surfaces de type general, Bull. Soc. Math. France 110 (1982), 319–346.

[Di]   A. C. Dixon, Note on the reduction of a ternary quantic to a symmetrical determinant, Proc. Cambridge Philos. Soc. 11 (1902), 350–351.

[Ei80]   D. Eisenbud, Homological algebra on a complete intersection, Trans. Amer. Math. Soc. 260 (980), 35–64.

[Ei95]   D. Eisenbud, Commutative Algebra with a view towards Algebraic Geometry, Grad. Texts in Math. 150, Springer-Verlag, New York 1995.

[E-F-S]   D. Eisenbud, G. Fløystad, F.-O. Schreyer, Sheaf cohomology and free fesolutions over exterior algebras, preprint, math.AG/0104203.

[E-S]   D. Eisenbud, F.-O. Schreyer, Resultants and Chow forms via exterior syzygies, preprint, math.AG/0111040.

[En]   F. Enriques, Le Superficie Algebriche, Zanichelli, Bologna 1949.

[Fit]   H. Fitting, Die Determinantenideale einer Moduls, Jahresber. Deutsch. Math.-Verein. 46 (1936), 195–220.

[Fuj]   T. Fujita, On Kähler fiber spaces over curves, J. Math. Soc. Japan 30 (1978), 779–794.

[Gra]   M. Grassi, Koszul modules and Gorenstein algebras, J. Algebra 180 (1996), 918–953.

[Har] R. Hartshorne, Algebraic Geometry, Grad. Texts in Math. 52, Springer-Verlag, New York 1977.

[Hilb] D. Hilbert, Über die Theorie der Algebraischen Formen, Math. Ann. 36 (1890), 473–534.

[Hor75] E. Horikawa, On deformation of quintic surfaces, Invent. Math. 31 (1975), 43–85.

[Hor1-5] E. Horikawa, Algebraic surfaces of general type with small c_1^2, I, Ann. of Math. (2) 104 (1976), 357–387; II, Invent. Math. 37 (1976), 121–155; III, Inv. Math. 47 (1978), 209–248; IV, Invent. Math. 50 (1979), 103–128; V, J. Fac. Sci. Univ. Tokyo, Sect. A. Math. 283 (1981), 745–755.

[dJ-vS] T. de Jong, D. van Straten Deformations of the normalization of hypersurfaces, Math. Ann. 288 (1990), 527–547.

[J-L-P] T. Josefiak, A. Lascoux, P. Pragacz, Classes of determinantal varieties associated with symmetric and skew-symmetric matrices, Izv. An. SSSR 45,9 (1981), 662–673; English translation in Math. USSR Izv. 18 (1982), 575–586.

[Kap] M. Kapranov, On the derived categories of coherent sheaves on some homogeneous spaces, Invent. Math. 92 (1988), 479–508.

[Kod] K. Kodaira, On characteristic systems of families of surfaces with ordinary singularities in a projective space, Amer. J. Math. 87 (1965), 227–256.

[Kon91] K. Konno, A note on surfaces with pencils of non-hyperelliptic curves of genus 3, Osaka J. Math. 28 (1991), 737–745.

[Kon96] K. Konno, A lower bound of the slope of trigonal fibrations, Internat. J. Math. 7 (1) (1996) 19–27.

[Kon93] K. Konno, Non-hyperelliptic fibrations of small genus and certain irregular canonical surfaces, Ann. Sc. Norm. Super. Pisa, Cl. Sci. (IV) 20 (4) (1993), 575–595.

[Miy] Y. Miyaoka, On the Chern numbers of surfaces of general type, Invent. Math. 42 (1977), 225–237.

[M-P] D. Mond, R. Pellikaan, Fitting ideals and multiple points of analytic mappings, Lecture Notes in Math. 1414, Springer-Verlag, New York–Berlin 1987.

[Mum] D. Mumford, On the equations defining Abelian varieties. I–III, Invent. Math. 1 (1966), 287–354; ibid. 3 (1967), 75–135, 215–244.

[Ran] Z. Ran, Hodge theory and deformations of maps. Compositio Math. 97 (1995), 309–328.

[Rao] P. Rao, Liaison among curves in $\mathbb{P}^3$, Invent. Math.50 (1979), 205–217.

[Reid] M. Reid, Math. Review 86c: 14027.

[Reid79] M. Reid, π_1 for surfaces with small c_1^2, in: Algebraic Geometry, Lecture Notes in Math. 732, Springer-Verlag, Berlin 1979, 534–544.

[Sch] F.-O. Schreyer, Small fields and constructive algebraic geometry, in: Moduli of vector bundles (M. Maruyama, ed.), Lecture Notes in Pure and Appl. Math. 179, Marcel Dekker Inc., New York 1996, 221–228

[Sern]   E. Sernesi, L'unirazionalita' della varieta' dei moduli delle curve di genere dodici, Ann. Scuola Norm. Sup. Pisa Cl. Sci. (4) 8 (1981), 405–439.

[Ser]   J. P. Serre, Faisceaux algebriques coherents, Ann. of Math. 61 2 (1955), 197–278.

[Sim]   C. Simpson, Subspaces of moduli spaces of rank one local systems. Ann. Sci. École Norm. Sup. (4) 26 (1993), 361–401.

[Taka96]   T. Takahashi, Certain algebraic surfaces of general type with irregularity one and their canonical mappings. Tohoku Math. Publ. 2, Tohoku Univ., Mathematical Institute, Sendai 1996.

[Taka98]   T. Takahashi, Certain algebraic surfaces of general type with irregularity one and their canonical mappings. Tohoku Math. J. (II) 50 (1998), 261–290.

[Xiao87]   Xiao,Gang Fibered algebraic surfaces with low slope. Math. Ann. 276 (1987), 449–466.

[Yau]   S. T. Yau, Calabi's conjecture and some new results in algebraic geometry, Proc. Natl. Acad. Sci. USA 74 (1977), 1798–1799.

[Zuc94]   F. Zucconi, Su alcune questioni relative alle superficie di tipo generale con mappa canonica composta con un fascio o di grado 3, Tesi di Dottorato, Università di Pisa, 1994

F. Catanese, Lehrstuhl Mathematik VIII, Universität Bayreuth, 95440 Bayreuth, Germany
E-mail: Fabrizio.Catanese@uni-bayreuth.de

F.-O. Schreyer, Mathematik und Informatik, Universität des Saarlandes, Im Stadtwald, 66123 Saarbrücken, Germany
E-mail: schreyer@math.uni-sb.de

On surfaces with $p_g = 2, q = 1$ and non-birational bicanonical map*

Ciro Ciliberto and Margarida Mendes Lopes

Abstract. In the present note we show that any surface of general type over $\mathbb{C}$ with $p_g = 2$, $q = 1$ and non birational bicanonical map has a pencil of curves of genus 2. Combining this result with previous ones, one obtains that an irregular surface S of general type with $\chi(S) \geq 2$ and non-birational bicanonical map has a pencil of curves of genus 2.

1. Introduction

If a smooth complex surface S of general type has a pencil of curves of genus 2, then the bicanonical map ϕ of S is not birational. We call this exception to the birationality of the bicanonical map ϕ the *standard case*. The classification of the non-standard cases has a long history and we refer to the expository paper [Ci] for information on this problem.

The classification of non-standard irregular surfaces has been first considered by Xiao Gang in [X]. He gives a list of numerical possibilities for the invariants of the cases which might occur. More precise results have been obtained in [CFM], [CCM] and [CM].

In this note we prove the following result.

Theorem 1.1. *Let S be an irregular surface of general type over $\mathbb{C}$ with $\chi(S) \geq 2$ and non-birational bicanonical map. Then S has a pencil of curves of genus 2.*

In view of the results contained in the aforementioned papers, in order to prove this theorem, it suffices to exclude the existence of non-standard cases with $p_g = 2, q = 1$. This we will do in §5 (see Theorem 5.1). In §2 we prove some vanishing results which we apply later. In §§3, 4 we study the paracanonical system of the surfaces in question, proving a few numerical and geometric properties of it which allow us to prove Theorem 5.1.

Acknowledgements. The present collaboration took place in the framework of the EC research project HPRN-CT-2000-00099, EAGER.

---
*2000 Mathematics Subject Classification: 14J29

The second author is a member of CMAF and of the Departamento de Matemática da Faculdade de Ciências da Universidade de Lisboa.

The present paper has been finished during a visit of the second author to Rome supported by GNSAGA of INDAM.

Finally the authors want to thank the referee for the careful reading of the paper and for useful comments.

2. A preliminary result

In this section we prove some results that we will need further on.

Lemma 2.1. *Let S be an irregular surface of general type.*

Assume that C is a 1-connected curve on S and $\eta \in \mathrm{Pic}^0(S)$ is a point such that $h^1(S, \mathcal{O}_S(K + \eta)) = 0$. Then $h^1(S, \mathcal{O}_S(K + \eta + C)) \leq 1$ and equality holds if and only if $\eta_{|C} \simeq \mathcal{O}_C$.

Furthermore, if $h^1(S, \mathcal{O}_S(K + C)) = 0$ and $h^1(S, \mathcal{O}_S(K + \eta + C)) = 1$, then η is a torsion point of $\mathrm{Pic}^0(S)$.

Proof. By Serre duality and the hypothesis, we have
$$h^1(S, \mathcal{O}_S(K + \eta)) = h^1(S, \mathcal{O}_S(-\eta)) = 0.$$

Considering the long exact sequence obtained from the exact sequence
$$0 \to \mathcal{O}_S(-\eta - C) \to \mathcal{O}_S(-\eta) \to \mathcal{O}_C(-\eta) \to 0$$
we obtain $h^1(S, \mathcal{O}_S(-\eta - C)) = h^0(C, \mathcal{O}_C(-\eta))$. The line bundle $\mathcal{O}_C(-\eta)$ has degree 0 on C, thus, by [CFM], corollary (A.2), $h^0(C, \mathcal{O}_C(-\eta)) \leq 1$ with equality holding if and only if $-\eta_{|C} \simeq \mathcal{O}_C$. Now the first assertion follows because, by Serre duality $h^1(S, \mathcal{O}_S(-\eta - C)) = h^1(S, \mathcal{O}_S(K + \eta + C))$.

The second assertion follows from the first if we recall that $h^1(S, \mathcal{O}_S(K + C))$ is the dimension of the kernel of the restriction map $\mathrm{Pic}^0(S) \to \mathrm{Pic}^0(C)$. □

Proposition 2.2. *Let S be an irregular surface of general type such that either $q = 1$ or the image of the Albanese map $a : S \to \mathrm{Alb}(S)$ is a surface.*

(i) If C is a 1-connected curve on S and $\eta \in \mathrm{Pic}^0(S)$ is a general point, then $h^1(S, \mathcal{O}_S(K + \eta + C)) \leq 1$.

(ii) If C is a 1-connected curve on S which is not contracted by the Albanese map $a : S \to \mathrm{Alb}(S)$, then $h^1(S, \mathcal{O}_S(K + C)) < q$ and if $\eta \in \mathrm{Pic}^0(S)$ is a general point, then $h^1(S, \mathcal{O}_S(K + \eta + C)) = 0$.

(iii) If C is a 1-connected curve on S which is contracted by the Albanese map $a : S \to \mathrm{Alb}(S)$, then $h^1(S, \mathcal{O}_S(K + C)) = q$ and if $\eta \in \mathrm{Pic}^0(S)$ is a general point, then $h^1(S, \mathcal{O}_S(K + \eta + C)) = 1$.

Proof. Assertion (i) is a direct consequence of Lemma 2.1 since $h^1(S, \mathcal{O}_S(K+\eta)) = 0$ for a general point $\eta \in \operatorname{Pic}^0(S)$. For $q = 1$ this is clear by upper semicontinuity of $h^0(S, \mathcal{O}_S(K + \eta))$ on $\eta \in \operatorname{Pic}^0(S)$, whereas for $q \geq 2$ it follows by the generic vanishing theorem from [GL] (see also [Be]).

Let us prove assertion (ii). The assertion about $h^1(S, \mathcal{O}_S(K + C))$ is well known (see [Ra] and [Ca], remark 6.8). Indeed, since C is not contracted by the Albanese map, if $\eta \in \operatorname{Pic}^0(S)$ is a general point, then $\eta_{|C}$ is non-trivial, i.e. the restriction map $\operatorname{Pic}^0(S) \to \operatorname{Pic}^0(C)$ is non-zero. Hence its tangent map $H^1(S, \mathcal{O}_S) \to H^1(C, \mathcal{O}_C)$ is non-zero, thus $h^1(S, \mathcal{O}_S(K + C))$, which is the dimension of its kernel, is smaller than $h^1(S, \mathcal{O}_S) = q$. The assertion about $h^1(S, \mathcal{O}_S(K + \eta + C))$ then follows also from Lemma 2.1.

The proof of (iii) is similar and therefore we omit it. □

3. Some properties of the paracanonical curves

Let S be an irregular surface of general type. We denote by K a canonical divisor of S. If $\eta \in \operatorname{Pic}^0(S)$ is any point, we will consider the linear system $|K + \eta|$. We denote by C_η a curve in $|K + \eta|$, called a *paracanonical curve* of S.

Assume now that $p_g = 2, q = 1$ for S. Then, for a general point $\eta \in \operatorname{Pic}^0(S)$, the linear system $|K + \eta|$ is a pencil. The curves C_η thus describe, for a general point $\eta \in \operatorname{Pic}^0(S)$, a continuous system $\mathcal{K}$ of dimension 2 of curves on S, called the *main paracanonical system* of S.

We will write $|K + \eta| = F_\eta + |M_\eta|$, where $F := F_\eta$ is the fixed part and $|M| := |M_\eta|$ is the movable part of $|K + \eta|$. Next we prove two lemmas about the main paracanonical system.

Lemma 3.1. *In the above setting, there is a not empty open Zariski subset $U \subseteq \operatorname{Pic}^0(S)$ such that either F_η or $|M_\eta|$ stays fixed as η varies in U.*

Proof. Let $U \subseteq \operatorname{Pic}^0(S)$ be the not empty open set such that F_η depends flatly on $\eta \in U$, so that the algebraic equivalence class of F_η is independent on $\eta \in U$.

Suppose F_η varies with $\eta \in U$. Fix a point $\epsilon \in \operatorname{Pic}^0(S)$. Then, when η varies, the system $F_\eta + |M_\epsilon|$ varies describing the general paracanonical system in $\mathcal{K}$. This proves that $|M_\epsilon|$ does not depend on η. □

Lemma 3.2. *Let S be a minimal surface of general type with $p_g = 2, q = 1$, with no pencil of curves of genus 2 and $K^2 \leq 8$. Then the general curve in $|M_\eta|$ is irreducible for any $\eta \in \operatorname{Pic}^0(S)$ with $h^0(S, \mathcal{O}_S(K + \eta)) = 2$. Furthermore:*

(i) *if $|M_\eta| = |M|$ does not depend on $\eta \in \operatorname{Pic}^0(S)$, then either F_η is 1-connected or $F_\eta = A_\eta + B$, where B is a fundamental cycle, $A := A_\eta$ is 1-connected, and $K_S^2 = 8$, $K_S \cdot M = 4, A \cdot B = 0, K \cdot A = 4, A^2 = 2, M \cdot A = M \cdot B = 2$;*

(ii) if $F_\eta = F$ does not depend on $\eta \in \mathrm{Pic}^0(S)$, then either $F = 0$, or $F \cdot M = 2$, F is 1-connected and $M^2 \geq 4$ or $K^2 = 8$, $M^2 = F \cdot M = 4$ and $K \cdot F = 0$. In this latter case either F is 1-connected or $F = A + B$ with A, B fundamental cycles such that $A \cdot B = 0$.

Proof. We start by proving that the general curve in $|M_\eta|$ is irreducible if $h^0(S, \mathcal{O}_S(K + \eta)) = 2$. Otherwise the pencil $|M_\eta|$ would be composed with a pencil $\mathcal{P}$. The pencil $\mathcal{P}$ cannot be rational, since $|M_\eta|$ has dimension 1. Since $q = 1$, then $\mathcal{P}$ would be an elliptic pencil, i.e. the Albanese pencil of S, and $M_\eta \equiv 2G$, where G is a curve in $\mathcal{P}$. Since $G^2 = 0$, we must have $F_\eta \neq 0$ and we cannot have $K \cdot F_\eta = 0$, otherwise F_η would be contained in curves of $\mathcal{P}$ and the canonical curves would be disconnected. Thus we have $8 \geq K^2 \geq K \cdot F + 2K \cdot G > 2K \cdot G$, hence $K \cdot G = 2$, i.e. $\mathcal{P}$ would be a pencil of curves of genus 2, a contradiction.

Now we prove (i). Suppose $|M_\eta| = |M|$ does not depend on $\eta \in \mathrm{Pic}^0(S)$. Then F_η moves with η. We claim that $K \cdot F_\eta \geq 4$ and therefore $K \cdot M \leq 4$. Indeed, let $G := G_\eta$ be an irreducible component of $F := F_\eta$ moving with η, so that G is nef. Notice that by Debarre's inequality [De] we have $K^2 \geq 2p_g = 4$. Then the index theorem tells us that $K \cdot G \geq 2$ and $K \cdot G = 2$ would imply that G moves in a pencil of curves of genus 2, a contradiction. If $K \cdot G = 3$, the index theorem again says that $G^2 = 1$ and we have a contradiction by proposition (0.18) from [CCM]. Therefore $K \cdot F \geq K \cdot G \geq 4$ and there is a unique component G of F moving with η, otherwise $K \cdot F \geq 8$ and therefore $K \cdot M \leq 0$, a contradiction.

If $F \cdot M = 2$, then F is 1-connected by A.4 from [CFM]. Assume $F \cdot M > 2$. Then we have $4 \geq K \cdot M = F \cdot M + M^2 \geq 4 + M^2$, yielding $F \cdot M = 4, M^2 = F^2 = 0, K \cdot M = K \cdot F = 4, K^2 = 8$. If F is not 1-connected, then $F = A + B$ with A, B effective, non-zero divisors, such that $A \cdot B \leq 0$. Then $(M + A) \cdot B \geq 2$, hence $M \cdot B \geq 2 - A \cdot B \geq 2$, and similarly $M \cdot A \geq 2$. Since $F \cdot M = 4$ we must have $M \cdot A = M \cdot B = 2$ and $A \cdot B = 0$. Thus F is 0-connected and A and B are also 0-connected by A.4 from [CFM]. Suppose the unique component G of F moving with η sits in A. Then $K \cdot A = 4$ and $K \cdot B = 0$. From $0 = K \cdot B = B \cdot A + B^2 + B \cdot M = B^2 + 2$, we deduce $B^2 = -2$, hence B is a fundamental cycle. Suppose A is not 1-connected and write $A = A' + A''$, with $A' \cdot A'' = 0$. We claim that $B \cdot A' = B \cdot A'' = 0$. Suppose indeed that $B \cdot A' > 0$. Since $0 = B \cdot A = B \cdot A' + B \cdot A''$, we would have $B \cdot A'' < 0$. But then $A'' \cdot (A' + B) = A'' \cdot B < 0$, contrary to the fact that F is 0-connected. Suppose G sits in A'. We claim that $M \cdot G \geq 2$. Indeed, clearly $M \cdot G > 0$. If $M \cdot G = 1$, since G is not rational, then $|M|$ would cut a fixed point on G, hence G would be properly contained in a curve of $|M|$, contrary to the fact that G moves on S. Since $M \cdot G \geq 2$, we have $M \cdot A' = 2, M \cdot A'' = 0$. But then $A'' \cdot (A' + B + M) = 0$, contrary to the 2-connectedness of the canonical divisors. This ends the proof of (i).

In case (ii) we set $M := M_\eta$. One has $M^2 > 0$ since M moves in a system $\mathcal{M}$ of curves of dimension 2 and the general curve M is irreducible. The case $M^2 = 1$ is excluded by proposition (0.14, iii) of [CCM]. The case $M^2 = 2$ is also excluded by theorem (0.20) of [CCM]. Thus we may assume $M^2 \geq 3$. Suppose that $M^2 = 3$, in which case $F \neq 0$, otherwise we would have $K^2 = M^2 = 3$ against [De]. Again by theorem (0.20) of [CCM], $\mathcal{M}$ has no base point. Fix $\eta \in \text{Pic}^0(S)$ general and a general curve $M \in |M_\eta|$. For a general $\epsilon \in \text{Pic}^0(S)$, the pencil $|M_\epsilon|$ cuts out on M a base point free g_3^1 which has to vary with ϵ by (1.6) of [CFM]. This is impossible, since $M \cdot (K + M) = M \cdot (2M + F) \geq 8$ and therefore $p_a(M) \geq 5$. Thus $M^2 \geq 4$. If $M \cdot F \geq 4$, then $8 \geq K^2 \geq K \cdot M = F \cdot M + M^2$, proving that $K^2 = 8$ and $M^2 = F \cdot M = 4$, $K \cdot F = 0$. Suppose F is not 1-connected and write $F = A + B$ with A, B effective, non-zero divisors, such that $A \cdot B \leq 0$. By the same argument we made before, we have $M \cdot A = M \cdot B = 2$ and $A \cdot B = K \cdot A = K \cdot B = 0$. Since $F^2 = -4$ we have that $A^2 = B^2 = -2$, and this concludes the proof of (ii). □

4. Non birationality of the bicanonical map and the paracanonical system

Now we will consider a minimal surface S of general type with $p_g = 2$, $q = 1$, with non-birational bicanonical map ϕ, presenting the non-standard case. Notice that, by [R] and by proposition 3.1 of [CM], one has then $K^2 \leq 8$. We also recall the following lemma already contained in [CCM] (see lemma (2.2) of that paper):

Lemma 4.1. *Fix $\eta \in \text{Pic}^0(S)$ and let x, y be points on S such that $\phi(x) = \phi(y)$, with y not lying in the base locus of $|K - \eta|$. In particular one may assume that x, y are general points such that $\phi(x) = \phi(y)$. Then x belongs to a curve C_η in $|K + \eta|$ if and only if $y \in C_\eta$.*

Let $\nu > 1$ be the degree of ϕ. In view of this lemma, ϕ restricts to the general curve M_η to a map of degree ν to its image. In particular, if ϕ has degree 2, then the bicanonical involution fixes the curves M_η and therefore also the curves F_η.

Now we are ready for the proof of the following proposition:

Proposition 4.2. *Let S be a minimal surface of general type with $p_g = 2$, $q = 1$, with non-birational bicanonical map, presenting the non-standard case. Then, with the same notation as in Lemma 3.2, we have that $F = F_\eta$ is independent of $\eta \in \text{Pic}^0(S)$, and either:*

(i) $F = 0$ or F is 1-connected, strictly contained in a fibre of the Albanese pencil of S and $F \cdot M = 2$, $M^2 \geq 4$, or;

(ii) $F = A + B$ with A, B fundamental cycles such that $A \cdot B = 0$ and $M^2 = F \cdot M = 4$, $K^2 = 8$.

In any case for a general $\eta \in \text{Pic}^0(S)$, the general curve $M \in |M_\eta|$ is bielliptic and for a general $\epsilon \in \text{Pic}^0(S)$, the linear system $|M_\epsilon|$ cuts out on M a complete linear series whose movable part is a g_4^1 composed with the bielliptic involution on M.

The bicanonical map has then degree 2 onto its image, and the bicanonical involution ι acts on the general curve M as the bielliptic involution.

Proof. First, assume that F is non-zero, 1-connected and not contained in a fibre of the Albanese pencil. We claim that this case is not possible.

Notice that $h^1(S, \mathcal{O}_S(2K - M_\eta)) = h^1(S, \mathcal{O}_S(K + F_\eta - \eta)) = 0$ for $\eta \in \text{Pic}^0(S)$ a general point. Indeed, if $|M_\eta|$ moves with η, then $F = F_\eta$ stays fixed by Lemma 3.1, and the assertion follows by Proposition 2.2, (ii). If $|M_\eta|$ stays fixed while η moves, the divisor $F_\eta - \eta$ stays also fixed, and we obtain the assertion by the second part of Lemma 2.1, because $h^1(S, \mathcal{O}_S(-F_\eta)) = 0$ and η is not a torsion point, for a general $\eta \in \text{Pic}^0(S)$.

Thus if $\eta \in \text{Pic}^0(S)$ is general, the bicanonical system $|2K|$ cuts out on the general curve $M \in |M_\eta|$ a complete, non special series of degree $2g - 2 + F \cdot M$ and dimension $g - 2 + F \cdot M$, where we set $g := p_a(M)$. By Lemma 4.1 this series is composite with an involution of degree $\delta \geq 2$. One has $2g - 2 + F \cdot M \geq \delta(g - 1) + \delta(F \cdot M - 1)$, hence $F \cdot M \geq (\delta - 2)(g - 1) + \delta(F \cdot M - 1) \geq 2(F \cdot M - 1)$, and therefore $F \cdot M = \delta = 2$ and the equality has to hold everywhere in the above inequalities. In particular M is hyperelliptic. Notice that two points $x, y \in M$ are conjugated in the hyperelliptic involution if and only if $\phi(x) = \phi(y)$, so that ϕ has degree 2 onto its image and we can talk about the bicanonical involution ι such that $\iota(x) = y$ if and only if $\phi(x) = \phi(y)$.

Suppose that $|M_\eta|$ moves with η. Let $\epsilon \in \text{Pic}^0(S)$ be a general point and consider the linear series $g_{\epsilon,\eta}$ cut out by $|M_\epsilon|$ on the general curve $M \in |M_\eta|$. This series is complete and its movable part is composite with the hyperelliptic involution on M by Lemma 4.1. Furthermore its fixed divisor does not depend on ϵ, since it is supported at the finitely many points of S which belong to every curve M_η. This would imply that $g_{\epsilon,\eta}$ is independent on ϵ, a contradiction to (1.6) of [CFM].

Suppose that $|M_\eta|$ does not move with η and let again M be the general curve in $|M_\eta|$. Let $\epsilon \in \text{Pic}^0(S)$ be a general point. By Lemma 4.1 the curve F_ϵ cuts out on M a divisor fixed by the hyperelliptic involution. This would imply that the restriction map $\text{Pic}^0(S) \to \text{Pic}^0(M)$ is the zero map, yielding $h^1(S, \mathcal{O}_S(-M)) = 1$, a contradiction. Our claim is thus proved.

Suppose next that F is 1-connected, contained in some fibre of the Albanese pencil of S. Remark that $h^1(S, \mathcal{O}_S(-F)) = 1$ and also $h^1(S, \mathcal{O}_S(-F + \eta)) = 1$, for $\eta \in \text{Pic}^0(S)$ general by Proposition 2.2. Now we will prove that (i) holds and therefore $|M_\eta|$ moves with $\eta \in \text{Pic}^0(S)$.

In this case $|2K|$ cuts out on M an incomplete series of degree $2g - 2 + F \cdot M$ and dimension $g - 3 + F \cdot M$, which is composite with an involution of degree $\delta \geq 2$. We claim that $\delta = 2$. Suppose indeed that $\delta \geq 3$. Then $2g - 2 + F \cdot M \geq 3(g - 3) + 3F \cdot M$, i.e. $g + 2F \cdot M \leq 7$, which yields $g = 3$, $F \cdot M = 2$, $\delta = 3$. Then $4 = (K + M) \cdot M = F \cdot M + 2M^2 = 2 + 2M^2$, and we find $M^2 = 1$. By theorem (0.20) of [CCM], we see that $|M|$ cannot vary with η, which means that F has to. Namely F has to be a fibre of the Albanese pencil. Then $K \cdot F = F^2 + F \cdot M = 2$, and we find a contradiction.

From $\delta = 2$ we deduce $2(g - 3 + F \cdot M) \leq 2g - 2 + F \cdot M$, hence $F \cdot M \leq 4$.

If $F \cdot M = 4$, then M has to be hyperelliptic. Arguing as above, we see that $|M|$ cannot move with η, hence F has to, so that F is a fibre of the Albanese pencil. In this case, by Lemma 4.1, the bicanonical involution has to fix every curve of the Albanese pencil. Thus, if $x, y \in M$ are in the g_2^1, i.e. if $\iota(x) = y$, then the fibre of the Albanese pencil through x also contains y. This would imply that the base of the Albanese pencil is rational, a contradiction.

If $F \cdot M = 2$, then $K \cdot F = F^2 + F \cdot M \leq 2$ and therefore F cannot move, otherwise the fibres of the Albanese pencil have genus 2. Then, by Proposition 3.2, $M^2 \geq 4$. This ends the proof of (i).

Suppose now that F is not 1-connected. Then we will prove that $|M_\eta|$ moves with $\eta \in \mathrm{Pic}^0(S)$, thus (ii) will follow by Lemma 3.2, (ii).

If $|M|$ does not depend on $\eta \in \mathrm{Pic}^0(S)$, then one has a decomposition $F = A + B$ as in part (i) of Lemma 3.2 and we claim that $h^1(S, \mathcal{O}_S(-F)) = 1$. Indeed $h^1(S, \mathcal{O}_S(-A)) = 0$ and $h^0(B, \mathcal{O}_B(A)) = 1$ and the claim follows by the exact sequence

$$0 \to \mathcal{O}_S(-F) \to \mathcal{O}_S(-A) \to \mathcal{O}_B(-A)) \to 0$$

Hence $|2K|$ cuts out on the general curve $M \in |M|$ a linear series of degree $2g + 2$ and dimension $r \geq g + 1$, which is composite with an involution of degree $\delta \geq 2$. This implies again that the involution in question is a g_2^1 on M. Arguing as before we see that the curves A should cut out on M divisors of this g_2^1, which leads to a contradiction. Thus we proved that $|M_\eta|$ moves with $\eta \in \mathrm{Pic}^0(S)$.

As for the final part of the statement, first suppose we are in case (i). As we saw, $|2K|$ cuts out on M a g_{2g}^{g-1} composed with an involution of degree 2. As usual, one sees that M cannot be hyperelliptic. Then it has to be bielliptic. The series cut out by $|M_\epsilon|$ on M is complete. Its fixed part does not depend on ϵ, whereas its movable part is composed with the bielliptic involution and it is complete, of dimension 1, hence it is a g_4^1.

Suppose now we are in case (ii). Recall that, by Lemma 4.1, the degree of the map $\phi_M : M \to \phi(M)$ is $\nu \geq 2$, which is also the degree of ϕ. Since $M \cdot (K + M) = 12$, then M has genus 7. The series cut out by $|M_\epsilon|$ on M is a complete g_4^1, varying with the parameter $\epsilon \in \mathrm{Pic}^0(S)$. Notice that its fixed part does not depend on ϵ, hence it is empty, since there are finitely many g_d^1's on M, with $d \leq 3$. Furthermore Lemma 4.1 yields that either $\nu = 2$ or $\nu = 4$. In the latter case however the g_4^1 would not vary with ϵ, a contradiction. Therefore $\nu = 2$ and the g_4^1 is composed with a fixed involution of degree 2, independent on ϵ. Since the g_4^1 is complete and varies with ϵ, we see that the involution on M has to be elliptic. □

Notice that the assertion concerning the degree of the bicanonical map contained in the above proposition, also follows by the results of [X].

5. The main theorem

In this section we will prove the following:

Theorem 5.1. *If S is a surface of general type with $p_g = 2$, $q = 1$, non-birational bicanonical map, then it presents the standard case.*

By Proposition 4.2, if S is a minimal surface of general type with $p_g = 2$, $q = 1$, such that the bicanonical map is not birational and S presents the non-standard case, then for $\eta \in \text{Pic}^0(S)$ a general point, the pencil $|K + \eta| = F + |M_\eta|$ has a fixed part F which does not depend on η and a movable part $|M| = |M_\eta|$ which depends on η. Notice that all components of F sit in fibres of the Albanese pencil.

We let $\mathcal{M}$ be the 2-dimensional system of curves which is the closure, in the Hilbert scheme, of the family of curves $|M_\eta|$, with $\eta \in \text{Pic}^0(S)$ a general point. If we let $\mu := M^2$, then $\mu \geq 4$ and the general curve $M \in \mathcal{M}$ is irreducible. Notice that $\mathcal{M}$ has a natural morphism $\mathcal{M} \to \text{Pic}^0(S)$ whose general fibre is a $\mathbb{P}^1$.

Fix a general point $x \in S$ and set $x' := \iota(x)$. Consider the system $\mathcal{M}_x$ of curves in $\mathcal{M}$ passing through x. We claim that the system $\mathcal{M}_x$ is irreducible, of dimension 1, parametrized by $\text{Pic}^0(S)$. Indeed any irreducible component of $\mathcal{M}_x$ has dimension 1 and no irreducible component of $\mathcal{M}_x$ can be contained in a fibre of $\mathcal{M} \to \text{Pic}^0(S)$. Since $\mathcal{M}_x$ cuts the general fibre of $\mathcal{M} \to \text{Pic}^0(S)$ in one point, the claim follows.

Notice that $\mathcal{M}_x = \mathcal{M}_{x'}$, so it is appropriate to denote this system by $\mathcal{M}_{x,x'}$.

Let us point out the following corollary of Proposition 4.2.

Corollary 5.2. *In the above setting, two general curves in $\mathcal{M}$ have intersection multiplicity $\mu - 4$ at fixed points of S, and intersect at 4 further distinct variable points, which are pairwise conjugated in the bicanonical involution ι. In particular, given a general point $x \in S$, two general curves in $\mathcal{M}_{x,x'}$ intersect at 2 distinct variable points, which are conjugated in the bicanonical involution.*

Recall that the *index* of a 1-dimensional system of curves on a surface is the number of curves of the system passing through a general point of the surface. Next we prove the following lemma:

Lemma 5.3. *In the above setting, the system $\mathcal{M}_{x,x'}$ has index $\nu = 2$.*

Proof. The index of $\mathcal{M}_{x,x'}$ cannot be 1. In this case, in fact, $\mathcal{M}_{x,x'}$ would be a pencil. Since $\mathcal{M}_{x,x'}$ has base points at x and x', it would be a rational pencil, whereas we know it is parametrized by $\text{Pic}^0(S)$. Thus $\nu \geq 2$.

Let y be a general point of S and set $y' := \iota(y)$. Let M be a curve in $\mathcal{M}_{x,x'}$ through y and therefore also through y'. Since $\nu \geq 2$, we know there is some other curve M' in $\mathcal{M}_{x,x'}$ through y and y'. Suppose there is a third one M''. Then M' and M'' would cut out on M the same divisor, a contradiction to (1.6) of [CFM]. □

Set $\text{Pic}^0(S) := A'$ and fix $x \in S$ a general point. Another general point $y \in S$ determines two point $m_{1,y}$, $m_{2,y}$ in A' corresponding to the two curves of $\mathcal{M}_{x,x'}$

containing y. Thus we can consider the map:

$$\alpha : S \to \text{Pic}^2(A') \simeq A'$$

which takes the general point $y \in S$ to the divisor class of $m_{1,y} + m_{2,y}$. This map is clearly surjective and it factors through the Albanese map. We denote by G the general fibre of α, which is composed of curves of the Albanese pencil.

We are finally in a position to give the:

Proof of Theorem 5.1. Assume S presents the non-standard case. Let us keep the above notation and let us set $n := G \cdot M$. By Proposition 4.2, one has $G \cdot K = G \cdot F + G \cdot M = G \cdot M = n$. Since we are in the non-standard case, we have $n \geq 4$.

Fix a general point $x \in X$ and the system $\mathcal{M}_{x,x'}$. Let M be a general curve in $\mathcal{M}_{x,x'}$. Let $M \cap G$ consist of the points $x_1, \ldots, x_n$. Notice that each of the points $x_1, \ldots, x_n$ is a general point of S, in particular it is different from x and x'. Let m be the point of A' corresponding to the curve M. By Lemma 5.3 and by the generality of $x_1, \ldots, x_n$, for each $i = 1, \ldots, n$ there is only another curve $M_i \in \mathcal{M}_{x,x'}$, different from M, containing x_i. Let m_i be the point of A' corresponding to M_i, $i = 1, \ldots, n$. One has $\alpha(x_i) = m + m_i$, $i = 1, \ldots, n$. On the other hand, by the meaning of G, one has $\alpha(x_1) = \cdots = \alpha(x_n)$, i.e. the divisor classes of $m + m_i$ on A' are the same for all $i = 1, \ldots, n$. This implies $m_1 = \cdots = m_n$, i.e. $M_1 = \cdots = M_n$. Let M' be this curve. Then $x_1, \ldots, x_n$ sit in the intersection of the two curves M and M' of $\mathcal{M}_{x,x'}$, off the points x, x'. By Corollary 5.2 we have $n + 2 \leq 4$, i.e. $n \leq 2$, a contradiction. □

References

[Be] A. Beauville, Annulation du H^1 et systémes paracanoniques sur les surfaces, J. Reine Angew. Math. 388 (1988), 149–157; Erratum: J. Reine Angew. Math. 418 (1991), 219–220.

[Ca] F. Catanese, On the moduli spaces of surfaces of general type, J. Differential Geom. 19 (1984), 483–515.

[CCM] F. Catanese, C. Ciliberto, M. Mendes Lopes, On the classification of irregular surfaces of general type with non birational bicanonical map, Trans. Amer. Math. Soc. 350 (1998), 275–308.

[Ci] C. Ciliberto, The bicanonical map for surfaces of general type, Proc. Sympos. Pure Math. 62 (1997), 57–84.

[CFM] C. Ciliberto, P. Francia, M. Mendes Lopes, Remarks on the bicanonical map for surfaces of general type, Math. Z. 224 (1997), 137–166.

[CM] C. Ciliberto, M. Mendes Lopes, On surfaces with $p_g = q = 2$ and non-birational bicanonical map, to appear in Adv. Geom.

[De] O. Debarre, Inégalités numériques pour les surfaces de type général, Bull. Soc. Math. France 110 (1982), 319–342.

[GL]   M. Green, R. Lazarsfeld, Deformation theory, generic vanishing theorems, and some conjectures of Enriques, Catanese and Beauville, Invent. Math. 90 (1987), 416–440.

[Ra]   C. P. Ramanujam, Remarks on the Kodaira vanishing theorem, J. Indian Math. Soc. (N.S.) 36 (1972), 41–51; suppl. ibidem 38 (1974), 121–124.

[R]   I. Reider, Vector bundles of rank 2 and linear systems on algebraic surfaces, Ann. of Math. 127 (1988), 309–316.

[X]   Xiao Gang, Degree of the bicanonical map of a surface of general type, Amer. J. Math. 112 (1990), 713–737.

C. Ciliberto, Dipartimento di Matematica, Università di Roma Tor Vergata, Via della Ricerca Scientifica, 00133 Roma, Italy

E-mail: `cilibert@mat.uniroma2.it`

M. Mendes Lopes, CMAF, Universidade de Lisboa, Av. Prof. Gama Pinto, 2, 1649-003 Lisboa, Portugal

E-mail: `mmlopes@lmc.fc.ul.pt`

On unirationality of double covers of fixed degree and large dimension; a method of Ciliberto

Alberto Conte, Marina Marchisio and Jacob P. Murre

To the memory of Paolo Francia

Abstract. Following an idea of Ciliberto we show that double covers of projective r-space branched over an hypersurface of degree $2d$ are unirational provided r is sufficiently big with respect to d.

2000 Mathematics Subject Classification: 14E08, 14J40

1. Introduction

The notion of *unirationality* plays an important role in classical algebraic geometry in the works of M. Noether, Enriques and especially by Fano; see for instance Chap. IV of the book by Roth [12]. At present the concept of *rationally connected* variety seems to become more and more important (see the recent book of Kollar [5]), clearly unirational varieties are rationally connected but whether the latter concept is more general than the former is not yet known (ibid., problem 55). Irrespectively of the answer to this question, it remains an interesting geometrical problem to decide whether certain types of varieties are unirational (or not!).

One the most striking results in this subject is a *theorem of U. Morin* from 1940 [6] saying that (always in characteristic zero) if $V = V_{r-1}(d) \subset \mathbb{P}^r$ is a *hypersurface* of degree d in a projective r-space then there exists a constant $c(d)$ such that if $r \geq c(d)$ and if V is "sufficiently general" then V is unirational (see Theorem 3.1 below for the precise statement). This theorem has been generalized to *complete intersections* by *Predonzan* in [10]. "Modern" treatments of the results of Morin and Predonzan were given in the papers of Ciliberto [1], Ramero [11], Paranjape–Srinivas [8] and we refer also to Chap. 10 in the book of Iskovskikh [4]. Recently the result of Morin has been improved by Harris, Mazur and Pandharipande [3] in the sense that "sufficiently general" has been relaxed to "smooth", this is an important improvement but – as far as we know – this result of [3] for hypersurfaces has not yet been extended to complete intersections.

A natural question is whether the result of Morin can be extended to *double covers* $\pi : W = W_r[2d, B] \longrightarrow \mathbb{P}^r$ of $\mathbb{P}^r$ ramified over an hypersurface $B = B_{r-1}(2d) \subset \mathbb{P}^r$

of degree $2d$; i.e., whether there exists a constant $\rho(d)$ such that if $r \geq \rho(d)$ and B is "sufficiently general" the variety $W = W_r[2d, B]$ is unirational. In the above quoted paper [1] Ciliberto has given a beautiful idea (osservazione 3.6) how to proceed to prove such a theorem (reducing it to a general criterion given by Morin in his Torino lecture of 1954 [7]). However, as Ciliberto remarks himself, the details of his outline depend upon a number of rather subtle verifications of "algebraic" nature. The purpose of this paper is to give these details and to prove the theorem for double covers; the precise statement is Theorem 4.1 below. Our starting point is the theorem of Morin–Predonzan in the version of Ciliberto, see for the precise statement Theorem 3.1 below.

There are in the theorems of Morin and Predonzan (at least) two important, but delicate, technical points: *firstly*, in order to specify the field over which the unirationality occurs one needs a "sufficiently large" linear space $L \simeq \mathbb{P}^q$ contained in V and *secondly* the pair (V, L) must be "sufficiently general". In the paper [8] and in the book of Iskovskikh one introduces the notion of "general pair". However it is difficult to control this notion of "general pair". On the other hand for the application of the results of Morin–Predonzan to the case of double covers these technical aspects play an important (and in fact crucial) role. Therefore we have preferred to work with the precise notion of "generic" in the sense of Weil [13] or Grothendieck [2] (although technically differently framed the notions "generic" of Weil and Grothendieck are – of course – essentially the same). Working with "generic" it is important to distinguish between: "V generic over K" (see Subsection 2.1) and in case V contains a linear space L "V generic over K subject to containing L" (see Subsection 2.2). We have given the precise definitions of these notions in Section 2.

2. Definition and preliminaries

2.1.

Let K be a field of characteristic zero. Let $V = V_n$ be an irreducible variety defined over K of dimension n. We recall that V is called *unirational* if there exists a rational *dominant* map $f : \mathbb{P}^n \longrightarrow V$ where $\mathbb{P}^n$ is projective n-space, f is defined over the algebraic closure $\overline{K}$ of K (or better: usually over a finite extension K' of K) and dominant means that the Zariski closure of the image $f(\mathbb{P}^n)$ in V is V itself. V is *unirational over K* if moreover f itself is also defined over K.

Let now $V = V_{r-1}(d) \subset \mathbb{P}^r$ be a *hypersurface* of degree d, i.e. V is defined in $\mathbb{P}^r$ by an equation $F(X_0, \ldots, X_r) = 0$ homogeneous of degree d. We shall say that V is *generic over K* if the coefficients of F are *independent transcendental over K*, i.e. in the parameter space $\mathbb{H}(r, d) = \mathbb{P}^N$ with $N = \binom{r+d}{d} - 1$ the V corresponds to a generic point $\vartheta(V)$ over K (in the terminology of Weil [13]). Similarly if $V = V_{r-m}(\underline{d}) \subset \mathbb{P}^r$ is a *complete intersection* of multidegree $\underline{d} = (d_1, d_2, \ldots, d_m)$ with

$0 < d_1 \leq d_2 \leq \cdots \leq d_m$ defined by equations $F_j(X_0, \ldots, X_r) = 0$ homogeneous of degree d_j ($j = 1, \ldots, m$) then we shall say that V is *generic over K* if the coefficients of the F_j are (mutually) independent transcendental over K (note that such $V(\underline{d})$ are parametrized by a product of projective spaces $\mathbb{H} = \mathbb{H}(r, \underline{d}) = \mathbb{P}^{N_1} \times \cdots \times \mathbb{P}^{N_m}$ with
$$N_j = \binom{r + d_j}{d_j} - 1).$$

2.2. Linear Spaces contained in $V = V_{r-m}(\underline{d})$

Let the field K and r and $\underline{d}$ be as above, and let q be an integer such that $0 < q < r$. We are interested in linear spaces $L = \mathbb{P}^q \subset \mathbb{P}^r$ which are contained in $V = V_{r-m}(\underline{d})$. There is the following theorem of Predonzan [9] (see also [1], thm 2.1).

Theorem 2.1. *If r and $\underline{d} = (d_1, d_2, \ldots, d_m)$ are as above and $\underline{d} \neq (1, 1, \ldots, 1, 2)$ then $V = V_{r-m}(\underline{d})$ contains a linear space $L = \mathbb{P}^q$ if*

$$(r - q)(q + 1) \geq \sum_i \binom{d_j + q}{q}. \tag{2.1}$$

Consider the incidence correspondence

$$I = I(r, \underline{d}, q) \xrightarrow{p_2} \mathbb{H}(r, \underline{d})$$

$$p_1 \downarrow$$

$$\mathbf{Gr}(q, r)$$

where, as usual, $\mathbf{Gr}(q, r)$ is the Grassmannian of the $\mathbb{P}^q \subset \mathbb{P}^r$, $\mathbb{H}(r, \underline{d})$ is a variety parametrizing the complete intersections of multidegree $\underline{d}$ in $\mathbb{P}^r$ and

$$I = I(r, \underline{d}, q) = \{(V, L); L \in \mathbf{Gr}(q, r), V \in \mathbb{H}(r, \underline{d}) \text{ and } L \subset V\}.$$

Given V, put $F(V, q) = p_2^{-1}(V) = \{L \in \mathbf{Gr}(q, r); L \subset V\}$; this is the *Fano variety* of $\mathbb{P}^q$'s in V.

It is well-known (and in fact very easy to see) that I is irreducible over K and of dimension

$$(r - q)(q + 1) + \sum_i \binom{d_i + r}{r} - \sum_i \binom{d_i + q}{q} - m$$

and p_2 is *onto* only if, and by the above Theorem 2.1 of Predonzan in fact if, the inequality (2.1) from above holds.

Given the field K, let $(V, L) \in I$. We shall say that *the pair (V, L) is generic over K* if (V, L) is a generic point of I. From the above we have immediately

Lemma 2.2. *Assume (2.1) holds. Let $(V, L) \in I$. Then the following are equivalent:*

1. *(V, L) is a generic pair over K;*

2. $L \in \mathbf{Gr}(q, r)$ is generic over K and V is generic over $K(l(L))$ in the fibre $p_1^{-1}(L)$;

3. $V \in \mathbb{H}(r, \underline{d})$ is generic over K, $F(V, q)$ is irreducible over $K(\vartheta(V))$ and L is generic over $K(\vartheta(V))$ in $F(V, q)$.

Remark 2.3. By $K(l(L))$, resp. $K(\vartheta(V))$, we denote the field of definition over K of L, resp. of V (i.e., obtained by adjoining to K the ratios of the Plücker coordinates, resp. the ratios of the coefficients of the equations).

Given the field K, let $L \in \mathbf{Gr}(q, r)$; let $(V, L) \in I$. We shall say that V is *generic over K subject to containing L* if V is generic over $K(l(L))$ in the fibre $p_1^{-1}(L)$. Note that this fibre itself is (isomorphic to) a product of projective spaces of type $\mathbb{P}^{M_1} \times \cdots \times \mathbb{P}^{M_m}$ with $M_i = \binom{d_i + r}{r} - \binom{d_i + q}{q} - 1$ (of course in general V will no longer by generic over K, i.e., no longer generic in $\mathbb{H}(r, \underline{d})$).

Given such $L = \mathbb{P}^q \subset \mathbb{P}^r$, say $L = L_0$, we can choose homogeneous coordinates $(Z_0, Z_1, \ldots, Z_q, Y_{q+1}, \ldots, Y_r)$ in $\mathbb{P}^r$ such that L_0 is given by

$$Y_{q+1} = Y_{q+2} = \cdots = Y_r = 0 \tag{2.2}$$

and we can use then $(Z_0, \ldots, Z_q)$ as homogeneous coordinates in L_0.

Note. If L_0 is defined over K then we can make this projective coordinate transformation over K itself, however if L_0 should not be defined over K (which of course is the general situation) we need to make base extension $K(l(L_0)) \supset K$ and perform this transformation over $K(l(L_0))$.

If we choose coordinates in this way then the equations of $V = V_{r-m}(\underline{d}) \subset \mathbb{P}^r$ take the following shape:

$$\mathcal{G}_j(Z_0, \ldots, Z_q, Y_{q+1}, \ldots, Y_r) = \sum_{\substack{I \\ 0 \leq |I| < d_j}} Z^I \mathcal{G}_{j,I}(Y) = 0 \quad (j = 1, \ldots, m), \tag{2.3}$$

where $I = (i_0, i_1, \ldots, i_q)$ and $Z^I = Z_0^{i_0} Z_1^{i_1} \ldots Z_q^{i_q}$, $|I| = \sum_{\rho=0}^{q} i_\rho$, i.e., the total degree in the Z_i's and $\mathcal{G}_{j,I}(Y)$ is homogeneous in the Y_i's of degree $d_j - |I|$. Note that the condition that the Z^I with $|I| = d_j$ do *not* occur is precisely the condition that $L_0 \subset V$. Also note that the coefficients of the $\mathcal{G}_j(Z, W)$ are in $K(l(L_0))$.

Now by counting the dimensions we have immediately the following lemma

Lemma 2.4. *Given K and $L_0 \in \mathbf{Gr}(q, r)$, let $(V, L_0) \in I(r, \underline{d}, q)$. Then the following are equivalent:*

1. *V is generic over K subject to containing L_0,*

2. *in the above equations (2.3) the (ratios of the) coefficients are independent transcendental over $K(l(L_0))$.*

Remark 2.5. Of course we mean not only for one separate j, but for all the j's together.

3. Theorems of Morin and Predonzan, in version of Ciliberto

Theorem 3.1 (Morin, Predonzan, Ciliberto). *Given a field K of characteristic zero and $\underline{d} = (d_1, d_2, \ldots, d_m)$ with $0 < d_1 \leq d_2 \leq \cdots \leq d_m$. Then there exist integers $c(\underline{d})$ and $q(\underline{d})$ such that if $r \geq c(\underline{d})$ and $L_0 \in \mathbf{Gr}(q(\underline{d}), r)$ and if $V = V_{r-m}(\underline{d}) \subset \mathbb{P}^r$ is generic over K subject to containing L_0 then V is unirational over the field $K(l(L_0), \vartheta(V))$.*

For $\underline{d} = d$ this is the theorem of Morin, for general $\underline{d}$ we get the theorem of Predonzan. The above theorem is Corollary 2.5 of Ciliberto [1], his notion "generica su $K(l(L_0))$" is indeed precisely what we call "$V = V_{r-m}(\underline{d}) \subset \mathbb{P}^r$ is generic over K subject to containing L_0" as we see from his remark on page 180 of his paper on the equations (2.7) in his paper. Finally in Ciliberto's notation $c(\underline{d}) = s_n$ and $q(\underline{d}) = s_{n-1}$ (see his Corollary 2.5); however remark that n in the paper of Ciliberto is determined by $\underline{d}$, and s_n and s_{n-1} are determined by $\underline{d}$, see his definitions on page 177.

We shall use the theorem of Morin and of Predonzan in the above precise version of Ciliberto.

4. Double covers of $\mathbb{P}^r$ and unirationality

Let, as above, K be a field of characteristic zero. Let $B = B_{r-1}(2d) \subset \mathbb{P}^r$ be a hypersurface of degree $2d$ in $\mathbb{P}^r$ and let $\pi : W = W_r[2d, B] \longrightarrow \mathbb{P}^r$ be the double cover of $\mathbb{P}^r$ branched over B. We want to prove the following:

Theorem 4.1. *Given K and $d > 2$, there exists a constant $\rho(d)$ such that if $r \geq \rho(d)$ and if $B_{r-1}(2d) \subset \mathbb{P}^r$ is generic over K then the double cover $W[2d, B]$ of $\mathbb{P}^r$ is unirational, and in fact unirational over a finite extension $K^{\#}$ of $K(\vartheta(B))$.*

5. Proof of the theorem

5.1. Preparations and beginning of proof

Let $d \geq 3$ and put $\underline{d} = (1, 2, \ldots, 2d - 2)$. Let furthermore $q(\underline{d})$ and $c(\underline{d})$ be the integers occurring in Theorem 3.1 (i.e., the integers s_{n-1} and s_n in Ciliberto's Corollary 2.5 of [1]).

Lemma 5.1. *For $d \geq 3$ then $q(\underline{d}) \geq 2d - 2$.*

Proof. Elementary, but cumbersome. We leave it to the reader but we make some remarks. The proof goes by induction and starting with $d = 3$. Examining the expressions (2.3) on page 177 of Ciliberto's paper we see that in our case his $n = 2d-2$. For $d = 3$ we get $q(3) = 25$ and since $25 > 4$ we can start. Next for the induction step, passing from $(d - 1)$ to d we see that the $q(\underline{d})$ increases by at least 2. $\square$

We take
$$q = q(\underline{d}) + 1 =: \rho''(d). \tag{5.1}$$

Now we *introduce a constant* $\rho_1(d)$ as follows: by Theorem 2.1 there exists a constant $c^*(2d, q)$ such that if $V = V_{n-1}(2d) \subset \mathbb{P}^n$ then V contains a linear space $L \simeq \mathbb{P}^q$ provided $n \geq c^*(2d, q)$. Now take
$$\rho_1(d) = c^*(2d, \rho''(d)). \tag{5.2}$$

Lemma 5.2. *Given the field K and $d \geq 3$, take q from (5.1). Let $r \geq \rho_1(d)$. Let $B = B_{r-1}(2d) \subset \mathbb{P}^r$ be generic over K. Then there exists a linear space $L_0 \simeq \mathbb{P}^q$ such that*

(i) *$L_0 \subset B$ and L_0 is defined over a finite extension $K^\#$ of $K(\vartheta(B))$,*

(ii) *B is generic over K subject to containing L_0 (recall that this notion is defined in Subsection 2.2 and it means that B is generic over $K(l(L_0))$ in the fibre $p_1^{-1}(L_0)$ of the diagram below).*

Proof. Consider the incidence diagram (like in Subsection 2.2)

$$I = \quad I(r, 2d, q) \quad \xrightarrow{p_2} \quad \mathbb{H}(r, 2d) \ni B$$
$$\quad \quad \quad \quad \quad p_1 \downarrow$$
$$\mathbf{Gr}(q, r) \supset F(B, q)$$

Consider the Fano variety $F(B, q) = p_2^{-1}(B) = \{L \in \mathbf{Gr}(q, r); L \subset B\}$, this variety is non-empty since $r \geq \rho_1(d)$, it is defined over $K(\vartheta(B))$ and as we have remarked in Subsection 2.2 it is irreducible over $K(\vartheta(B))$ and of dimension

$$\dim F(B, q) = (q + 1)(r - q) - \binom{q + 2d}{2d} =: b(r, d).$$

Now intersect this Fano variety with the linear section (via hyperplanes in the Plücker embedding) $\mathcal{L}$ of $\mathbf{Gr}(q, r)$ of codimension $b(r, d)$, i.e., of dimension $\binom{q + 2d}{2d}$, *defined over K and sufficiently general such that the intersection consists* of points (this is possible, K has infinitely many elements hence we can choose the hyperplane "sufficiently general"). These points are defined over the algebraic closure

$\overline{K(\vartheta(B))}$ of $K(\vartheta(B))$; take one of them, say, L_0 so $L_0 \in F(B,q) \cap \mathcal{L}$. Now we claim that B is (still) generic in $p_1^{-1}(L_0)$. In fact we have

$$\binom{r+2d}{2d} - 1 = \mathrm{trdeg}_K K(\vartheta(B)) = \mathrm{trdeg}_K K(\vartheta(B), l(L_0))$$

$$= \mathrm{trdeg}_K K(l(L_0)) + \mathrm{trdeg}_{K(l(L_0))} K(l(L_0), \vartheta(B))$$

$$\leq \binom{q+2d}{2d} + \dim p_1^{-1}(L_0)$$

$$= \binom{q+2d}{2d} + \left\{\binom{r+2d}{2d} - 1 - \binom{q+2d}{2d}\right\}.$$

(Note: we use $L_0 \in \mathcal{L}$, $\dim \mathcal{L} = \binom{q+2d}{2d}$). Hence we must have equality, in particular $\mathrm{trdeg}_{K(l(L_0))} K(l(L_0), \vartheta(B)) = \dim p_1^{-1}(L_0)$, i.e., B is generic over $K(l(L_0))$ in the fibre $p_1^{-1}(L_0)$. This completes the proof of the lemma. □

5.2. Continuation. Choice of $\rho(d)$

Take

$$\rho(d) := \max\{c(\underline{d}), \rho_1(d)\} + 1, \qquad (5.3)$$

where $c(\underline{d})$ is the constant in Theorem 3.1 with $\underline{d} = (1, 2, \ldots, 2d-2)$ and $\rho_1(d)$ is from (5.2).

Now let $r \geq \rho(d)$, take $B = B_{r-1}(2d) \subset \mathbb{P}^r$ generic over K and let $W_r[2d, B]$ be a *double cover* of $\mathbb{P}^r$ branched over B. We must prove that $W_r[2d, B]$ is *unirational* and in fact unirational over a finite extension of $K(\vartheta(B))$.

Let q be the integer from (5.1). Since $r \geq \rho_1(d)$ we can, by Lemma 5.2, find a linear space $L_0 \subset B$ of dimension q satisfying the conditions of Lemma 5.2, in particular B is generic over K subject to containing L_0 (also: L_0 is defined over a finite extension of $K(\vartheta(B))$).

Fix moreover in L_0 a linear space $M_0 \subset L_0$ of dimension $(2d-2)$ (possible since $q \geq (2d-2)$ by Lemma 5.1) and take M_0 to be defined over $K(l(L_0))$.

After we have choosen such $L_0 \subset B = B_{r-1}(2d) \subset \mathbb{P}^r$ we make a projective coordinate transformation, defined over $K(l(L_0))$, such that we have homogeneous coordinates $Z_0, \ldots, Z_q, Y_{q+1}, \ldots, Y_r$ such that L_0 is given by (see Subsection 2.2)

$$Y_{q+1} = \cdots = Y_r = 0 \qquad (5.4)$$

and we use $Z_0, Z_1, \ldots, Z_q$ as homogeneous coordinates in L_0.

We can take for $M_0 \subset L_0$ the space defined by

$$Z_{2d-1} = \cdots = Z_q = Y_{q+1} = \cdots = Y_r = 0. \qquad (5.5)$$

The equation of $B_{r-1} = B_{r-1}(2d) \subset \mathbb{P}^r$ has now the following form (see in Section 2 the equation (2.3))

$$g(Z, Y) = \sum_{0 \leq |I| < 2d} Z^I g_I(Y) = 0 \tag{5.6}$$

where $I = (i_0, \ldots, i_q)$, $|I| = \sum_{\sigma=0}^{q} i_\sigma$ and $g_I(Y)$ is homogeneous in the Y_j's of degree $2d - |I|$. The coefficients of $g_{i_0 \ldots i_q}(Y)$ are $b_{i_0 \ldots i_q j_{q+1} \ldots j_\sigma}$ (the sum of all the indices is 2d). Since B is generic over K subject to containing L_0 we have, by Lemma 2.4 that the

$$b_{i_0 \ldots i_q j_{q+1} \ldots j_\sigma} \text{ are independent transcendentals over } K(l(L_0)). \tag{5.7}$$

Next we choose a hyperplane H_0 in $\mathbb{P}^r$ not containing M_0 and, a fortiori, therefore not containing L_0. Let this hyperplane be $\{Z_0 = 0\}$. Put $L_0^* = L_0 \cap H_0$ and $M_0^* = M_0 \cap H_0$. So the equations of L_0^* are

$$Z_0 = 0, Y_{q+1} = Y_{q+2} = \cdots = Y_r = 0 \tag{5.8}$$

and similar for the M_0^*.

5.3. The varieties F_η and F_η^*

Take in M_0 a point η generic over $K(l(L_0), \vartheta(B))$, so η has coordinates (see the equations (5.5))

$$\eta = (\eta_0 = 1, \eta_1, \ldots, \eta_{2d-2}, 0, \ldots, 0, \ldots, 0) \tag{5.9}$$

with $\eta_1, \ldots, \eta_{2d-2}$ independent transcendentals over $K(l(L_0), \vartheta(B))$. For the sake of simplicity of notation we sometimes write

$$\eta = (\eta_0 = 1, \eta_1, \ldots, \eta_{2d-2}, \eta_{2d-1} = 0, \ldots, \eta_q = 0, 0, \ldots, 0). \tag{5.10}$$

Now consider the (higher) *polar varieties* of η with respect to B, up to the $(2d - 2)$-th one, i.e., the varieties

$$\Delta_\eta^1(B) = 0, \Delta_\eta^2(B) = 0, \ldots, \Delta_\eta^{2d-2}(B) = 0, \tag{5.11}$$

where $\Delta_\eta^j(B)$ is defined by the equation

$$\Delta_\eta^j(B) = \left(Z_0 \frac{\partial}{\partial Z_0} + \cdots + Z_q \frac{\partial}{\partial Z_q} \right. \tag{5.12}$$

$$\left. + Y_{q+1} \frac{\partial}{\partial Y_{q+1}} + \cdots + Y_r \frac{\partial}{\partial Y_r} \right)^{(j)} g(Z, Y)\bigg|_\eta = 0.$$

So in particular $\Delta_\eta^1(B) = 0$ is the tangent space to B in η, $\Delta_\eta^1(B) \cap \Delta_\eta^2(B)$ is the tangent cone to B in η, etc.

Let
$$F_\eta = \Delta_\eta^1(B) \cap \Delta_\eta^2(B) \cap \cdots \cap \Delta_\eta^{2d-2}(B). \tag{5.13}$$
So F_η consists of the lines m through η such that
$$m \cap B = (2d-1)\eta + \beta \tag{5.14}$$
with $\beta \in B \cap F_\eta := B_\eta^*$. Also put $F_\eta^* = F_\eta \cap H_0$ and clearly F_η^* is the projection of F_η from the point η and is birationally equivalent with F_η over the field $K(l(L_0), \vartheta(B), \eta)$. Also clearly F_η^* is a variety of type $V_{r-2d+1}(1, 2, \ldots, 2d-2) \subset H_0$ and it contains the $(q-1)$-dimensional linear space $L_0^* (= L_0 \cap H_0)$.

The following lemma is crucial (but subtle!).

Lemma 5.3. *F_η^* is a variety of type $V_{r-2d+1}(1, 2, \ldots, 2d-2)$ in H_0 and F_η^* is generic over $K(l(L_0), \eta)$ subject to containing L_0^* (in the sense of Subsection 2.2).*

Note. This lemma plays in our case a role analogous to the assertion i) on page 186 of Ciliberto's paper [1]. Since the lemma is crucial for our purpose we give a full proof (with all details).

Proof. We start with two remarks.

Remark 5.4. Working first (for simplicity) over the field $K' := K(l(L_0))$, we can consider *over this field* a projective coordinate transformation T in the linear space M_0 of the type $\tilde{Z}_0 = Z_0$, $\tilde{Z}_l = \sum_{i=0}^{2d-2} a_{il} Z_i = T(Z_0, \ldots, Z_{2d-2})$ with $a_{il} \in K'$. Of course we can also consider – if we prefer – the T as a linear coordinate transformation in $\mathbb{P}^r \supset L_0 \supset M_0$ itself (leaving the other coordinates unchanged). (Note also that it leaves the equation $Z_0 = 0$ of H_0 unchanged.)

T operates now on the parameter space $\mathbb{H}(r, 2d)$ of the hypersurfaces of degree $2d$ in $\mathbb{P}^r$ (see the diagram in the proof of Lemma 5.2), leaving the space $p_1^{-1}(L_0)$ fixed. It transforms the point $B \in p_1^{-1}(L_0)$ into a point $B^T \in p_1^{-1}(L_0)$ corresponding to the (new) equation obtained from (5.6):
$$\mathcal{G}^T(\tilde{Z}, Y) = \mathcal{G}(T^{-1}(\tilde{Z}), Y) = \sum_{\substack{I \\ |I|<2d}} T^{-1}(\tilde{Z})^I \mathcal{G}_I(Y) = \sum_{\substack{I \\ |I|<2d}} \tilde{Z}^I \mathcal{G}_I^T(Y). \tag{5.15}$$

Claim. *The B^T is (still) generic over K subject to containing L_0 (in the sense of Section 2, i.e., $B^T \in p_1^{-1}(L_0)$ is a generic point of $p_1^{-1}(L_0)$ over $K(l(L_0))$).*

Proof of the Claim. The coefficients $b_{i_0 i_1 \ldots i_q j_{q+1} \ldots j_T}^T$ of $\mathcal{G}^T(\tilde{Z}, Y)$ are linear over $K' = K(l(L_0))$ in the coefficients of $\mathcal{G}^T(Z, Y)$, hence $K'(\vartheta(B^T)) \subset K'(\vartheta(B))$. Applying the inverse of T we get $K'(\vartheta(B)) = K'(\vartheta(B^T))$.

Remark 5.5. Let $\eta \in M_0$ be a generic point of M_0 over $K(l(L_0), \vartheta(B))$, then "conversely" B is still generic over the field $K(\eta)$ subject to containing L_0 (count transcendence degrees).

Hence if we replace the field $K' = K(l(L_0))$ by the field $K'' := K(\eta, l(L_0)) = K'(\eta)$ then we can apply the above remark 5.4 to B and the field K''.

After the two remarks we proceed as follows.

Apply over the field $K'' = K(\eta, l(L_0))$ in M_0 the projective coordinate transformation T such that $\widetilde{Z}_0 = Z_0$ and such that the point $\eta = (1, \eta_1, \ldots, \eta_{2d-2}, 0, \ldots, 0)$ becomes the point $\eta_0 = (1, 0, \ldots, 0)$.

Then the hypersurface $B = B_{r-1}(2d) \subset \mathbb{P}^r$ corresponds in the parameter space $\mathbb{H}(r, 2d)$ with the point B^T corresponding to the equation

$$g^T(\widetilde{Z}, Y) = \sum_{\substack{I \\ |I|<2d}} \widetilde{Z}^I g_I^T(Y) = \sum_{\substack{I \\ |I|<2d,\ I+J=2d}} \widetilde{b}_{I,J} \widetilde{Z}^I Y^J \quad (5.16)$$

and by Remark 5.4 the coefficients $\widetilde{b}_{I,J} (\in K''(\vartheta(B)))$ are all *independent transcendental* over $K'' = K(\eta, l(L_0))$ (and moreover also note that in *the allowed range*, i.e., $0 \leq |I| < 2d$, *all coefficients occur*).

Now we are going to rearrange them according to *decreasing* powers of $\widetilde{Z}_0 = Z_0$. (Note since $\eta_0 \in B$ Z_0^{2d} does not occur.)

We get

$$\begin{aligned} g^T(\widetilde{Z}, Y) &= Z_0^{2d-1} \Phi_1(Y_{q+1}, \ldots, Y_r) \\ &+ Z_0^{2d-2} \Phi_2(\widetilde{Z}_1, \ldots, \widetilde{Z}_q, Y_{q+1}, \ldots, Y_r) \quad (5.17) \\ &+ \cdots + Z_0^{2d-s} \Phi_s(\widetilde{Z}, Y) + \cdots \end{aligned}$$

where the $\Phi_s(\widetilde{Z}, Y)$ are homogeneous polynomials in the $\widetilde{Z}_1, \ldots, \widetilde{Z}_q, Y_{q+1}, \ldots, Y_r$ of degree s, subject to the condition that they do not contain terms in the $\widetilde{Z}_1, \ldots, \widetilde{Z}_q$ alone (this $\Longleftrightarrow |I| < 2d \Longleftrightarrow B$ contains L_0; in particular $\Phi_1(\widetilde{Z}, Y) = \Phi_1(Y)$). The coefficients in the above equation are the $\widetilde{b}_{I,J}$ from above, hence by what we did say above they are independent transcendental over $K'' = K(l(L_0), \eta)$ and all of them occur in the allowed range.

However now (and compare with Ciliberto page 187) the s-th polar of the point $\eta_0 = (1, 0, \ldots, 0)$ with respect to B is precisely given by

$$\Delta_{\eta_0}^s(B^T) = \Phi_s(\widetilde{Z}_1, \ldots, \widetilde{Z}_q, Y_{q+1}, \ldots, Y_r) = 0$$

and the $F_\eta^* = (F^T)_{\eta_0}^*$ is given by $\Phi_1 = \Phi_2 = \cdots = \Phi_{2d-2} = 0$.

Hence by what we have said above about the coefficients $\widetilde{b}_{I,J}$ in (5.17) it follows that the F_η^* is indeed a $V(1, 2, \ldots, 2d-2) \subset H_0$ generic over $K'' = K(\eta, l(L_0))$ subject to containing L_0^* (this latter condition is precisely equivalent to the condition that there are in the $\Phi_s(s = 1, \ldots, 2d-2)$ no terms in the $\widetilde{Z}_i$'s alone). $\square$

Corollary 5.6. *If $r \geq \rho(d) = \max\{c(\underline{d}), \rho_1(d)\}+1$ (see (5.3)) then F_η^* is unirational over the field $K(l(L_0), \eta, \vartheta(B))$.*

Proof. By Lemma 5.3 and since $r - 1 \geq c(d)$ we can apply the theorem of Predonzan–Ciliberto, i.e. Theorem 3.1. □

Next, in order to simplify notation, let us write

$$K_1 := K(\vartheta(B)) = K(\vartheta(W)) \subset K_1' := K(\vartheta(B), l(L_0)). \tag{5.18}$$

From Corollary 5.6 we have that the function field $K_1'(\eta)(F_\eta^*)$ of F_η^* over $K_1'(\eta)$ is contained in a purely transcendental extension. Therefore if $\xi \in F_\eta^*$ is generic over $K_1'(\eta)$ then we have

$$K_1'(\eta, \xi) \cong K_1'(\eta)(F_\eta^*) \subset K_1'(\eta, t_1, \ldots, t_{r-2d+1}) \tag{5.19}$$

where the $t_1, \ldots, t_{r-2d+1}$ are independent transcendental over $K_1'(\eta)$.

5.4. Construction of a unirational family of lines in $\mathbb{P}^r$

Let $S \subset M_0 \times H_0$ be the Zarisky closure over K_1' of the point (η, ξ) where $\eta \in M_0$ is generic over K_1' and $\xi \in F_\eta^*$ is generic over $K_1'(\eta)$. From (5.19) it follows that S is a variety which is *unirational* over K_1', since we have

$$K_1'(S) :\cong K_1'(\eta, \xi) \subset K_1'(\eta_1, \ldots, \eta_{2d-2}, t_1, \ldots, t_{r-2d+1}) \tag{5.20}$$

and the right hand side of (5.20) is a purely transcendental extension of K_1'. Moreover, clearly dim $S = r - 1$.

Now we consider the line $m = \langle \eta, \xi \rangle$ spanned by η and ξ. Then $\langle \eta, \xi \rangle \subset F_\eta$ and by (5.14) we have

$$\langle \eta, \xi \rangle \cap B = (2d - 1)\eta + \beta \tag{5.21}$$

with $\beta \in B_\eta^* = B \cap F_\eta$.

Lemma 5.7. *When $\eta \in M_0$ is generic over K_1' and $\xi \in F_\eta^*$ is generic over $K_1'(\eta)$, then $\beta \in B$ is generic over K_1'.*

Proof. This is assertion ii) of page 186 of Ciliberto [1]. For the convenience of the reader we repeat – in our language and notation – the argument here.

Let us denote during the proof of this lemma by K' the field $K' := K(l(L_0))$.

With the notations and points from our lemma we have the following inclusions of fields

$$\begin{array}{ccc} K'(\vartheta(B)) & \longrightarrow & K'(\vartheta(B), \eta) \\ \downarrow & & \downarrow \\ K'(\vartheta(B), \beta) & \longrightarrow & K'(\vartheta(B), \eta, \beta) = K'(\vartheta(B), \eta, \xi) \end{array}$$

We have $\mathrm{trdeg}(K'(\vartheta(B), \eta) : K'(\vartheta(B))) = 2d - 2$ and $\mathrm{trdeg}(K'(\vartheta(B), \eta, \xi) : K'(\vartheta(B), \eta))) = r - 2d + 1$, hence $\mathrm{trdeg}(K'(\vartheta(B), \eta, \beta) : K'(\vartheta(B))) = r - 1$.

Furthermore clearly $\operatorname{trdeg}(K'(\vartheta(B), \beta) : K'(\vartheta(B))) \leq r - 1$ and hence in order *to prove the lemma* we have to prove that $K'(\vartheta(B), \eta, \beta)$ is an *algebraic extension* of $K'(\vartheta(B), \beta)$.

The proof goes by contradiction. Suppose that

$$(*) \operatorname{trdeg}(K'(\vartheta(B), \eta, \beta) : K'(\vartheta(B), \beta)) > 0.$$

Now comes the nice *specialization argument* of Ciliberto! Consider the following variety

$$J \subset p_1^{-1}(L_0) \times M_0 \times \mathbb{P}^r \subset \mathbb{H}(r, 2d) \times M_0 \times \mathbb{P}^r$$

with J the Zariski locus over K' of the point $(\vartheta(B), \eta, \beta)$ and let $J_{13} = \operatorname{pr}_{13} J$. By construction these are irreducible varieties over K'.

Now our assumption $(*)$ about the positive transcendence degree means that over the generic point $(\vartheta(B), \beta)$ of J_{13} the fibre $p_{13}^{-1}(\vartheta(B), \beta)$ of J has positive dimension. So then *the fibres over all points of J_{13} have positive dimension*. Consider now the following special point $(\vartheta(B^*), \eta^*, \beta^*)$ of J with (with the coordinates of (5.5)): $B^* \subset \mathbb{P}^r$ is the hypersurface with equation

$$Z_{2d-3}^{2d-1} Y_r + Z_0^2 Y_r^{2d-2} + Z_1^3 Y_r^{2d-3} + \cdots + Z_{2d-4}^{2d-2} Y_r^2 = 0$$

(note that $B^* \supset L_0$, in fact $B^* \supset (Y_r = 0)$)

$$\eta^* = (0, \ldots, 0, 1, 0, \ldots, 0)$$

place $\uparrow 2d - 2$

$$\beta^* = (0, \ldots, 0, 1).$$

Then one checks that indeed $\beta^* \in B^* \cap F_{\eta^*}$ (i.e. that β^* is on the polars of η^* with respect to B^*), and hence indeed $(\vartheta(B^*), \eta^*, \beta^*) \in J$ and hence $(\vartheta(B^*), \beta^*) \in p_{13}(J) = J_{13}$. However now one also checks by direct computation (via the equations $\Delta_\eta^j(B)$) that η^* is the *only* point in the fibre $p_{13}^{-1}(\vartheta(B^*), \beta^*)$! (counted in fact $(2d-1)!$ times). However this contradicts our assumption $(*)$, which proves the lemma. □

Corollary 5.8. *Let η and ξ be as above and let α be a point on $\langle \eta, \xi \rangle$ generic over $K'_1(\eta, \xi)$. Then α is a generic point of $\mathbb{P}^r$ over K'_1.*

Proof. By Lemma 5.7 the Zariski closure of α over K'_1 contains B; since it clearly contains B *strictly* it must be $\mathbb{P}^r$. □

5.5. Construction of a unirational family of rational curves on W. End of the proof

Returning to the double cover $\pi : W_r[2d, B] \longrightarrow \mathbb{P}^r$ for $r \geq \rho(d)$ (from (5.3)), consider on $W = W_r[2d, B]$ the curve $C_{\eta\xi} = \pi^{-1}(\langle \eta, \xi \rangle)$ where η and ξ are as in Corollary 5.8.

Lemma 5.9. *The curve $C_{\eta\xi}$ is a rational curve, rational over the field $K'_1(\eta, \xi)$.*

Proof. The curve is a double cover of the line $\langle \eta, \xi \rangle$ branched only over the two points η and β (the β from (5.21)). So it is a rational curve. Moreover it is rational over the field $K'_1(\xi, \eta)$ since it contains a rational point $\widetilde{\eta}$ over this field where $\widetilde{\eta} \in W$ such that $\pi(\widetilde{\eta}) = \eta$. $\square$

Let $\omega \in C_{\eta\xi}$ be a generic point of this curve over $K'_1(\eta, \xi)$ and consider in $M_0 \times H_0 \times W$ the Zariski closure W' over K'_1 of the point (η, ξ, ω). Then we have a diagram

$$\begin{array}{ccc} W' & \xrightarrow{p_3} & W \\ {\scriptstyle p_{12}}\downarrow & & \\ S & & \end{array}$$

where S is the variety from Subsection 5.4.

Lemma 5.10. *W' is unirational over K'_1 and p_3 is onto.*

Proof. As we have seen already above $p_{12} : W' \longrightarrow S$ has a *section* $s : S \longrightarrow W'$ defined by $s(\eta, \xi) = \widetilde{\eta}$ with $\pi(\widetilde{\eta}) = \eta$. Since the curve $C_{\eta\xi}$ is rational over $K'_1(\eta, \xi)$ we have $K'_1(\eta, \xi)(\omega) \simeq K'_1(\eta, \xi)(C_{\eta\xi}) = K'_1(\eta, \xi)(\tau)$ with τ transcendental over $K'_1(\eta, \xi)$. Therefore we get from (5.20)

$$K'_1(W') \simeq K'_1(\eta, \xi, \omega) \subset K'_1(\eta_1, \ldots, \eta_{2d-1}, t, \ldots, t_{r-2d+1}, \tau) \qquad (5.22)$$

and since the field on the right hand side of (5.22) is purely transcendental over K'_1 we have that W' is unirational over K'_1. Finally p_3 is surjective, for putting $p_3(\omega) = \alpha$ then α is generic on $\mathbb{P}^r$ over K'_1 by Corollary 5.8, hence ω is generic on W over K'_1. $\square$

Finally since p_3 is surjective we get the following corollary which concludes the proof of Theorem 4.1 (with $K^\# = K'_1$, the field from (5.18)).

Corollary 5.11. *The variety W is unirational over K'_1.*

References

[1] C. Ciliberto, Osservazioni su alcuni classici teoremi di unirazionalità per ipersuperficie e complete intersezioni algebriche proiettive, Ricerche Mat. 29 (1980), 175–191.

[2] A. Grothendieck and J. Dieudonné, Eléments de Géometrie Algébrique I, Springer-Verlag, Berlin–Heidelberg 1971.

[3] J. Harris, B. Mazur and R. Pandharipande, Hypersurfaces of Low Degree, Duke Math. J. 95 (1998), 125–160.

[4] V. A. Iskovskikh and Yu. G. Prokhorov, Fano Varieties, Encyclopaedia Math. Sci. 47, Springer-Verlag, Berlin–Heidelberg 1999, 1–245.

[5] J. Kollár, Rational Curves on Algebraic Varieties, Ergeb. Math. Grenzgeb. (3) 32, Springer-Verlag, Berlin–Heidelberg 1996.

[6] U. Morin, Sull'irrazionalità dell'ipersuperficie algebrica di qualunque ordine e dimensione sufficientemente alta, Atti II Congr. Un. Mat. Ital., Bologna, 277 (1940), 298–302.

[7] U. Morin, Alcuni problemi di unirazionalità, Rend. Sem. Mat. Univ. Politec. Torino 14 (1954/55), 39–53.

[8] K. Paranjape and V. Srinivas, Unirationality of Complete Intersections, in: Flips and Abundance for Algebraic Threefolds, Astérisque 211 (1992), 241–247.

[9] A. Predonzan, Intorno agli S_k giacenti sulla varietà intersezione completa di più forme, Rend. Accad. Naz. Lincei VIII, Ser. 5 (1948), 238–242.

[10] A. Predonzan, Sull'unirazionalità delle varietà intersezione completa di più forme, Rend. Sem. Mat. Padova 18 (1949), 161–176.

[11] L. Ramero, Effective Estimates for Unirationality, Manuscripta Math. 68 (1990), 435–445.

[12] L. Roth, Algebraic Threefolds, Ergeb. Math. Grenzgeb., Reihe Algebraische Geometrie, Heft 6, Springer-Verlag, Berlin–Göttingen–Heidelberg 1955.

[13] A. Weil, Foundations of Algebraic Geometry, Amer. Math. Soc. Colloq. Publ. 39, Amer. Math. Soc., Providence, RI, 1946 and 1962.

A. Conte, Dipartimento di Matematica, Università di Torino, Via Carlo Alberto 10, 10123 Torino, Italy
E-mail: conte@dm.unito.it

M. Marchisio, Dipartimento di Matematica, Università di Torino, Via Carlo Alberto 10, 10123 Torino, Italy
E-mail: marchisio@dm.unito.it

J. P. Murre, Department of Mathematics, University of Leiden, P.O. Box 9512, 2300 RA Leiden, The Netherlands
E-mail: murre@math.leidenuniv.nl

Weighted Grassmannians

Alessio Corti and Miles Reid

In memory of Paolo Francia

Abstract. Many classes of projective algebraic varieties can be studied in terms of graded rings. Gorenstein graded rings in small codimension have been studied recently from an algebraic point of view, but the geometric meaning of the resulting structures is still relatively poorly understood. We discuss here the weighted projective analogs of homogeneous spaces such as the Grassmannian Gr(2, 5) and orthogonal Grassmannian OGr(5, 10) appearing in Mukai's linear section theorem for Fano 3-folds, and show how to use these as ambient spaces for weighted projective constructions. This is a first sketch of a subject that we expect to have many interesting future applications.

1. Introduction

We are interested in describing algebraic varieties explicitly in terms of graded rings and, conversely, in algebraic varieties for which such an explicit description is possible. Our varieties X always come with a polarisation A, usually the canonical class or an integer submultiple of it. Our favourites include the following:

(1) canonical curves, K3 surfaces, Fano 3-folds. A Fano 3-fold V is canonically polarised by its anticanonical class $A = -K_V$. We consider K3 surfaces with Du Val singularities polarised by a Weil divisor. A canonical curve C is a curve of genus $g \geq 2$ with its canonical polarisation by K_C; but we will more often be concerned with subcanonical curves, polarised by a divisor A that is a submultiple of $K_C = kA$, and variants on orbifold curves also occur naturally (see Altınok, Brown and Reid [ABR]).

(2) Regular canonical surfaces, Calabi–Yau 3-folds.

(3) Regular canonical 3-folds.

If $X, \mathcal{O}(1)$ is a polarised n-fold with $K_X = \mathcal{O}(k)$, Mukai defines the *coindex* of X to be $n + 1 + k$; the above are varieties of coindex 3, 4 and 5.

Remark 1.1. (a) Describing a variety explicitly means embedding it into a suitable ambient space and writing down its equations. This is closely related to the problem of finding generators and relations for the graded ring

$$R(X, A) = \bigoplus_{n=0}^{\infty} H^0(X, nA).$$

In all the examples we consider, this is a Gorenstein ring; this property is one of the most powerful general tools we have in studying X and its deformations. It seems to us that this point is not adequately appreciated.

(b) Varieties often come in ladders of successive hyperplane sections. Thus, in good situations, a general elephant $S \in |-K_V|$ on a nonsingular Fano 3-fold V is a K3 surface polarised by $A = -K_V|_S$, and a general $C \in |A|_S$ a canonical curve. Finding the equations of V is closely related to finding the equations of S or C, and often practically equivalent to it.

(c) The natural context to study Fano 3-folds is the Mori category of projective varieties with terminal singularities. The key examples of these are cyclic quotient singularities

$$\frac{1}{r}(1, a, -a) = \mathbb{C}^3/(\mathbb{Z}/r\mathbb{Z}),$$

where the notation signifies that the cyclic group $\mathbb{Z}/r\mathbb{Z}$ acts diagonally with weights $1, a, -a$, and $\mathrm{hcf}(a, r) = 1$. We are thus led to consider K3 surfaces with singularities $\frac{1}{r}(a, -a)$ polarised by ample *Weil* divisors, and, one further step down the ladder, *orbifold* canonical curves; compare [ABR].

Our original motivation is Mukai's description of a prime Gorenstein Fano 3-fold of genus $7 \leq g \leq 10$ as a *linear section* of a special projective homogeneous space, that is, the quotient G/P of a (semisimple) Lie group G by a maximal parabolic subgroup P. For example, consider $V = \mathbb{C}^{2n}$ endowed with a complex symmetric quadratic form q; it is traditional to take

$$q = \begin{pmatrix} 0 & I \\ I & 0 \end{pmatrix}, \quad \text{where } I = I_{n \times n},$$

so that $V = U \oplus U^\vee$ where $U = \langle e_1, \ldots, e_n \rangle$. It is well known that the space of isotropic n-dimensional vector subspaces of V splits into two components for reasons of "spin". The O'Grassmann or orthogonal Grassmann variety $\mathrm{OGr}(n, 2n)$ is one connected component; we take the component containing the reference subspace U. It is a homogeneous space for the group $G = \mathrm{SO}(2n, \mathbb{C})$, and has a natural Plücker style spinor embedding in the projective space $\mathbb{P}(S^+)$ of the spinor representation $S^+ = \bigwedge^{\mathrm{even}} U$. Mukai proves the following result:

Theorem 1.2 (Mukai [Mu]). *A prime Gorenstein Fano 3-fold of genus 7 is a linear section of* $\mathrm{OGr}(5, 10)$ *in its spinor embedding. In other words, there are 7 hyperplanes*

$H_1, \ldots, H_7$ of $\mathbb{P}(S^+) = \mathbb{P}^{15}$ such that

$$(V_{12} \subset \mathbb{P}^8) = \mathrm{OGr}(5, 10) \cap H_1 \cap \cdots \cap H_7.$$

We wanted to see how far these ideas of Mukai generalise. In this note, we define weighted Grassmann and orthogonal Grassmann varieties, and study some examples of their linear sections.

Acknowledgements. The treatment of weighted projective homogeneous spaces in Section 3 is based on notes of Ian Grojnowski [G], whom we had hoped to involve as coauthor. We have also benefitted from discussion with Jorge Neves, whose forthcoming paper [N] takes these ideas further. Gavin Brown and Roberto Pignatelli helped us with computer algebra calculations.

2. Weighted Gr(2, 5)

The affine Grassmannian aGr(2, 5). Weighted versions of a projective homogeneous variety Σ arise on dividing out the affine cone over Σ by different $\mathbb{C}^\times$ actions. The construction is particularly transparent for $\mathrm{Gr}(2, 5)$. Set $V = \mathbb{C}^5$; the affine Grassmann variety

$$\mathrm{aGr}(2, 5) \subset \bigwedge^2 V$$

can be defined in any of the following equivalent ways:

(i) The variety of skew tensors of rank ≤ 2, that is, the image of $V \times V \to \bigwedge^2 V$ given by $(a, b) \mapsto a \wedge b$. In coordinates, we write

$$\bigwedge^2 \begin{pmatrix} a_1 & a_2 & a_3 & a_4 & a_5 \\ b_1 & b_2 & b_3 & b_4 & b_5 \end{pmatrix} = \begin{pmatrix} c_{12} & c_{13} & c_{14} & c_{15} \\ & c_{23} & c_{24} & c_{25} \\ & & c_{34} & c_{35} \\ & & & c_{45} \end{pmatrix}, \quad (2.1)$$

where $c_{ij} = \det \begin{vmatrix} a_i & a_j \\ b_i & b_j \end{vmatrix}$. Our convention is to write out only the upper diagonal entries of the 5×5 skew matrix (c_{ij}).

(ii) The closed orbit of the highest weight vector $e_{12} = (1, 0, \ldots) \in \bigwedge^2 V$ under the action of $\mathrm{GL}(5) = \mathrm{GL}(V)$. In other words, any tensor of rank 2 is in the $\mathrm{GL}(5)$ orbit of $(1, 0, 0, 0, 0) \wedge (0, 1, 0, 0, 0)$.

(iii) The quotient of the variety $M(2, 5)$ of 2×5 matrices by $\mathrm{SL}(2)$ acting on the left: indeed, the ring of invariant functions is generated by the 2×2 minors $c_{ij} = \det \begin{vmatrix} a_i & a_j \\ b_i & b_j \end{vmatrix}$.

(iv) The variety defined by the 4×4 Pfaffians of the generic 5×5 skew matrix, that is, the Plücker equations

$$\mathrm{Pf}_{ijkl} = x_{ij}x_{kl} - x_{ik}x_{jl} + x_{il}x_{jk} = 0,$$

where x_{ij} for $1 \leq i < j \leq 5$ are coordinates on $\bigwedge^2 V$. The point is just that setting the Pfaffians of a skew matrix (x_{ij}) equal to zero enforces rank ≤ 2.

(v) In other words, $\mathrm{aGr}(2, 5)$ has affine coordinate ring

$$R = \mathbb{C}[\mathrm{aGr}(2, 5)] = \mathbb{C}[(x_{ij})]/I, \qquad (2.2)$$

where I is the ideal $I = \langle \mathrm{Pf}_1, \ldots, \mathrm{Pf}_5 \rangle$, and $\mathrm{aGr}(2, 5) = \mathrm{Spec}\, R$.

Equivariant resolution. As a prelude to introducing weights and defining wGr, it is convenient to explain the symmetry group of $\mathrm{aGr}(2, 5) \subset \bigwedge^2 V$ and to write its equations and syzygies in their full symmetry. Under the induced action of $\mathrm{GL}(V)$ on $\bigwedge^2 V$, the scalar matrices $\lambda \cdot I$ act by λ^2. However, the *straight* Grassmannian $\mathrm{Gr}(2, 5) \subset \mathbb{P}^9$ is the quotient of $\mathrm{aGr}(2, 5)$ by $\mathbb{C}^\times$ acting on $\bigwedge^2 V$ by overall scalar multiplication by $\mu \in \mathbb{C}^\times$, and this is not covered by the $\mathrm{GL}(5)$ action; the full symmetry group is thus a double cover of $\mathrm{GL}(V)$ (an index 2 central extension). Rather than introducing notation for the double cover, we write $L = \mathbb{C}$ with the usual action of $\mathbb{C}^\times$, and view the Plücker embedding as $\mathrm{aGr}(2, 5) \hookrightarrow \bigwedge^2 V \otimes L$, where $\mathrm{GL}(5)$ acts on the first factor and $\mathbb{C}^\times$ on the second. We also write

$$D = \det V \otimes L^2 = \bigwedge^5 V \otimes L^2, \qquad (2.3)$$

a 1-dimensional representation of $\mathrm{GL}(V) \times \mathbb{C}^\times$. It is useful to bear in mind the straight homogeneous case, when D pushed forward to $\mathbb{P}^9$ corresponds to $\mathcal{O}(-2)$, and L to $\mathcal{O}(-1)$.

Proposition 2.1. *There are universal maps of vector bundles over the affine space* $\mathbb{A}^{10} = \bigwedge^2 V \otimes L$

$$M: \bigwedge^2 V \otimes L \to \mathbb{C} \quad (\text{that is, } V \otimes L \to V^\vee)$$

and

$$\mathrm{Pf} = \mathrm{Pf}\, M: V^\vee \otimes D = \bigwedge^4 V \otimes L^2 \to \mathbb{C}.$$

Note that interpreted intrinsically, Pf is the second wedge of $M: V \otimes L \to V^\vee$.

Now write $\mathcal{O}$ for the structure sheaf of $\mathbb{A}^{10}$, and $M: \mathcal{O} \otimes \bigwedge^2 V \otimes L \to \mathcal{O}$, etc., for the above universal maps viewed as sheaf homomorphisms. Then the structure sheaf

$\mathcal{O}_{\mathrm{aGr}}$ *of* $\mathrm{aGr}(2,5)$ *has a* $\mathrm{GL}(5) \times \mathbb{C}^\times$ *equivariant projective resolution of the form*

$$0 \leftarrow \mathcal{O} \xleftarrow{\mathrm{Pf}} \mathcal{O} \otimes V^\vee \otimes D \xleftarrow{M} \mathcal{O} \otimes V \otimes L \otimes D \xleftarrow{{}^t\mathrm{Pf}} \mathcal{O} \otimes L \otimes D^2 \leftarrow 0$$

$$\downarrow$$

$$\mathcal{O}_{\mathrm{aGr}}.\qquad\qquad\qquad (2.4)$$

Proof. In coordinates x_{ij} on $\bigwedge^2 V \otimes L$, the map M is the generic 5×5 skew matrix (x_{ij}) and $\mathrm{Pf} = (\mathrm{Pf}_1, \ldots, \mathrm{Pf}_5)$ its vector of Pfaffians. Thus (2.4) follows at once from the well known fact that the ideal of $\mathrm{aGr}(2,5)$ is generated by the 5 Pfaffians, and M is the matrix of syzygies between them. $\square$

Remark 2.2. Each term in (2.4) is a G-equivariant bundle, where $G = \mathrm{GL}(5) \times \mathbb{C}^\times$, and the complex gives the projective resolution of $\mathcal{O}_{\mathrm{aGr}}$ in terms of G-equivariant vector bundles on the ambient space $\bigwedge^2 V \otimes L$. We write out the definitions for completeness.

Let G be a group and Y a space with a left G-action $G \times Y \to Y$; write $l_g : Y \to Y$ for the action of $g \in G$. A G-equivariant sheaf is a sheaf $\mathcal{F}$ on Y, together with isomorphisms

$$\alpha_g : \mathcal{F} \to l_g^* \mathcal{F} \quad \text{satisfying} \quad \alpha_{g_2 g_1} = l_{g_2}^*(\alpha_{g_1}) \circ \alpha_{g_2}. \qquad (2.5)$$

The collection of maps $\{\alpha_g\}$ is called a *G-linearisation* or *descent data* for $\mathcal{F}$; the cocycle condition in (2.5) ensures that G acts on the pushforward $\pi_* \mathcal{F}$, where $\pi : Y \to X = Y/G$ is the quotient morphism. A quasicoherent sheaf $\mathcal{F}$ over an affine scheme $Y = \mathrm{Spec}\, A$ is the associated sheaf $\mathcal{F} = \widetilde{F}$ for an A-module $F = \Gamma(Y, \mathcal{F})$; a G-equivariant sheaf arises in the same way from a module F over the twisted group ring $A * G$. That is, F is an A-module with a representation of G such that $g(am) = g(a)g(m)$ for all $a \in A$ and $m \in F$, where G has the left action on A by $g(a) = l_{g^{-1}}^\#(a)$.

If G acts freely with quotient $X = Y/G$, taking pushforward and invariant sections identifies a G-equivariant sheaf with a sheaf on X; if G has fixed points, the same construction only gives an orbi-sheaf (or a sheaf on the *quotient stack* $[Y/G]$ of which X is the coarse moduli space). With a little common sense, we can mostly ignore this point, and pretend that we get a genuine sheaf on the space Y/G.

Remark 2.3. The $\bigwedge^2 V \otimes L$ occurring here tells us how to define $\mathrm{Gr}(2,5)$-bundles over an arbitrary base scheme S, more or less as for conic bundles: choose a rank 5 vector bundle $\mathcal{V}$, a line bundle $\mathcal{L}$ and a morphism $\mu : \bigwedge^2 \mathcal{V} \otimes \mathcal{L} \to \mathcal{O}_S$, and take the locus rank $\mu \leq 2$ defined by the relative equations $\mathrm{Pf}(\mu) = \bigwedge^2 \mu = 0$. If $\mathcal{L}$ is a square, say $\mathcal{L} = \mathcal{L}_0^2$, we can get rid of it by replacing $\mathcal{V} \mapsto \mathcal{V} \otimes \mathcal{L}_0$.

The definition of wGr(2,5). Choosing weights on $\mathrm{aGr}(2,5)$ is equivalent to specifying a 1-parameter subgroup $\mathbb{C}^\times \hookrightarrow \mathrm{GL}(5) \times \mathbb{C}^\times$. Up to conjugacy, we can choose it in the maximal torus, that is, diagonal of the form

$$\left(\operatorname{diag}(\lambda^{w_1}, \lambda^{w_2}, \lambda^{w_3}, \lambda^{w_4}, \lambda^{w_5}); \lambda^u\right) \subset \operatorname{GL}(5) \times \mathbb{C}^\times.$$

To put weights on $\operatorname{aGr}(2,5) \subset \bigwedge^2 V$, we thus specify integer weights $(w_1, \ldots, w_5)$ on V, and a separate overall weight u on $\bigwedge^2 V$. The ambient space $\bigwedge^2 V$ thus has coordinates

$$x_{ij} \quad \text{with} \quad \operatorname{wt} x_{ij} = w_i + w_j + u.$$

Replacing $w_i \mapsto w_i - [\frac{u}{2}]$, we can always take $u = 0$ or 1. In fact, for brevity in calculations, we usually use the trick of absorbing the weight u into the w_i by $w_i \mapsto w_i + \frac{u}{2}$, at the cost of working with half-integers w_i. For odd u this is formally incorrect, but completely harmless, and hardly ever leads to confusion.

Definition 2.4. Let $w = (w_1, \ldots, w_5)$ and u be weights such that $w_i + w_j + u > 0$ for all i, j. We define

$$\operatorname{wGr}(2,5) = \left(\operatorname{aGr}(2,5) \setminus 0\right)/\mathbb{C}^\times,$$

where $\mathbb{C}^\times$ acts on $\operatorname{aGr}(2,5) \subset \bigwedge^2 V$ by $x_{ij} \mapsto \lambda^{w_i+w_j+u} x_{ij}$. Clearly

$$\operatorname{wGr}(2,5) = \operatorname{Proj} R$$

where $R = \mathbb{C}[\operatorname{aGr}(2,5)]$ is the affine coordinate ring as in (v) above, graded by $\operatorname{wt} x_{ij} = w_i + w_j + u$. By definition, $\operatorname{wGr}(2,5)$ comes with a Plücker embedding in weighted projective space (wps) $\mathbb{P}^9(\{w_i + w_j + u\})$, and is defined by the usual Plücker equations, the 4×4 Pfaffians of the generic 5×5 skew matrix (x_{ij}).

The elementary properties of $\operatorname{wGr}(2,5)$ are easy enough to figure out. We get affine charts by setting $x_{ij} \neq 0$, where (say)

$$x_{12} = \det \begin{vmatrix} a_1 & a_2 \\ b_1 & b_2 \end{vmatrix}.$$

This chart is the quotient $\mathbb{C}^6/(\mathbb{Z}/\operatorname{wt} x_{12})$ of $\mathbb{C}^6$ by the cyclic group of order $\operatorname{wt} x_{12} = w_1 + w_2 + u$ acting on coordinates $b_3, b_4, b_5, a_3, a_4, a_5$ with weights

$$w_1 + w_3 + u, \quad w_1 + w_4 + u, \quad w_1 + w_5 + u,$$
$$w_2 + w_3 + u, \quad w_2 + w_4 + u, \quad w_2 + w_5 + u.$$

This formula shows the point of our shorthand setting $u = 0$, allowing the w_i to be half-integers. As with weighted projective spaces, we usually impose "well formed" conditions to ensure that the cyclic group acts effectively and without ramification in codimension 1. We omit the details, but compare [Fl], Definition 6.9.

The Hilbert numerator of $\operatorname{wGr}(2,5) \subset \mathbb{P}(\{w_i + w_j + u\})$ is

$$\prod_{i,j}(1 - t^{w_i+w_j+u}) P(t) = 1 - \sum_{i=1}^{5} t^{d-w_i} + \sum_{j=1}^{5} t^{d+w_j+u} - t^{2d+u},$$

where $d = \sum w_i + 2u$. This formula is a numerical version of (2.4), and essentially equivalent to it by the splitting principle. Multiplying by $(1-t)^3$, we deduce that

$$\deg \text{wGr} = \frac{\sum \binom{d-w_i}{3} - \sum \binom{d+w_i+u}{3} + \binom{2d+u}{3}}{\prod(w_i + w_j + u)}.$$

If wGr(2, 5) is well formed, its canonical class is $K_{\text{Gr}(2,5)} = \mathcal{O}(-2d - u)$. In fact the wps has $K = -\det(\bigwedge^2 V \otimes L)$, which has degree

$$-\sum \text{wt}\, x_{ij} = -4\sum w_i - 10u = -4d - 2u,$$

and wGr(2, 5) $\subset \mathbb{P}(\bigwedge^2 V \otimes L)$ has the adjunction number $\deg(L \otimes D^2) = 2d + u$ by (2.4).

Tautological sequences

Tautological vector bundles over aGr(2, 5) can be discussed in several ways, parallel to the different treatments of aGr(2, 5). Taking invariants of the $\mathbb{C}^\times$ action gives rise to tautological (orbi-)bundles on wGr(2, 5), as in the case of the straight Grassmannian. These sheaves on wGr(2, 5) can also be understood in terms of the well known Serre correspondence

$$\mathcal{E} \mapsto E_* = \bigoplus_{k \geq 0} H^0(\mathcal{E}(k)).$$

We describe the Serre module of the tautological bundles explicitly as modules over the affine coordinate ring of aGr.

First, aGr is locally a codimension 3 complete intersection wherever the matrix of syzygies M has rank 2, that is, at every point of aGr $\setminus 0$. At any such point, we can use two rows of M to express 2 of the 5 Pfaffians as linear combinations of the others, so that the ideal sheaf $\mathcal{I}_{\text{aGr}}$ is locally generated by 3 Pfaffians. Thus the conormal sheaf to aGr(2, 5) is a vector bundle of rank 3 outside the origin with 5 sections. In more detail, consider (2.4) as a resolution of the ideal sheaf $\mathcal{I}_{\text{aGr}}$:

$$0 \leftarrow \mathcal{I}_{\text{aGr}} \xleftarrow{\text{Pf}} \mathcal{O} \otimes V^\vee \otimes D \xleftarrow{M} \mathcal{O} \otimes V \otimes L \otimes D \leftarrow \cdots.$$

Tensoring with $\mathcal{O}_{\text{aGr}} = \mathcal{O}/\mathcal{I}_{\text{aGr}}$ gives the exact sequence

$$0 \leftarrow \mathcal{I}/\mathcal{I}^2 \xleftarrow{\text{Pf}} \mathcal{O}_{\text{aGr}} \otimes V^\vee \otimes D \xleftarrow{M} \mathcal{O}_{\text{aGr}} \otimes V \otimes L \otimes D \leftarrow \cdots; \quad (2.6)$$

twisting back by D^{-1} gives a tautological exact sequence

$$0 \leftarrow \mathcal{F} \xleftarrow{\text{Pf}} \mathcal{O}_{\text{aGr}} \otimes V^\vee \leftarrow \mathcal{E} \otimes L \leftarrow 0 \quad (2.7)$$

of vector bundles over aGr $\setminus 0$, where $\mathcal{F} = \mathcal{I}/\mathcal{I}^2 \otimes D^{-1}$ and the identification ker Pf = im $M = \mathcal{E} \otimes L$ is justified below.

Next, M has rank 2 at every point of $\mathrm{aGr}(2, 5) \setminus 0$, so that if we restrict the sheaf homomorphism $M: \mathcal{O} \otimes V \otimes L \otimes D \to \mathcal{O} \otimes V^\vee \otimes D$ to $\mathrm{aGr}(2, 5)$, this restriction maps onto a $\mathrm{GL}(5) \times \mathbb{C}^\times$ equivariant sheaf over $\mathrm{aGr}(2, 5)$ that is a rank 2 vector bundle on $\mathrm{aGr}(2, 5) \setminus 0$. We twist it back by $L^{-1}D^{-1}$ for convenience, obtaining a second tautological exact sequence:

$$0 \leftarrow \mathcal{E} \leftarrow \mathcal{O}_{\mathrm{aGr}} \otimes V \leftarrow \mathcal{K} \leftarrow 0. \tag{2.8}$$

Here $\mathcal{E}$ is the same sheaf as in (2.7), up to the indicated twist, because the sequence in (2.6) is exact. By playing with determinant bundles in (2.7) and (2.8) one sees that $\det \mathcal{E} = L = \det \mathcal{F}$ so that $\mathcal{E}^\vee = \mathcal{E} \otimes L$, and then the two sequences are dual, which determines the kernel in (2.8):

$$0 \leftarrow \mathcal{E} \leftarrow \mathcal{O}_{\mathrm{aGr}} \otimes V \leftarrow \mathcal{F}^\vee \leftarrow 0. \tag{2.9}$$

We can concatenate the exact sequences (2.9) and (2.7) to obtain the following explicit description of the module $E_* = H^0(\mathrm{aGr}(2, 5), \mathcal{E})$ over the affine coordinate ring $R = \mathbb{C}[(x_{ij})]/I = \mathbb{C}[\mathrm{aGr}(2, 5)]$: it is generated by 5 sections $s_1, \ldots, s_5$ that one identifies either with the columns of M, or with the 5 columns $s_i = \binom{a_i}{b_i}$ subject to the 10 relations

$$x_{ij}s_k - x_{ik}s_j + x_{jk}s_i \quad \text{for } 1 \leq i < j < k \leq 5. \tag{2.10}$$

We can say the same thing in invariant terms by taking global sections in the exact sequence

$$0 \leftarrow \mathcal{E} \leftarrow \mathcal{O}_{\mathrm{aGr}} \otimes V \leftarrow \mathcal{O}_{\mathrm{aGr}} \otimes \bigwedge^2 V^\vee.$$

Our 4th and final treatment of the bundle $\mathcal{E}$ is intrinsic and starts from the model $(\mathrm{aGr}(2, 5) \setminus 0) = M(2, 5)^*/\mathrm{SL}(2)$. Consider the given representation of $\mathrm{SL}(2)$ on $\mathbb{C}^2$ and the diagonal action of $\mathrm{SL}(2)$ on the trivial bundle $M(2, 5) \times \mathbb{C}^2$; the quotient is the total space of a rank 2 vector bundle $\mathcal{E}$ on $\mathrm{aGr}(2, 5) \setminus 0$. The sections of $\mathcal{E}$ are functions $f: M(2, 5)^* \to \mathbb{C}^2$ that transform as

$$f(gM) = gf(M) \quad \text{for all } g \in \mathrm{SL}(2) \text{ and } M \in M(2, 5).$$

We can identify this bundle $\mathcal{E}$ with any of the above constructions: the 5 columns of M give global sections of $\mathcal{E}$, and they satisfy the same relations as in (2.10), leading to the same presentation of $\mathcal{E}$ by $\mathrm{GL}(5) \times \mathbb{C}^\times$-equivariant free sheaves on $\mathrm{aGr}(2, 5) \setminus 0$. The advantage of this construction is that, since it involves the 2-planes parametrised by points of $\mathrm{aGr}(2, 5)$, it really relates to the functor represented by $\mathrm{aGr}(2, 5)$, and thus to the traditional tautological bundle of a Grassmannian.

Examples

Example 2.5. Take $w = (\frac{1}{2}, \frac{1}{2}, \frac{1}{2}, \frac{1}{2}, \frac{3}{2})$; then wGr(2, 5) $\subset \mathbb{P}(1^6, 2^4)$ is given by a skew matrix

$$M = \begin{pmatrix} x_{12} & x_{13} & x_{14} & y_1 \\ & x_{23} & x_{24} & y_2 \\ & & x_{34} & y_3 \\ & & & y_4 \end{pmatrix} \quad \text{with weights} \quad \begin{pmatrix} 1 & 1 & 1 & 2 \\ & 1 & 1 & 2 \\ & & 1 & 2 \\ & & & 2 \end{pmatrix}.$$

This has 5 Pfaffians of degrees 2, 3, 3, 3, 3. The section $V = \text{wGr}(2, 5) \cap (2)^3$ by 3 general forms of weight 2 is a Fano 3-fold with

$$h^0(V, -K) = 6, \quad -K^3 = 6 + \frac{1}{2};$$

that is, $g = 4$ and V has in general a singular point $\frac{1}{2}(1, 1, 1)$; the corresponding family of K3 surfaces is No. 2 in Altınok's list, `Altinok3(2)` in the Magma database.

Because the 3 equations of degree 2 are general, they involve 3 of the weight 2 coordinates y_1, y_2, y_3 with nonzero coefficients, and we can use them to eliminate y_i as generators. Thus we say that V is a *quasilinear section* of wGr(2, 5). By analogy with Mukai's results, we want to call this a *linear section theorem*, but we keep the "quasi" for the moment to keep away the unclean spirit.

The section $S = \text{wGr}(2, 5) \cap (2)^4$ by 4 forms of weight 2 has been studied in detail by Neves [N]; we can use the 4 equations to write $y_i = q_i(x)$ for $i = 1, \ldots, 4$, giving a canonical surface $S \subset \mathbb{P}^5$ with $p_g = 6$, $K^2 = 13$ defined by the Pfaffians of

$$\begin{pmatrix} l_{12} & l_{13} & l_{14} & q_1 \\ & l_{23} & l_{24} & q_2 \\ & & l_{34} & q_3 \\ & & & q_4 \end{pmatrix}$$

with l_{ij} linear and q_i quadratic forms on $\mathbb{P}(1^6, 2^4)$. Conversely (and slightly more generally), Neves [N] shows that a surface S with $p_g = 6$, $K^2 = 13$ satisfying appropriate generality assumptions has a *nongeneral* canonical curve $C \in |K_S|$ for which the restricted linear system splits as $|K_S|_C = g_6^1 + g_7^1$. Following Mukai's strategy, Neves shows how to derive the "tautological" rank 2 vector bundle E over S and the embedding of S into wGr(2, 5) or a cone over it from this Brill–Noether data on C. The linear entries l_{ij} of M may be linearly dependent, corresponding to a model of S as a section of a cone over wGr(2, 5).

Example 2.6. Taking $w = (\frac{1}{2}, \frac{1}{2}, \frac{1}{2}, \frac{3}{2}, \frac{3}{2})$ gives wGr(2, 5) $\subset \mathbb{P}(1^3, 2^6, 3)$ defined by a matrix with weights

$$\begin{pmatrix} 1 & 1 & 2 & 2 \\ & 1 & 2 & 2 \\ & & 2 & 2 \\ & & & 3 \end{pmatrix},$$

having Pfaffians of degrees 3, 3, 4, 4, 4.

(a) Write $\mathcal{C}\,\mathrm{wGr}(2, 5) \subset \mathbb{P}(1^4, 2^6, 3)$ for the projective cone over $\mathrm{wGr}(2, 5)$; this means that we add one extra variable of degree 1 to the homogeneous coordinate ring, not involved in any relation. Then a general quasilinear section $S = \mathcal{C}\,\mathrm{wGr}(2, 5) \cap (2)^5$ of the cone by 5 general forms of degree 2 is a K3 surface with ample Weil divisor D satisfying

$$h^0(S, D) = 4, \quad D^2 = 4 + \frac{2}{3}$$

that is, $g = 3$ and S has a singular point $\frac{1}{3}(1, 2)$. This family of K3 surfaces is Altinok3(3).

The new phenomenon in this example is that S has $h^0(S, D) = 4$, so that the graded ring $R(S, D)$ has 4 generators $x_1, \ldots, x_4$ of degree 1. On the other hand, the matrix only has 3 entries of weight 1, so that not all the x_i can appear as degree 1 terms. Thus S is obtained from the cone over $\mathrm{wGr}(2, 5)$.

(b) On the other hand a general quasilinear section $S = \mathrm{wGr}(2, 5) \cap (2)^3 \cap (3)$ is a K3 with ample D such that

$$h^0(S, D) = 3 \quad \text{and} \quad D^2 = 2 + 3 \times \frac{1}{2}$$

that is, $g = 2$ and S has three singular points $\frac{1}{2}(1, 1)$; this is Altinok3(5).

The following is due to Selma Altınok.

Theorem 2.7 (Altınok [Al]). *There are precisely 69 families of K3 surfaces with cyclic singularities $\frac{1}{r}(a, -a)$ whose general element is a codimension 3 subvariety in weighted projective space given by the 4×4 Pfaffians of a skew 5×5 matrix.*

The next result is a nice structural description of these surfaces; unfortunately, we don't know how to prove it in an entirely conceptual way.

Proposition 2.8. *All K3 surfaces of Altınok are quasilinear sections of a weighted Grassmannian $\mathrm{wGr}(2, 5)$ or a cone over $\mathrm{wGr}(2, 5)$.*

Proof. Ultimately, this is based on a case by case check against Altınok's list. By Buchsbaum–Eisenbud, $S \subset \mathbb{P}(a_1, \ldots, a_6)$ is defined (scheme theoretically) by the 4×4 Pfaffians of a 5×5 skew matrix

$$\begin{pmatrix} f_{12} & f_{13} & f_{14} & f_{15} \\ & f_{23} & f_{24} & f_{25} \\ & & f_{34} & f_{35} \\ & & & f_{45} \end{pmatrix}$$

where the entry $f_{ij}(y_1, \ldots, y_6)$ is a weighted homogeneous form of degree d_{ij} in the coordinates y_i (the condition for S to be a K3 implies that, if $b_i = \deg \mathrm{Pf}_i$ is the degree of the ith Pfaffian, then $\sum b_i = 2\sum a_i$). Now an easy combinatorial argument shows that the 5 Pfaffians are weighted homogeneous if and only if $d_{ij} = w_i + w_j$ for some $w_i, i = 1, \ldots, 5$. The idea is to map S to the weighted Grassmannian wGr(2, 5) with weights $w = (w_1, \ldots, w_5)$, immersed in $\mathbb{P}(x_{ij})$, by setting

$$f_{ij} = x_{ij}$$

and check that this is an embedding and maps to a quasilinear section. As far as we can tell, these conclusions must be checked explicitly on each of the 69 families of Altınok. Intuitively, the key point is that, for S to be a K3, the degrees f_{ij} must be "small". Slightly more precisely, the formula $\sum b_i = 2\sum a_i$ implies in practice that many of the d_{ij} equal an a_i, which is to say that f_{ij} is linear in one of the variables. □

Remark 2.9. If $S \subset$ wGr(2, 5) is a K3 quasilinear section, it is tempting to try to reconstruct the embedding from intrinsic data on S, by analogy with Mukai's constructions and Neves [N]. It is easy to see that $E_{|S}$ is a rigid simple vector bundle, hence stable and uniquely characterised by its Chern classes and local nature at the singularities. We know many ad hoc constructions but no unified way to produce the bundle directly on S, and no a priori reason why it must exist. In the case of an (orbifold) canonical curve C, the vector bundles $E_{|C}$ arising from embeddings in wGr(2, 5) are often interesting and rather exceptional from the point of view of higher rank Brill–Noether theory.

3. Weighted homogeneous spaces

This section is based on a close reading of part of Ian Grojnowski's notes [G]. We give the general definition of weighted projective homogeneous spaces under an algebraic group G and describe an explicit atlas of coordinate charts on them. A homogeneous variety that is projective is of course homogeneous under a semisimple group G; however, weighted homogeneous spaces always involve central extensions, as we saw with SL(5), GL(5) and GL(5) $\times \mathbb{C}^\times$ in the preceding section. Thus we work from now on with a reductive group G.

Notation. Let G be a reductive complex algebraic group. We fix a maximal torus and Borel subgroup $T \subset B \subset G$ and write $\mathfrak{B} = G/B$ for the maximal flag variety. Let $X = \text{Hom}(T, \mathbb{C}^\times)$ be the lattice of weights (or characters), and $Y = \text{Hom}(\mathbb{C}^\times, T)$ the dual lattice of 1-parameter subgroups, with the perfect pairing $\langle \, , \, \rangle : X \times Y \to \mathbb{Z}$.

Recall that the *roots* of G are defined as the weights of T appearing in the adjoint action of T on the Lie algebra $\mathfrak{g}$. We write $\Delta \subset X$ for the set of these. A root $\alpha \in \Delta$ determines an involution of the maximal torus T, and hence a reflection r_α of X; these reflections generate the Weyl group $W(G)$. The *negative roots* $-\Delta_+$ are the roots appearing in $\mathfrak{b}/\mathfrak{t}$. Let $S \subset \Delta_+$ be the set of simple roots.

Projective homogeneous spaces and parabolic subgroups. A projective homogeneous space under G is a quotient space $\Sigma = G/P$ by a parabolic subgroup P. Every such is conjugate to a *standard* parabolic subgroup, that is, one containing B. A standard parabolic subgroup P corresponds to the subset of simple roots

$$I = \{\alpha \in S \mid r_\alpha(B) \subset P\} \subset S.$$

We recover P as follows: let $W_I \subset W(G)$ be the subgroup generated by r_α for $\alpha \in I$; then $P = P_I = BW_I B$. We write $\Sigma_I = G/P_I$ for the corresponding projective homogeneous space or *generalised flag variety*.

Dominant weights. A weight $\chi \in X$ extends to a unique character $B \to \mathbb{C}^*$, and hence gives rise to a line bundle $\mathcal{O}(\chi)$ on $\mathfrak{B}$. A weight χ is *dominant* if $V_\chi = H^0(\mathfrak{B}, \mathcal{O}(\chi)) \neq 0$; thus the cone X^+ of dominant weights is the effective cone of $\mathfrak{B}$. If χ is a dominant weight, V_χ is an irreducible representation of G with highest weight vector v_χ. The linear system $|V_\chi|$ is free, and defines an equivariant morphism $G \to \mathbb{P}(V_\chi)$ whose image $\Sigma = G \cdot \mathbb{C}v_\chi$ is the orbit of the highest weight line $\mathbb{C}v_\chi$. Therefore $\Sigma = G/P$ is a projective homogeneous space, where $P = \text{Stab}(\mathbb{C}v_\chi)$ is a parabolic subgroup. Then $P = P_I$ as above and $\Sigma = \Sigma_I \subset \mathbb{P}(V_\chi)$ is a *generalised Plücker embedding*.

Definition of weighted homogeneous spaces. Let $\rho \in Y = \text{Hom}(\mathbb{C}^\times, T)$ be a 1-parameter subgroup, and $u \geq 0$ an overall weight. We use ρ and u to make V_χ into a representation of $G \times \mathbb{C}^\times$, with the second factor acting by

$$\lambda \colon v \mapsto \lambda^u \rho(\lambda) \cdot v.$$

We assume from now on that this action has only positive weights.

Remark 3.1. This never happens if, say, G is semisimple and $u = 0$. However, we can always make it happen by taking u large enough; more precisely, the weights are all positive if and only if

$$N_w = \langle \chi, w\rho \rangle + u > 0 \quad \text{for every } w \in W(G).$$

Here we assume that this condition is satisfied.

Then the quotient $\mathbb{P}(V_\chi)(\rho, u) = (V_\chi \setminus 0)/\mathbb{C}^\times$ is a weighted projective space. From the description $a\Sigma_I \subset V_\chi = G \cdot \mathbb{C}v_\chi$, we see that $a\Sigma_I$ is invariant under the $\mathbb{C}^\times$-action.

Definition 3.2. The weighted homogeneous variety associated to this data is the quotient

$$w\Sigma_I = \bigl(a\Sigma_I \setminus 0\bigr)/\mathbb{C}^\times \subset \mathbb{P}(V_\chi)(\rho, u).$$

To stress the choices of the data χ, ρ, u, we write $w\Sigma_I = w\Sigma_I(\chi, \rho, u)$.

Lemma 3.3. *We have* $w\Sigma_I = w\Sigma_I(\chi, \rho, u) = w\Sigma_I(\chi, w\rho, u)$ *for all* $w \in W(G)$.

Proof. Almost obvious, but see the explicit coordinatisation given below. □

Coordinate charts. We write down explicit T-invariant coordinate charts on weighted homogeneous varieties as quotients of affine spaces by a cyclic group. This explicit coordinate atlas is useful in studying various properties of $w\Sigma_I$.

Let U^- be the unipotent radical of the opposite Borel subgroup B^-. Choose a T-equivariant isomorphism $\mathbb{C}^{\Delta_+} \cong U^-$, where T acts on $\mathbb{C}^{\Delta_+}$ by $x \cdot s_\alpha = \alpha(x^{-1})s_\alpha$. Thus a 1-parameter subgroup $\rho \colon \mathbb{C}^\times \to G$ gives rise to an action of $\mathbb{C}^\times$ on $\mathbb{C}^{\Delta_+}$ by

$$\lambda \cdot s_\alpha = \lambda^{-\langle \alpha, \rho \rangle} s_\alpha.$$

As a warm up, we start with the maximal flag variety $\mathfrak{B} = G/B$. Then for each $w \in W(G)$, the image of $wU^- v_\chi$ in $\mathbb{P}(V_\chi)$ is an open set of $\mathfrak{B}$, isomorphic to the affine space $\mathbb{C}^{\Delta_+}$ with T-action twisted by w^{-1}. Moreover, the union of these open sets is all of $\mathfrak{B}$. Thus we get a covering of the weighted flag variety $w\Sigma$ by $|W(G)|$ open subsets, each isomorphic to $\mathbb{C}^{\Delta_+}/\mu_{N_w}$, where $N_w = u + \langle \chi, w\mu \rangle$ as before, and $\lambda \in \mu_{N_w}$ acts by

$$\lambda \cdot s_\alpha = \lambda^{-\langle \alpha, w\rho \rangle}.$$

We now treat the general case of $\Sigma_I = G/P_I$, for $I \subset S$. Write Δ^I for the roots that can be written as linear combinations of the roots in I, and $\Delta^I_+ = \Delta^I \cap \Delta_+$. We set

$$U^- = U^J \times U_J^- \quad \text{where} \quad U_J^- \cong \mathbb{C}^{\Delta^I_-} \quad \text{and} \quad U^J \cong \mathbb{C}^{\Delta_+ \setminus \Delta^I_+}$$

(T-equivariantly). Then the weighted homogeneous space $w\Sigma_I(\chi, \rho, k)$ admits a cover by $|W(G)/W_I|$ open charts, each a cyclic quotient of affine space. The chart corresponding to w is the image of $wU^J(v_\chi)$ in $\mathbb{P}(V_\chi)(\rho, u)$; it is isomorphic to U^J/μ_{N_w} where $\lambda \in \mu_{N_w}$ acts by

$$\lambda \cdot s_\alpha = \lambda^{-\langle \alpha, u\rho \rangle}.$$

Problem 3.4. As with weighted projective spaces, to use weighted homogeneous spaces $w\Sigma$ as ambient spaces in which to construct varieties, we need to study questions such as when a subvariety $X \subset \Sigma$ is well formed (that is, no orbifold behaviour in codimension 0 or 1, no quasireflections), or quasismooth (that is, the affine cone over X is nonsingular); for $w\Sigma$ itself, it seems reasonable to expect that the $\mathbb{C}^\times$ action of V_χ is well formed if and only if its restriction to $a\Sigma_I \subset V_\chi$ is. By analogy with the toric case, there must be straightforward adjunction formulas for the canonical class of weighted homogeneous spaces $w\Sigma$, together with criteria to determine whether the affine cone $a\Sigma$ is Gorenstein or $\mathbb{Q}$-Gorenstein, and has terminal or canonical singularities. The results of the preceding section on $\mathrm{Gr}(2, 5)$ raise the interesting question of writing down the projective resolution of $a\Sigma_I \subset V_\chi$ in equivariant terms; Lascoux [La] has related results in some important cases that might serve as a model.

Since Σ_I has the status of a generalised flag variety, it is also interesting to study the corresponding tautological structures over aΣ_I.

4. Weighted orthogonal Grassmannian OGr(5, 10)

As in the introduction, let $V = \mathbb{C}^{10}$ with a nondegenerate quadratic form q; a change of basis puts q in the normal form

$$q = \begin{pmatrix} 0 & I \\ I & 0 \end{pmatrix}, \quad \text{that is, } V = U \oplus U^\vee, \text{ where } U = \langle e_1, \ldots, e_5 \rangle.$$

We write $f_1, \ldots, f_5$ for the dual basis of $U^\vee$. A vector subspace $F \subset V$ is *isotropic* if q is identically zero on F. For example, U is an isotropic 5-space. Since q is nondegenerate, it is clear that an isotropic subspace $F \subset V$ has dimension ≤ 5. We say that a maximal isotropic subspace is a *generator* of q, or of the quadric hypersurface $Q: (q = 0) \subset \mathbb{P}(V)$. The parity $\dim F_\lambda \cap U \mod 2$ is known to be locally constant in a continuous family of generators F_λ. Thus parity splits the generators into two connected components. We choose the component containing the reference subspace U.

Thus we define the *orthogonal Grassmann* or *O'Grassmann variety* OGr(5, 10) by

$$\text{OGr}(5, 10) = \left\{ F \in \text{Gr}(5, V) \,\middle|\, \begin{array}{l} F \text{ is isotropic for } q \\ \text{and } \dim F \cap U \text{ is odd} \end{array} \right\}$$

The Weyl group $W(D_5)$. The study of the algebraic group $\text{SO}(10, \mathbb{C})$, its double cover $\text{Spin}(10) \to \text{SO}(10, \mathbb{C})$, and their representations is governed by the Weyl group $W(D_5)$, which acts as a permutation group on every combinatoric set in the theory. We are particularly interested in two permutation representations of $W(D_5)$ that base respectively the given representation V of $\text{SO}(10)$, and the space of spinor $S^+ = \bigwedge^{\text{even}} U$, which is a representation of $\text{Spin}(10)$.

The given representation $V = U \oplus U^\vee$ has basis $e_1, \ldots, e_5, f_1, \ldots, f_5$, and $W(D_5)$ acts by permuting the indices $\{1, \ldots, 5\}$ on the e_i and f_i simultaneously, and by swapping evenly many e_i with f_i. For example, the permutations

$$(e_1 f_1)(e_2 f_2) \quad \text{and} \quad (e_1 f_1)(e_2 f_2)(e_3 f_3)(e_4 f_4)$$

are elements of $W(D_5)$. One checks that in this permutation representation, $W(D_5)$ is the Coxeter group generated by the 5 involutions

$$(12) \quad \text{---} \quad (23) \quad \text{---} \quad (34) \quad \text{---} \quad (45)$$
$$|$$
$$(e_4 f_5)(e_5 f_4)$$

with the Coxeter relations indicated by the Dynkin diagram.

The spinor representation S^+ has basis the 16 nodes of the graph

$$\Gamma = \text{5-cube modulo antipodal identification.}$$

The action of $W(D_5)$ on Γ has 5 involutions parallel to the facets of the 5-cube, whose product is the antipodal involution, and thus acts trivially on Γ. These define a normal subgroup $(\mathbb{Z}/2)^5/(\text{diag}) \triangleleft W(D_5)$, the quotient by which is the symmetric group S_5 permuting the 5 orthogonal directions of the 5-cube. Thus $W(D_5)$ is the extension $(\mathbb{Z}/2)^4 \triangleleft W(D_5) \twoheadrightarrow S_5$.

To introduce notation for the nodes of Γ, we break the symmetry by choosing a preferred node $x = x_\emptyset \in \Gamma$ and an order 1, 2, 3, 4, 5 on the 5 edges $xx_1, \ldots, xx_5$ out of x. Then Γ consists of x_I, where $I \subset \{1, 2, 3, 4, 5\}$, and $x_I = x_{\mathcal{C}I}$ (where $\mathcal{C}I$ is the set complement $\mathcal{C}I = \{1, 2, 3, 4, 5\} \setminus I$). The *short* representatives are x, x_i, x_{ij}, with $x = x_\emptyset = x_{12345}$, $x_1 = x_{2345}$, $x_{12} = x_{345}$, etc. We use this below to work out the equations and syzygies of $\mathrm{OGr}(5, 10)$.

The symmetry here is the same as that of the 16 lines on the del Pezzo surface of degree 4, see Reid [R].

Notation. We take the construction $S^+ = \mathbb{C} \oplus \bigwedge^2 U \oplus \bigwedge^4 U$ as the definition of S^+, without attempting to deal with it intrinsically (which can be done in terms of the even Clifford algebra). This construction depends on the choice of U or of the decomposition $V = U \oplus U^\vee$, and S^+ is a representation of the double cover $\mathrm{Spin}(10)$, not of $\mathrm{SO}(10)$ itself.

We write $(e, M, P) \in S^+$ for an element of S^+, where $e \in \mathbb{C}$, $M = (x_{ij})$ is a skew 5×5 matrix and P a 5×1 column vector. If $M = (x_{ij})$ is a skew 5×5 matrix then $\mathrm{Pf}\, M$ is the column vector of its Pfaffians, that is,

$$\mathrm{Pf}\, M = \begin{pmatrix} x_{23}x_{45} - x_{24}x_{35} + x_{25}x_{34} \\ -x_{13}x_{45} + x_{14}x_{35} - x_{15}x_{34} \\ x_{12}x_{45} - x_{14}x_{25} + x_{15}x_{24} \\ -x_{12}x_{35} + x_{13}x_{25} - x_{15}x_{23} \\ x_{12}x_{34} - x_{13}x_{24} + x_{14}x_{23} \end{pmatrix}.$$

Affine cover of $\mathrm{OGr}(5, 10)$. Since $V = U \oplus U^\vee$, a 5-plane $F \subset V$ near U is the graph of a linear map $\varphi \colon U \to U^\vee$, so that $U \in \mathrm{Gr}(5, 10)$ has an affine neighbourhood parametrised by $\mathrm{Hom}(U, U^\vee)$: in other words, F has a basis of 5 vectors in $V = \mathbb{C}^{10}$ that we can write as a matrix (I, M) with $I = I_5$ and M a 5×5 matrix. One sees that F is isotropic for q if and only if the linear map φ or the matrix M is skew. Thus an affine neighbourhood of U in $\mathrm{OGr}(5, 10)$ is given by (I, M) with M a skew 5×5 matrix.

There are 16 standard affine pieces of $\mathrm{OGr}(5, 10)$. Each is obtained from this one by acting on the basis of V by a permutation of $W(D_5)$. That is, take matrices (I, M)

with M skew, and swap evenly many of the first 5 columns with the corresponding columns from the last 5.

The spinor embedding aOGr(5, 10) $\subset S^+$. Corresponding to the different treatment of tensors of rank 2 in $\bigwedge^2 V$ in Section 2, we can write down 4 characterisations of *simple* spinors. For a spinor $s \in S^+$, we have the following equivalent conditions:

(i) Explicit: s is in the $W(D_5)$-orbit of a spinor of the form

$$e(1, M, \operatorname{Pf} M) \quad \text{with } e \in \mathbb{C} \text{ and } M \text{ a skew } 5 \times 5 \text{ matrix.}$$

(ii) Orbit of highest weight vector: s is in the Spin(10)-orbit of $(1, 0, \ldots, 0) \in S^+$.

(iii) Quotient by SL(5): consider all 5×10 matrices N whose rows base a generator $\Pi \in \operatorname{OGr}(5, 10)$ of q, and take the quotient by SL(5) acting by left multiplication. To explain this briefly, suppose that $N = (A, B)$ with nonsingular first 5×5 block A. Then up to the SL(5) action, N is of the form $e(I, M)$. The affine embedding into S^+ is then given by $e(1, M, \operatorname{Pf} M)$. Every other N is of this form up to the action of the Weyl group $W(D_5)$.

(iv) Equations: $s = (e, M, P) \in S^+$ satisfies the 10 equations

$$eP = \operatorname{Pf} M, \quad MP = 0$$

(see below). The first set of 5 equations with $e \neq 0$ describes the embedding of the first affine open in spinor space. On the other hand, as we discuss next, this set of equations is $W(D_5)$-invariant.

Remark 4.1. We use the following point of view in (iii): it is well known that the spinor embedding $\operatorname{OGr} \hookrightarrow \mathbb{P}(S^+)$ is the Veronese square root of the Plücker embedding $\operatorname{OGr} \hookrightarrow \operatorname{Gr}(5, 10) \hookrightarrow \mathbb{P}(\bigwedge^5 \mathbb{C}^{10})$. In other words, up to a straightforward (!) change of coordinates, the set of 5×5 minors of N is the second symmetric power of the set of spinor coordinate functions $e, x_{ij}, \operatorname{Pf}_k$.

Equations of aOGr(5, 10)

As described above, S^+ has a basis indexed by the graph Γ. A pair x_I, x_J is an *edge* of Γ (that is, x_I is joined to x_J) if and only if I and J or I and $\mathcal{C}J$ differ by one element. Because of this definition, edges of Γ fall into 5 sets of 8 parallel edges, with *directions* given by adding the same i: for example, the 8 edges

$$xx_1, \quad x_i x_{1i}, \quad x_{ij} x_{kl} \text{ with } \{i, j, k, l\} = \{2, 3, 4, 5\}$$

are all of the form $x_I x_{I+1}$, so in the 1 direction. Two edges of Γ are *remote* if no edge of Γ joins either end of one to either end of the other. Any two parallel edges either form two sides of a square, or are remote. Each set of 8 parallel edges breaks up into two *remote quads*. For example, the 8 edges in the 1 direction give

$$xx_1, x_{23}x_{45}, x_{24}x_{35}, x_{25}x_{34} \quad \text{and} \quad x_1x_{12}, x_1x_{13}, x_1x_{14}, x_1x_{15},$$

The 10 equations of $OGr(5, 10)$ in (iv) are sums of these quads with appropriate choice of signs:

$$xx_1 - x_{23}x_{45} + x_{24}x_{35} - x_{25}x_{34} = 0 \quad \text{and} \quad x_1x_{12} - x_1x_{13} + x_1x_{14} - x_1x_{15} = 0,$$

and permutations.

The 10 equations *centred at* x are written in terms of the matrices

$$M = \begin{pmatrix} x_{12} & x_{13} & x_{14} & x_{15} \\ & x_{23} & x_{24} & x_{25} \\ & & x_{34} & x_{35} \\ -\text{sym} & & & x_{45} \end{pmatrix} \quad \text{and} \quad \mathbf{v} = \begin{pmatrix} x_1 \\ x_2 \\ x_3 \\ x_4 \\ x_5 \end{pmatrix}.$$

They take the form

$$\begin{pmatrix} N_1 \\ N_2 \\ N_3 \\ N_4 \\ N_5 \end{pmatrix} = x\mathbf{v} - \text{Pf}\, M = 0 \quad \text{and} \quad \begin{pmatrix} N_{-1} \\ N_{-2} \\ N_{-3} \\ N_{-4} \\ N_{-5} \end{pmatrix} = M\mathbf{v} = 0.$$

The 16 first syzygies are also indexed by the 16 vertices of Γ. Each is a 5 term syzygy involving the 5 neighbouring vertices:

	x	x_1	x_2	x_3	x_4	x_5	x_{12}	x_{13}	x_{14}	x_{15}	x_{23}	x_{24}	x_{25}	x_{34}	x_{34}	x_{45}
	0	0	x_{12}	x_{13}	x_{14}	x_{15}	x_2	x_3	x_4	x_5	0	0	0	0	0	0
	0	$-x_{12}$	0	x_{23}	x_{24}	x_{25}	$-x_1$	0	0	0	x_3	x_4	x_5	0	0	0
	0	$-x_{13}$	$-x_{23}$	0	x_{34}	x_{35}	0	$-x_1$	0	0	$-x_2$	0	0	x_4	x_5	0
	0	$-x_{14}$	$-x_{24}$	$-x_{34}$	0	x_{45}	0	0	$-x_1$	0	0	$-x_2$	0	$-x_3$	0	x_5
	0	$-x_{15}$	$-x_{25}$	$-x_{35}$	$-x_{45}$	0	0	0	0	$-x_1$	0	0	$-x_2$	0	$-x_3$	$-x_4$
x_1	x	0	0	0	0	0	0	0	0	0	$-x_{45}$	x_{35}	$-x_{34}$	$-x_{25}$	x_{24}	$-x_{23}$
x_2	0	x	0	0	0	0	0	0	x_{45}	$-x_{35}$	x_{34}	0	0	0	x_{15}	$-x_{14}$	x_{13}
x_3	0	0	x	0	0	0	$-x_{45}$	0	x_{25}	$-x_{24}$	0	$-x_{15}$	x_{14}	0	0	$-x_{12}$
x_4	0	0	0	x	0	x_{35}	$-x_{25}$	0	x_{23}	x_{15}	0	$-x_{13}$	0	x_{12}	0
x_5	0	0	0	0	x	$-x_{34}$	x_{24}	$-x_{23}$	0	$-x_{14}$	x_{13}	0	$-x_{12}$	0	0

The 2nd syzygies are likewise indexed by the 16 monomials; they form a 16×16 symmetric matrix with typical columns

$S(x)$	$S(x_1)$	$S(x_{12})$
x^2	$xx_1 - 2N_1$	xx_{12}
$xx_1 - 2N_1$	x_1^2	$x_1x_{12} + 2N_{-2}$
$xx_2 - 2N_2$	x_1x_2	$x_2x_{12} - 2N_{-1}$
$xx_3 - 2N_3$	x_1x_3	x_3x_{12}
$xx_4 - 2N_4$	x_1x_4	x_4x_{12}
$xx_5 - 2N_5$	x_1x_5	x_5x_{12}
xx_{12}	$x_1x_{12} + 2N_{-2}$	x_{12}^2
xx_{13}	$x_1x_{13} + 2N_{-3}$	$x_{12}x_{13}$
xx_{14}	$x_1x_{14} + 2N_{-4}$	$x_{12}x_{14}$
xx_{15}	$x_1x_{15} + 2N_{-5}$	$x_{12}x_{15}$
xx_{23}	x_1x_{23}	$x_{12}x_{23}$
xx_{24}	x_1x_{24}	$x_{12}x_{24}$
xx_{25}	x_1x_{25}	$x_{12}x_{25}$
xx_{34}	x_1x_{34}	$x_{12}x_{34} + 2N_5$
xx_{35}	x_1x_{35}	$x_{12}x_{35} - 2N_4$
xx_{45}	x_1x_{45}	$x_{12}x_{45} + 2N_3$

Numerology

From this we get the following numerology and representation theory. Write $U = \mathbb{C}^5$ with weights $w_1, \ldots, w_5$, where SO(10) acts on $V = U \oplus U^\vee$. As in Section 3, to ensure that all the weights are positive, we introduce a further overall weight u on S^+. To keep track of this, we introduce the bigger group $G = \mathrm{Spin}(10) \times \mathbb{C}^\times$, and replace S^+ by $S^+ \otimes L$, where Spin(10) acts on the first factor in the usual way, and $\mathbb{C}^\times$ acts on $L = \mathbb{C}$ with weight u. Set

$$S^+ = \bigwedge^{\text{even}} U \otimes L = (\mathbb{C} \oplus \bigwedge^2 U \oplus \bigwedge^4 U) \otimes L$$

for the 16 dimensional spinor space. The only representations we need are V, S^+, its dual S^- and their twists by line bundles. By analogy with Proposition 2.1 and (2.3), we define[1]

$$D = \bigwedge^5 U \otimes L^2, \quad s = \sum w_i = \mathrm{wt} \bigwedge^5 U \quad \text{and} \quad d = \mathrm{wt}\, D = s + 2u,$$

---

[1] As in (2.3–2.4), this move cleans up the formulas below in a most miraculously way. However, to be quite honest, at the time of writing, we have absolutely no idea what representation D is, or for what group it is supposed to be an equivariant line bundle on $\mathbb{C}^{16} = S^+$. Cf. Problem 4.2.

Then the generators have weights
$$\text{wt } x = u,$$
$$\text{wt } x_{ij} = u + w_i + w_j,$$
$$\text{wt } x_i = u + w_j + w_k + w_l + w_m = u + s - w_i.$$

The average of the 16 weights is $\frac{1}{2}d$.

The 10 relations with their representative terms have weights
$$N_i = xx_i + \cdots \mapsto 2u + w_j + w_k + w_l + w_m = d - w_i,$$
$$N_{-i} = \sum x_j x_{ij} \mapsto d + w_i,$$

which have average d. The 16 first syzygies have weights
$$T(x) = \sum x_i N_{-i} \mapsto 2d - u,$$
$$T(x_i) = x N_{-i} + \cdots \mapsto 2d - u - s + w_i,$$
$$T(x_{ij}) = x_i N_j + \cdots \mapsto 2d - u - w_i - w_j,$$

which have average $\frac{3}{2}d$, and the 16 second syzygies
$$S(x) = x^2 T(x) + \cdots \mapsto 2d + u,$$
$$S(x_i) = xx_i T(x) + \cdots \mapsto 2d + u + s - w_i,$$
$$S(x_{ij}) = xx_{ij} T(x) + \cdots \mapsto 2d + u + w_i + w_j,$$

which have average $\frac{5}{2}s$.

To write out the Hilbert series of wOGr(5, 10) with the above weights, introduce the Laurent polynomials
$$Q_V = \sum_i t^{w_i} + \sum_i t^{-w_i}$$
$$Q_{S^+} = 1 + \sum_{i,j} t^{w_i + w_j} + \sum_i t^{s - w_i}$$
$$Q_{S^-} = 1 + \sum_{i,j} t^{-w_i - w_j} + \sum_i t^{-s + w_i}$$

Then wOGr(5, 10) has Hilbert series
$$P(t) = 1 - t^d Q_V + t^{2d-u} Q_{S^-} - t^{2d+u} Q_{S^+} + t^{3d} Q_V - t^{4d}.$$

This numerology implies that the spaces of relations, first syzygies, etc., in the resolution are the following representations of Spin(10):

$$\mathcal{O} \leftarrow V \otimes D \leftarrow S^- \otimes D^2 \otimes L^{-1} \leftarrow S^+ \otimes D^2 \otimes L \leftarrow V \otimes D^3 \leftarrow D^4 \leftarrow 0. \quad (4.1)$$

Providing it is well formed, wOGr(5, 10) has canonical class $K_{\text{wGr}} = \mathcal{O}(-4d)$. In fact the wps $\mathbb{P}(S^+ \otimes L)$ has $K_{\mathbb{P}} = -8d$ (the sum of weights of the coordinates), and by (4.1), the adjunction number of wGr $\subset \mathbb{P}$ equals $4d$.

Problem 4.2. We believe that the affine O'Grassmannian aOGr(5, 10) has an equivariant resolution of the form (4.1), in complete analogy with Proposition 2.1. We have written out the maps in this sequence in explicit coordinate expressions in our treatment, with the right $W(D_5)$ symmetry and weights. It should be possible to specify them intrinsically in terms of Clifford multiplication.

5. Examples

We have searched in vain for examples of Fano 3-folds, K3 surfaces or canonical surfaces as quasilinear sections of wGr(5, 10), and we believe that there are very few, or even none, apart from the well known straight cases. In this section, we construct nice examples of a canonical 3-fold and a Calabi–Yau 3-fold having isolated cyclic quotient singularities.

Example 5.1. Let V be a regular 3-fold of general type with $p_g = 7$, $K^3 = 21$ and $2 \times \frac{1}{2}(1, 1, 1)$ singularities. The plurigenus formula of Fletcher and Reid [YPG] states that

$$p_n = \begin{cases} 1 \\ p_g \\ \frac{n(n-1)(2n-1)}{12} K^3 + (2n-1)(p_g - 1) + l(n) & \text{for } n \geq 2, \end{cases}$$

where $l(n)$ is a sum of the local orbifold contributions

$$l(n) = \begin{cases} \frac{n}{4} & \text{if } n \text{ is even} \\ \frac{n-1}{4} & \text{if } n \text{ is odd} \end{cases}$$

from each of the $\frac{1}{2}(1, 1, 1)$ singularities. One easily calculates the Hilbert function $H(t) = \sum p_n t^n$ from this:

$$H(t) = 1 + t + \frac{t+t^2}{(1-t)^2} \times (p_g - 1) + \frac{t^2 + t^3}{(1-t)^4} \times \frac{K^3}{2} + \frac{1}{4} \times \frac{t^2}{(1-t)(1-t^2)}$$

$$= 1 + 7t + 29t^2 + 83t^3 + 190t^4 + 370t^5 + 645t^6 + 1035t^7 + 1562t^8 + \cdots$$

$$= \frac{1 + 4t + 10t^2 + 12t^3 + 10t^4 + 4t^5 + t^6}{(1-t)^3(1-t^2)}.$$

We need seven generators in degree 1 (since $p_g = 7$) and at least two in degree 2 to accommodate the two $\frac{1}{2}(1, 1, 1)$ singularities; the simplest possibility is that V has codimension 5 in $\mathbb{P}(1^7, 2^2)$, with Hilbert numerator

$$(1-t)^7(1-t^2)^2 H(t) = 1 - t^2 - 8t^3 + 7t^4 + 8t^5 - 8t^7 - \cdots$$

We easily recognise this as the Hilbert numerator of the weighted orthogonal Grassmannian wOGr(5, 10) with weights $\mathbf{w} = (0, 0, 0, 0, 1)$, $u = 1$ and $s = 1$, therefore $d = 3$. With $w_1 = \cdots = w_4 = 0$ and $w_5 = 1$, it follows that x, x_{ij} for $1 \leq i < j \leq 4$ and x_{1234} have weight 1 and anything involving x_5 has weight 2. The spinor embedding takes wOGr(5, 10) into $\mathbb{P}(1^8, 2^8)$ and we construct V as a general quasilinear section

$$V = \text{wOGr}(5, 10) \cap (1) \cap (2)^6.$$

We check that the canonical class adds up: either $V \subset \mathbb{P}(1^7, 2^2)$, with adjunction number $4d = 12$ gives $-7 \times 1 - 2 \times 2 + 12 = 1$, or $V = (1) \cap (2)^6 \subset$ wOGr has $K_V = \mathcal{O}(-4d + 1 + 6 \times 2) = \mathcal{O}(1)$.

Example 5.2. Let V be a Calabi–Yau 3-fold polarised by a divisor A with $A^3 = \frac{6}{5}$ and $A \cdot c_2 = \frac{108}{5}$, and having singular points $P' = \frac{1}{3}(1, 1, 1)$, $P'' = \frac{1}{3}(2, 2, 2)$, and $Q = \frac{1}{5}(3, 3, 4)$ (we are writing these so that $A = \mathcal{O}(1)$).

The orbifold Riemann–Roch formula of Fletcher and Reid [YPG] states that

$$p_n = \frac{A^3}{6} n^3 + \frac{A \cdot c_2}{12} n + c_{P'}(n) + c_{P''}(n) + c_Q(n)$$

where $c_\bullet(n)$ is a local contribution from the singularity that can be calculated explicitly using the instructions in [YPG]. Following the instructions, we discover that $c_{P'}(n) + c_{P''}(n) = 0$ for all n, and

$$c_Q(n) = \begin{cases} 0 & \text{if } n \equiv 0 \bmod 5; \\ 0 & \text{if } n \equiv 1 \bmod 5; \\ \frac{-1}{5} & \text{if } n \equiv 2 \bmod 5; \\ \frac{1}{5} & \text{if } n \equiv 3 \bmod 5; \\ 0 & \text{if } n \equiv 4 \bmod 5. \end{cases}$$

From this it is easy to calculate the Hilbert function

$$H(t) = 1 + \frac{A^3}{6} \times \frac{(1 + 4t + t^2)t}{(1-t)^4} + \frac{A \cdot c_2}{12} \times \frac{t}{(1-t)^2} + \frac{1}{5} \times \frac{-t^2 + t^3}{1 - t^5}$$

$$= \frac{1 - 2t + 3t^2 - t^3 - t^4 + t^5 + t^6 - 3t^7 + 2t^8 - t^9}{(1-t)^4(1-t^5)}$$

$$= 1 + 2t + 5t^2 + 11t^3 + 20t^4 + 34t^5 + 54t^6 + 81t^7 + 117t^8 + \cdots$$

We see that we need to multiply by $(1-t)^2(1-t^2)^2$:

$$(1-t)^2(1-t^2)^2(1-t^5)H(t) = 1 + 3t^3 - 2t^5 + 2t^6 - 3t^8 - t^{11}$$

Then we need three generators in degree 3:

$$(1-t)^2(1-t^2)^2(1-t^3)^3(1-t^5)H(t) = 1 - 2t^5 - 4t^6 + 3t^8 + 2t^9 + 2t^{11} + \cdots$$

At first sight this looks like a plausible 6×10 codimension 4 format; the typical example of this is a nonspecial canonical curve C of genus 6, that is known to be a quadric section of a cone over $\mathrm{Gr}(2, 5)$. We might hope to find V as a nonlinear section of a cone over a weighted $\mathrm{wGr}(2, 5)$. Indeed the polynomial in the last displayed equation is the Hilbert numerator of $\mathcal{C}\,\mathrm{wGr}_{2,6} \cap (6) \subset \mathbb{P}(1, 2^3, 3^6, 4)$. However, this is a mirage of a fairly typical type: although it would have the correct Hilbert function, a quasilinear section of this variety can't have a $\frac{1}{5}(3, 3, 4)$ singularity. The simplest assumption is that V is codimension 5; the easiest guess is that there is an additional generator (and relation) in degree 4, giving

$$(1-t)^2(1-t^2)^2(1-t^3)^3(1-t^4)(1-t^5)H(t)$$
$$= 1 - t^4 - 2t^5 - 4t^6 + 3t^8 + 4t^9 + 4t^{10} + 2t^{11} - 2t^{13} - \cdots$$

We easily recognise this as the Hilbert numerator of the weighted orthogonal Grassmannian wOGr(5, 10) with weights $\mathbf{w} = (0, 0, 1, 1, 2)$, $u = 1$ and $s = 4$, therefore $d = 6$, embedded in $\mathbb{P}(1^2, 2^4, 3^4, 4^4, 5^2)$, with canonical class $\mathcal{O}(-4d) = \mathcal{O}(-24)$. We can construct V as a general quasilinear section

$$V = \mathrm{wOGr}(5, 10) \cap (2)^2 \cap (3) \cap (4)^3 \cap (5)$$

(the calculation that V has the correct singularities is a bit tedious but can be done by hand).

References

[Al] S. Altınok, Graded rings corresponding to polarised K3 surfaces and $\mathbb{Q}$-Fano 3-folds, Univ. of Warwick Ph.D. thesis, Sep. 1998, vii+93 pp., available from www.maths.warwick.ac.uk/~miles/doctors/Selma.

[ABR] S. Altınok, G. Brown and M. Reid, Fano 3-folds, K3 surfaces and graded rings, in: Singapore International Symposium in Topology and Geometry (A. J. Berrick, M. C. Leung and X. W. Xu, eds.), Singapore 2001, Contemp. Math., Amer. Math. Soc., to appear, math.AG/0202092.

[FH] W. Fulton and J. Harris, Representation theory. A first course, Grad. Texts in Math. 129, Springer-Verlag, New York 1991.

[Fl] A. R. Iano-Fletcher, Working with weighted complete intersections, in: Explicit birational geometry of 3-folds (Alessio Corti and Miles Reid, eds.), London Math. Soc. Lecture Note Ser. 281, Cambridge University Press., Cambridge 2000, 101–173.

[G] Ian Grojnowski, Weighted flag varieties, get from www.maths.warwick.ac.uk/~miles/3folds/wfv.ps.

[La] Alain Lascoux, Syzygies des variétés déterminantales, Adv. Math. 30 (1978), 202–237.

[Mu] MUKAI Shigeru, Curves and symmetric spaces. I, Amer. J. Math. 117 (1995), 1627–1644.

[N] J. Neves, A note on regular surfaces of general type with $K^2 = 13$ and $p_g = 6$, work in progress.

[PR] Stavros Papadakis and Miles Reid, Kustin–Miller unprojection without complexes, J. Algebraic Geom., to appear, preprint math.AG/0011094.

[R] Miles Reid, The complete intersection of two or more quadrics, Cambridge Ph.D. thesis, June 1972, 84 pp., www.maths.warwick.ac.uk/~miles/3folds/qu.ps.

[YPG]   Miles Reid, Young person's guide to canonical singularities, in: Algebraic geometry, Bowdoin, 1985 (Brunswick, Maine, 1985), Proc. Sympos. Pure Math. 46, Part 1, Amer. Math. Soc., Providence, RI, 1987, 345–414.

[Ki]   Miles Reid, Graded rings and birational geometry, in: Proceedings of algebraic geometry symposium (K. Ohno, ed.) , Kinosaki 2000, 1–72; also available from www.maths.warwick.ac.uk/~miles/3folds/Ki/Ki.ps.

A. Corti, DPMMS, University of Cambridge, Centre for Mathematical Sciences, Wilberforce Road, Cambridge CB3 0WB, England
E-mail: a.corti@dpmms.cam.ac.uk
Web: can.dpmms.cam.ac.uk/~corti

M. Reid, Mathematical Institute, University of Warwick, Coventry CV4 7AL, England
E-mail: miles@maths.warwick.ac.uk
Web: www.maths.warwick.ac.uk/~miles

Resolution of indeterminacy of pairs

*Tommaso de Fernex and Lawrence Ein**

In memory of Paolo Francia

Introduction

Suppose that G is a finite subgroup of the group of birational transformations $\mathrm{Bir}(Y)$ of a projective variety Y. Then a resolution of indeterminacy of the pair (Y, G) consists of a smooth variety X, birationally equivalent to Y, and a birational map $\phi : X \dashrightarrow Y$ such that for every $\tau \in G$ the composite map $\phi^{-1}\tau\phi$ is an automorphism of X.

One motivation for finding resolutions is the study of the group $\mathrm{Bir}(Y)$ itself. The general philosophy is that, by resolving the indeterminacy, we reduce to study isomorphisms of smooth varieties. See for instance [1] and [5], where this process of resolution is applied to classify cyclic subgroups of $\mathrm{Bir}(\mathbb{P}^2)$ up to conjugation.

The first non-trivial example of resolution of indeterminacy of pairs occurs when resolving the fundamental locus of a birational involution τ of a smooth surface by a minimal sequence of monoidal transformations $f : X \to Y$. Indeed one can show that in this case $f^{-1}\tau f \in \mathrm{Aut}(X)$. On the other hand, if for instance we consider birational transformations τ of any order greater that two, then in general this is not true, even in dimension two. Clearly the picture becomes more complicated if we consider non-cyclic groups or increase the dimension.

One can see that resolutions of indeterminacy of pairs (Y, G) always exist. This is probably well known to the specialists. We give an elementary proof of this basic result in Section 1.

In this paper, we study the two dimensional case in detail. In Section 2 we formalize the terminology concerning infinitely near points in the language of algebraic valuations. We use a theorem of Zariski on algebraic valuations to define a topology on the set of algebraic valuations of a given smooth projective surface, and establish a correspondence between birational morphisms of smooth surfaces and finite closed subsets of algebraic valuations. In Section 3, we apply these results to show that a minimal resolution of a pair (Y, G) exists, when Y is a smooth projective surface.

---

*Research of first author partially supported by MURST of Italian Government, National Research Project (Cofin 2000) "Geometry on Algebraic Varieties". Research of second author partially supported by NSF Grant DMS 99-70295.

In the last section we introduce a birational invariant of each subgroup of prime order of Bir(Y) when Y is a projective surface. We use this invariant to distinguish whether two subgroups of the same prime order of Bir($\mathbb{P}^2$) are conjugate.

We would like to mention Cheltsov's paper [4], where an explicit construction of resolution of indeterminacy of pairs (called there *regularization*) is given in dimension three using the Minimal Model Program.

We are very grateful to Professor A. Lanteri for precious suggestions. We would like to thank also Professor I. Dolgachev for useful conversations.

Throughout this paper, all varieties are assumed to be defined over an algebraically closed field k of characteristic zero.

1. The resolution of pairs in its general form

Let Y be a projective variety defined over an algebraically closed field k of characteristic zero, and let G be a group.

Definition 1.1. We say that G *acts birationally* on Y if a homomorphism $\eta : G \to$ Bir(Y) is assigned; G is said to *act faithfully* if η is injective. The action of G on Y is said *biregular* if $\eta(G) \subset$ Aut(Y).

We fix a group G, and consider pairs (Y, G), where Y is a projective variety and G acts birationally on Y. We will always assume that the action of G is faithful, and look at G as a subgroup of Bir(Y).

Definition 1.2. Let (X, G) and (Y, G) two pairs. A dominant rational map $\phi : X \dashrightarrow Y$ is said to be a *G-equivariant rational map*, or *rational map of pairs*, if the birational actions of G on X and Y commute with ϕ. We also write $\phi : (X, G) \dashrightarrow (Y, G)$. Two pairs (X, G) and (Y, G) are said to be *birationally equivalent* if there is a birational map of pairs $\phi : (X, G) \dashrightarrow (Y, G)$. In this case, we write $(X, G) \sim_\phi (Y, G)$.

Definition 1.3. Given a pair (Y, G), a second pair (X, G) is called *resolution of indeterminacy of* (Y, G) if the following three conditions are satisfied:

(i) X is a smooth projective variety,

(ii) there is a birational map $\phi : X \dashrightarrow Y$ such that $(X, G) \sim_\phi (Y, G)$,

(iii) $G \subset$ Aut(X) inside Bir(X).

We also say that ϕ *resolves* the indeterminacy of (Y, G).

Theorem 1.4 (Existence of resolution – General form). *Let Y be a variety and G be a finite group acting birationally on Y. Then there exists a smooth projective variety X and a birational map $\phi : X \dashrightarrow Y$ which resolves the indeterminacy of (Y, G).*

Proof. Choose W to be a smooth projective variety such that $K(W) = K(Y)^G$. Let Z be the normalization of W in the function field $K(Y)$. Now observe that G acts biregularly on Z. Then a G-equivariant resolution of the singularities of Z (cf. [8]) gives a resolution of indeterminacy (X, G) of (Y, G). $\square$

2. Algebraic valuations and birational morphisms

Throughout this section Y will denote a smooth projective surface over an algebraically closed field k of characteristic zero.

Since M. Noether, mathematicians realized the importance of understanding the geometry of a surface Y at the level of infinitely near points in order to study rational maps upon it. Classically, an *infinitely near point* of Y is a (reduced closed) point lying on some exceptional divisor over Y. For our purpose, it is more convenient to describe an infinitely near point as the discrete valuation along the exceptional divisor obtained by blowing up such point. So, we recall the following definition:

Definition 2.1. An *algebraic valuation ring* R of Y is a discrete valuation ring in $K(Y)$ which is determined, through a birational map $\phi : X \dashrightarrow Y$, by the local ring of an irreducible divisor E of a smooth projective surface X. An *algebraic valuation* v of Y is the discrete valuation determined by an algebraic valuation ring R of Y. The center on Y of v is the image of the closed point of Spec R in Y. The set of algebraic valuations of Y is denoted by $\mathrm{Val}(Y)$. We denote by $\mathrm{Val}_0(Y)$ the set of valuations along irreducible divisors of Y, and set $\mathrm{Val}_+(Y) := \mathrm{Val}(Y) - \mathrm{Val}_0(Y)$.[1]

In this section we will introduce a natural topology on the set $\mathrm{Val}(Y)$ of algebraic valuations of Y. This is probably well known to the experts, but we would like to set up some standard notation which will be used in Section 3.

An important consequence in dimension two of Zariski's Theorem [7, Lemma 2.45] is the following property:

Lemma 2.2. *For any algebraic valuation $v \in \mathrm{Val}_+(Y)$ of a smooth surface Y there is a unique minimal sequence of blowups of centers of v such that v is the valuation along the exceptional divisor E of the last blow up of the sequence.*

Definition 2.3. We will refer to the minimal sequence of blowups mentioned in Lemma 2.2 as the *minimal extraction* of v.

Lemma 2.2 allows us to define a partial ordering in $\mathrm{Val}_+(Y)$.

Definition 2.4. Let $v, w \in \mathrm{Val}_+(Y)$. We put $v \geq w$, and say that v *dominates* w, if w is the valuation along one of the prime exceptional divisors occurring in the minimal

---
[1] In [7, p. 50], algebraic valuations of a variety Y are defined as the valuations along irreducible exceptional divisors over Y. These correspond to the elements in $\mathrm{Val}_+(Y)$ in our notation.

extraction of v. The *level* (over Y) of v is defined as the number i of valuations $w \in \text{Val}_+(Y)$ dominated by v (in other words, i is the number of prime exceptional divisors occurring in the minimal extraction of v).

Remark 2.5. The level 1 elements of $\text{Val}_+(Y)$ are precisely the minimal elements with respect to the partial order defined above.

If we denote by $\text{Val}_i(Y)$ the set of valuations of Y of level i, then

$$\text{Val}(Y) = \text{Val}_0(Y) \sqcup \text{Val}_+(Y) = \sqcup_{i \geq 0} \text{Val}_i(Y).$$

Definition 2.6. Let $V \subset \text{Val}_+(Y)$ be a subset. The *support* of V in Y is the union of the centers on Y of the elements of V. An element v of V is said to be *maximal* in V if it is not dominated by any other element of V.

We can define a topology on $\text{Val}_+(Y)$ by choosing as closed sets the subsets $V \subset \text{Val}_+(Y)$ which satisfy the following condition: if $v \in V$ and $w < v$, then $w \in V$. It is immediate to verify indeed that

Lemma 2.7. *The closed subsets of* $\text{Val}_+(Y)$, *as defined above, satisfy the axioms for a topology on* $\text{Val}_+(Y)$.

Note that, in this topology, if V is a closed set in $\text{Val}_+(Y)$ containing all maximal elements of another set $W \subset \text{Val}_+(Y)$, then $V \supset W$. Note also that the support of a closed set $V \subset \text{Val}_+(Y)$ is the union of the centers on Y of the elements of level 1 of V.

Remark 2.8. The topology of $\text{Val}_+(Y)$ extends to a topology of $\text{Val}(Y)$ by choosing as closed subsets the sets of the form $\text{Val}_0(Y) \sqcup V$, where V is closed in $\text{Val}_+(Y)$, and the empty set. See also [9] for other natural topology on the set of valuations.

Let $\mathcal{V}$ be the category whose objects are finite closed subsets $V \subset \text{Val}_+(Y)$ and whose morphisms are inclusions of sets. Let $\mathcal{B}$ be the category of smooth projective Y-surfaces birational to Y. An object in $\mathcal{B}$ is a birational morphism $f : X \to Y$, where X is a smooth projective surface, and, for $f, f' \in \text{Obj}(\mathcal{B})$, a morphism $\alpha : f \to f'$ is a commutative diagram

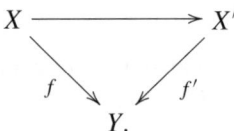

Then we define

$$F : \mathcal{V} \to \mathcal{B}$$

by associating to any $V \in \text{Obj}(\mathcal{V})$ the morphism

$$f_V = f_h \cdots f_0 : X := Y_{h+1} \to Y_0 := Y,$$

where $f_i : Y_i \to Y_{i-1}$ is recursively defined as the blow up of the centers on Y_{i-1} of the elements of V of level i. Conversely, consider

$$G : \mathcal{B} \to \mathcal{V}$$

which associates to any $f \in \mathrm{Obj}(\mathcal{B})$ the set $V(f)$ of valuations of the prime exceptional divisors of f.

Theorem 2.9. *F and G establish an equivalence of categories $\mathcal{V} \cong \mathcal{B}$.*

Proof. It is immediate to verify that F and G are well defined contravariant functors, and that each one is the inverse of the other. □

We observe that, if $\phi : X \dashrightarrow Y$ is a birational map of smooth projective surfaces, then ϕ induces a bijection $\phi_* : \mathrm{Val}(X) \to \mathrm{Val}(Y)$. Moreover, if T is another smooth projective surface and $\psi : Y \dashrightarrow T$ is a birational map, then $\psi_* \phi_* = (\psi \phi)_*$ (as a bijection $\mathrm{Val}(X) \to \mathrm{Val}(T)$). In particular, we see that $(\phi_*)^{-1} = (\phi^{-1})_* =: \phi_*^{-1}$ (as a bijection $\mathrm{Val}(Y) \to \mathrm{Val}(X)$).

Now we consider a birational morphism of smooth surfaces

$$f : X \to Y.$$

Let $V(f) \subset \mathrm{Val}_+(Y)$ be the associated closed set. We introduce the following notation:

$$V(Y, f) := \mathrm{Val}_0(Y) \sqcup V(f).$$

We observe that

$$f_*^{-1} V(Y, f) = \mathrm{Val}_0(X).$$

In other words, $V(Y, f)$ is the set of algebraic valuations of Y which correspond, through f_*^{-1}, to valuations of X along effective divisors (on X). Note that this set depends on the particular morphism f from X to Y. We have the following characterization of resolution of birational transformations:

Proposition 2.10. *Let f, X and Y as above, and let $\tau \in \mathrm{Bir}(Y)$. Then $f^{-1} \tau f \in \mathrm{Aut}(X)$ if and only if*

$$\tau_*^{-1} V(Y, f) = V(Y, f).$$

Before proving this proposition, we investigate some properties, concerning resolutions of birational maps of surfaces, which follow from Theorem 2.9. Let $\phi : X \dashrightarrow Y$ be a birational map of smooth projective surfaces, and let

(∗)

be a resolution of indeterminacy of ϕ.

Lemma 2.11. (1) *With the above notation,*
$$V(X, p) = \phi_*^{-1} V(Y, q) \supset \phi_*^{-1} \mathrm{Val}_0(Y).$$

(2) *If $E \subset Z$ is an irreducible divisor and $v_E \in \mathrm{Val}(X)$ and $w_E \in \mathrm{Val}(Y)$ are the respective valuations determined by E, then $\phi_*(v_E) = w_E$. Moreover, if E is both p-exceptional and q-exceptional, then v_E is maximal in $V(p)$ if and only if w_E is maximal in $V(q)$.*

Proof. The first part follows from
$$p_*^{-1} V(X, p) = \mathrm{Val}_0(Z) = q_*^{-1} V(Y, q).$$
Regarding the second part, we see that $w_E(h) = v_E(\phi^*(h))$ for every function $h \in K(Y)$. The last assertion follows by observing that the valuation along E is maximal in either direction if and only if E is a (-1)-curve on Z. $\square$

Now we review the definition of minimal resolution of indeterminacy:

Definition 2.12. Diagram $(*)$ is a *minimal resolution* (of indeterminacy) of ϕ if there are no maximal elements v of $V(p)$ such that $\phi_*(v)$ is a maximal element of $V(q)$.

This definition clearly coincides with the usual notion of minimal resolution.

Lemma 2.13. *Consider the resolution given by the commutative diagram $(*)$.*

(1) *The diagram is a minimal resolution of ϕ if and only if it is a minimal resolution of ϕ^{-1}.*

(2) *The resolution is minimal if and only if*
$$V(X, p) = \overline{\phi_*^{-1} \mathrm{Val}_0(Y)}.$$
In particular, the minimal resolution of ϕ exists and is unique.

(3) *If the resolution is minimal, then every maximal element v of $V(p)$ is in $\phi_*^{-1} \mathrm{Val}_0(Y)$.*

Proof. The definition of minimal resolution is symmetric by Lemma 2.11, thus (1) follows. To prove (2), assume that $V(X, p) \ne \overline{\phi_*^{-1} \mathrm{Val}_0(Y)}$. Note that
$$\mathrm{Val}_0(X) \subset \overline{\phi_*^{-1} \mathrm{Val}_0(Y)} \subset V(X, p).$$
Thus we can find a maximal element v of $V(p)$ such that $\phi_*(v) \in V(q)$. Then, by Lemma 2.11 (2), $\phi_*(v)$ is maximal in $V(q)$, thus the resolution is not minimal. Note that the morphism p which gives the minimal resolution of ϕ is uniquely determined by the condition in (2). Now, (3) follows directly from (2) and the way the topology is defined. $\square$

The following lemma characterizes morphisms and isomorphisms among birational maps of smooth projective surfaces.

Lemma 2.14. $\mathrm{Val}_0(X) \supset \phi_*^{-1} \mathrm{Val}_0(Y)$ *if and only if ϕ is a morphism, and equality holds if and only if ϕ is an isomorphism.*

Proof. One direction is clear, so assume that $\mathrm{Val}_0(X) \supset \phi_*^{-1} \mathrm{Val}_0(Y)$. Then
$$\mathrm{Val}_0(X) = \overline{\phi_*^{-1} \mathrm{Val}_0(Y)}.$$
We conclude that ϕ is a morphism by Lemma 2.13 (2). The last statement follows by the same argument applied to ϕ^{-1}. □

Proof of Proposition 2.10. Since $f_*^{-1} V(Y, f) = \mathrm{Val}_0(X)$, the assertion follows from Lemma 2.14 applied to $\phi := f^{-1}\tau f$. □

3. Explicit resolution of pairs in dimension two

Let G be a finite group acting birationally on a smooth projective surface Y. For every $\tau \in G$ we consider the minimal resolution

$$\begin{array}{ccc}
 & Z_\tau & \\
{}^{p_\tau}\swarrow & & \searrow{}^{q_\tau} \\
Y \dashrightarrow^{\tau} & & Y
\end{array}$$

of the indeterminacy of τ. Let $V(p_\tau)$ be the closed subset of $\mathrm{Val}_+(Y)$ associated to p_τ by Theorem 2.9.

Definition 3.1. The subset of $\mathrm{Val}_+(Y)$ given by
$$V_G := \bigcup_{\tau \in G} V(p_\tau).$$
is called *set of indeterminacy of the pair* (Y, G). We also set
$$V(Y, G) := \mathrm{Val}_0(Y) \sqcup V_G.$$

Theorem 3.2 (Construction of resolution). *Let Y, G and V_G be as above. Let $f : X \to Y$ be the birational morphism associated to V_G by Theorem 2.9. Then f resolves the indeterminacy of (Y, G). In other words, G acts biregularly on X via f.*

Proof. Note that $V(Y, f) = V(Y, G)$. By Proposition 2.10, we need to prove that, for every $\sigma \in G$,
$$\sigma_*^{-1} V(Y, G) = V(Y, G).$$

Since $(\sigma_*^{-1})^{-1} = (\sigma^{-1})_*^{-1}$, it is enough to show only one inclusion. In fact, since
$$V(Y, G) = \bigcup_{\tau \in G} V(Y, p_\tau),$$
we reduce to show that, for every $\sigma, \tau \in G$,
$$\sigma_*^{-1} V(Y, p_\tau) \subset V(Y, G).$$
We consider the birational morphism $g : Z \to Y$ associated to the closed set
$$V(p_\sigma) \cup V(p_{\tau\sigma}).$$
We observe that g resolves the indeterminacy of both σ and $\tau\sigma$. We have the following commutative diagram

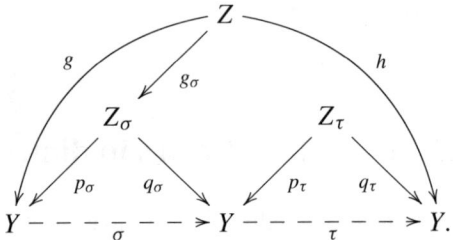

Then, by the universal property applied to the minimal resolution of τ, there is a morphism $h_\tau : Z \to Z_\tau$ such that $h = q_\tau h_\tau$. Now, Lemma 2.11 (1) applied to the diagram

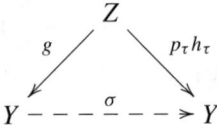

gives
$$\sigma_*^{-1} V(Y, p_\tau h_\tau) \subset V(Y, g).$$
Then, observing that $V(Y, p_\tau) \subset V(Y, p_\tau h_\tau)$ and recalling that $V(Y, g) = V(Y, p_\sigma) \cup V(Y, p_{\tau\sigma})$, we conclude that
$$\sigma_*^{-1} V(Y, p_\tau) \subset \sigma_*^{-1} V(Y, p_\tau h_\tau) \subset V(Y, G).$$
Therefore the theorem is proved. □

The following theorem implies that the resolution constructed in Theorem 3.2 is minimal among all resolutions of indeterminacy of pairs determined by birational morphisms.

Theorem 3.3 (Universal property of resolution of pairs). *In the notation of Theorem 3.2, assume that there is another smooth projective surface X' and a birational* morphism $f' : X' \to Y$ which resolves the indeterminacy of (Y, G). Then f' factors through f.

Proof. All $\tau \in G$ lift to automorphisms of X'. Then, for every $\tau \in G$, $V(p_\tau)$ is contained in $V(f')$, hence f' factors through f. □

The following theorem follows directly from the existence of equivariant resolution of singularities. On the other hand, it is interesting to observe that, for finite groups acting on surfaces, equivariant resolution of singularities follows from Theorem 3.2 (and, of course, usual resolution of singularities).

Theorem 3.4 (Strong equivariant factorization). *Let G be a finite group acting biregularly on two smooth projective surfaces X and Y, and $\phi : X \dashrightarrow Y$ be a G-equivariant birational map. Then there exists a smooth projective surface Z, a biregular action of G on Z, and G-equivariant birational morphisms p and q, giving the following commutative diagram of pairs:*

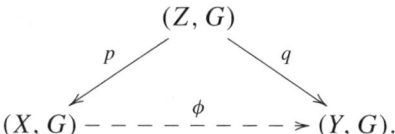

Proof. Let Z_0 be the closure of the graph of ϕ inside $X \times Y$. Since ϕ is G-equivariant, the componentwise action of G on $X \times Y$ induces a biregular action on Z_0. Then the theorem follows by taking a G-equivariant resolution of singularities of Z_0. □

Remark 3.5. Assume that, in the set up introduced at the beginning of the section, G is a cyclic group, and let τ be a generator of G. Then, in order to construct the minimal resolution of (Y, G), we can follow an explicit procedure that involves recursive resolutions and lifts of the generator τ. We start with $Y_0 := Y$ and $\tau_0 := \tau$, and then define recursively:

(1) $f_i : Y_i \to Y_{i-1}$ to be the morphism giving the minimal resolution of indeterminacy of τ_{i-1}, and

(2) $\tau_i \in \mathrm{Bir}(Y_i)$ to be the lift of $\tau_{i-1} \in \mathrm{Bir}(Y_{i-1})$ to Y_i via f_i.

One can see that the composition $g_i := f_1 \cdots f_i : Y_i \to Y_0$ determines resolutions of indeterminacy for $\tau, \tau^2 \ldots, \tau^i$. On the other hand, if Z is a smooth surface and $h : Z \to Y$ is a birational morphism which determines resolutions of indeterminacy for both τ and τ^{-1}, then it follows, by the universal property of the blow up for smooth surfaces, that τ lifts to an automorphism on Z via h. Therefore, if n denotes the order of G, the above sequence of recursive resolutions stops after a number $l \leq n-1$ of steps, producing a surface Y_l and an automorphism $\tau_l \in \mathrm{Aut}(Y_l)$, so that (Y_l, τ_l) is a resolution of indeterminacy of (Y, τ). In fact, it turns out that this is the minimal resolution, i.e., $Y_l = X$ in the notation of Theorem 3.2.

Note that, if τ and τ^{-1} have the same set of indeterminacy, this process stops at the first step, giving $Y_1 = X$. However, in general the resolution of the indeterminacy of the generator of the group does not produce a resolution of indeterminacy of the pair. For example, consider the pair $(\mathbb{P}^2, \tau)$, where τ is the birational transformation of order 5 given, in coordinates (x, y, z) of $\mathbb{P}^2$, by

$$\tau : (x, y, z) \to (x(z-y), z(x-y), xz). \qquad (\dagger)$$

In this case, one can check that $V(p_\tau) \not\supset V(p_{\tau^2})$, thus the resolution of indeterminacy of τ would not resolve the birational action of the group generated by τ.

4. A birational invariant of pairs

Throughout this section (X, σ) and (Y, τ) will denote two pairs consisting respectively of smooth projective surfaces X and Y, defined over an algebraically closed field k of characteristic 0, and automorphisms of finite order σ and τ of X and Y, respectively.

Definition 4.1. A birational morphism of pairs $f : (X, \sigma) \to (Y, \tau)$ is said to be a *minimal equivariant blow up* if any times it is written as a composition $f = f_1 f_2$ of two equivariant birational morphisms f_i, then either f_1 or f_2 (but not both) is an isomorphism.

Lemma 4.2. *Let $f : (X, \sigma) \to (Y, \tau)$ be a minimal equivariant blow up. Assume that the order of σ (and thus of τ) is finite. Then f is the blow up of Y along the τ-orbit of a point q. Moreover, if $g : (X, \sigma) \to (Y, \tau)$ is a birational morphism of pairs, then g factors as a composition of minimal equivariant blowups (and an automorphism).*

Proof. Let $q \in Y$ be a point where f^{-1} is not defined. Since f is equivariant, f^{-1} is not defined at $\tau^k q$ for all $k = 1, \ldots, n-1$, where n is the order of τ. Then, by the universal property of the blow up for smooth surfaces, f factors through the blow up $h : \mathrm{Bl}_\Sigma Y \to Y$, where Σ is the τ-orbit of q. $\square$

We can attach to a pair (X, σ) a number which is birational invariant modulo the order of σ. Let $r_i(X, \sigma)$ be the number of connected components of dimension i (for $i = 0, 1$) of the locus of points of X which are fixed by σ.

Proposition 4.3. *Assume that (X, σ) and (Y, τ) are birational equivalent pairs. Assume moreover that the order of σ and τ is a prime number p. Then*

$$\rho(X) - r_0(X, \sigma) - 2r_1(X, \sigma) \equiv \rho(Y) - r_0(Y, \tau) - 2r_1(Y, \tau) \qquad (\ddagger)$$

modulo p, where $\rho(X)$ and $\rho(Y)$ are the ranks of the respective Neron–Severi groups.

Proof. By Theorem 3.4 and Lemma 4.2, we can reduce to the case when (X, σ) is a minimal equivariant blow up of (Y, τ). Let $f : (X, \sigma) \to (Y, \tau)$ denote the blowing up.

If f is the blow up of Y along an orbit consisting of p distinct points, then $\rho(X) = \rho(Y) + p$ and $r_i(X, \sigma) = r_i(Y, \tau)$ for $i = 1, 2$. Thus ($\ddagger$) is satisfied.

Otherwise f is the blow up of Y at a fixed point q. In this case $\rho(Y) = \rho(X) + 1$. Note that the exceptional curve $E = f^{-1}(q)$ is σ-invariant. If $\sigma|_E \neq 1_E$, then there are two fixed points contained in E, which correspond to the two distinct eigenspaces of the action of τ on the tangent space of Y at q. Then, independently of the fact that q is an isolated fixed point or is contained in a fixed curve, we have

$$r_0(X, \sigma) = r_0(Y, \tau) + 1 \quad \text{and} \quad r_1(X, \sigma) = r_1(Y, \tau).$$

Suppose now that $\sigma|_E = 1_E$. In the complete local ring $\widehat{O}_{Y,q}$, the action of τ can be linearized [3, Lemma 2]. Hence, after been linearized, the action is given by $\lambda \cdot \text{Id}$, for some $\lambda \neq 1$. Therefore q is an isolated fixed point, and

$$r_0(X, \sigma) = r_0(Y, \tau) - 1 \quad \text{and} \quad r_i(X, \sigma) = r_i(Y, \tau) + 1.$$

We see then that in both cases ($\ddagger$) is satisfied. $\square$

We can apply this invariant to distinguish conjugacy classes of given birational transformations. For instance, we can consider the birational transformations occurring in the classification, up to conjugation, of the elements of prime order in $\text{Bir}(\mathbb{P}^2)$ (cf. [2], [6], [1], [5]). The previous method, used to distinguish the conjugacy classes of elements of $\text{Bir}(\mathbb{P}^2)$ is based on the consideration of the geometric genus of the fixed curve. Consider the following examples of birational transformations of $\mathbb{P}^2$:

E1. Let $\tau \in \text{Bir}(\mathbb{P}^2)$ be the transformation induced by the order 3 Galois automorphism σ of the cyclic covering given by a cubic $X \subset \mathbb{P}^3$ over $\mathbb{P}^2$.

E2. Let $\tau \in \text{Bir}(\mathbb{P}^2)$ be the transformation induced by the order 3 Galois automorphism σ of the cyclic covering given by a sextic $X \subset \mathbb{P}(1, 1, 2, 3)$ of equation of the form $z^3 = F(x, y, w)$ over $\mathbb{P}(1, 1, 3)$. In other words, σ is given by the diagonal action on the coordinates sending $(x, y, z, w) \to (x, y, e^{2\pi i/3} z, w)$.

E3. Let $\tau \in \text{Bir}(\mathbb{P}^2)$ be the transformation induced by the order 5 Galois automorphism σ of the cyclic covering given by a sextic $X \subset \mathbb{P}(1, 1, 2, 3)$ of equation of the form $xy^5 = F(x, z, w)$ over $\mathbb{P}(1, 2, 3)$. In other words, σ is given by the diagonal action on the coordinates sending $(x, y, z, w) \to (x, e^{2\pi i/5} y, z, w)$.

We refer to [5] for more details concerning these transformations. In E2 and E3, (x, y, z, w) denote the weighted coordinates of $\mathbb{P}(1, 1, 2, 3)$, and all cyclic coverings mentioned in these examples are the ones induced by the obvious linear projections of the ambient spaces. The pairs (X, σ) coincide with cases A1–A3 of [5, Theorem A].

Proposition 4.4. *All three examples E1–E3 are not birational equivalent to linear automorphisms of $\mathbb{P}^2$. Moreover, E1 is not birational equivalent to E2.*

Proof. By considering the corresponding eigenspaces, we see that, if α is any linear automorphism of finite order of $\mathbb{P}^2$, then

$$\rho(\mathbb{P}^2) - r_0(\mathbb{P}^2, \alpha) - 2r_1(\mathbb{P}^2, \alpha) = -2.$$

On the other hand, if τ is as in one of examples E1–E3, then the pair $(\mathbb{P}^2, \tau)$ is birational equivalent to the corresponding pair (X, σ). Then we see that

$$\rho(X) - r_0(X, \sigma) - 2r_1(X, \sigma) = 5 \not\equiv -2 \quad \text{mod } 3 \quad \text{for case E1}$$
$$\rho(X) - r_0(X, \sigma) - 2r_1(X, \sigma) = 6 \not\equiv -2 \quad \text{mod } 3 \quad \text{for case E2}$$
$$\rho(X) - r_0(X, \sigma) - 2r_1(X, \sigma) = 6 \not\equiv -2 \quad \text{mod } 5 \quad \text{for case E3}.$$

Therefore these transformations can not be birationally equivalent to any linear automorphism α. Moreover, $5 \not\equiv 6$ modulo 3 shows that the two transformations in examples E1 and E2 are not birationally equivalent. □

Remark 4.5. Unfortunately, both the above method and the method of considering the genus of the fixed curve are not sufficient to determine whether the birational transformation defined by (†) in Remark 3.5 is conjugate to a linear automorphism of $\mathbb{P}^2$.

References

[1] L. Bayle and A. Beauville, Birational involutions of $\mathbb{P}^2$, Asian J. Math. 4 (2000), 11–18.

[2] E. Bertini, Ricerche sulle trasformazioni univoche involutorie nel piano, Annali di Mat. 8 (1877), 244–286.

[3] H. Cartan, Quotient d'un espace analytique par un groupe d'automorphismes, in: Algebraic geometry and topology (R. H. Fox, D. C. Spencer and A. W. Tucker, eds.), A symposium in honor of S. Lefschetz, Princeton Math. Ser. 12, Princeton University Press, Princeton, NJ, 1957, 90–102.

[4] I. Cheltsov, Regularization of birational automorphisms, Preprint, 1999.

[5] T. de Fernex, Birational transformations of prime order of the projective plane, Preprint, 2001.

[6] S. Kantor, Theorie der endlichen Gruppen von eindeutigen Transformationen in der Ebene, Mayer & Müller, Berlin 1895.

[7] J. Kollár and S. Mori, Birational geometry of algebraic varieties (with the collaboration of C. H. Clemens and A. Corti), translated from the 1998 Japanese original, Cambridge Tracts in Math. 134, Cambridge University Press, Cambridge 1998.

[8] Z. Reichstein and B. Youssin, Essential dimensions of algebraic groups and a resolution theorem for G-varieties (with an appendix by János Kollár and Endre Szabó, Fixed points of group actions and rational maps), Canad. J. Math. 52 (2000), 1018–1056.

[9] O. Zariski and P. Samuel, Commutative algebra I, II, Van Nostrand, Princeton 1958, 1960.

T. de Fernex, University of Illinois at Chicago, Department of Mathematics, Statistics and Computer Science, 851 South Morgan, Chicago, IL 60607-7045, U.S.A.

E-mail: defernex@math.uic.edu

L. Ein, University of Illinois at Chicago, Department of Mathematics, Statistics and Computer Science, 851 South Morgan, Chicago, IL 60607-7045, U.S.A.

E-mail: ein@uic.edu

The Chow motive of the Godeaux surface

*Vladimir Guletskiĭ and Claudio Pedrini**

Abstract. Let X be the Godeaux surface obtained as a quotient of the Fermat quintic in $\mathbb{P}^3_{\mathbb{C}}$ under the appropriate action of $\mathbb{Z}/5$. We show that its Chow motive $h(X)$ splits as $1 \oplus 9\mathbb{L} \oplus \mathbb{L}^2$ where $\mathbb{L}$ is the Lefschetz motive. This provides a purely motivic proof of the Bloch conjecture for X, which is a surface of general type. Our results also give a motivic proof of the Bloch conjecture for those surfaces considered in [BKL], i.e. all surfaces with $p_g = 0$ which are not of general type.

1. Introduction

Let $\mathcal{V}$ be the category of smooth projective varieties over a field k. For an object $X \in \mathrm{Ob}(\mathcal{V})$ let $A^i(X) = CH^i(X) \otimes \mathbb{Q}$ be the Chow-group of codimension i cycles with $\mathbb{Q}$-coefficients on X. For any $X, Y \in \mathrm{Ob}(\mathcal{V})$ the group $A^0(X, Y)$ of correspondences (of degree zero) from X into Y is, by definition, the direct sum $\oplus_j A^{d_j}(X_j \times Y)$ where X_j are the irreducible components of X and $d_j = \dim(X_j)$ for any j. If $f \in A^0(X, Y)$ and $g \in A^0(Y, Z)$ for three varieties X, Y and Z then the composition $g \circ f \in A^0(X, Z)$ can be defined by the usual formula $g \circ f = p_{XZ*}(p^*_{XY}(f) \cdot p^*_{YZ}(g))$ where p_{XZ}, p_{XY} and p_{YZ} are the appropriate projections. Varieties and correspondences form an additive category. Its pseudoabelian envelope $\mathcal{M}$ is called the *category of Chow motives* over k with coefficients in $\mathbb{Q}$. Objects in $\mathcal{M}$ are pairs (X, p) where $X \in \mathrm{Ob}(\mathcal{V})$ and p is a projector of X, that is a correspondence $p \in A^0(X, X)$, such that $p \circ p = p$. For any two motives $M = (X, p)$ and $N = (Y, q)$ a morphism $f : M \to N$ is a correspondence $f \in A^0(X, Y)$, such that $q \circ f = f = f \circ p$. Since the category $\mathcal{M}$ is pseudo-abelian, every projector on $M \in \mathrm{Ob}(\mathcal{M})$ induces a decomposition of M as the sum of $\ker(p)$ and $\mathrm{im}(p)$. If $f \in A^*(X \times Y)$ we write its transpose as f^t. The diagonal Δ_X of a variety X is a projector of X and $h(X) = (X, \Delta_X)$ is called the *motive of the variety* X. If $M = (X, p)$ is a motive then its identity morphism $1_M : M \to M$ is given by the projector p. In particular, Δ_X can be viewed as the

---
*The authors acknowledge the support of TMR ERB FMRX CT-97-0107. The first named author also acknowledges SFB 343 "Diskrete Strukturen in der Mathematik" and the hospitality of the University of Bielefeld. The second named author is a member of GNSAGA of CNR.

identity morphism of $h(X)$ for any variety X. If $f : Y \to X$ is a morphism of varieties then the transpose of its graph $\Gamma_f^t : h(X) \to h(Y)$ is a morphism of Chow motives. Hence $X \to h(X)$, $f \to \Gamma_f^t$ is a contravariant functor from the category of varieties $\mathcal{V}$ into the category of Chow motives $\mathcal{M}$. The Chow groups of a motive M are defined by $A^i(M) = p_*(A^i(X))$ (see [Mu1]).

By $1 = h(\text{Spec}(k))$ we will denote the *unit motive* and by $\mathbb{L} = (\mathbb{P}_k^1, [\mathbb{P}_k^1 \times p])$, where p is a rational point of $\mathbb{P}_k^1$, the *Lefschetz motive*. $\mathbb{L}$ can also be defined by the equality: $h(\mathbb{P}^1) = 1 \oplus \mathbb{L}$ (see [Ful]). In general: if X is irreducible of dimension d and p is a rational point on X then one has: $(X, [p \times X]) \cong 1$ and $(X, [X \times p]) \cong \mathbb{L}^d$, where $\mathbb{L}^d$ is the d-fold tensor product of the motive $\mathbb{L}$ by itself.

Now let k be an algebraically closed field of characteristic zero and let H be a *Weil cohomology theory* on the category of smooth projective varieties over k. Then for every variety X there exists a graded algebra $H^*(X)$ and the usual axioms hold (see [Kl]). In particular there is a cycle map $cl_X^i : A^i(X) \to H^{2i}(X)$. For example, if $k = \mathbb{C}$ one can take for $H^*(X)$ the singular cohomology groups $H^*(X, \mathbb{Q})$ of the underlying manifold $X(\mathbb{C})$ with coefficients in $\mathbb{Q}$.

The group of codimension i cycles homologically equivalent to 0 is then defined to be $A^i_{\text{hom}}(X) = \{\alpha \in A^i(X) \mid cl_X^i(\alpha) = 0\}$. For every motive $M = (X, p)$ its cohomology groups are defined by $H^i(M) = p_*(H^i(X))$. As the cycle maps of varieties commute with correspondences, one can similarly define the cycle class of a motive. We will write $b_i(X)$ for the Betti number $\dim(H^i(X, \mathbb{Q}))$ of a complex variety X.

One of the crucial questions that arise in this context is to ask how much of the properties of the Chow graded ring $A(X) = \oplus A^i(X)$ can be recovered by looking at the Chow motive $h(X)$. In particular Beilinson conjectured the existence of a filtration $\{F^*\}$ on $A(X)$ (see [Ja], §2), such that $F^0 A^j(X) = A^j(X)$, $F^1 A^j(X) = A^j(X)_{\text{hom}}$ and the associated graded groups $Gr_F^v A^j(X) = F^v A^j(X)/F^{v+1} A^j(X)$ depend only on the motive of X modulo homological equivalence (i.e. the correspondences in $A^i(X \times X)_{\text{hom}}$ act as 0 on $Gr_F^v A^j(X)$).

In this paper we will consider the case of a particular surface X over $\mathbb{C}$ of general type with $p_g = 0$, the so called *classical Godeax surface*. We prove (Corollary 3.3) that the motive $h(X)$ is isomorphic to the direct sum of "trivial" motives in the sense that $h(X) \cong 1 \oplus 9 \cdot \mathbb{L} \oplus \mathbb{L}^2$, where $9 = b_2(X)$. This shows that $h(X)$ is isomorphic to the motive of a rational surface having the same Betti numbers as X. Therefore Corollary 3.3 appears as a natural extension of a known result for Enriques surfaces: if X is an Enriques surface then $h(X)$ is isomorphic to the motive of a rational surface J (see [Co]). In this case J is the Jacobian fibration associated to the elliptic fibration $\pi : X \to \mathbb{P}^1$. From this one immediately deduce the so called *Bloch Conjecture* for the Godeaux surface X because $A^2(X) \cong A^2(\mathbb{L}^2) = \mathbb{Q}$. The computation of $A^2(X)$ for the Godeaux surface has been done in different ways by Inose and Mizukami ([IM]) and by C. Voisin ([Vo]), while the problem of proving Bloch's conjecture for all surfaces of general type with $p_g = 0$ remains still open. Our motivic proof is along the line of the ideas of Bloch and Beilinson: let $T(X) = \ker(A_0(X) \to \text{Alb}(X))$

be the *Albanese kernel* of X, where $A_0(X)$ denotes the group of 0-cycles of degree 0. Then, if a Beilinson's filtration $\{F^*\}$ exists for a smooth projective variety X of dimension d, we must have $F^2 A^d(X) = T(X)$ ([Ja], 2.10). Bloch ([Bl], 1.11) proves that if X is a surface with $p_g = 0$, such that the action of correspondences on $T(X)$ factors trough homological equivalence, then $T(X)$ is 0. We show that the motive of the Godeaux surface X is finite dimensional in the sense of S. Kimura (see [Ki]). According to Kimura's result (see Proposition 7.5 in [Ki] or Theorem 2.9 below) this implies that if a correspondence acts as 0 on the cohomology of X then it is a nilpotent as a morphism of $h(X)$.

More generally we prove (Theorem 2.11) that, if X is a complex surface such that $h(X)$ is finite dimensional and $p_g(X) = 0$, then $T(X) = 0$. This result yields (Corollary 2.12) a uniform motivic proof of the Bloch conjecture for all surfaces with $p_g = 0$ which are not of general type.

We also compare our result with a theorem of J. Murre on the so called *Chow–Künneth decomposition* of a surface X. In general Murre conjectured that for a smooth projective variety X of dimension d there exists a decomposition of the motive $h(X)$:

$$h(X) = \bigoplus_{0 \le i \le 2d} h^i(X),$$

where $h^i(X) = (X, \pi_i)$, π_i are pairwise orthogonal projectors, such that $cl(\pi_i)$ coincides with the $(2d - i, i)$-component of Δ_X in the Künneth decomposition of $H^{2d}(X \times X, \mathbb{Q})$ and $\pi_0, \ldots, \pi_{j-1}$ and $\pi_{2j+1}, \ldots, \pi_{2d}$ act trivially on $A^j(X)$. Also $h^0(X) = (X, \pi_0) \cong 1$ and $h^{2d}(X) = (X, \pi_{2d}) \cong \mathbb{L}^d$.

By setting $F^0 A^j(X) = A^j(X)$ and $F^k A^j(X) = \ker(\pi_{2j+1-k} \mid_{F^{k-1}})$ one inductively defines a filtration on $A^*(X)$. Janssen proved in [Ja], §5, that the existence of a Chow–Künneth decomposition with $F^1 A^j(X) = A^j(X)_{\text{hom}}$ is, in fact, equivalent to the existence of a Beilinson filtration on $A^*(X)$.

Murre [Mu1] proved that such a decomposition exists when $\dim(X) = 2$. In this case the problem of understanding $h(X)$ reduces to the case of $h^2(X)$, a motive which carries at the same time the Neron–Severi group of X, the Albanese kernel $T(X)$ and the group of transcendental cycles. If $X = \mathbb{P}^2$ then $h^2(X)$ is "trivial", i.e. $h^2(X) \cong \mathbb{L}$. Here we prove (Theorem 2.14) that, if X is a smooth projective surface with $p_g(X) = q(X) = 0$ and $h(X)$ is finite dimensional, then $h^2(X)$ is a direct sum of trivial motives, i.e. $h^2(X) \cong b_2(X) \cdot \mathbb{L}$. These results notably applies to a class of surfaces of general type with $p_g = 0$ and $K_X^2 = 8$, recently classified by R. Pardini in [Pa], see the example in the end of Section 2.

The paper is organized as follows: in Section 2 we recall some definitions and results on finite dimensional motives from [Ki]. In particular, the class of finite dimensional motives is closed under direct sums and tensor products (Proposition 2.2), and subobjects and quotient objects of a finite dimensional motives are finite dimensional (Proposition 2.6). It follows (Theorem 2.8) that the finite dimensionality of the motive $h(X)$ is a birational invariant for a smooth projective surface X. In particular, the motive of a rational surface is finite dimensional. Kimura proved (see [Ki]) that, if M is a finite dimensional motive and $H^*(M) = 0$, then $M = 0$ (Theorem 2.10). These results imply that if X is a smooth projective surface over $\mathbb{C}$ with $p_g = 0$ and $\dim h(X) < \infty$, then the Albanese kernel $T(X)$ vanishes (Theorem 2.11). Then we prove the decomposition of $h^2(X)$ for surfaces with $p_g = q = 0$ and $\dim h(X) < \infty$ and consider the example of surfaces of general type with $p_g = 0$ and $K_X^2 = 8$.

In Section 3 we recall some geometrical properties of the Godeaux surface, defined as the quotient of a quintic surface in $\mathbb{P}^3$ invariant under the action of the group μ_5 of 5th roots of 1. X has an involution β such that $S = X/\beta$ is a rational surface with 5 nodes. Then we show that the motive $h(X)$ is finite dimensional and that $h(\tilde{S}) = h(X) \oplus 5 \cdot \mathbb{L}$, where $\tilde{S}$ is the desingularization of S.

The authors wish to thank I. Panin for many helpful conversations over the subject of this paper. They also thank C. Weibel for useful comments on an early version of the paper.

2. Finite dimensional motives

In this section we recall some definitions and results on finite dimensional motives from [Ki]. In particular, any finite dimensional motive, which is homologically trivial, is zero in $\mathcal{M}$ (Theorem 2.10).

Let $M = (X, p)$ and $N = (Y, q)$ be motives. The direct sum $M \oplus M$ is the motive $(X \coprod Y, p + q)$ where $X \coprod Y$ is the disjoint union of the two varieties and $p + q \in A^0(X, X) \oplus A^0(Y, Y)$.

The tensor product $M \otimes N$ is defined to be the motive $(X \times Y, p \otimes q)$ where $X \times Y$ is the fiber product of the varieties and the tensor product of correspondences $p \otimes q$ is the intersection product $\pi_X^*(p) \cdot \pi_Y^*(q)$ where $\pi_X : (X \times Y) \times (X \times Y) \to X \times X$ and $\pi_Y : (X \times Y) \times (X \times Y) \to Y \times Y$. In particular this gives: $h(X) \otimes h(Y) = h(X \times Y)$.

For any variety X let X^n be n-times fiber product $X \times \cdots \times X$ over the base field k. If $f_i : X \to Y, i = 1, \ldots, n$, are correspondences, that is $f_i \in A^*(X \times Y)$, then their tensor product $f_1 \otimes \cdots \otimes f_n$ can be defined as follows. Let $p_i : X^n \times Y^n \to X \times Y$ be the projection defined by the rule $((x_1, \ldots, x_n), (y_1, \ldots, y_n)) \to (x_i, y_i)$. Then $f_1 \otimes \cdots \otimes f_n$ is the intersection of correspondences $p_1^*(f_1) \ldots p_s^*(f_s)$. For a correspondence $f : X \to Y$ we will write $f^{(n)}$ for its n-times tensor power $f \otimes \cdots \otimes f$. Then $f^{(n)} \in A^*(X^n \times Y^n)$: we will use $f^{(n)}$ instead of f^n because the last symbol means the power of f in the sense of compositions of correspondences, i.e. as en element of $A^*(X \times Y)$.

If M is a motive by M^n we will denote the n-th fold tensor product of M, that is $M^n = M \otimes \cdots \otimes M$. A morphism $f : M \to N$ between motives is called *smash nilpotent* if $f^{(n)} = 0$ for some $n > 0$. One can also define exterior and symmetric powers of a motive M in the following way. Let Σ_n be the symmetric group of permutations of n elements. Then there is a one-to-one correspondence between all irreducible representations of Σ_n (over $\mathbb{Q}$) and all partitions of the integer n, where a partition of n is an ordered collection of integers $\lambda_1 \geq \cdots \geq \lambda_s \geq 0$, such that $\sum_{i=1}^{s} \lambda_i = n$.

Let V_λ be the irreducible representation corresponding to a partition λ of n and let χ_λ be the character of the representation V_λ. Let

$$d_\lambda = \frac{\dim(V_\lambda)}{n!} \sum_{\sigma \in \Sigma_n} \chi_\lambda(\sigma) \cdot \Gamma_\sigma,$$

where Γ_σ is the correspondence $(x_1, \ldots, x_n) \mapsto (x_{\sigma(1)}, \ldots, x_{\sigma(n)})$.

$\{d_\lambda\}$ is a set of pairwise orthogonal idempotents in $\mathrm{End}_\mathcal{M}(h(X^n))$ such that $\sum d_\lambda = \Delta_{X^n}$. Hence they give a decomposition of the motive $h(X)^n = (X, \Delta_X)^n$:

$$h(X)^n = h(X^n) = \bigoplus_\lambda (X^n, d_\lambda).$$

If $M = (X, p)$ is a motive then $p^{(n)}$ is a projector on X^n and $d_\lambda \circ p^{(n)} = p^{(n)} \circ d_\lambda$. Therefore the correspondences $p_\lambda^{(n)} = d_\lambda \circ p^{(n)}$ are pairwise orthogonal idempotents in $\mathrm{End}_\mathcal{M}(M^n)$, such that $\sum_\lambda p_\lambda^{(n)} = p^{(n)}$. In conclusion,

$$M^n = (X^n, p^{(n)}) = \bigoplus_\lambda (X^n, p_\lambda^{(n)})$$

where λ runs over all partitions of n.

Definition 2.1 (see [Ki], 3.5, 3.7). The n-th symmetric power of a motive $M = (X, p)$ is the motive $S^n M = (X^n, p_\lambda^{(n)})$ where λ is the partition (n) and the n-th exterior power of M is the motive $\wedge^n M = (X^n, p_\lambda^{(n)})$ for $\lambda = (1, \ldots, 1)$. A motive M is evenly finite dimensional if $\wedge^n M = 0$ for some n and it is called oddly finite dimensional if $S^n M = 0$ for some n.

We say that M is finite dimensional if it can be decomposed as $M = M_+ \oplus M_-$ where M_+ is evenly finite dimensional and M_- is oddly finite dimensional. We also define $\dim(M_+) = m$ if $\wedge^m M_+ \neq 0$, $\wedge^{m+1} M_+ = 0$ and similarly for M_-. For a finite dimensional motive M we let $\dim(M) = \dim(M_+) + \dim(M_-)$.

Examples. 1) If M is isomorphic to the motive of a point then it is evenly one dimensional. In fact the group Σ_2 acts trivially on $M \otimes M$, so $\wedge^2 M = 0$.

2) The Lefschetz motive $\mathbb{L} = (\mathbb{P}_k^1, [\mathbb{P}_k^1 \times p])$ is evenly finite dimensional. In fact $\wedge^2 \mathbb{L} = 0$ because the symmetric group Σ_2 acts trivially on $\mathrm{End}_\mathcal{M}(h(\mathbb{P}^1) \otimes h(\mathbb{P}^1)) = A^2((\mathbb{P}^1)^4) \cong \mathbb{Q}^6$ and hence also on the factor $\mathbb{L} \otimes \mathbb{L}$. From Proposition 2.2 below it follows that $\mathbb{L}^i$ is evenly finite dimensional for all i.

3) The motive $M = h(\mathbb{P}^n)$ is evenly $(n+1)$ dimensional because it is isomorphic to $\bigoplus_{0 \leq i \leq n} \mathbb{L}^i$.

4) Let C be a smooth projective curve of genus g and let $p \in C$ be a point. Then we have:

$$h(C) = h^0(C) \oplus h^1(C) \oplus h^2(C)$$

where $h^i(C) = (C, \pi_i)$, $\pi_0 = [p \times C]$, $\pi_2 = [C \times p]$ and $\pi_1 = \Delta_C - \pi_0 - \pi_2$. By a result of Shermenev [Sh] (see also [Ki], 4.2) the motive $h^1(C)$ is oddly $2g$-dimensional and each of the motives $h^0(C)$ and $h^2(C)$ is evenly finite dimensional, being isomorphic to 1 and to the motive $\mathbb{L}$. Then the motive $h(C)$ is finite dimensional as a sum of finite dimensional motives (see Proposition 2.2).

Proposition 2.2. *If the motives M and N are finite dimensional then $M \oplus N$ and $M \otimes N$ are finite dimensional. Moreover if M and N are both evenly or oddly finite dimensional motives then $M \otimes N$ is evenly finite dimensional; if M and N have a different parity, then $M \otimes N$ is oddly finite dimensional. In both cases we have $\dim(M \otimes N) \leq \dim(M) \cdot \dim(N)$.*

Proof. For $M \oplus N$ the result immediately follows from the definitions, for $M \otimes N$ the proof is given in [Ki], Proposition 5.10 and Corollary 5.11 □

Proposition 2.3. *Let $\tilde{X}$ be the blow up of a variety X along a subvariety $Y \subset X$ of pure codimension r. If the Chow motives $h(Y)$ and $h(X)$ are finite dimensional then $h(\tilde{X})$ is finite dimensional.*

Proof. Let $\mathbb{L} = (\mathbb{P}^1, [\mathbb{P}^1 \times pt])$ be the Lefschetz motive. Then

$$h(\tilde{X}) \cong h(X) \oplus \left(\bigoplus_{i=1}^{r-1} h(Y) \otimes \mathbb{L}^i \right) \tag{2.1}$$

(see [Ma]). The motive $\mathbb{L}$ is finite dimensional (see Example 2 above) and the motives $h(X)$ and $h(Y)$ are finite dimensional by assumption. By Proposition 2.2 $h(\tilde{X})$ is also finite dimensional. □

The converse of 2.3 is also true, in the sense that if the Chow motive of $\tilde{X}$ is finite dimensional so it is $h(X)$. To prove this result we use Kimura's theorem asserting that if a motive N is isomorphic to a submotive of a finite dimensional motive M then N is finite dimensional too (Proposition 2.6).

Definition 2.4. Let $f : M \to N$ be a morphism in $\mathcal{M}$ (or in any pseudoabelian category): f is said to be surjective if there exists a morphism $g : N \to M$, such that $f \circ g = 1_N$. A morphism $g : N \to M$ is said to be injective if there exists a morphism $f : M \to N$ such that $f \circ g = 1_N$. If $f : M \to N$ is surjective then N is canonically a quotient object of M and non canonically a direct factor of M. If $g : N \to M$ is injective then it has an image which is (non canonically) a direct factor of M and $N \to \text{im}(g)$ is an isomorphism (see [Sch], 1.7).

If a morphism in $\mathcal{M}$ is at the same time injective and surjective then it is an isomorphism.

Remark 2.5. Let $M = h(X)$ be the motive of a variety X and N the motive of a variety Y: let $f : X \to Y$ be a proper surjective morphism of varieties. Then the graph $\Gamma_f : h(X) \to h(Y)$ is a surjective morphism of motives (see [Ki], Remark 6.6).

Proposition 2.6. *Let M and N be two Chow motives. Suppose that there exists a surjective morphism $M \to N$ or, equivalently, an injective morphism $N \to M$. Then, if M is finite dimensional, N is also finite dimensional.*

Proof. Let $f : M \to N$ be a surjective morphism. It is clear that $\wedge^n M \xrightarrow{\wedge^n f} \wedge^n N$ and $S^n M \xrightarrow{S^n f} S^n N$ are also surjective morphisms for any n. Hence, if M is oddly or evenly finite dimensional then N enjoys the same property. Let M be finite dimensional, i.e. $M = M_+ \oplus M_-$ where M_+ is evenly finite dimensional and M_- is oddly finite dimensional. Then it is enough to show that there exists a decomposition $N = N_+ \oplus N_-$ and two surjective morphisms $f_+ : M_+ \to N_+$ and $f_- : M_- \to N_-$, such that $f = f_+ + f_-$. This is proved in [Ki], Proposition 6.9. □

Corollary 2.7. *Let $\tilde{X}$ be the blow up of a variety X along a smooth subvariety Y. Assume that the motive $h(\tilde{X})$ is finite dimensional. Then $h(X)$ is finite dimensional.*

Proof. From (2.1) it follows that $h(X)$ is a direct summand of $h(\tilde{X})$. By 2.6 it follows that $h(X)$ is finite dimensional. □

The following theorem shows that finite dimensionality of the motive $h(X)$ is a birational invariant for a smooth projective surface X.

Theorem 2.8. *Let X and Y be smooth projective surfaces which are birationally equivalent. Assume $h(X)$ is finite dimensional. Then $h(Y)$ is finite dimensional. In particular: the Chow motive of a rational surface is finite dimensional.*

Proof. Let T be a birational transformation between X and Y. Then T can be factored into a finite sequence of monoidal transformations (i.e. blow-ups of points) and their inverses. More precisely there exits a surface Z and two birational morphisms $f : Z \to X$ and $g : Z \to Y$, such that: $T = g \circ f^{-1}$, f and g can be factored into a composition of finitely many blow-ups of points. Therefore by Proposition 2.3 $h(Z)$ is finite dimensional. Similarly: $h(Y)$ is isomorphic to a submotive of the finite dimensional motive $h(Z)$. Hence by Proposition 2.6 it is finite dimensional. The last statement immediately follows for the finite dimensionality of the motive of $\mathbb{P}^2$. □

Suppose now that $\mathcal{M}$ is the category of Chow motives over an algebraically closed field of characteristic 0 and let H^* be a Weil cohomology theory for smooth projective varieties over k. If $M = (X, p)$ and $f : M \to M$ is a morphism in $\mathcal{M}$ then f is a cycle on $X \times X$ and its cohomology class $cl_{X \times X}(f)$ can be considered (using Künneth formula and Poincaré duality) as an endomorphism of the vector space $H^*(X)$.

Theorem 2.9. *Let $M = (X, p)$ be a finite dimensional Chow-motive and let $f : M \to M$ be an endomorphism of M. If $cl_{X \times X}(f) = 0$, then f is nilpotent in the ring $\mathrm{End}_{\mathcal{M}}(M)$.*

Proof. See [Ki], Proposition 7.5. □

Theorem 2.10. *Let $M = (X, p)$ be a Chow-motive. Assume that M is finite dimensional and $H^*(M) = 0$, i.e. $cl_{X \times X}(p) = 0$ in $H^*(X \times X)$. Then $M = 0$.*

Proof. See [Ki], Corollary 7.3. □

The following theorem shows that Bloch's conjecture holds for a smooth projective surface X with $p_g = 0$ if $h(X)$ is finite dimensional.

Theorem 2.11. *Let X be a smooth projective surface over $\mathbb{C}$ with $p_g = 0$. Assume that the Chow motive $M = h(X)$ is finite dimensional. Then the Albanese kernel $T(X)$ vanishes.*

Proof. Let $[\Delta_X] = \pi_0 + \pi_1 + \pi_2 + \pi_3 + \pi_4$ be a Chow–Künneth decomposition in $A^2(X \times X)$ ([Mu1]). Here $\pi_0 = [pt \times X]$ and $\pi_4 = [X \times pt]$ are the trivial parts, π_1 is the Picard projector, $\pi_3 = \tilde{\pi}_3 - \pi_1 \circ \tilde{\pi}_3$ where $\tilde{\pi}_3$ is the Albanese projector, i.e. the transpose π_1^t of π_1 (see Remark 6.5 in [Mu1]), and $\pi_2 = \Delta - \pi_0 - \pi_1 - \pi_3 - \pi_4$. Then the π_i's are orthogonal idempotents and the image of π_i under the cycle class map in $H^4(X \times X, \mathbb{Q})$ is the Künneth component $\Delta(4 - i, i)$.

Let $\{\alpha_1, \ldots, \alpha_s\}$, where $s = b_2(X)$, be a basis of the $\mathbb{Q}$-vector space $H^2(X, \mathbb{Q})$ and $\{\tilde{\alpha}_1, \ldots, \tilde{\alpha}_s\}$ the dual basis by the cup product identifying $H^2(X)$ with $H_2(X)$. Then $\alpha_i \cup \tilde{\alpha}_j = 0$ in $H^4(X, \mathbb{Q})$ for $i \neq j$ and $\sum q_i(\alpha_i \otimes \tilde{\alpha}_i) = \Delta(2, 2)$ in $H^4(X \times X, \mathbb{Q})$ for some $q_i \in \mathbb{Q}$.

The map $cl_X^1 : A^1(X) \to H^2(X, \mathbb{Q})$ being surjective (because $p_g(X) = 0$) there exist divisors $\{a_i\}$ and $\{\tilde{a}_i\}$ in $A^1(X)$, such that $cl_X^1(a_i) = \alpha_i$ and $cl_X^1(\tilde{a}_i) = \tilde{\alpha}_i$ for any i. Then $cl_{X \times X}^2 (\sum q_i(a_i \otimes \tilde{a}_i)) = \Delta(2, 2)$. Let $\sigma_2 = \sum q_i(a_i \otimes \tilde{a}_i)$: then we have $\pi_0 \circ \sigma_2 = \sigma_2 \circ \pi_0 = 0$ and similarly $\pi_4 \circ \sigma_2 = \sigma_2 \circ \pi_4 = 0$. Moreover, $(\sigma_2)_*(x) = 0$ for every $x \in A^2(X)$ because σ_2 is generated by divisors.

Let
$$f = \pi_0 + \pi_1 + \sigma_2 + \pi_3 + \pi_4.$$

Then f is an endomorphism of the motive $M = h(X) = (X, \Delta_X)$ and $f - \Delta_X$ is homologically equivalent to 0. By Theorem 2.9 $f - \Delta_X$ is nilpotent, i.e. $(f - \Delta_X)^n = 0$ in $A^2(X \times X)$ for some integer n. $A^2(X \times X)$ is an associative ring with unit Δ_X. Therefore we get:

$$(f - \Delta_X)^n = \sum \binom{n}{i} f^i + (-1)^n \Delta_X = 0.$$

Let $x \in A^2(X)$: then, according to [Mu1], we have: $\pi_0(x) = \pi_1(x) = 0$, $\pi_4(x) = \deg(x)$ and $\ker(\pi_3) = T(X) = \ker(A_0(X) \to \text{Alb}(X))$ where $A_0(X)$ is the group of 0-cycles of degree zero. If $x \in A_0^2(X)$ then $f(x) = \pi_3(x)$ and $(f - \Delta_X)^n(x) = 0$. Therefore we get:

$$(-1)^{n+1} x = \sum \binom{n}{i} (\pi_3(x))^i.$$

This shows that $\ker(\pi_3) = T(X) = 0$. □

The following result yields a motivic proof of a theorem in [BKL]:

Corollary 2.12. *Let X be a complex surface with Kodaira dimension less than 2. If $p_g(X) = 0$ then Bloch's conjecture holds for X.*

Proof. By Theorem 2.11 it is enough to show that $h(X)$ is finite dimensional. By Theorem 2.8 the finite dimensionality is a birational invariant, so we can just look at the birational classification of surfaces with Kodaira dimension $\kappa < 2$. If $\kappa < 0$ then X is either rational or birational to $C \times \mathbb{P}^1$ where C is a smooth curve. So, in both cases $h(X)$ is finite dimensional. If $0 \leq \kappa < 2$ then X is elliptic: if $p_g = q = 0$ then X is an Enriques surface and $h(X)$ is isomorphic in $\mathcal{M}$ to the motive of a rational surface (see [Co]). By Theorem 2.8 $h(X)$ is finite dimensional. If $p_g = 0, q = 1$ and $\kappa = 0$ then X is hyperelliptic, i.e. it is isomorphic to the quotient of the product of two elliptic curves by the action of a finite group. If $\kappa = 1$ then X is still the quotient of the product of 2 curves (see [BKL]). By Proposition 2.6 and by Remark 2.5 the motive $h(X)$ is finite dimensional. This concludes the proof. □

Lemma 2.13. *Let X and Y be smooth projective varieties over a field k and let $f : X \to Y$ be a morphism of finite degree. Then $h(Y)$ is isomorphic to a submotive of $h(X)$.*

Proof. Let f be of degree d: then the following equality between correspondences holds ([Ma], p. 450)

$$\Gamma_f^t \circ \Gamma_f = d \circ 1_{h(Y)}.$$

Therefore the morphism $1/d \cdot \Gamma_f : h(Y) \to h(X)$ is a right inverse to the morphism $\Gamma_f^t : h(X) \to h(Y)$. Hence $h(Y)$ is isomorphic to a submotive of $h(X)$. □

The next result (Theorem 2.14) shows that if X is a smooth projective surface over $\mathbb{C}$, such that $p_g = q = 0$ and the motive $h(X)$ is finite dimensional, then $h^2(X)$ is isomorphic to a finite direct sum of copies of the Lefschetz motive $\mathbb{L}$. In order to prove this result we need to "enlarge" the category $\mathcal{M}$ of Chow effective motives over a field k by adjoining the *Tate twists* $M(r)$ for every $M \in \mathrm{Ob}(\mathcal{M})$ and every $r \in \mathbb{Z}$. We then obtain a category $\bar{\mathcal{M}}$ whose objects are triples $M = (X, p, m)$ where p is a projector and $m \in \mathbb{Z}$ (see [Sch], 1.4). Morphisms from $M = (X, p, m)$ into $N = (Y, q, n)$ in the category $\bar{\mathcal{M}}$ are given by correspondences $f \in A^{n-m}(X, Y)$ of degree $n - m$ such that $f \circ p = q \circ f = f$, where $A^r(X, Y) = A^{d+r}(X \times Y)$ if X is purely d-dimensional. Clearly, $\mathcal{M}$ is a full subcategory of $\bar{\mathcal{M}}$. The Lefschetz motive $\mathbb{L}$ is then isomorphic in $\bar{\mathcal{M}}$ to $(\mathrm{Spec}(k), \mathrm{id}, -1)$. For every motive $M = (X, p, m)$ one defines the Tate twist $M(r)$ to be the motive $M \otimes \mathbb{L}^{-r} = (X, p, m+r)$, where $\mathbb{L}^r = \mathbb{L}^{\otimes r}$ for a positive integer r, $\mathbb{L}^0 = 1$ and $\mathbb{L}^r = \mathbb{L}^{\otimes -r}$ for a negative r (see [Sch], 1.9).

$\bar{\mathcal{M}}$ is a pseudoabelian $\mathbb{Q}$-linear tensor category. It is then easy to see that all definitions and formal properties of finite dimensionality for $\mathcal{M}$ also hold in $\bar{\mathcal{M}}$ (or, more generally, in every pseudoabelian $\mathbb{Q}$-linear tensor category). In particular direct sums and tensor products of finite dimensional motives in $\bar{\mathcal{M}}$ are finite dimensional. If $f : M \to N$ is a surjective morphism in $\bar{\mathcal{M}}$ (or, equivalently, if $N \to M$ is injective) and if M is finite dimensional then N is finite dimensional too.

If H^* is a Weil cohomology theory for smooth projective varieties over k, then one defines a functor H^i on $\bar{\mathcal{M}}$ for every $i \in \mathbb{Z}$ by $H^i((X, p, m)) = p_* H^{i+2m}(X)$.

Theorem 2.14. *Let X be a smooth projective surface over $\mathbb{C}$, such that $p_g(X) = q(X) = 0$. Assume that the motive $h(X)$ is finite dimensional. Then:*

$$h(X) \cong 1 \oplus b_2(X) \cdot \mathbb{L} \oplus \mathbb{L}^2,$$

where $b_2(X) = \dim(H^2(X, \mathbb{Q}))$ and $b_2(X) \cdot \mathbb{L}$ is a direct sum of $b_2(X)$ copies of $\mathbb{L}$.

Proof. By [Mu1] the motive $h(X)$ has a Chow–Künneth decomposition $h(X) = \sum h^i(X)$. Since $H^1(X) = H^3(X) = 0$, we get:

$$h(X) = h^0(X) \oplus h^2(X) \oplus h^4(X)$$

where $h^i(X) = (X, \pi_i, 0)$ and $\Delta_X = \pi_0 + \pi_2 + \pi_4$. $h^0(X)$ and $h^4(X)$ are the trivial parts of $h(X)$: in $\bar{\mathcal{M}}$ there are isomorphisms $h^0(X) \cong 1$ and $h^4(X) \cong \mathbb{L}^2$ (see [Sch], 1.13). By [Sch] 2.1 we also have:

$$\mathrm{Hom}_{\bar{\mathcal{M}}}(\mathbb{L}, h^2(X)) = \mathrm{Hom}_{\bar{\mathcal{M}}}((\mathrm{Spec}(\mathbb{C}), \mathrm{id}, -1), (X, \pi_2, 0))$$
$$= \pi_2 \circ A^1(\mathrm{Spec}(\mathbb{C}), X) = \pi_{2*} A^1(X).$$

Since $\pi_2 = \Delta_X - \pi_0 - \pi_4$ and projectors π_0 and π_4 acts as zero on $A^1(X)$ (see [Mu1], Theorem 3), we get $(\pi_2)_* A^1(X) = (\Delta_X)_* A^1(X) = A^1(X)$. Hence,

$$\mathrm{Hom}_{\bar{\mathcal{M}}}(\mathbb{L}, h^2(X)) = (\pi_2)_* A^1(X) = A^1(X).$$

Analogously,

$$\mathrm{Hom}_{\bar{\mathcal{M}}}(h^2(X), \mathbb{L}) = \mathrm{Hom}_{\bar{\mathcal{M}}}((X, \pi_2, 0), (\mathrm{Spec}(\mathbb{C}), \mathrm{id}, -1)) = A^1(X).$$

Note that $A^1(X) \cong H^2(X, \mathbb{Q})$, because $p_g(X) = q(X) = 0$. Let $\alpha \in H^2(X, \mathbb{Q})$: then $\alpha = \sum q_i [e_i]$ where $[e_i]$ for $i = 1, \ldots, b_2$ is a basis of the $\mathbb{Q}$-vector space $H^2(X, \mathbb{Q}) \cong A^1(X)$. Let $f_\alpha : \mathbb{L} \to h^2(X)$ be the corresponding morphism in $\bar{\mathcal{M}}$. Then $f_\alpha = \sum q_i [\mathrm{Spec}(\mathbb{C}) \times e_i]$. The transpose f_α^t is a morphism $h^2(X) \to \mathbb{L}$ ([Sch], 2.1) and

$$f_\alpha^t \circ f_\alpha = \sum q_i [\mathrm{Spec}(\mathbb{C}) \times \mathrm{Spec}(\mathbb{C})] \in \mathrm{Hom}_{\bar{\mathcal{M}}}(\mathbb{L}, \mathbb{L})$$
$$\cong A^0(\mathrm{Spec}(\mathbb{C}) \times \mathrm{Spec}(\mathbb{C})) \cong \mathbb{Q}.$$

Therefore, if we take $\alpha = e_i$ then the corresponding morphism $f_i = f_{e_i} : \mathbb{L} \to h^2(X)$ is injective. Let $f = \sum f_i : b_2 \cdot \mathbb{L} \to h^2(X)$: then f is an injective morphism in $\bar{\mathcal{M}}$. By [Sch], Remark 1.7, there exists a motive $N = (Y, q, n)$ in $\bar{\mathcal{M}}$, such that $h^2(X) \cong b_2 \cdot \mathbb{L} \oplus N$. $h(X)$ being finite dimensional, $h^2(X)$ is also finite dimensional. Hence N is finite dimensional.

By taking cohomology we get:
$$H^i(h^2(X)) = H^i(b_2 \cdot \mathbb{L}) \oplus H^i(N).$$
It follows from the results in [Mu1] that $H^i(h^2(X)) = 0$ unless $i = 2$ and $H^2(h^2(X)) = H^2(X, \mathbb{Q})$, and therefore $\dim(H^*(h^2(X))) = b_2(X)$. On the other hand we have $H^i(\mathbb{L}) = H^i((\operatorname{Spec}(\mathbb{C}), \operatorname{id}, -1)) = H^{i-2}(\operatorname{Spec}(\mathbb{C})) = 0$ unless $i = 2$ in which case $H^2(\mathbb{L}) = \mathbb{Q}$. Therefore, $\dim(H^*(b_2(X) \cdot \mathbb{L})) = b_2(X)$. It follows that $\dim(H^*(N)) = 0$, that is $H^i(N) = 0$ for all i. Let n be an integer, such that $N(-n)$ lies in $\mathcal{M}$. Since N is finite dimensional, $N(-n)$ is also finite dimensional, and since $H^*(N) = 0$, it follows that $H^*(N(-n)) = 0$. Then, by Theorem 2.10, $N(-n) = 0$. Therefore $N = 0$ in the category $\bar{\mathcal{M}}$ and $h^2(X) \cong b_2 \cdot \mathbb{L}$. This implies $h(X) \cong 1 \oplus b_2 \cdot \mathbb{L} \oplus \mathbb{L}^2$. □

Example. Let X be a minimal complex projective surface of general type with $p_g = 0$ (and therefore also $q = 0$) and $K_X^2 = 8$, such that there exists an involution β on X with X/β a rational surface. These surfaces have been classified in [Pa]: they are isomorphic to a free quotient $F \times C/G$, where F is a curve, C a hyperelliptic curve and G is a finite group. The motive $h(F \times C)$ is finite dimensional because it is a tensor product of motives of curves (see Example 4 above and use Proposition 2.2). By Proposition 2.6 $h(X)$ is also finite dimensional. From Theorem 2.14 it follows that $h^2(X)$ is isomorphic to a finite direct sum of copies of $\mathbb{L}$. This implies $A^2(X) \cong \mathbb{Q}$.

3. The Chow motive of the Godeaux surface

We recall the classical construction of the Godeaux surface. Let μ_5 be the group of 5-th roots of unity generated by the primitive root ϵ. Let μ_5 act on $\mathbb{P}^3$ by the rule
$$(x_1 : x_2 : x_3 : x_4) \to (\epsilon x_1 : \epsilon^2 x_2 : \epsilon^3 x_3 : \epsilon^4 x_4).$$
Let $f = f(x_1, \ldots, x_4)$ be a μ_5-invariant quintic form. If $f(P_i) \neq 0$ where P_i are the four coordinate points $(0, \ldots, 1, \ldots, 0)$ then the quintic surface in $\mathbb{P}^3$ defined by $f = 0$ is invariant under μ_5 and Y is non singular. For the sake of simplicity we will take Y to be the Fermat quintic defined by the equation
$$x_1^5 + x_2^5 + x_3^5 + x_4^5 = 0.$$
The quotient $X = Y/\mu_5$ is the Godeaux surface. Let $\eta : Y \to X$ be the quotient map.

Theorem 3.1. *Let Y be the Fermat quintic in $\mathbb{P}^3$, and let $X = Y/\mu_5$ be the Godeaux surface. The Chow motive $h(X)$ is finite dimensional.*

Proof. We will use the "inductive structure" on Fermat hypersurfaces proved in [SK]. It follows that the Fermat quintic Y in $\mathbb{P}^3$ can be obtained from the product of two copies of the Fermat quintic plane curve C defined by the equation $x_1^5 + x_2^5 + x_3^5 = 0$ in the following way. Let $c_i = (1 : -\epsilon^i : 0) \in C$, where $0 \leq i \leq 4$ and ϵ is a generator of μ_5. Let Z be the blow up of the product $C \times C$ in the 25 points $c_i \times c_j$. Then μ_5 acts on Z and the quotient surface Z/μ_5 is smooth (see Lemma 1.4 in [SK]). Moreover the Fermat quintic Y is obtained by blowing down two projective lines on Z/μ_5). The Chow-motive of a curve being finite dimensional, $h(C)$ is finite dimensional. By Proposition 2.2 the motive $h(C \times C) = h(C) \otimes h(C)$ is also finite dimensional. Since Z is a blow up of $C \times C$ in a finite collection of points, the motive $h(Z)$ is finite dimensional by Proposition 2.3. The quotient map $Z \to Z/\mu_5$ induces a surjective morphism of Chow motives $h(Z) \to h(Z/\mu_5)$ (see Remark 2.5). Hence $h(Z/\mu_5)$ is finite dimensional by Proposition 2.6. By Corollary 2.7 $h(Y)$ is finite dimensional too. $h(Y)$ being finite dimensional, the motive $h(X)$ of the Godeaux surface X is finite dimensional. □

Remark 3.2. Actually one can prove a more general result. Let

$$Y_d^{n-1} : x_1^d + \cdots + x_{n+1}^d = 0$$

be the Fermat hypersurface of degree d in $\mathbb{P}^n$. Then its Chow motive $h(Y_d^{n-1})$ is finite dimensional because we can use Shioda/Katsura's inductive systems to approximate Y_d^{n-1} by $n-1$ copies of the curve Y_d^1.

The above result, together with Theorem 2.14, yields an isomorphism $h(X) \cong 1 \oplus b_2(X) \cdot \mathbb{L} \oplus \mathbb{L}^2$, where $b_2(X)$ is the second Betti number of the Godeaux surface X. Since $b_2(X) = 9$ we get the following:

Corollary 3.3. *Let X be the Godeaux surface: then*

$$h(X) \cong 1 \oplus 9 \cdot \mathbb{L} \oplus \mathbb{L}^2.$$

In the next theorem we show that there exists a smooth rational surface $\tilde{S}$ such that: $h(\tilde{S}) = h(X) \oplus 5\mathbb{L}$.

We first recall some geometrical properties of the Godeaux surface X.
Consider the 3 lines in $\mathbb{P}^3$ lying on the Fermat quintic Y:

$$L_1: \quad x_1 = -x_4, \quad x_2 = -x_3$$

$$L_2: \quad x_1 = -x_3, \quad x_2 = -x_4$$

$$L_3: \quad x_1 = -x_2, \quad x_3 = -x_4$$

We have $L_s \cdot L_t = 1$ for $s \neq t$. Since Y is a degree five hypersurface in $\mathbb{P}^3$, the canonical invertible sheaf ω_Y is equal to $\mathcal{O}_Y(1)$. The canonical class K_Y can be cut by a hyperplane, e.g. $x_1 = 0$. Any L_s meets this hyperplane transversally: hence $K_Y \cdot L_s = 1$ and, by the adjunction formula, $L_s^3 = -3$ for any $s \in \{1, 2, 3\}$. For any s, the orbit of the action of μ_5 on L_s contains five different lines on Y. In other words a nontrivial element $\alpha \in \mu_5$ shifts any line L_s inside a set of lines on Y. It follows that η maps L_s isomorphically into X for any s. Therefore, $M_s = \eta(L_s) \simeq \mathbb{P}^1$. If

$s, t \in \{1, 2, 3\}$ and $s \neq t$, then L_t intersects only one line in the orbit $\sum_{\alpha \in \mu_5} \alpha(L_s)$, so that, for $s \neq t$,

$$M_s \cdot M_t = 1.$$

Since the map η is unramified of degree 5 we have: $K_X = \frac{1}{5} \eta_* K_Y$. The class K_Y is very ample and the quotient morphism η is a surjective finite map. Therefore (see [Ha], p. 25) K_X is ample and

$$M_s \cdot K_X = 1$$

for $s = 1, 2, 3$. The equality $K_Y^2 = 5$ implies $K_X^2 = 1$. By the adjunction formula $M_s^2 = -3$ for $s = 1, 2, 3$.

X is a surface of general type with $p_g = q = 0$, and therefore the cycle map $cl : A^1(X) \to H^2(X, \mathbb{Q})$ is an isomorphism and we have $b_2(X) = \dim H^2(X, \mathbb{Q}) = \dim A^1(X) = 9$.

Let's consider the linear system $\Phi = |3K_X - M_1|$ on X (see [Re]): Φ is a dimension one linear system without fixed points. Therefore it induces a pencil $\Phi : X \to \mathbb{P}^1$. This fibration has five reducible fibers and a genus two curve as a general fiber. The line $L : x_1 = x_4$, $x_2 = x_3$ intersects the quintic Y in five points $(1 : -\epsilon^i : -\epsilon^i : 1)$, $i \in \{0, 1, 2, 3, 4\}$. Let

$$Q_i = \eta(1 : -\epsilon^i : -\epsilon^i : 1)$$

be the corresponding points on the surface X. For any i there exist two elliptic curves, say F_i^+ and F_i^-, on the surface X such that

$$F_i^+ \cap F_i^- = \{Q_i\},$$
$$F_i^+ \cdot F_i^- = 1$$

and the divisors $F_i^+ + F_i^-$ form five reducible fibers of the pencil $\Phi : X \to \mathbb{P}^1$ (see [Re], p. 316). The nine divisors

$$M_1, \quad M_2, \quad M_3, \quad K_X, \quad F_0^+, \quad F_1^+, \quad F_2^+, \quad F_3^+, \quad F_4^+ \qquad (3.2)$$

have the following non degenerate intersection matrix:

	M_1	M_2	M_3	K_X	F_0^+	F_1^+	F_2^+	F_3^+	F_4^+
M_1	-3	1	1	1	3	3	3	3	3
M_2	1	-3	1	1	1	0	2	2	0
M_3	1	1	-3	1	1	2	0	0	2
K_X	1	1	1	1	1	1	1	1	1
F_0^+	3	1	1	1	-1	0	0	0	0
F_1^+	3	0	2	1	0	-1	0	0	0
F_2^+	3	2	0	1	0	0	-1	0	0
F_3^+	3	2	0	1	0	0	0	-1	0
F_4^+	3	0	2	1	0	0	0	0	-1

whose determinant is equal to -64. Hence the cohomology classes of the nine divisors in (3.2) form a basis in the $\mathbb{Q}$-vector space $H^2(X, \mathbb{Q})$ (one can also use F_i^- instead of F_i^+).

Now we recall some known facts about the involution on the Godeaux surface. Let the dihedral group D_{10} act on $\mathbb{P}^3$ by adjoining to the action of a generator α of μ_5 an involution β defined by: $\beta : x_i \to x_{5-i}$. If Y is the Fermat quintic then D_{10} acts on Y and β induces an involution (which we will again call β) on the Godeaux surface $X = Y/\mu_5$. It is easy to see that the three lines L_1, L_2 and L_3 are β-invariant. Therefore, M_1, M_2 and M_3 are β-invariant as well. Of course, the canonical class K_X is invariant under β. Moreover using the results in [Re] one can show that the curves F_i^+ are also β-invariant. Indeed, the five cubic forms

$$ax_1^2 x_3 - bx_1 x_2^2 + cx_3^2 x_4 - dx_2 x_4^2 = 0,$$

$$b\sigma x_2^3 - a\sigma x_1 x_2 x_3 + a\tau x_1^2 x_4 + c\tau x_3 x_4^2 = 0,$$

$$d\tau x_4^3 + b\tau x_1 x_2 x_4 + b\sigma x_2^2 x_3 - a\sigma x_1 x_3^2 = 0,$$

$$a\tau x_1^3 + c\tau x_1 x_3 x_4 + c\sigma x_2 x_3^2 - d\sigma x_2^2 x_4 = 0,$$

$$c\sigma x_3^3 - d\sigma x_2 x_3 x_4 + d\tau x_1 x_4^2 + b\tau x_1^2 x_2 = 0$$

define a μ_5-invariant elliptic curve in $\mathbb{P}^3$ (loc. cit., p. 362). For any $i \in \{0, \ldots, 4\}$ let E_i^+ be the elliptic curve defined by the specialization

$$a = \epsilon^i, \quad b = -1, \quad c = 1, \quad d = -\epsilon^i, \quad \sigma = 1 + \sqrt{5}, \quad \tau = 2\epsilon^{2i}.$$

A simple computation shows that the curves E_i^+ lie on Y and the images under η of E_i^+ are the elliptic curves F_i^+ on X. The explicit equations for E_i^+ show that the curves F_i^+ are β-invariant. Hence the algebraic basis for the $\mathbb{Q}$-vector space $A^i(X) = H^2(X, \mathbb{Q})$ appearing in (3.2) is invariant under the involution β.

The involution β on X has five isolated fixed points $Q_i = \eta(1 : -\epsilon^i : -\epsilon^i : 1)$ and a fixed curve $M_1 = \eta(L_1)$ with $M_1^2 = -3$ and $M_1 \simeq \mathbb{P}^1$. The quotient surface S is a rational surface with five singular points $\theta(Q_i)$ (see [Re], p. 314). They are of type $(2, 1)$ and locally isomorphic to the cone $z^2 = xy$ (see [BPV], 5.1, 5.3). Therefore they can be resolved with a blow up $\mu : \tilde{S} \to S$ having exceptional -2-curves. Let $\lambda : \tilde{X} \to X$ be the blow up of the surface X at the 5 points Q_i, $i \in \{0, \ldots, 4\}$. Since Q_i are smooth points, all exceptional fibers over Q_i are -1-curves.

Lemma 3.4. *For any $s \in \{0, 1, 2, 3, 4\}$ we have that $b_s(\tilde{X}) = b_s(\tilde{S})$.*

Proof. The surface $\tilde{X}$ being birationally isomorphic to the surface X we have: $p_g(\tilde{X}) = p_g(X) = 0$ and $q(\tilde{X}) = q(X) = 0$. Therefore, $b_1(\tilde{X}) = b_3(\tilde{X}) = 0$. Analogously, since $\tilde{S}$ is birationally isomorphic to S, we have: $b_1(\tilde{S}) = b_3(\tilde{S}) = 0$. Moreover, $b_2(\tilde{X}) = b_2(\tilde{S})$ if and only if $\dim_{\mathbb{Q}} A^1(\tilde{X}) = \dim_{\mathbb{Q}} A^1(\tilde{S})$. Since the $\mathbb{Q}$-vector space

$A^1(X)$ has a β-invariant algebraic basis we get $A^1(X)^\beta = A^1(X)$ which implies that $\dim_{\mathbb{Q}} A^1(S) = \dim_{\mathbb{Q}} A^1(X) = 9$. Hence $\dim_{\mathbb{Q}} A^1(\tilde{S}) = \dim_{\mathbb{Q}} A^1(\tilde{X}) = 9+5 = 14$. This shows that: $b_2(\tilde{S}) = b_2(\tilde{X}) = 14$. □

Theorem 3.5. *Let X be the Godeaux surface and let $\tilde{S}$ be a resolution of singularities of the rational surface $S = X/\beta$, β being the involution on X. Then there exists an isomorphism of Chow motives:*

$$h(\tilde{S}) \cong h(X) \oplus 5\mathbb{L} \, .$$

Proof. Consider the following commutative diagram

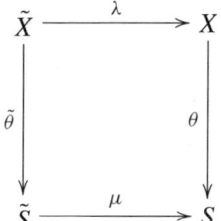

The map $\tilde{\theta}$ induces a morphism of motives $\Gamma_{\tilde{\theta}} : h(\tilde{S}) \to h(\tilde{X})$. Since $\theta : X \to S$ is a degree five map, the map $\tilde{\theta}$ has a degree five as well. Hence

$$p = \frac{1}{5} \Gamma_{\tilde{\theta}} \circ \Gamma_{\tilde{\theta}}^t$$

is a projector of the surface $\tilde{X}$ (see [Ma]). Moreover,

$$h(\tilde{X}) = (\tilde{X}, p) \oplus (\tilde{X}, \Delta_{\tilde{X}} - p)$$

and

$$(\tilde{X}, p) \cong h(\tilde{S}).$$

Let $M = (\tilde{X}, \Delta_{\tilde{X}} - p)$: then $h(\tilde{X}) \cong h(\tilde{S}) \oplus M$. By Theorem 3.1 and Proposition 2.3 $h(\tilde{X})$ is finite dimensional. By Proposition 2.6 the motive M is also finite dimensional. The decomposition $h(\tilde{X}) = h(\tilde{S}) \oplus M$ yields a similar the decomposition for the cohomology groups:

$$H^*(h(\tilde{X})) \cong H^*(h(\tilde{S})) \oplus H^*(M).$$

From Lemma 3.4 it follows that $b_s(\tilde{X}) = b_s(\tilde{S})$: hence $\dim(H^*(M)) = 0$, i.e. $H^*(M, \mathbb{Q}) = 0$. From Theorem 2.10 we get $M = 0$. Therefore, $h(\tilde{S}) = h(\tilde{X})$. On the other hand, according to Manin's theorem [Ma, §9], we have that $h(\tilde{X}) \cong h(X) \oplus 5\mathbb{L}$. Therefore, $h(\tilde{S}) \cong h(X) \oplus 5\mathbb{L}$. □

References

[BPV] W. Barth, C. Peters, A. Van de Ven, Compact Complex Surfaces, Ergeb. Math. Grenzgeb. (3) 4, Springer-Verlag, New-York, 1984.

[Bl] S. Bloch, Lectures on algebraic cycles, Duke Univ. Math. Series IV, 1980.

[BKL] S. Bloch, A. Kas, D. Lieberman, Zero cycles on surfaces with $p_g = 0$. Compositio Math. 33 (1976), 135–145.

[Co] K. Coombes, The K-cohomology of Enriques surfaces, Contemp. Math. 126 (1992) 47–57.

[Ful] W. Fulton, Intersection theory, Ergeb. Math. Grenzgeb. (3) 2, Springer-Verlag, New-York, 1984.

[FH] W. Fulton, J. Harris, Representation theory: a first course, Springer-Verlag, 1991.

[Ha] R. Hartshorne, Ample subvarieties of algebraic varieties, Lecture Notes in Math. 156, Springer-Verlag, Berlin, 1970.

[IM] H. Inose, M. Mizukami, Rational equivalence of zero-cycles on some surfaces with $p_g = 0$, Math. Ann. 244 (1979) 205–217.

[Ja] U. Jannsen, Motivic Sheaves and Filtratins on Chow Groups, in: Motives, Proc. Sympos. Pure Math. 55, Part 1, Amer. Math. Soc., Providence, RI, 1994, 245–302.

[Ki] S.-I. Kimura, Chow groups can be finite dimensional, in some sense, to appear in J. Algebraic Geom.

[Kl] S. L. Kleiman, Algebraic cycles and the Weil conjectures, Dix Exposés sur la Cohomologie des Schémas, North-Holland, Amsterdam, 1968.

[Ma] Yu. I. Manin, Correspondences, motives and monoidal transformations, Math. USSR Sb. 6 (1968) 439–470.

[Mu1] J. P. Murre, On the motive of an algebraic surface, J. Reine Angew. Math. 409 (1990), 190–204.

[Mu2] J. P. Murre, On a conjectural filtration on the Chow groups of an algebraic surface – I, II, Indag. Math. 4 (2) (1990), 177–188, 189–201.

[Pa] R. Pardini, The classification of double planes of general type with $K^2 = 8$ and $p_g = 0$, math.AG/0107100.

[Re] M. Reid, Campedelli versus Godeaux, in: Problems in the theory of surfaces and their classification (Cortona, 1988), Academic Press, 1991, 309–365.

[Sh] A. M. Shermenev, The motive of an abelian variety, Funct. Anal. 8 (1974), 47–53.

[Sch] A. J. Scholl, Classical motives, in: Motives, Proc. Sympos. Pure Math. 55, Part 1 (1994), 163–187.

[SK] T. Shioda, T. Katsura, On Fermat varieties, Tôhoku Math. J. 31 (1979), 97–115.

[Vo] C. Voisin, Sur les zero-cycles de certaine hypersurfaces munies d'un automorphisme, Ann. Scuola Norm. Sup. Pisa Ck. Si. (4) 19 (1992), 473–492.

V. Guletskiĭ, Institute of Mathematics, Surganova 11, Minsk 220072, Belarus
E-mail: guletskii@im.bas-net.by

C. Pedrini, Dipartimento di Matematica, Università di Genova, Via Dodecaneso 35, 16146 Genova, Italy
E-mail: pedrini@dima.unige.it

Francia's flip and derived categories

Yujiro Kawamata

Abstract. We extend some of the results of Bondal–Orlov on the equivalence of derived categories to the case of orbifolds by using the category of coherent orbifold sheaves.

1. Introduction

We consider an approach to the problems on flips and flops in the birational geometry from the point of view of the theory of derived categories. The purpose of this paper is to extend some of the existing results for smooth varieties to the case of varieties having only quotient singlarities.

The idea of using the derived categories can be explained in the following way. The category of sheaves on a given variety is directly related to the biregular geometry of the variety. But the category derived from the category of complexes of sheaves by adding the inverses of quasi-isomorphisms and dividing modulo chain homotopy equivalences acquires more symmetry, and is beleived to reflect more essential properties of the variety, namely the birational geometry of the variety. More precisely, the varieties which have the same level of the canonical divisors (so called K-equivalent varieties) are beleived to have equivalent derived categories.

Bondal–Orlov [1] considered a smooth variety X which contains a subvariety E isomorphic to the projective space $\mathbb{P}^m$ such that the normal bundle is isomorphic to $\mathcal{O}(1)^{n+1}$. If we blow up X with center E, then the exceptional divisor can be contracted to another direction to yield another smooth variety X' which contains a subvariety E' isomorphic to the projective space $\mathbb{P}^n$. The induced birational map $X - \to X'$ is a flip if $m > n$ and a flop if $m = n$. In other words, the K-level of X is higher than that of X' if $m > n$ and they are equal if $m = n$. Then [1] proved that the natural functor between the derived categories of bounded complexes of coherent sheaves $D^b_{\text{coh}}(X') \to D^b_{\text{coh}}(X)$, called the Fourier–Mukai transform after [12], is fully faithful if $m > n$ and an equivalence if $m = n$.

[1] (see also [2]) also proved a reconstruction theorem in the following sense: if there exists an equivalence of derived categories $D^b_{\text{coh}}(X) \to D^b_{\text{coh}}(X')$ for smooth projective varieties X and X' such that either the canonical divisor K_X or its negative $-K_X$ is ample, then there exists an isomorphism $X \to X'$.

We shall extend these results for varieties having quotient singularities in this paper. The main results are Theorem 5.2 and Theorem 6.1. We consider some toric flips and flops defined in §4 after [14] and [15]. We first remark in Example 5.1 of §5 that this kind of extension does not work if we consider the usual derived categories of bounded complexes of coherent sheaves. In order to overcome this difficulty, we introduce the concept of coherent orbifold sheaves in §2. In Theorem 5.2, we prove that only the level of K deternimes the equivalence class of the derived categories, though the varieties with the same level of K may have very different geometric outlook. For example, the dimensions of the exceptional loci may be different. Unlike the smooth case, there is no obvious geometric order between the varieties, though there is order of canonical divisors, and the derived categories follow the latter.

According to the minimal model program, we should deal with varieties which admit mild singularities, and results for smooth varieties should be extended to such varieties (cf. [10]). Our extension for varieties with quotient singularities can be regarded as the first step toward the general case of varieties with arbitrary terminal singularities. A recent result by Yasuda [16] on the motivic integration for orbifolds is also one of such extensions.

The existence of the flips for arbitrary small contraction with relatively negative canonical divisor is one of the most important but difficult conjectures in the minimal model program. It is proved only in dimension 3 by Mori [11]. In §3, we recall a result of the author [9] which reduces the existence problem of the flips to that of the flops (Theorem 3.3). The reason is that the flops seem to be better suited to the categorical argument than the flips, because the flop corresponds to the equivalence of categories while the flip to the fully faithful embedding.

Bridgeland [4] constructed the flop for any small crepant contraction of a smooth 3-dimensional variety by using only the categorical argument as in [5]. It is remarkable that the existence of the flop and the equivalence of derived categories are simultaneously proved. While preparing this manuscript, the author learned that Chen [6] obtained a result which extends the above result [4] to the flops of 3-dimensional varieties with Gorenstein terminal singularities. One might even extend this to the case of 3-folds having arbitrary terminal singularities by combining with our method since such singularities can be deformed to quotient singularities. We hope that we could eventually prove the existence of flips in this way.

The author would like to thank Akira Ishii and Adrian Langer for the useful discussions on the derived categories and orbifold sheaves, respectively. We work over the complex number field $\mathbb{C}$.

2. Orbifold sheaf

We begin with recalling the definition of the quasi-projective orbifolds (or Q-varieties) and coherent orbifold sheaves (or Q-sheaves) from [13] §2.

Definition 2.1. Let X be a quasi-projective variety. An *orbifold structure* on X consisits of the data $\{\pi_i : X_i \to X, G_i\}_{i \in I}$, where the X_i are smooth quasi-projective varieties, the π_i are quasi-finite morphisms, and the G_i are finite groups acting faithfully on the X_i, such that $X = \bigcup_{i \in I} \pi_i(X_i)$, the π_i induce etale morphisms $\pi'_i : X_i/G_i \to X$, and that, if $X_{ij} = (X_i \times_X X_j)^\nu$ denotes the normalization of the fiber product, then the projections $p_1 : X_{ij} \to X_i$ and $p_2 : X_{ij} \to X_j$ are etale for any i and j, where i and j may be equal.

In this case, X has only quotient singularities. Conversely, if X is a quasi-projective variety having only quotient singularities, then there exists an orbifold structure on X such that the π_i are etale in codimension 1. We call such a structure *natural*.

A *global cover* $\tilde{X}$ is the normalization of X in a Galois extension of the function field $k(X)$ which contains all the extensions $k(X_i)$.

Definition 2.2. An *orbifold sheaf* F is a collection of sheaves F_i of $\mathcal{O}_{X_i}$-modules on the X_i together with the gluing isomorphisms $g_{ji} : p_1^* F_i \to p_2^* F_j$ on the X_{ij}, such that the compatibility conditions $(p_{23}^* g_{kj}) \circ (p_{12}^* g_{ji}) = p_{13}^* g_{ki}$ hold on the triple overlaps $X_{ijk} = (X_i \times_X X_j \times_X X_k)^\nu$, where ν denotes the normalization.

For example, we define the *orbifold structure sheaf* $\mathcal{O}_X^{\mathrm{orb}}$ by the $\mathcal{O}_{X_i}$.

Let $\tilde{X}_i$ be the normalization of X_i in the function field $k(\tilde{X})$ of the global cover, and $H'_i = \mathrm{Gal}(\tilde{X}_i/X'_i)$, where $X'_i = X_i/G_i$. Then an orbifold sheaf F on X is in a one-to-one correspondense to a sheaf $\tilde{F}$ of $\mathcal{O}_{\tilde{X}}$-modules on $\tilde{X}$ such that the action of the Galois group $\mathrm{Gal}(\tilde{X}/X)$ lifts to $\tilde{F}$ and that the restriction $\tilde{F}|_{\tilde{X}_i}$ with its H'_i-action is isomorphic to the pull-back of a sheaf of $\mathcal{O}_{X_i}$-modules on X_i.

A *homomorphism* $h : F \to F'$ of orbifold sheaves is a collection of G_i-equivariant $\mathcal{O}_{X_i}$-homomorphisms $h_i : F_i \to F'_i$ which are compatible with gluing isomorphisms. A *tensor product* $F \otimes_{\mathcal{O}_X^{\mathrm{orb}}} F'$ is given by the sheaves $F_i \otimes_{\mathcal{O}_{X_i}} F'_i$. The category of orbifold sheaves $Sh(X^{\mathrm{orb}})$ on X thus defined becomes an abelian category.

An orbifold sheaf is said to be *coherent* (resp. *locally free*) if each F_i is coherent (resp. locally free). For example, the *orbifold sheaf of differential p-forms* $\Omega_X^{p,\mathrm{orb}}$ is a locally free coherent orbifold sheaf given by the sheaves $\Omega_{X_i}^p$. In particular, the *dualizing orbifold sheaf* ω_X^{orb} is the invertible orbifold sheaf consisting of the dualizing sheaves ω_{X_i}. We note that even if ω_{X_i} is isomorphic to $\mathcal{O}_{X_i}$, the action of G_i on them may be different.

Let $D^b(X_{\mathrm{coh}}^{\mathrm{orb}})$ (resp. $D_c^b(X_{\mathrm{coh}}^{\mathrm{orb}})$) be the derived category of bounded complexes of coherent orbifold sheaves on X (resp. with compact supports).

Proposition 2.3. *Let X be a quasi-projective variety with an orbifold structure.*

(1) *The category $Sh(X^{\mathrm{orb}})$ has enough injectives.*

(2) *([13] Proposition 2.1.) If X has a Cohen–Macaulay global cover, then any coherent orbifold sheaf F has a finite locally free resolution.*

Proposition 2.4. Let $f : X \to Y$ be a generically surjective morphism of quasi-projective varieties with orbifold structures $\{\pi_i : X_i \to X, G_i\}$ and $\{\rho_\alpha : Y_\alpha \to Y, H_\alpha\}$. Assume that the natural morphism $p_1^{i\alpha} : (X_i \times_Y Y_\alpha)^\nu \to X_i$ is etale for any i and α, where ν denotes the normalization. Then one can define the direct image functor $f_* : Sh(X^{\mathrm{orb}}) \to Sh(Y^{\mathrm{orb}})$ and the inverse image functor $f^* : Sh(Y^{\mathrm{orb}}) \to Sh(X^{\mathrm{orb}})$ which are adjoints each other.

Proof. When we fix α and vary i, then we obtain a covering of the normalized fiber product $(X \times_Y Y_\alpha)^\nu$ by the morphisms $\sigma_{i\alpha} : (X_i \times_Y Y_\alpha)^\nu \to (X \times_Y Y_\alpha)^\nu$ induced from the π_i. For an orbifold sheaf E on X, we define a sheaf E_α on $(X \times_Y Y_\alpha)^\nu$ as the kernel

$$E_\alpha \to \bigoplus_i \sigma_{i\alpha*} p_1^{i\alpha*} E_i \rightrightarrows \bigoplus_{i,j} \sigma_{ij\alpha*} p_1^{ij\alpha*} E_i$$

where $p_1^{ij\alpha} : (X_i \times_X X_j \times_Y Y_\alpha)^\nu \to X_i$ and $\sigma_{ij\alpha} : (X_i \times_X X_j \times_Y Y_\alpha)^\nu \to (X \times_Y Y_\alpha)^\nu$ are natral morphisms. Then we define an orbifold sheaf f_*E on Y by $(f_*E)_\alpha = p_{2*}E_\alpha$. Since the $p_1^{i\alpha}$ are etale, f_* is left exact.

When we fix i and vary α, then we obtain an etale covering of X_i by $p_1^{i\alpha} : (X_i \times_Y Y_\alpha)^\nu \to X_i$. For an orbifold sheaf F on Y, we define a sheaf $(f^*F)_i$ on X_i as the kernel

$$(f^*F)_i \to \bigoplus_\alpha p_{1*}^{i\alpha} p_2^{i\alpha*} F_\alpha \rightrightarrows \bigoplus_{\alpha,\beta} p_{1*}^{i\alpha\beta} p_2^{i\alpha\beta*} F_\alpha$$

where $p_1^{i\alpha\beta} : (X_i \times_Y Y_\alpha \times_Y Y_\beta)^\nu \to X_i$ and $p_2^{i\alpha\beta} : (X_i \times_Y Y_\alpha \times_Y Y_\beta)^\nu \to Y_\alpha$ are natural morphisms. Since the $p_1^{i\alpha}$ are etale, f^* is right exact. $\square$

Corollary 2.5. *In addition to the assumptions of Proposition 2.4, assume that f is proper and Y has a Cohen–Macaulay global cover. Then functors f_* and f^* induces exact functors*

$$Rf_*^{\mathrm{orb}} : D^b(X_{\mathrm{coh}}^{\mathrm{orb}}) \to D^b(Y_{\mathrm{coh}}^{\mathrm{orb}})$$

$$Lf_{\mathrm{orb}}^* : D^b(Y_{\mathrm{coh}}^{\mathrm{orb}}) \to D^b(X_{\mathrm{coh}}^{\mathrm{orb}}).$$

We remark that we need to consider Q-stacks instead of Q-varieties in order to deal with general morphisms $f : X \to Y$ of orbifolds.

We have the Serre functor as follows:

Proposition 2.6. *Let X be a quasi-projective variety with an orbifold structure. Assume that X has a Cohen–Macaulay global cover $\tilde{X}$. Then there exists a Serre functor $S = S_X$ for the derived categories $D = D^b(X_{\mathrm{coh}}^{\mathrm{orb}})$ and $D_c = D_c^b(X_{\mathrm{coh}}^{\mathrm{orb}})$ defined by*

$$S(u) = u \otimes_{\mathcal{O}_X^{\mathrm{orb}}} \omega_X^{\mathrm{orb}}[\dim X]$$

for $u \in D$. There are bifunctorial isomorphisms

$$\mathrm{Hom}_D(u, v) \cong \mathrm{Hom}_D(v, S(u))^*$$

for $u \in D$ and $v \in D_c$ or $u \in D_c$ and $v \in D$.

Proof. We first assume that u is a locally free orbifold sheaf and v is an orbifold sheaf with compact support. If we replace v by $u^* \otimes v$, we may assume that $u = \mathcal{O}_X^{\mathrm{orb}}$. We denote by $\tilde{v}$ the sheaf on $\tilde{X}$ corresponding to v with the action of $G = \mathrm{Gal}(\tilde{X}/X)$. Let $\pi : \tilde{X} \to X$ denote the natural morphism. We have

$$\mathrm{Hom}_D(u, v[k]) \cong H^k(\tilde{X}, \tilde{v})^G \cong H^k(X, (\pi_*\tilde{v})^G).$$

On the other hand, the Zariski sheaves $v'_i = (\pi_{i*} v_i)^{G_i}$ on the $X'_i = X_i/G_i$ define an etale sheaf on X. Indeed, we have $v'_i = (\pi'_i)^*(\pi_*\tilde{v})^G$, where $\pi'_i : X'_i \to X$ is induced from π_i. By the relative duality for the morphism $X_i \to X'_i$, we have $R\,\mathrm{Hom}(v_i, \omega_{X_i})^{G_i} \cong R\,\mathrm{Hom}(v'_i, \omega_{X'_i})$. Since $\omega_{X'_i} = (\pi'_i)^*\omega_X$, we have

$$\mathrm{Hom}_D(v[k], S(u)) \cong \mathrm{Ext}^{d-k}((\pi_*\tilde{v})^G, \omega_X)$$

where $d = \dim X$. Hence our assertion is reduced to the usual duality theorem on X.

If v is a locally free orbifold sheaf and u is an orbifold sheaf with compact support, then by the first part

$$\mathrm{Hom}_D(u, v[k]) \cong \mathrm{Hom}_D(S(u), S(v)[k]) \cong \mathrm{Hom}_D(v[k], S(u))^*.$$

The general case is obtained by taking the locally free resolutions. $\square$

3. Flip to flop

We shall reduce the existence problem of the flips to that of flops. For this purpose, we consider the total space KX of the $\mathbb{Q}$-bundle K_X:

Lemma 3.1. *Let X be a variety of dimension n with only log terminal singularities. Then*

$$KX = \mathrm{Spec}(\bigoplus_{m=0}^{\infty} \mathcal{O}_X(-mK_X))$$

is a variety of dimension $n+1$ with only rational Gorenstein singularities and trivial canonical bundle.

Proof. We take a small open subset U of X which has an index 1 cover $\pi : U_1 \to U$ with the Galois group G. We have a commutative diagram

$$\begin{array}{ccc} KU_1 & \xrightarrow{K\pi} & KU \\ \downarrow & & \downarrow \\ U_1 & \xrightarrow{\pi} & U. \end{array}$$

U_1 has only rational Gorenstein singularities and KU_1 is the total space of the line bundle K_{U_1}, hence KU_1 has only rational Gorenstein singularities. Since $\mathcal{O}_X(-mK_U) = (\pi_*\mathcal{O}_{U_1}(-mK_{U_1}))^G$, we have $KU_1 = KU/G$.

Let ω be a generating section of K_{U_1}. Then ω^{-1} gives a fiber coordinate along the fiber of K_{U_1}, and $\tilde{\omega} = \omega \wedge d\omega^{-1}$ is a generating section of K_{KU_1}. Since $\tilde{\omega}$ is G-invariat, K_{KX} is again invertible.

Let U' be another small open subset of X which has an index 1 cover $\pi' : U'_1 \to U'$ and ω' a generating section of $K_{U'_1}$. We can write $\omega' = u\omega$ for an invertible function u on $U_1 \times_X U'_1$. Then $\tilde{\omega}' = \omega' \wedge d\omega'^{-1} = u\omega \wedge u^{-1}d\omega^{-1} = \omega \wedge d\omega^{-1} = \tilde{\omega}$. Therefore, we obtain a global generating section of K_{KX_1}. $\square$

Question 3.2. If X has only terminal singularities, so has KX?

Theorem 3.3. *Let n be an integer such that $n \geq 3$. Assume that the Gorenstein canonical flop always exists in dimension $n + 1$ in the following sense: for any projective birational small crepant morphism $\psi : \tilde{X} \to \tilde{Z}$ of $(n + 1)$-dimensional varieties with only Gorenstein canonical singularities and a ψ-negative effective $\mathbb{Q}$-Cartier divisor D on $\tilde{X}$, there exists another projective birational small crepant morphism $\psi^+ : \tilde{X}^+ \to \tilde{Z}$ from a $(n + 1)$-dimensional variety with only Gorenstein canonical singularities such that the strict transform D^+ of D on $\tilde{X}^+$ is ψ^+-ample. Then the canonical flip always exists in dimension n in the following sense: for any projective birational small morphism $\phi : X \to Z$ from an n-dimensional variety with only canonical singularities such that K_X is ϕ-negative, there exists another projective birational small morphism $\phi^+ : X^+ \to Z$ from an n-dimensional variety with only canonical singularities such that K_{X^+} is ϕ^+-ample.*

Proof. Let $\phi : X \to Z$ be a small projective birational morphism from a variety of dimension n with only canonical singularities to a normal variety such that the canonical divisor K_X is ϕ-negative. Let E be the exceptional locus of ϕ. Let KX be as in Lemma 3.1 with a natural morphism $\xi : KX \to X$. We have an embedding $i : X \to KX$ to the zero section. We also define

$$KZ = \operatorname{Spec}(\bigoplus_{m=0}^{\infty} \mathcal{O}_Z(-mK_Z)).$$

Since

$$\mathcal{O}_Z(-mK_Z) \cong \phi_*(\mathcal{O}_X(-mK_X))$$

the direct sum on the right hand side of the formula for KZ gives a finitely generated sheaf of $\mathcal{O}_Z$-algebras.

We claim that the induced morphism $K\phi : KX \to KZ$ is a projective birational morphism whose exceptional locus coincides with $i(E)$. In order to prove this, we may assume that Z is affine. In this case, $K\phi$ is given by the linear system Λ generated by divisors of the form $mi(X) + \xi^*D$ for $D \in |-mK_X|$. Since $|-mK_X|$ is very ample on X for sufficiently large m, so is Λ on $KX \setminus i(X)$. Since the restriction of $K\phi$ to $i(X)$ coincides with ϕ, we have our assertion. Since K_{KX} is globally trivial, $K\phi$ is crepant.

By the assuption, there exists a flop $(K\phi)^+ : (KX)^+ \to KZ$ with respect to $i(X)$. Let X^+ be the strict transform of $i(X)$ on $(KX)^+$. By definition, X^+ is a $\mathbb{Q}$-Cartier divisor on $(KX)^+$ which is ample for $(K\phi)^+$.

Hence X^+ is $\mathbb{Q}$-Gorenstein though X^+ may not be normal. We shall prove that X^+ is regular in codimension 1 and the induced morphism $\phi^+ : X^+ \to Z$ is small. Then X^+ is normal, since it is Cohen–Macaulay, and $K_{X^+} = K_{(KX)^+} + X^+|_{X^+}$ is ϕ^+-ample, so that ϕ^+ is the flip of ϕ.

Let $\tilde{Y}$ be a common desingularization of KX and $(KX)^+$ with projective birational morphisms $\tilde{\mu} : \tilde{Y} \to KX$ and $\tilde{\mu}^+ : \tilde{Y} \to (KX)^+$ with normal crossing exceptional locus $F = \sum F_j$. We write $\tilde{\mu}^*i(X) = Y + \sum_j r_j F_j$ and $\tilde{\mu}^*K_{KX} + \sum_j a_j F_j = K_{\tilde{Y}}$, where Y is the strict transform of $i(X)$. Since X has only canonical singularities, we have $r_j \leq a_j$ for any j. We write also $(\tilde{\mu}^+)^*X^+ = Y + \sum_j r_j^+ F_j$ and $(\tilde{\mu}^+)^*K_{(KX)^+} + \sum_j a_j^+ F_j = K_{\tilde{Y}}$. Since $\tilde{\mu}^+$ is the flop of $\tilde{\mu}$ with respect to $i(X)$, we have $r_j^+ < r_j$ for any divisor F_j which lies above $i(E)$. Since we have also $a_j^+ = a_j$ for any j, we conclude that $r_j^+ < a_j^+$ for such j.

Let P be any codimension 2 point of $(KX)^+$ contained in the exceptional locus of $\tilde{\mu}^+$. Since $(KX)^+$ has only canonical singularities, there exists an exceptional divisor F_{j_0} above P such that $a_{j_0}^+ = 0$ if $(KX)^+$ is singular at P or $= 1$ if $(KX)^+$ is smooth at P. Thus $r_{j_0}^+ = 0$, and X^+ does not contain P. Therefore, X^+ is regular in codimension 1 and the induced morphism $\phi^+ : X^+ \to Z$ is small. □

Example 3.4. In the notation of §4, if $X^- = X^-(a; b)$ and $c = \sum_i a_i - \sum_j b_j > 0$, then we have $KX^- = X^-(a; b, c)$.

Let a and b be coprime positive integers. The sequene of integers

$$(1, a, b; 1, a + b)$$

is obtained in the above way from the following sequences: (1) $(1, a, b; a + b)$, (2) $(1, a, b; 1)$, (3) $(1, a + b; a, b)$, (4) $(1, a + b; 1, a)$, (5) $(1, a + b; 1, b)$. (1) and (2) correspond to divisorial contractions of 3-folds with only terminal quotient singularities to the quotient singularity of type $\frac{1}{a+b}(1, a, b)$ and to a smooth point, respectively. (3), (4) and (5) correspond to flips from 3-folds with only terminal quotient singularities.

4. Toric flip and flop

According to [14] and [15], we consider the following toric varieties:

Definition 4.1. Let $(a; b) = (a_1, \ldots, a_m; b_1, \ldots, b_n)$ be a sequence of positive integers. We let the multiplicative group G_m act on $\mathbb{A} = \mathbb{A}(a; b) = \operatorname{Spec} R \cong \mathbb{A}^{m+n}$ for $R = \mathbb{C}[x_1, \ldots, x_m, y_1, \ldots, y_n]$ by

$$\lambda_t(x_1, \ldots, x_m, y_1, \ldots, y_n) = (t^{a_1} x_1, \ldots, t^{a_m} x_m, t^{-b_1} y_1, \ldots, t^{-b_n} y_n)$$

for $t \in G_m$. We consider GIT quotients

$$X^- = X^-(a; b) = (\mathbb{A} \setminus \{x_1 = \cdots = x_m = 0\})/G_m$$
$$X^+ = X^+(a; b) = (\mathbb{A} \setminus \{y_1 = \cdots = y_n = 0\})/G_m$$
$$X^0 = X^0(a; b) = \mathbb{A}//G_m = \operatorname{Spec} R^{G_m}.$$

We also define $Y = Y(a; b)$ to be the fiber product $X^- \times_{X^0} X^+$. Let $\phi^\pm : X^\pm \to X^0$ and $\mu^\pm : Y \to X^\pm$ be the induced morphisms as in the following commutative diagram:

$$\begin{array}{ccc} Y & \xrightarrow{\mu^+} & X^+ \\ \mu^- \downarrow & & \downarrow \phi^+ \\ X^- & \xrightarrow{\phi^-} & X^0 \end{array}$$

Let $A_i^\pm$ and $B_j^\pm$ be prime divisors on $X^\pm$ corresponding to the x_i and the y_j, and let $U_i^- = X^- \setminus A_i^-$ and $U_j^+ = X^+ \setminus B_j^+$. Thus $X^- = \bigcup_i U_i^-$ and $X^+ = \bigcup_j U_j^+$. Let A_i and B_j be the strict transforms of $A_i^\pm$ and $B_j^\pm$ on Y, respectively. Let $U_{i,j} = Y \setminus (A_i \cup B_j)$ so that $Y = \bigcup_{i,j} U_{i,j}$.

Example 4.2. If $n = 0$, then X^- is nothing but the weighted projective space $\mathbb{P}(a)$. In this case, X^0 is a point and $X^+ = Y = \emptyset$. If $n = 1$, then ϕ^- is a divisorial contraction and $X^+ = X^0$. If $m = n = 2$ and $a_1 = a_2 = b_1 = b_2 = 1$, then this is Atiyah's flop. If $m = n = 2$ and $a_1 = 2, a_2 = b_1 = b_2 = 1$, then this is Francia's flip.

Proposition 4.3. *If $m, n \geq 2$, then the following hold.*

(1) *The morphisms $\phi^\mp$ are projective and birational whose exceptional loci $E^\mp$ are isomorphic to the weighted projective spaces $\mathbb{P}(a_1, \ldots, a_m)$ and $\mathbb{P}(b_1, \ldots, b_n)$, respectively.*

(2) $E = (\mu^\pm)^{-1}(E^\pm)$ *is a prime divisor on Y isomorphic to the product $E^- \times E^+$.*

(3) *The divisors $\mp A_i^\pm$ and $\pm B_j^\pm$ are $\phi^\pm$-ample.*

with Galois group

$$G \cong \prod_i \mathbb{Z}_{a_i} \times \prod_j \mathbb{Z}_{b_j}.$$

There are induced Galois morphisms $\pi^{\pm} : \tilde{X}^{\pm} \to X^{\pm}$, $\pi_X^0 : \tilde{X}^0 \to X^0$ and $\pi_Y : \tilde{Y} \to Y$ with the same Galois group G.

$\tilde{U}_1^- = \tilde{X}^- \setminus \tilde{A}_1^-$ is isomorphic to $\mathbb{A}^{m+n-1}$ with coordinates

$$\tilde{x}_2/\tilde{x}_1, \ldots, \tilde{x}_m/\tilde{x}_1, \tilde{x}_1\tilde{y}_1, \ldots, \tilde{x}_1\tilde{y}_n.$$

The quotient $(U_1^-)' = \tilde{U}_1^-/G'$ for $G' = \prod_{i>1} \mathbb{Z}_{a_i} \times \prod_j \mathbb{Z}_{b_j}$ is again isomorphic to $\mathbb{A}^{m+n-1}$ with coordinates

$$(\tilde{x}_2/\tilde{x}_1)^{a_2}, \ldots, (\tilde{x}_m/\tilde{x}_1)^{a_m}, (\tilde{x}_1\tilde{y}_1)^{b_1}, \ldots, (\tilde{x}_1\tilde{y}_n)^{b_n}.$$

Hence the first assertion.

If the action is not small, then there exists an integer a_1' such that $0 < a_1' < a_1$, $a_1' | a_1$, and $a_1' a_2 \equiv \cdots \equiv a_1' b_n \equiv 0 \bmod a_1$ except possibly one of the $a_2, \ldots, b_n$. Then $c > 1$, $c_i > 1$ or $c_{m+j} > 1$ for some i or j, a contradiction. □

Proposition 4.6. *Let $a = \gcd(a_1, \ldots, a_m)$ and $b = \gcd(b_1, \ldots, b_n)$. Set $a_i = aa_i'$ and $b_j = bb_j'$. Then the open subset $U_{1,1}$ is isomorphic to the quotient of the affine space $\mathbb{A}^{m+n-1}$ by the group $\mathbb{Z}_{a_1'} \times \mathbb{Z}_{b_1'}$ whose action is given by the weights*

$$\frac{1}{a_1'}(-a_2', \ldots, -a_m', b, 0, \ldots, 0), \quad \frac{1}{b_1'}(0, \ldots, 0, a, -b_2', \ldots, -b_n').$$

Moreover, the action of the group is small in the sense that the induced morphism $\mathbb{A}^{m+n-1} \to U_{1,1}$ is etale in codimension 1.

Proof. $\tilde{U}_{1,1} = \tilde{Y} \setminus (\tilde{A}_1 \cup \tilde{B}_1)$ is isomorphic to $\mathbb{A}^{m+n-1}$ with coordinates

$$\tilde{x}_2/\tilde{x}_1, \ldots, \tilde{x}_m/\tilde{x}_1, \tilde{x}_1\tilde{y}_1, \tilde{y}_2/\tilde{y}_1, \ldots, \tilde{y}_n/\tilde{y}_1.$$

The quotient $U'_{1,1} = \tilde{U}_{1,1}/G'$ for $G' = \prod_{i>1} \mathbb{Z}_{a_i} \times \prod_{j>1} \mathbb{Z}_{b_j}$ is again isomorphic to $\mathbb{A}^{m+n-1}$ with coordinates

$$(\tilde{x}_2/\tilde{x}_1)^{a_2}, \ldots, (\tilde{x}_m/\tilde{x}_1)^{a_m}, \tilde{x}_1\tilde{y}_1, (\tilde{y}_2/\tilde{y}_1)^{b_2}, \ldots, (\tilde{y}_n/\tilde{y}_1)^{b_n}.$$

The group $\mathbb{Z}_{a_1} \times \mathbb{Z}_{b_1}$ acts on $U'_{1,1}$ with weights

$$\frac{1}{a_1}(-a_2, \ldots, -a_m, 1, 0, \ldots, 0), \quad \frac{1}{b_1}(0, \ldots, 0, 1, -b_2, \ldots, -b_n).$$

The quotient $U''_{1,1} = U'_{1,1}/G''$ for $G'' = \mathbb{Z}_a \times \mathbb{Z}_b$ is still isomorphic to $\mathbb{A}^{m+n-1}$ with coordinates

$$(\tilde{x}_2/\tilde{x}_1)^{a_2}, \ldots, (\tilde{x}_m/\tilde{x}_1)^{a_m}, (\tilde{x}_1\tilde{y}_1)^{ab}, (\tilde{y}_2/\tilde{y}_1)^{b_2}, \ldots, (\tilde{y}_n/\tilde{y}_1)^{b_n}.$$

We have the following reduction for the sequence of integers in a similar way to the case of the weighted projective spaces ([7] and [8]).

Proposition 4.4. (1) *Let* $c = \gcd(a_1, \ldots, a_m, b_1, \ldots, b_n)$ *be the greatest common divisor, and let* $(\boldsymbol{a}'; \boldsymbol{b}') = (a_1/c, \ldots, a_m/c; b_1/c, \ldots, b_n/c)$. *Then* $X^{\pm}(\boldsymbol{a}'; \boldsymbol{b}') \cong X^{\pm}(\boldsymbol{a}; \boldsymbol{b})$ *and* $X^0(\boldsymbol{a}'; \boldsymbol{b}') \cong X^0(\boldsymbol{a}; \boldsymbol{b})$.
(2) *Assume* $c = 1$. *Let*

$$c_i = \gcd(a_1, \ldots, \widehat{a_i}, \ldots, a_m, b_1, \ldots, b_n)$$
$$c_{m+j} = \gcd(a_1, \ldots, a_m, b_1, \ldots, \widehat{b_j}, \ldots, b_n)$$
$$d_k = \mathrm{lcm}(c_1, \ldots, \widehat{c_k}, \ldots, c_{m+n})$$
$$(\boldsymbol{a}''; \boldsymbol{b}'') = (a_1/d_1, \ldots, a_m/d_m; b_1/d_{m+1}, \ldots, b_n/d_{m+n}).$$

Then $X^{\pm}(\boldsymbol{a}''; \boldsymbol{b}'') \cong X^{\pm}(\boldsymbol{a}; \boldsymbol{b})$ *and* $X^0(\boldsymbol{a}''; \boldsymbol{b}'') \cong X^0(\boldsymbol{a}; \boldsymbol{b})$.

We may therefore assume that $c_i = c_{m+j} = 1$ for any i and j from now on.

Proof. (1) The action of G_m factors through a homomorphism $G_m \to G_m$ given by $t \mapsto t^c$.

(2) Let $d = \mathrm{lcm}(c_1, \ldots, c_{m+n})$. Since $c = 1$, we have $\gcd(c_i, d_i) = 1$ and $d = c_i d_i$. The ring of invariants R^{G_m} for the sequence $(\boldsymbol{a}; \boldsymbol{b})$ is generated by monomials $x^m y^n$ such that $\sum_i a_i m_i = \sum_j b_j n_j$, and the corresponding ring of invariants $(R'')^{G_m}$ for the new sequence $(\boldsymbol{a}''; \boldsymbol{b}'')$ is generated by monomials $(x'')^{m''}(y'')^{n''}$ such that $\sum_i a_i c_i m_i'' = \sum_j b_j c_{m+j} n_j''$. If $\sum_i a_i m_i = \sum_j b_j n_j$, then it follows that $c_i | m_i$ and $c_{m+j} | n_j$. Hence there is an isomorphism $f : (R'')^{G_m} \to R^{G_m}$ given by $f(x_i'') = x_i^{c_i}$ and $f(y_j'') = y_j^{c_{m+j}}$. Thus we have $X^0(\boldsymbol{a}''; \boldsymbol{b}'') \cong X^0(\boldsymbol{a}; \boldsymbol{b})$. The assertions for $X^{\pm}$ follow from this isomorphism. □

Proposition 4.5. *The open subset* U_1^- *is isomorphic to the quotient of the affine space* $\mathbb{A}^{m+n-1}$ *by the group* $\mathbb{Z}_{a_1}$ *whose action is given by the weights*

$$\frac{1}{a_1}(-a_2, \ldots, -a_m, b_1, \ldots, b_n).$$

Moreover, the action of the group is small in the sense that the induced morphism $\mathbb{A}^{m+n-1} \to U_1^-$ *is etale in codimension 1.*

Proof. Let $(\boldsymbol{1}, \boldsymbol{1}) = (1, \ldots, 1; 1, \ldots, 1)$, and denote $\tilde{\mathbb{A}} = \mathbb{A}(\boldsymbol{1}, \boldsymbol{1})$, $\tilde{X}^{\pm} = X^{\pm}(\boldsymbol{1}, \boldsymbol{1})$, $\tilde{X}^0 = X^0(\boldsymbol{1}, \boldsymbol{1})$ and $\tilde{Y} = Y(\boldsymbol{1}, \boldsymbol{1})$. Denote the coordinates of $\tilde{\mathbb{A}}$ by $\tilde{x}_1, \ldots, \tilde{x}_m$, $\tilde{y}_1, \ldots, \tilde{y}_n$, and define $\tilde{A}_i^{\pm}$ and so on. Then there is a G_m-equivariant finite Galois morphism $\pi_{\mathbb{A}} : \tilde{\mathbb{A}} \to \mathbb{A}$ given by

$$\tilde{x}_i \mapsto x_i = \tilde{x}_i^{a_i}, \quad \tilde{y}_j \mapsto y_j = \tilde{y}_j^{b_j}$$

Here we note that $\gcd(a, b) = 1$. We can check that there are at least 2 numbers which are coprime to a'_1 among $a'_2, \ldots, a'_m$ and b by Proposition 4.4. We also check a similar statement for the b'_j and a. Hence the assertion. $\square$

Definition 4.7. We put natural orbifold structures on the toric varieties $X^\pm(a, b)$, and we take the fiber product $Y(a, b)$ in the sense of orbifolds. More precisely, the orbifold structure of the latter is not the natural one given by the minimal coverings $U''_{i,j} \to U_{i,j}$ but by the coverings $U'_{i,j} \to U_{i,j}$.

The ideal sheaves $\mathcal{O}^{\mathrm{orb}}_{X^-}(-A_i^-)$ and $\mathcal{O}^{\mathrm{orb}}_{X^-}(-B_j^-)$ have the structure of invertible orbifold sheaves. Indeed, on the covering $(U_1^-)'$ of the affine open subset U_1^-, the sheaves $\mathcal{O}^{\mathrm{orb}}_{X^-}(-A_i^-)$ for $i = 2, \ldots, m$ and $\mathcal{O}^{\mathrm{orb}}_{X^-}(-B_j^-)$ for $j = 1, \ldots, n$ are generated by the coordinates

$$(\tilde{x}_2/\tilde{x}_1)^{a_2}, \ldots, (\tilde{x}_m/\tilde{x}_1)^{a_m}, (\tilde{x}_1\tilde{y}_1)^{b_1}, \ldots, (\tilde{x}_1\tilde{y}_n)^{b_n}.$$

The sheaves of invariants under the Galois group action coincide with the usual reflexive ideal sheaves:

$$\mathcal{O}^{\mathrm{orb}}_{X^-}(-A_i^-)|_{X^-} = \mathcal{O}_{X^-}(-A_i^-), \quad \mathcal{O}^{\mathrm{orb}}_{X^-}(-B_j^-)|_{X^-} = \mathcal{O}_{X^-}(-B_j^-).$$

We can define invertible orbifold sheaves $\mathcal{O}^{\mathrm{orb}}_{X^\pm}(k)$ for $k \in \mathbb{Z}$ on $X^\pm$ and $\mathcal{O}^{\mathrm{orb}}_Y(k_1, k_2)$ for $k_1, k_2 \in \mathbb{Z}$ on Y so that we have isomorphisms

$$\mathcal{O}^{\mathrm{orb}}_{X^\pm}(A_i^\pm) \cong \mathcal{O}^{\mathrm{orb}}_{X^\pm}(\pm a_i), \quad \mathcal{O}^{\mathrm{orb}}_{X^\pm}(B_j^\pm) \cong \mathcal{O}^{\mathrm{orb}}_{X^\pm}(\mp b_j),$$

$$\mathcal{O}^{\mathrm{orb}}_Y(A_i) \cong \mathcal{O}^{\mathrm{orb}}_Y(a_i, 0), \quad \mathcal{O}^{\mathrm{orb}}_Y(B_j) \cong \mathcal{O}^{\mathrm{orb}}_Y(0, b_j)$$

$$\mathcal{O}^{\mathrm{orb}}_Y(\bar{E}) \cong \mathcal{O}^{\mathrm{orb}}_Y(-1, -1)$$

where $\bar{E}$ is the exceptional prime divisor on the Galois covers $U'_{i,j}$ so that

$$\pi^*_{i,j} E = ab\bar{E}$$

for $\pi_{i,j} : U'_{i,j} \to U_{i,j}$. Indeed, the coordinates

$$(\tilde{x}_2/\tilde{x}_1)^{a_2}, \ldots, (\tilde{x}_m/\tilde{x}_1)^{a_m}, \tilde{x}_1\tilde{y}_1, (\tilde{y}_2/\tilde{y}_1)^{b_2}, \ldots, (\tilde{y}_n/\tilde{y}_1)^{b_n}.$$

on $U'_{1,1}$ correspond to the prime divisors $A_2, \ldots, A_m, \bar{E}$, and $B_2, \ldots, B_n$. We have the following equalities

$$(\mu^-)^* A_i^- = A_i, \qquad (\mu^-)^* B_j^- = B_j + b_j \bar{E}$$

$$(\mu^+)^* A_i^+ = A_i + a_i \bar{E}, \qquad (\mu^+)^* B_j^- = B_j$$

$$(\mu^-)^* \mathcal{O}^{\mathrm{orb}}_{X^-}(k) = \mathcal{O}^{\mathrm{orb}}_Y(k, 0), \qquad (\mu^+)^* \mathcal{O}^{\mathrm{orb}}_{X^+}(k) = \mathcal{O}^{\mathrm{orb}}_Y(0, k).$$

Since $K_{X^\pm} + \sum_i A_i^\pm + \sum_j B_j^\pm \sim 0$, we have

$$\omega^{\mathrm{orb}}_{X^\pm} \cong \mathcal{O}^{\mathrm{orb}}_{X^\pm}\left(\pm\left(\sum_i a_i - \sum_j b_j\right)\right).$$

Hence
$$(\mu^-)^* K_{X^-} \sim (\mu^+)^* K_{X^+} + \left(\sum_i a_i - \sum_j b_j\right)\bar{E}.$$

On the other hand, though we have $K_Y + \sum_i A_i + \sum_j B_j + E \sim 0$, we have
$$\omega_Y^{\text{orb}} \cong \mathcal{O}_Y^{\text{orb}}\left(-\sum_i a_i + 1, -\sum_j b_j + 1\right)$$

because we have additional ramification along E.

5. Flip and derived categories

The following example shows that we should consider the derived categories of orbifold sheaves instead of ordinary sheaves.

Example 5.1. We consider Francia's flop; we take a sequence of integers $(a; b) = (1, 2; 1, 1, 1)$.

X^- has only one singular point $P_0 \in U_1^-$ which is a quotient singularity of type $\frac{1}{2}(1, 1, 1, 1)$, while X^+ is smooth. Y has 2-dimensional singular locus of type A_1. The exceptional loci $E^- \subset X^-$, $E^+ \subset X^+$ and $E \subset Y$ are respectively isomorphic to $\mathbb{P}^1$, $\mathbb{P}^2$ and $\mathbb{P}^1 \times \mathbb{P}^2$.

By direct calculation, we observe that the image of a sheaf under the Fourier–Mukai transform
$$R\mu_*^+ L(\mu^-)^* \mathcal{O}_{X^-}(-A_2^-) \in D^-((X^+)_{\text{coh}})$$

has unbounded cohomology sheaves.

On the other hand, we can calculate
$$R\mu_*^- L(\mu^+)^*(\Omega_{E^+}^1(-1)) = R\mu_*^-(\mu^+)^!(\Omega_{E^+}^1(1)) = 0$$

in $D^b((X^-)_{\text{coh}})$. Indeed, we have
$$R(\mu^-)_*^{\text{orb}} L(\mu^+)_{\text{orb}}^*(\Omega_{E^+}^1(-1)) = R(\mu^-)_*^{\text{orb}}(\mu^+)_{\text{orb}}^! \Omega_{E^+}^1(1) = \mathcal{O}_{\tilde{P}_0}^-[-1]$$

in $D^b((X^-)_{\text{coh}}^{\text{orb}})$, where $\mathcal{O}_{\tilde{P}_0}^-$ is the structure sheaf $\mathcal{O}_{\tilde{P}_0}$ of the point $\tilde{P}_0 \in \tilde{U}_1^-$ above P_0 with the non-trivial action of $\text{Gal}(\tilde{U}_1^-/U_1^-)$.

The following theorem is the main result of this section. This is an extension of the result of Bondal–Orlov [1] Theorem 3.6 to the orbifold case. First we introduce the notation. Fixing a sequence of positive integers $(a; b)$, we consider the following cartesian diagram of quasi-projective toroidal varieties

$$
\begin{array}{ccc}
\mathcal{Y} & \xrightarrow{\hat{\mu}^+} & \mathcal{X}^+ \\
{\scriptstyle \hat{\mu}^-}\downarrow & & \downarrow{\scriptstyle \hat{\phi}^+} \\
\mathcal{X}^- & \xrightarrow{\hat{\phi}^-} & \mathcal{X}^0
\end{array}
\qquad (5.1)
$$

whose local models are the product of the toric varieties defined in §4 and a fixed smooth closed subvariety W of $\mathcal{X}^0$:

$$
\begin{array}{ccc}
Y(a;b) \times W & \xrightarrow{\mu^+ \times \mathrm{Id}_W} & X^+(a;b) \times W \\
{\scriptstyle \mu^- \times \mathrm{Id}_W}\downarrow & & \downarrow{\scriptstyle \phi^+ \times \mathrm{Id}_W} \\
X^-(a;b) \times W & \xrightarrow{\phi^- \times \mathrm{Id}_W} & X^0(a;b) \times W.
\end{array}
\qquad (5.2)
$$

We assume that the base change of the diagram 5.1 by the completion of $\mathcal{X}^0$ at any point $w \in W$ is isomorphic to that of the diagram 5.2 by the completion of $X^0(a;b) \times W$ at (P_0, w), where $P_0 = \phi^\pm(E^\pm)$. We put natural orbifold structures on $\mathcal{X}^\pm$ and the orbifld structure of the fiber product on $\mathcal{Y}$.

Theorem 5.2. *In the situation above, assume that the orbifolds $\mathcal{X}^\pm$ have Cohen–Macaulay global covers. Asume moreover that $\sum a_i \leq \sum b_j$, i.e.,*

$$(\hat{\mu}^-)^* K_{\mathcal{X}^-} \leq (\hat{\mu}^+)^* K_{\mathcal{X}^+}.$$

Then the Fourie–Mukai functors

$$\mathcal{F} = R(\hat{\mu}^+)_* L(\hat{\mu}^-)^* : D^b_{\mathrm{coh}}(\mathcal{X}^-) \to D^b_{\mathrm{coh}}(\mathcal{X}^+)$$
$$\mathcal{F}' = R(\hat{\mu}^+)_* (\hat{\mu}^-)^! : D^b_{\mathrm{coh}}(\mathcal{X}^-) \to D^b_{\mathrm{coh}}(\mathcal{X}^+)$$

are fully faithful. In particular, if $\sum a_i = \sum b_j$, then they are equivalences of categories.

We recall the definition of the spanning class.

Definition 5.3. A set of objects Ω of a triangulated category A is said to be a *spanning class* if the following hold for any $a \in A$:

(1) $\mathrm{Hom}_A(a, \omega[k]) = 0$ for any $\omega \in \Omega$ and $k \in \mathbb{Z}$ implies $a \cong 0$.
(2) $\mathrm{Hom}_A(\omega[k], a) = 0$ for any $\omega \in \Omega$ and $k \in \mathbb{Z}$ implies $a \cong 0$.

Lemma 5.4. *Let $f : A \to B$ be an exact functor between triangulated categories with a right adjoint g and a left adjoint h. Let Ω be a spanning class of A. Assume that $gf(\omega) \cong hf(\omega) \cong \omega$ for any $\omega \in \Omega$. Then $gf(a) \cong hf(a) \cong a$ for any $a \in A$ and f is fully faithful.*

Proof. If $a \in A$, then

$$\text{Hom}_A(\omega, gf(a)) \cong \text{Hom}_A(hf(\omega), a) \cong \text{Hom}(\omega, a)$$
$$\text{Hom}_A(hf(a), \omega) \cong \text{Hom}_A(a, gf(\omega)) \cong \text{Hom}(a, \omega)$$

hence the natural morphisms $a \to gf(a)$ and $hf(a) \to a$ are isomorphisms. Thus

$$\text{Hom}(f(a), f(a')) \cong \text{Hom}(hf(a), a') \cong \text{Hom}(a, a')$$

for any $a' \in A$. □

Example 5.5. (1) Let X be a quasi-projective variety with an orbifold structure having a Cohen–Macaulay global cover. For any point $x \in X$, there exists a finite group G_x such that, if $x \in \pi_i(X_i)$, then the stabilizer subgroup of G_i at any point $\tilde{x} \in \pi_i^{-1}(x)$ is isomorphic to G_x. Let V be any irreducible representation of G_x. Then the sheaf

$$Z_{x,V,i} = \bigoplus_{\tilde{x} \in \pi_i^{-1}(x)} V \otimes_{\mathbb{C}} \mathcal{O}_{\tilde{x}}$$

on X_i glue together to define an orbifold sheaf $Z_{x,V}$ on X.

Let $\mathcal{P}_X$ be the set of all the orbifold sheaves of the form $Z_{x,V}$ for the points $x \in X$ and irreducible representations V of G_x. Then $\mathcal{P}_X$ is a spanning class for $D^b(X_{\text{coh}}^{\text{orb}})$. The proof is similar to [3] Example 2.2.

(2) Let $X^{\pm} = X^{\pm}(a; b)$ be as in §4 and fix a positive integer k_0. Then the set of orbifold sheaves

$$\mathcal{Q}_{X^{\pm}} = \{\mathcal{O}_{X^{\pm}}^{\text{orb}}(k) \mid k \in \mathbb{Z} \text{ and } k \geq k_0\}$$

is a spanning class for $D^b((X^{\pm})_{\text{coh}}^{\text{orb}})$. Indeed, any orbifold sheaf in $\mathcal{P}_{X^{\pm}}$ can be resolved into a complex of orbifold sheaves which are direct sums of the orbifold sheaves in $\mathcal{Q}_{X^{\pm}}$.

Lemma 5.6. *Let* $X^{\pm} = X^{\pm}(a; b)$ *be as in* §4. *Let*

$$F = R(\mu^+)_*^{\text{orb}} \circ L(\mu^-)_{\text{orb}}^* : D^b((X^-)_{\text{coh}}^{\text{orb}}) \to D^b((X^+)_{\text{coh}}^{\text{orb}})$$
$$G = R(\mu^-)_*^{\text{orb}} \circ (\otimes \omega_{Y/X^+}^{\text{orb}}) \circ L(\mu^+)_{\text{orb}}^* : D^b((X^+)_{\text{coh}}^{\text{orb}}) \to D^b((X^-)_{\text{coh}}^{\text{orb}})$$
$$H = R(\mu^-)_*^{\text{orb}} \circ (\otimes \omega_{Y/X^-}^{\text{orb}}) \circ L(\mu^+)_{\text{orb}}^* : D^b((X^+)_{\text{coh}}^{\text{orb}}) \to D^b((X^-)_{\text{coh}}^{\text{orb}})$$
$$F' = R(\mu^+)_*^{\text{orb}} \circ (\otimes \omega_{Y/X^-}^{\text{orb}}) \circ L(\mu^-)_{\text{orb}}^* : D^b((X^-)_{\text{coh}}^{\text{orb}}) \to D^b((X^+)_{\text{coh}}^{\text{orb}})$$
$$G' = R(\mu^-)_*^{\text{orb}} \circ (\otimes \omega_{Y/X^+}^{\text{orb}} + \omega_{Y/X^-}^{\text{orb}-1}) \circ L(\mu^+)_{\text{orb}}^* : D^b((X^+)_{\text{coh}}^{\text{orb}}) \to D^b((X^-)_{\text{coh}}^{\text{orb}})$$
$$H' = R(\mu^-)_*^{\text{orb}} \circ L(\mu^+)_{\text{orb}}^* : D^b((X^+)_{\text{coh}}^{\text{orb}}) \to D^b((X^-)_{\text{coh}}^{\text{orb}}).$$

Then (H, F, G) *and* (H', F', G') *are adjoint triples of functors.*

We need a simple lemma in commutative algebra:

Lemma 5.7. *Let $R = \mathbb{C}[x_1, \ldots, x_m]$ be a polynomial ring with graded ring structure defined by $\deg x_i = a_i$. Let I_k be the graded ideal of R consisting of elements of degree greater than or equal to k. Then there exists a graded free resolution*

$$0 \to \bigoplus_{\lambda \in \Lambda_m} R(-e_\lambda^{(m)}) \to \cdots \to \bigoplus_{\lambda \in \Lambda_1} R(-e_\lambda^{(1)}) \to I_k \to 0$$

given by matrices with monomial entries such that

$$k \le e_\lambda^{(l)} < k + \sum_{i=1}^m a_i$$

for any $1 \le l \le m$ and any λ.

Proof. From the Koszul complex

$$0 \to R(-\sum_{i=1}^m a_i) \to \cdots \to \bigoplus_{i=1}^m R(-a_i) \to R \to \mathbb{C} \to 0$$

we obtain

$$\mathrm{Ext}_R^l(\mathbb{C}, \mathbb{C}) \cong \bigwedge^l (\bigoplus_{i=1}^m \mathbb{C}(a_i)).$$

We can express the R-module R/I_k as extensions of the R-modules $\mathbb{C}(-e)$ such that $0 \le e < k$. Since

$$\mathrm{Ext}_R^l(R/I_k, \mathbb{C}) \cong \bigoplus_{\lambda \in \Lambda_l} \mathbb{C}(e_\lambda^{(l)})$$

we obtain our assertion. $\square$

Let $\mathcal{I}_k^-$ ($k \ge 0$) be the orbifold ideal sheaf on $X^- = X^-(a; b)$ generated by monomials of order k on $\tilde{y}_1, \ldots, \tilde{y}_n$. By the vanishing theorem, we have the following:

Lemma 5.8.

$$R(\mu^-)_*^{\mathrm{orb}} \mathcal{O}_Y^{\mathrm{orb}}(k\bar{E}) = \mathcal{O}_{X^-}^{\mathrm{orb}}$$

for $0 \le k \le \sum_j b_j - 1$ and

$$R(\mu^-)_*^{\mathrm{orb}} \mathcal{O}_Y^{\mathrm{orb}}(-k\bar{E}) = \mathcal{I}_k^-$$

for $0 \le k$

Proposition 5.9. *Under the notation of Lemma 5.6, let $u = \mathcal{O}_{X^-}^{\mathrm{orb}}(k)$. Assume that $\sum_i a_i \le \sum_j b_j$.*

(1) *If $k \ge 0$, then $GF(u) \cong HF(u) \cong u$.*

(2) *If $k \ge \sum_j b_j - 1$, then $G'F'(u) \cong H'F'(u) \cong u$.*

Proof. Since
$$L(\mu^-)^*_{\mathrm{orb}} \mathcal{O}^{\mathrm{orb}}_{X^-}(k) \cong \mathcal{O}^{\mathrm{orb}}_Y(0, -k)(-k\bar{E})$$
we have
$$F(\mathcal{O}^{\mathrm{orb}}_{X^-}(k)) \cong I_k^+(-k).$$
We have a locally free resolution
$$0 \to \bigoplus_{\lambda \in \Lambda_m} \mathcal{O}^{\mathrm{orb}}_{X^+}(e_\lambda^{(m)}) \to \cdots \to \bigoplus_{\lambda \in \Lambda_1} \mathcal{O}^{\mathrm{orb}}_{X^+}(e_\lambda^{(1)}) \to I_k^+ \to 0$$
where the maps are given by matrices whose entries are monomials in the x_i. Therefore
$$(\otimes \omega^{\mathrm{orb}}_{Y/X^+}) \circ L(\mu^+)^*_{\mathrm{orb}} \circ F(\mathcal{O}^{\mathrm{orb}}_{X^-}(k))$$
$$\cong (0 \to \bigoplus_{\lambda \in \Lambda_m} \mathcal{O}^{\mathrm{orb}}_Y(k - e_\lambda^{(m)}, 0)((k - e_\lambda^{(m)} + \sum_i a_i - 1)\bar{E}) \to$$
$$\cdots \to \bigoplus_{\lambda \in \Lambda_1} \mathcal{O}^{\mathrm{orb}}_Y(k - e_\lambda^{(1)}, 0)((k - e_\lambda^{(1)} + \sum_i a_i - 1)\bar{E}) \to 0).$$

Since $\sum_i a_i \leq \sum_j b_j$ and $k \leq e_\lambda^{(l)} < k + \sum_i a_i$ for any $1 \leq l \leq m$ and any λ, we obtain
$$G \circ F(\mathcal{O}^{\mathrm{orb}}_{X^-}(k))$$
$$\cong (0 \to \bigoplus_{\lambda \in \Lambda_m} \mathcal{O}^{\mathrm{orb}}_{X^-}(k - e_\lambda^{(m)}) \to \cdots \to \bigoplus_{\lambda \in \Lambda_1} \mathcal{O}^{\mathrm{orb}}_{X^-}(k - e_\lambda^{(1)}) \to 0)$$
$$\cong \mathcal{O}^{\mathrm{orb}}_{X^-}(k)$$
where the latter isomorphism is obtained because I_k in Lemma 5.7 is primary to the maximal ideal I_1.

The other isomorphisms are proved similarly. □

Corollary 5.10. *If $\sum_i a_i \leq \sum_j b_j$, then*
$$GF(u) \cong HF(u) \cong G'F'(u) \cong H'F'(u) \cong u$$
for any $u \in D^b((X^-)^{\mathrm{orb}}_{\mathrm{coh}})$, and the functors F and G are fully faithful.

Corollary 5.11. *If $\sum_i a_i = \sum_j b_j$, then the functor F is an equivalence of categories whose inverse is given by G.*

Proof of Theorem 5.2. We can extend the result of Corollary 5.10 by replacing $X^\pm$ and Y by $X^\pm \times W$ and $Y \times W$. If we define $\mathcal{F}, \mathcal{G}$, and so on as in Lemma 5.6, then we have
$$\mathcal{G}\mathcal{F}(u) \cong \mathcal{H}\mathcal{F}(u) \cong \mathcal{G}'\mathcal{F}'(u) \cong \mathcal{H}'\mathcal{F}'(u) \cong u$$
for any $u \in \mathcal{P}_{X^-}$, hence the result. □

6. Reconstruction

We extend the reconstruction theorem by Bondal–Orlov to the orbifold case in this section.

Theorem 6.1. *Let X and X' be projective varieties with only quotient singularities. Assume the following conditions for X and X':*

(1) *The natural orbifold structure has a Cohen–Macaulay global cover.*

(2) *The canonical divisor generates local class groups at any point.*

Suppose that K_X or $-K_X$ is ample, and there is an equivalence of categories $D^b(X_{\text{coh}}^{\text{orb}}) \to D^b((X')_{\text{coh}}^{\text{orb}})$ which is compatible with shifting functors. Then there exists an isomorphism $X \to X'$.

Proof. We follow closely the proof of [1] Theorem 4.5. Denoting X or X' by Y, we let $D(Y) = D^b(Y_{\text{coh}}^{\text{orb}})$ and S_Y its Serre functor.

Step 1. We define a *point object of codimension s* to be an object $P \in D(Y)$ satisfying the following conditions:

(1) $S_Y^r(P) \cong P[rs]$ for some positive integer r. Let r_P be the smallest such r.
(2) $\text{Hom}^{<0}(P, S_Y^m(P)[-ms]) = 0$ for any integer m.
(3) $\text{Hom}^0(P, P) = \mathbb{C}$ and $\text{Hom}^0(P, S_Y^m(P)[-ms]) = 0$ for $0 < m < r_P$.

Step 1.1. We claim that any point object P of $D(X)$ is of the form $\mathcal{O}_x \otimes (\omega_X^{\text{orb}})^m[t]$ for some $x \in X$ and $m, t \in \mathbb{Z}$.

Indeed, it follows from (1) that $H^i(P) \otimes (\omega_X^{\text{orb}})^r \cong H^i(P)$ and $s = \dim X$, hence the supports of the cohomology orbifold sheaves $H^i(P)$ are 0-dimensional and ω_X^r is invertible there. Then P can be represented by a complex of orbifold sheaves whose supports are 0-dimensional. By (3), the support of P is a single point. Let i_0 and i_1 be the minimum and the maximum of the i such that $H^i(P) \neq 0$. Then we have $\text{Hom}^{i_0-i_1}(P, P \otimes (\omega_X^{\text{orb}})^m) \neq 0$ for some m, hence $i_0 = i_1$ by (2), i.e., P is an orbifold sheaf. If the length of P is more than 1, then there is a non-invertible homomorphism $P \to P \otimes (\omega_X^{\text{orb}})^m$ for some m, a contradiction to (3), hence P has the claimed form.

Step 1.2. We claim that any point object P of $D(X')$ is also of the form $\mathcal{O}_{x'} \otimes (\omega_{X'}^{\text{orb}})^m[t]$ for some $x' \in X'$ and $m, t \in \mathbb{Z}$.

Indeed, since $D(X)$ and $D(X')$ are equivalent, it follows from Step 1.1 that, for any point objects P and Q of $D(X')$, either $Q = S_{X'}^m(P)[t]$ for some $m, t \in \mathbb{Z}$ or $\text{Hom}^i(P, Q) = 0$ for any i holds. If P is not of the claimed form, then $\text{Hom}^i(P, \mathcal{O}_{x'} \otimes (\omega_{X'}^{\text{orb}})^m) = 0$ for any i, m and x', hence $P = 0$.

Step 2. An *invertible object* $L \in D(Y)$ is defined by the following condition: If P is any point object of codimension s, then there exist uniquely integers $m_0 \in [0, r_P - 1]$

and t_0 such that $\text{Hom}^{t_0}(L, S_Y^{m_0}(P)) = \mathbb{C}$ and $\text{Hom}^i(L, S_Y^m(P)) = 0$ for $i \neq t_0$ or $m \not\equiv m_0 \mod r_P$.

We claim that any invertible object of $D(Y)$ is of the form $L[t]$ for some invertible orbifold sheaf L on Y and some $t \in \mathbb{Z}$. Indeed, we consider a spectral sequence

$$E_2^{p,q} = \text{Hom}^p(H^{-q}(L), \mathcal{O}_y \otimes (\omega_Y^{\text{orb}})^m) \Rightarrow \text{Hom}^{p+q}(L, \mathcal{O}_y \otimes (\omega_Y^{\text{orb}})^m)$$

for $y \in Y$ and $m \in \mathbb{Z}$. If i_1 is the maximum of the i such that $H^i(L) \neq 0$, then $E_2^{p,-i_1}$ for $p = 0, 1$ survive at E_∞. On the other hand, for any $y \in \text{Supp}(H^{i_1}(L))$, there exists an integer m_y such that $\text{Hom}^0(H^{i_1}(L), \mathcal{O}_y \otimes (\omega_Y^{\text{orb}})^{m_y}) \neq 0$. Thus

$$\text{Hom}^0(H^{i_1}(L), \mathcal{O}_y \otimes (\omega_Y^{\text{orb}})^m) = \begin{cases} \mathbb{C} & \text{if } m \equiv m_y \mod r_y \\ 0 & \text{otherwise} \end{cases}$$

$$\text{Hom}^1(H^{i_1}(L), \mathcal{O}_y \otimes (\omega_Y^{\text{orb}})^m) = 0$$

for any m, hence $H^{i_1}(L)$ is an invertible orbifold sheaf. Then $E_2^{p,-i_1} = 0$ for $p \neq 0$, hence $E_2^{0,-i_1+1}$ survives at E_∞. Thus $\text{Hom}^0(H^{i_1-1}(L), \mathcal{O}_y \otimes (\omega_Y^{\text{orb}})^m) = 0$ for any $y \in Y$ and m, and $H^{i_1-1}(L) = 0$. Continuing this process, we obtain that $H^i(L) = 0$ for $i \neq i_1$, and conclude that $L[i_1]$ is an invertible orbifold sheaf.

Step 3. We fix an invertible orbifold sheaf L_0 on X. By Step 2, there exists an invertible orbifold sheaf L_0' on X' such that $L_0 \in D(X)$ corresponds to $L_0'[t_0] \in D(X')$ for some t_0. If we compose the shift functor to the given equivalence functor $D(X) \to D(X')$, we may assume that $t_0 = 0$. The set of point objects $P \in D(X)$ such that $\text{Hom}(L_0, P) \cong \mathbb{C}$ corresponds bijectively to the set of those $P' \in D(X')$ such that $\text{Hom}(L_0', P') \cong \mathbb{C}$. They correspond respectively to the sets of points $x \in X$ and $x' \in X'$ by the isomorphisms $P \cong P_x = \mathcal{O}_x \otimes (\omega_X^{\text{orb}})^{m_x}$ and $P' \cong P_{x'} = \mathcal{O}_{x'} \otimes (\omega_{X'}^{\text{orb}})^{m_{x'}}$, where m_x and $m_{x'}$ are integers depending on the points x and x'. Therefore we obtain a bijection of sets of points on X and X'. From this it follows that the sets of invertible orbifold sheaves on X and X' also correspond bijectively.

Step 4. If L_1 and L_2 are invertible orbifold sheaves on X and $u \in \text{Hom}(L_1, L_2)$, then the set $U(L_1, L_2, u)$ of points $x \in X$ such that the map $u^* : \text{Hom}(L_2 \otimes (\omega_X^{\text{orb}})^m, P_x) \to \text{Hom}(L_1 \otimes (\omega_X^{\text{orb}})^m, P_x)$ is bijective for any m is an affine open subset of X. The subsets $U(L_1, L_2, u)$ for all the L_1, L_2 and u form a basis of the Zariski topology of X. Hence the Zariski topologies on X and X' coincide under the bijection given in Step 3.

Step 5. Let $L_m = L_0 \otimes (\omega_X^{\text{orb}})^m$ and $L_m' = L_0' \otimes (\omega_{X'}^{\text{orb}})^m$. We set $\epsilon = \pm 1$ such that ϵK_X is ample. Then the subsets $U(L_0, L_{m\epsilon}, u)$ for all the positive integers m and all the non-zero sections $u \in \text{Hom}(L_0, L_m)$ form a basis of the Zariski topology of X. Since the same statement holds for X', it follows that $\epsilon K_{X'}$ is also ample. Since $\text{Hom}(L_i, L_{i+m}) \cong H^0(X, mK_X)$ for any i, the multiplication on the (anti-)canonical ring $R(X) = \bigoplus_{m=0}^\infty H^0(X, m\epsilon K_X)$ is given by the composition of morphisms in $D(X)$. Hence X and X' have isomorphic (anti-) canonical rings. □

Added in Proof. B. Totaro proved in "The resolution property for schemes and stacks" (to appear in J. Reine Angew. Math.) that any coherent sheaf on a smooth orbifold whose coarse moduli space is a separated scheme has a finite locally free resolution. Therefore, the assumption on the Cohen–Macaulay property in Proposition 2.3 (2) should be removed as well as in Corollary 2.5, Proposition 2.6, Theorem 5.2, Example 5.5 and Theorem 6.1.

References

[1] A. I. Bondal and D. O. Orlov, Semiorthogonal decompositions for algebraic varieties, math.AG/9506012.

[2] A. I. Bondal and D. O. Orlov, Reconstruction of a variety from the derived category and groups of autoequivalences, Compositio Math. 125 (2001), 327–344.

[3] T. Bridgeland, Equivalences of triangulated categories and Fourier-Mukai transforms, Bull. London Math. Soc. 31 (1999), 25–34.

[4] T. Bridgeland, Flops and derived categories, math.AG/9809114.

[5] T. Bridgeland, A. King and M. Reid, Mukai implies McKay: the McKay correspondence as an equivalence of derived categories, J. Amer. Math. Soc. 14 (2001), 535–554.

[6] J. C. Chen, Flops and equivalence of derived categories for threefolds with only terminal singularities, math.AG/0202005.

[7] C. Delorme, Espaces projectifs anisotropes, Bull. Soc. Math. France 103 (1975), 203–223.

[8] I. Dolgachev, Weighted projective varieties, in: Group actions and vector fields (J. B. Carrell, ed.), Lecture Notes in Math. 956, Springer-Verlag, Berlin 1982, 34–71.

[9] Y. Kawamata, Canonical and minimal models of algebraic varieties, Proc. Internat. Congr. Math. Kyoto 1990, Math. Soc. Japan, 1991, 699–707.

[10] Y. Kawamata, K. Matsuda and K. Matsuki, Introduction to the minimal model problem, in: Algebraic Geometry Sendai 1985, Adv. Stud. Pure Math. 10, Kinokuniya and North-Holland, 1987, 283–360.

[11] S. Mori, Flip theorem and the existence of minimal models for 3-folds,. J. Amer. Math. Soc. 1 (1988), 117–253.

[12] S. Mukai, Duality between $D(X)$ and $D(\hat{X})$ with its application to Picard sheaves, Nagoya Math. J. 81 (1981), 153–175.

[13] D. Mumford, Towards an enumerative geometry of the moduli space of curves, in: Arithmetic and Geometry II, Prog. Math. 36, Birkhäuser, Basel 1983, 271–328.

[14] M. Reid, What is a flip?, unpublished.

[15] M. Thaddeus, Geometric invariant theory and flips, J. Amer. Math. Soc. 9(1996), 691–723.

[16] T. Yasuda, Twisted jet, motivic measure and orbifold cohomology, math.AG/0110228.

Y. Kawamata, Department of Mathematical Sciences, University of Tokyo, Komaba, Meguro, Tokyo, 153-8914, Japan
E-mail: kawamata@ms.u-tokyo.ac.jp

On the quadric hull of a canonical surface

Kazuhiro Konno

Abstract. A minimal surface of general type with birational canonical map is called a canonical surface, and the quadric hull is defined as the intersection of all hyperquadrics through the canonical image. An upper bound is given for the dimension of the quadric hull of a canonical surface. Also all canonical surfaces on a certain extremal line are classified and their defining equations are given.

2000 Mathematics Subject Classification: 14J10

0. Introduction

Let S be a minimal projective surface of general type defined over the complex number field $\mathbb{C}$, and let K_S be the canonical bundle. We call S a *canonical surface* if the canonical map, i.e. the rational map associated with the linear system $|K_S|$, is a birational map of S onto its image X. We denote by $\mathrm{Quad}(S)$ the intersection of all hyperquadrics through X and call it the *quadric hull* of X. The dimension of $\mathrm{Quad}(S)$ is defined as the maximum of the dimensions of irreducible components of $\mathrm{Quad}(S)$ containing X.

Recall that the quadric hull of canonical curves plays an important role in the curve theory: Enriques–Petri's theorem states that the dimension of the quadric hull of a canonical curve is either 1 or 2, and if it is 2 the quadric hull is a surface of minimal degree whose ruling induces either a g_3^1 or a g_5^2 on the curve. In analogy, one naturally expects that a canonical surface has a particular structure when its quadric hull is of dimension strictly greater than 2. In [15], Reid conjectured that $\dim \mathrm{Quad}(S) = 3$ holds for any canonical surfaces with $K_S^2 < 4p_g - 12$. This would imply that any such surface has a flavor similar to a trigonal (or a plane quintic) curve. Indeed, it is true when S is an even surface (see, [11]). We would be able to say something even when $K_S^2 \geq 4p_g - 12$ provided that $\dim \mathrm{Quad}(S) > 2$. But our knowledge is very poor in this direction.

As the begining of the general programme to understand such special canonical surfaces, the present article aims to give an answer to the naive question: Can $\dim \mathrm{Quad}(S)$ become arbitrarily large? With the help of Miyaoka–Yau's inequality [14], we show that there is a universal upper bound 19 and that $\dim \mathrm{Quad}(S)$ is at most 9 when p_g is large enough (see Theorem 1.6 for the precise statement). Though the proof is rather primitive, it can introduce several lines in the zone of the existence, including Castelnuovo's line $K_S^2 = 3p_g - 7$, which may have a particular importance also in future study. We can find among them the line $K_S^2 = 4p_g - 11$ which is next to the above mentioned Reid's line. Hence it is natural to try to clarify the possible structure of surfaces on such an extremal line. This is the second purpose of the paper. Our classification result can be found in several propositions in §3. Through the exploration, we get a naive feeling that there should be a bound sharper than that given in Theorem 1.6, at least when p_g is sufficiently large.

Another important problem left is to determine the structure of the quadric hull itself. An ideal conclusion modeled on Enriques–Petri's theorem expects that Quad(S) is geometrically simple and has, for example, a pencil easy to handle with. An inequality given in §1 shows that there is a degree bound, which leads us to study varieties of small degree. In §2, we exibit some partial results for them as well as their applications to canonical surfaces, although most of which should be well-known.

It is a pleasure to thank the organizers of the conference in memory of Paolo Francia for inviting me to contribute.

1. Counting quadrics

A *prepolarized variety* (V, L) is a pair of an irreducible projective variety V and a line bundle L on V. We denote by $\Phi_L : V \to \mathbb{P}^{\dim |L|}$ the rational map associated with $|L|$.

Definition 1.1. The intersection of all hyperquadrics through $\Phi_L(V)$ is called the *quadric hull* of (V, L) and we denote it by Quad(V, L). An irreducible component W of Quad(V, L) is called an *essential* component if $\Phi_L(V) \subset W$. Let $\{W_1, \cdots, W_N\}$ be the set of all essential components of Quad(V, L). We put

$$\dim \text{Quad}(V, L) := \max_i \dim W_i$$

and sometimes call it the *quadric dimension* of (V, L). We clearly have $\dim \Phi_L(V) \leq \dim \text{Quad}(V, L) \leq \dim |L|$. If $\dim \Phi_L(V) = \dim \text{Quad}(V, L)$, then we say that $\Phi_L(V)$ is *generically cut out by quadrics*. If $\dim \text{Quad}(V, L) = \dim |L|$, then there are no hyperquadrics through $\Phi_L(V)$ and Quad$(V, L) = \mathbb{P}^{\dim |L|}$. When V is a Gorenstein variety of general type, we write Quad(V) instead of Quad(V, K_V).

Let $W \subset \mathbb{P}^n$ be a non-degenerate variety. The *Hilbert function* of W is defined as

$$h_W(m) := \text{rank}\{H^0(\mathbb{P}^n, \mathcal{O}(m)) \to H^0(W, \mathcal{O}_W(m))\}$$

for any non-negative integer m. For the properties of the Hilbert function, consult [8].

Lemma 1.2. *Let $W \subset \mathbb{P}^n$ be an irreducible non-degenerate variety of dimension w. Then*

$$h_W(2) \geq (w+1)(n+1) - \frac{1}{2}w(w+1) + \min\{\Delta(W), n-w\},$$

where $\Delta(W) = w + \deg W - n - 1$ is the Δ-genus of $(W, \mathcal{O}_W(1))$.

Proof. We choose a general flag $\mathbb{P}^n \supset \mathbb{P}^{n-1} \supset \cdots \supset \mathbb{P}^{n-w}$ and put $W_{w-i} = W \cap \mathbb{P}^{n-i}$ for $0 \leq i \leq w$. Since W_{w-i-1} is a general hyperplane section of the irreducible non-degenerate variety W_{w-i}, we have $h_{W_{w-i}}(2) \geq h_{W_{w-i}}(1) + h_{W_{w-i-1}}(2)$ and $h_{W_{w-i}}(1) = n - i + 1$ for each i. Since W_0 is a set of $\deg W$ distinct points in uniform position, we have $h_{W_0}(2) \geq \min\{\deg W, 2(n-w)+1\}$. It follows

$$h_W(2) \geq \sum_{i=0}^{w-1} h_{W_{w-i}}(1) + h_{W_0}(2) \geq \sum_{i=0}^{w-1}(n-i+1) + \min\{\deg W, 2(n-w)+1\}.$$

Hence we get the inequality. $\square$

Proposition 1.3. *Let (V, L) be a prepolarized variety with $\dim \text{Quad}(V, L) = w$ and let W be a w-dimensional essential component of $\text{Quad}(V, L)$. Then*

$$h^0(V, 2L) \geq (w+1)h^0(V, L) - \frac{1}{2}w(w+1) + \min\{\Delta(W), h^0(V, L) - w - 1\}. \quad (1.1)$$

If the equality sign holds, then the multiplication map $\text{Sym}^2 H^0(V, L) \to H^0(V, 2L)$ is surjective.

Proof. Put $V' = \Phi_L(V)$. Since V' is non-degenerate, so is W. Furthermore, it follows from $V' \subset W \subset \text{Quad}(V, L)$ that $h_{V'}(2) = h_W(2)$. Since $h^0(V, 2L) \geq h_{V'}(2)$, we get (1.1) by Lemma 1.2. If the equality holds in (1.1), then $h^0(V, 2L) = h_{V'}(2)$ which implies the surjectivity of the multiplication. $\square$

Recall that this can recover a part of Enriques–Petri's theorem as follow:

Corollary 1.4. *If C is a canonical curve of genus g, then $\dim \text{Quad}(C) \leq 2$. If the equality holds, then $\text{Quad}(C)$ is an irreducible surface of minimal degree.*

Proof. We apply Proposition 1.3 for $V = C$, $L = K_C$. Since $h^0(C, 2K_C) = 3g - 3$, (1.1) gives us

$$(w-2)(2g-w-3) + 2\min\{\Delta(W), g-w-1\} \leq 0.$$

It follows that $w = \dim \text{Quad}(C) \leq 2$ and $\Delta(W) = 0$ when $w = 2$. The last assertion follows from the fact that the homogeneous ideal of W is generated by quadrics when $\Delta(W) = 0$. $\square$

Corollary 1.5. *Let S be a canonical surface with* $\dim \mathrm{Quad}(S) = w$, *and let W be a w-dimensional essential component of* $\mathrm{Quad}(S)$. *Then*

$$K_S^2 \geq wp_g + q - \frac{1}{2}w(w+1) - 1 + \min\{\Delta(W), \ p_g - w - 1\} \quad (1.2)$$

If the equality holds here, then the multiplication map

$$\mathrm{Sym}^2 H^0(S, K_S) \to H^0(S, 2K_S)$$

is surjective and $\mathrm{Bs}|K_S| = \emptyset$.

Proof. The first assertion follows from Proposition 1.3, since we have $h^0(S, 2K_S) = K_S^2 + \chi(\mathcal{O}_S)$ by the pluri-genus formula. If the equality holds in (1.2), then the multiplication map $\mathrm{Sym}^2 H^0(S, K_S) \to H^0(S, 2K_S)$ is surjective. Since $\mathrm{Bs}|2K_S| = \emptyset$ (see e.g. [7]), we get $\mathrm{Bs}|K_S| = \emptyset$. □

Remarks. (1) Let $\sigma : \tilde{S} \to S$ be a minimal composite of blowing-ups which eliminates the base points of the variable part of $|K_S|$, and let $|\sigma^* K_S| = |M| + Z$ be the decomposition into the variable and fixed parts. We have $h_X(2) \leq h^0(2M) \leq h^0(K_{\tilde{S}} + M)$. By Ramanujam's vanishing theorem, we have

$$h^0(K_{\tilde{S}} + M) = M^2 + \frac{1}{2}M(Z + E) + \chi(\mathcal{O}_S)$$

where E is an exceptional divisor for σ with $K_{\tilde{S}} = \sigma^* K_S + [E]$. Hence, we get

$$M^2 + \frac{1}{2}M(Z + E) \geq wp_g + q - \frac{1}{2}w(w+1) - 1 + \min\{\Delta(W), \ p_g - w - 1\},$$

which is slightly stronger than (1.2).

(2) The inequality (1.2) is valid even when we drop the assumption that S is canonical. For example, we get $K_S^2 \geq 2p_g + q - 4$ whenever the canonical image is a surface (a version of Noether's inequality). We can also rediscover Castelnuovo's inequality $K_S^2 \geq 3p_g + q - 7$ for a canonical surface: When $\dim \mathrm{Quad}(S) \geq 3$, it is clear from (1.2) and, when $\dim \mathrm{Quad}(S) = 2$, we get it since $\deg \Phi_{K_S}(S) > 2p_g - 5$.

Now, Miyaoka–Yau's inequality [14] and (1.2) give us the following:

Theorem 1.6. *Let S be a canonical surface. Then* $\dim \mathrm{Quad}(S) \leq 19$ *with equality holding only if* $p_g = 20$, $q = 0$, $K_S^2 = 189$ *and there are no hyperquadric through the canonical image. Furthermore, the following hold.*

(1) *If $p_g + 10q > 65$, then* $\dim \mathrm{Quad}(S) \leq 9$.

(2) *If $q > 5$, then* $\dim \mathrm{Quad}(S) \leq 8$.

Proof. Let the notation be as before. Since $\Delta(W) \geq 0$ and $w \leq p_g - 1$, it follows from (1.2) and Miyaoka–Yau's inequality $K_S^2 \leq 9\chi(\mathcal{O}_S)$ that

$$10\chi(\mathcal{O}_S) \geq \frac{1}{2}(w+1)(2p_g - w).$$

We have $2p_g - w \geq p_g + 1 \geq \chi(\mathcal{O}_S)$. Hence we get $w \leq 19$. We can show (1) and (2) in a similar way. □

In the rest of the section, we assume that S is a regular canonical surface with $\dim \text{Quad}(S) = w$ whose numerical invariants attain the bound in (1.2), and we give a few comments on the structure of the canonical ring.

A special case of the following was used to study surfaces near Castelnuovo's line $K_S^2 = 3p_g - 7$ (see e.g., [1] and [10]).

Lemma 1.7. *Let S be a regular canonical surface with $\dim \text{Quad}(S) = w \geq 3$, and assume that the equality sign holds in (1.2). Then the canonical ring $R(S, K_S) = \bigoplus_{m \geq 0} H^0(S, mK_S)$ is generated in degree 1. In particular, $\Phi_{K_S}(S)$ is isomorphic to the canonical model of S.*

Proof. Put $n = p_g - 1$ and $X = \Phi_{K_S}(S)$. We choose a general flag $\mathbb{P}^n \supset \mathbb{P}^{n-1} \supset \mathbb{P}^{n-2}$ and put $X_{2-i} = X \cap \mathbb{P}^{n-i}$, $W_{w-i} = W \cap \mathbb{P}^{n-i}$ for $1 \leq i \leq 2$. We have $X_{2-i} \subset W_{w-i}$. Then $h_{X_0}(1) = n - 1$ and $h_{X_0}(2) = h_{W_{w-2}}(2) = K_S^2 - (n-1)$. Since $h_{X_0}(i+1) \geq \min\{K_S^2, h_{X_0}(i) + n - 2\}$, we have $h_{X_0}(3) \geq K_S^2 - 1$ and $h_{X_0}(j) = K_S^2$ for $j \geq 4$. We have $h_{X_1}(1) = n$. By the Riemann–Roch theorem, we have

$$h_{X_1}(i) \leq h^0(K_S|_{X_1}) = \begin{cases} K_S^2 + 1, & \text{if } i = 2, \\ (i-1)K_S^2, & \text{if } i \geq 3, \end{cases}$$

where we identified X_1 with its preimage in $|K_S|$. Hence we have $h_{X_1}(i+1) = h_{X_1}(i) + h_{X_0}(i+1)$ for any i. Similarly, one can check that $h_X(i+1) = h_X(i) + h_{X_1}(i+1)$ holds for any i. It follows that X is projectively normal and $R(S, K_S)$ is generated in degree 1. Hence X is isomorphic to the canonical model. □

Let $\Omega = \text{Sym}(H^0(K_S))$ be the symmetric algebra on $H^0(K_S)$. Since $R(S, K_S)$ is the homogeneous coordinate ring of $\Phi_{K_S}(S)$, its minimal free resolution as a graded Ω-module can be determined by looking at the Koszul cohomology groups due to Green [5]. Put $R_m = H^0(S, mK_S)$ and consider the Koszul complex

$$\cdots \to \wedge^{i+1} R_1 \otimes R_{j-1} \xrightarrow{d_{i+1,j-1}} \wedge^i R_1 \otimes R_j \xrightarrow{d_{i,j}} \wedge^{i-1} R_1 \otimes R_{j+1} \to \cdots$$

We put $\mathcal{K}_{i,j} = \text{Ker}(d_{i,j})/\text{Im}(d_{i+1,j-1})$. Then

$$\cdots \to \bigoplus_{j \geq 0} \mathcal{K}_{2,j} \otimes \Omega(-j-2) \to \bigoplus_{j \geq 0} \mathcal{K}_{1,j} \otimes \Omega(-j-1) \to \Omega$$

gives us the resolution. The following is a special case of Theorem(4.c.1) in [5].

Lemma 1.8. *Let S be a regular canonical surface with* $\dim \mathrm{Quad}(S) = w \geq 3$ *and* $K_S^2 = wp_g - w(w+1)/2 - 1 + \min\{\Delta(W), p_g - w - 1\}$, *where W is an essential component of dimension w. Put $n = p_g - 1$ and $\kappa_{i,j} = \kappa_{i,j}(S) := \dim \mathcal{K}_{i,j}$. Then the following hold.*

(1) $\kappa_{i,j} = \kappa_{n-2-i,4-j}$.

(2) *Except when* $(w, \Delta(W)) = (3, 0)$, $\kappa_{1,j} = 0$ *for* $j \geq 3$, *i.e., the homogeneous ideal of* $\Phi_{K_S}(S)$ *in* $\mathbb{P}^n$ *is generated by quadrics and cubics.*

2. Varieties of small degree

In this section, we shall study the structure of the quadric hull under some conditions, and show how it can be used to see the existence of a special linear system on a canonical surface.

Let W be an essential irreducible component of $\mathrm{Quad}(S)$ and put $w = \dim W$. Let V be a non-singular model of W and L the pull-back of the hyperplane bundle.

Lemma 2.1. *Let (V, L) be as above. Then $h^0(V, L) = p_g$ and the natural morphism $V \to W$ is induced from $|L|$.*

Proof. Put $N = h^0(V, L) - 1$ and suppose that $N \geq p_g$. Then the canonical map of S can be lifted to the map into $\mathbb{P}^N$. This is impossible, because the canonical map is induced from the complete linear system $|K_S|$. □

Since $|L|$ is free from base points, we have a smooth ladder

$$V_1 \subset V_2 \subset \cdots \subset V_{w-1} \subset V_w = V,$$

that is, each V_i is an irreducible non-singular member of $|L_{i+1}|$, where L_{i+1} is the restriction of L to V_{i+1}. By the adjunction formula, the canonical bundle of V_i is induced by $K_V + (w - i)L$. We say that (V, L) (or W) is a variety of small degree if V_2 is a ruled surface. Then as is well-known, we have the following:

Lemma 2.2. *If $L^w < 2h^0(V, L) - 2w$, then (V, L) is a variety of small degree.*

Proof. If L_1 were special, then Clifford's theorem implies that $2h^0(V_1, L_1) - 2 \leq \deg L_1 = L^w$. Since $h^0(L_1) \geq h^0(L_2) - 1 \geq \cdots \geq h^0(L) - w + 1$, we would get $L^w \geq 2h^0(L) - 2w$, which is impossible. Hence L_1 is non-special and it follows from the Riemann–Roch theorem on V_1 that $h^0(V_1, L_1) = L^w + 1 - g(V_1)$. We have $2g(V_1) - 2 = (K_V + (w-1)L)L^{w-1}$. Since $2h^0(V_1, L_1) \geq 2h^0(V, L) - 2w + 2 > L^w + 2$, we get $(K_V + (w-2)L)L^{w-1} < -2$. It follows that $\kappa(V_i) = -\infty$ for $i \geq 2$. □

Lemma 2.3. *If (V, L) is a variety of small degree, then $q_1(V) = q(V_2)$ and $q_2(V) = 0$, where $q_j(V) = h^j(V, \mathcal{O}_V)$.*

Proof. Consider the long exact sequence for
$$0 \to \mathcal{O}(-L_{i+1}) \to \mathcal{O}_{V_{i+1}} \to \mathcal{O}_{V_i} \to 0$$
for each i, $2 \leq i \leq w - 1$. It follows from Kawamata–Viewseg vanishing theorem that $H^p(-L_{i+1}) = 0$ for $p < i + 1$. Hence we get $q_1(V) = q(V_2)$ by induction. Since V_2 is ruled, we have $H^2(V_2, \mathcal{O}_{V_2}) = 0$. Hence $q_2(V) = 0$. □

The following is a special case of [4], Theorem 4.1.

Lemma 2.4. *Assume that $q_1(V) = 0$. If $L^w < 2h^0(L) - 2w$, then $H^j(V, L) = 0$ for $i > 0$. If $L^w < 2h^0(L) - 2w - 1$, then the graded ring $R(V, L) := \oplus_{m \geq 0} H^0(V, mL)$ is generated in degree 1 and the homogeneous ideal of $W = \Phi_L(V)$ is generated by quadrics.*

Proof. Consider the long exact sequence for
$$0 \to \mathcal{O}_{V_{i+1}} \to \mathcal{O}_{V_{i+1}}(L_{i+1}) \to \mathcal{O}_{V_i}(L_i) \to 0$$
for $1 \leq i \leq w - 1$. Since L_1 is non-special by Lemma 2.2, we have $H^1(L_1) = 0$. Therefore, we see that $H^1(L) = 0$ by induction. Similarly $H^i(L) = 0$ for $i > 0$. The second assertion follows from the corresponding assertion for V_1. □

Proposition 2.5. *Let S be a canonical surface and let W be a w-dimensional essential component of $\mathrm{Quad}(S)$. If $q(W) = 0$ and $\deg W \leq 2p_g - 2w - 2$, then $\mathrm{Quad}(S)$ is irreducible, that is, $\mathrm{Quad}(S) = W$.*

Proof. By the previous lemma, we see that W is cut out by quadrics which by definition generate the degree 2 part of the homogeneous ideal of the canonical image of S. □

Corollary 2.6. *Let S be a regular canonical surface with $\dim \mathrm{Quad}(S) = w \geq 3$ and*
$$wp_g - \frac{1}{2}w(w+1) - 1 \leq K_S^2 < (w+1)p_g - \frac{1}{2}(w+1)(w+2) - 1$$
Then $\mathrm{Quad}(S)$ is irreducible.

Proof. Let W be an essential component of dimension w. In the above range of K_S^2, we have $\deg W \leq 2p_g - 2w - 2$. Furthermore, we must have $q(W) = 0$, because S is a regular surface. □

Lemma 2.7. *Let (V, L) be a w-fold of small degree with $q_1(V) > 0$. Let $\alpha : V \to \mathrm{Alb}(V)$ be the Albanese map and put $B := \alpha(V)$. Then B is a curve. Furthermore, if Γ denotes a general fibre of $\alpha : V \to B$, then (Γ, L_Γ) induces a $(w-1)$-fold of minimal degree, where L_Γ is the restriction of L to Γ.*

Proof. Assume that the image of α is not a curve. Then, by Lemma 10.1 in [16], we would have $q_2(V) > 0$ which is impossible by Lemma 2.3. Therefore, the albanese image B is a non-singular curve of genus $q(V)$. Let Γ be a general fibre of α, which we can assume irreducible and non-singular. Since L is birationally very ample, so is L_Γ. Furthermore $Bs|L_\Gamma| = \emptyset$. It follows that any general member of it is irreducible and non-singular.

We proceed by induction on the dimension. The assertion is known for $w = 2$. We let $\alpha_{w-1} : V_{w-1} \to \text{Alb}(V_{w-1})$ be the Albanese map. We can assume that the image B_{w-1} is a curve and that $(\Gamma_{w-1}, L_{\Gamma_{w-1}})$ gives a $(w-2)$-fold of minimal degree for a general fibre Γ_{w-1} of α_{w-1}. Recall that we have the canonical isomorphism $H^1(\mathcal{O}_V) \simeq H^1(\mathcal{O}_{V_{w-1}})$. The inclusion $V_{w-1} \hookrightarrow V$ induces a morphism $\iota : B_{w-1} \to B$ by the universality of the Albanese map. We know that B_{w-1} and B have the same genus and that ι is non-constant. It follows that ι is an isomorphism except when $q_1(V) = 1$ and ι is an unramified covering of degree > 1. If we were in this exceptional case, α_{w-1} is the Stein factorization of $\alpha|_{V_{w-1}}$, and we can find a member of $|L_\Gamma|$ consisting of several disjoint irreducible $(w-2)$-folds. This is impossible, since a general member of $|L_\Gamma|$ is irreducible. Therefore, α_{w-1} is the restriction of α to V_{w-1}.

Put $\delta = \Gamma L^{w-1}$. We may assume that $\Gamma_{w-1} \in |L_\Gamma|$ which can be regarded as a $(w-2)$-fold of degree δ. We have $H^1(\Gamma, -L_\Gamma) = 0$ by Kawamata–Viewheg vanishing theorem. From the cohomology long exact sequence for

$$0 \to \mathcal{O}_\Gamma(-L_\Gamma) \to \mathcal{O}_\Gamma \to \mathcal{O}_{\Gamma_{w-1}} \to 0,$$

we have $H^1(\mathcal{O}_\Gamma) \hookrightarrow H^1(\mathcal{O}_{\Gamma_{w-1}}) = 0$. Thus, it follows from the cohomology long exact sequence for

$$0 \to \mathcal{O}_\Gamma \to \mathcal{O}_\Gamma(L_\Gamma) \to \mathcal{O}_{\Gamma_{w-1}}(L_{\Gamma_{w-1}}) \to 0$$

that $h^0(L_\Gamma) = h^0(L_{\Gamma_{w-1}}) + 1 = \delta + w - 1$. Hence $\Delta(\Gamma, L_\Gamma) = (w-1) + \delta - (\delta + w - 1) = 0$, and (Γ, L_Γ) induces a $(w-1)$-fold of minimal degree. $\square$

Let (V, L) and $\alpha : V \to B$ be as in the above lemma. Though there are various choice of birational models, we can put it in a natural (singular) model as follows. Put $\mathcal{E}_L = \alpha_* \mathcal{O}_V(L)$. Then the natural sheaf homomorphism $\alpha^* \mathcal{E}_L \to \mathcal{O}_V(L)$ induces a morphism $V \to \mathbb{P}_B(\mathcal{E}_L)$ which is birational onto the image $\tilde{W}$. Note that $\tilde{W}$ is ruled by $(w-1)$-fold of minimal degree and the natural map $V \to W$ factors through $\tilde{W}$.

The following is closely related to Severi's conjecture that the Albanese image of a minimal irregular surface S of general type is a curve provided that $K_S^2 < 4\chi(\mathcal{O}_S)$.

Proposition 2.8. *Let S be an irregular canonical surface and assume that* Quad(S) *has a three dimensional essential component W which is an irregular variety of small degree. Then S has a pencil of trigonal curves or plane quintic curves. If furthermore $K_S^2 < 4\chi(\mathcal{O}_S)$ holds, then the albanese image of S is a curve.*

Proof. The ruling of W induces on S a morphism to B. Taking Stein factorization if necessary, we get a fibration $f : S \to C$. (We indeed have $C = B$, since S is canonical.) Consider the relative canonical map for f, i.e., the rational map associated with the sheaf homomorphism $f^* f_* \omega_{S/C} \to \omega_{S/C}$. By the assumption, the image is contained in a threefold whose general fibre is a surface of minimal degree. It follows that a general fibre of f has either a g_3^1 or a g_5^2.

If $q(S) > g(C)$, then we have $K^2_{S/C} \geq 4 \deg f_* \omega_{S/C}$ by a result of Xiao [17]. It is impossible when $K_S^2 < 4\chi(\mathcal{O}_S)$. Then we have $q(S) = g(C)$ and see that C itself is the albanese image of S. □

Remarks. (1) We do not know whether a rough structure theorem as in Lemma 2.7 holds also for regular varieties of small degree. In view of [9], we can expect a similar result.

(2) The presence of a Petri-special pencil gives us a further restriction on the numerical invariants of S. Indeed, we showed in [12] that $K^2_{S/C} \geq 14(g-1)/(3g+1) \deg f_* \omega_{S/C}$, where g denotes the genus of a general fibre.

3. Regular surfaces on an extremal line

Let S be a canonical surface with $\dim \text{Quad}(S) = w$ and $K_S^2 = wp_g - w(w+1)/2 - 1$. Our main concern is in the case $w = 4$, that is, $K_S^2 = 4p_g - 11$ which is the line next to one appears in Reid's conjecture.

Note that S is automatically regular and $\text{Quad}(S)$ is an irreducible non-degenerate w-fold of degree $n - w + 1$ in $\mathbb{P}^n$ ($n = p_g - 1$). Hence, as is well-known, it is one of the following varieties with obvious polarization (see e.g., [3]):

(A) $\mathbb{P}^w$ ($n = w$).

(B) A hyperquadric in $\mathbb{P}^{w+1}$ ($n = w + 1$).

(C) A generalized cone over the Veronese surface ($n = w + 3$).

(D) A rational normal scroll of dimension w ($n > w + 1$).

We say that S is of type (A), (B), (C) or (D) according to the types of $\text{Quad}(S)$. Since the case $w = 3$ is well-known (see, e.g., [1]), we assume that $w \geq 4$ in what follows.

3.1. Surfaces of types (A) and (B)

We retain the notation in Section 1 and study surfaces of types (A), (B) by looking at the minimal free resolution of the ideal of $X = \Phi_{K_S}(S)$.

Lemma 3.1. *Let S be a canonical surface with* $\dim \operatorname{Quad}(S) = w$ *satisfying* $K_S^2 = wp_g - w(w+1)/2 - 1$, $w \geq 4$. *Then*

$$\kappa_{i,1}(S) = \kappa_{i,1}(W) = i\binom{n-w+1}{i+1}$$

and hence for $1 \leq i \leq n-3$,

$$\kappa_{i,2}(S) = \sum_{j=0}^{i+2}(-1)^{i+j}\binom{n+1}{j}\dim R_{i+2-j}$$
$$+ (n-1-i)\binom{n-w+1}{n-i} + (i+1)\binom{n-w+1}{i+2}.$$

Proof. Since we have $R_m \simeq H^0(W, \mathcal{O}(m))$ for $m \leq 2$, we get $\kappa_{i,1} = \kappa_{i,1}(W)$. Recall that the latter is the same as that of a rational normal curve of degree $n - w + 1$. Hence an Eagon–Northcott complex gives us the value of $\kappa_{i,1}$. Then looking at the complex

$$0 \to \bigwedge^{i+2} R_1 \to \bigwedge^{i+1} R_1 \otimes R_1 \to \bigwedge^{i} R_1 \otimes R_2 \to \bigwedge^{i-1} R_1 \otimes R_3 \to \cdots$$

we can compute $\kappa_{i,2}$, because we have $\kappa_{i+j,j} = 0$ for $j \neq 1, 2$. □

Type (A). When $W = \mathbb{P}^w$, we have $p_g = w + 1$ and $K_S^2 = w(w+1)/2 - 1$, $w \leq 19$. We have $\kappa_{i,1} = 0$ by Lemma 3.1. Hence the resolution takes the form

$$0 \to \Omega(-w-2) \to \kappa_{w-3,2}\Omega(-w+1) \to \cdots \to \kappa_{i,2}\Omega(-i-2) \to \cdots \to \Omega,$$

where $\kappa_{i,2}$ can be calculated by the formula in Lemma 3.1 as

$$\kappa_{i,2} = \frac{i(w-i-2)(w-1)(w+2)}{2(i+2)(w-i)}\binom{w-2}{i}. \quad (3.1)$$

Though it seems difficult to check the existence of S directly from it, we can say more at least when w is small.

Proposition 3.2. *Let S be a surface of type* (A) *and X its canonical image.*

(1) *If $w = 4$, then X is a complete intersection of two hypercubics.*

(2) *If $w = 5$, then X is the common zero locus of the seven diagonal 6×6 Pfaffians of a skew-symmetric matrix*

$$\begin{pmatrix} 0 & l_1 & l_2 & l_3 & l_4 & l_5 & l_6 \\ -l_1 & 0 & l_7 & l_8 & l_9 & l_{10} & l_{11} \\ -l_2 & -l_7 & 0 & l_{12} & l_{13} & l_{14} & l_{15} \\ -l_3 & -l_8 & -l_{12} & 0 & l_{16} & l_{17} & l_{18} \\ -l_4 & -l_9 & -l_{13} & -l_{16} & 0 & l_{19} & l_{20} \\ -l_5 & -l_{10} & -l_{14} & -l_{17} & -l_{19} & 0 & l_{21} \\ -l_6 & -l_{11} & -l_{15} & -l_{18} & -l_{20} & -l_{21} & 0 \end{pmatrix},$$

where the l_i's are linear forms in six variables.

Proof. (1) can be found in [10]. (2): Since $X \subset \mathbb{P}^5$ is Gorenstein in codimension 3, its defining equation is given by Pfaffians of a skew-symmetric matrix giving $7\Omega(-4) \to 7\Omega(-3)$ in the resolution, by Buchsbaum–Eisenbud's theorem [2]. □

Type (B). When W is a hyperquadric, we have $p_g = w+2$ and $K_S^2 = w(w+3)/2-1$, $w \leq 18$. Lemmas 1.9 and 3.1 show that $\kappa_{1,1} = \kappa_{w-2,3} = 1$ and $\kappa_{i,1} = \kappa_{w-1-i,3} = 0$ for $i > 1$. Hence the resolution takes the following form:

$$0 \to \Omega(-w-3) \to \begin{matrix} \Omega(-w-1) \\ \oplus \\ \kappa_{w-2,2}\Omega(-w) \end{matrix} \to \cdots \to \kappa_{i,2}\Omega(-i-2) \to \cdots$$

$$\to \begin{matrix} \kappa_{1,2}\Omega(-3) \\ \oplus \\ \Omega(-2) \end{matrix} \to \Omega$$

where $\kappa_{i,2}$ is given by

$$\kappa_{i,2} = \frac{i(w-i-1)(w^2+3w-2) - 4(w+1)}{2(i+2)(w-i+1)} \binom{w-1}{i}. \tag{3.2}$$

Again Buchsbaum–Eisenbud's theorem gives us for $w = 4$ the following:

Proposition 3.3. *Let S be a regular canonical surface with $p_g = 6$, $K_S^2 = 13$. If $\dim \mathrm{Quad}(S) = 4$, then the canonical image is the common zero locus of the five diagonal 4×4 Pfaffians of a skew-symmetric matrix*

$$\begin{pmatrix} 0 & q_1 & q_2 & q_3 & q_4 \\ -q_1 & 0 & l_1 & l_2 & l_3 \\ -q_2 & -l_1 & 0 & l_4 & l_5 \\ -q_3 & -l_2 & -l_4 & 0 & l_6 \\ -q_4 & -l_3 & -l_5 & -l_6 & 0 \end{pmatrix}$$

where the q_i's are quadratic forms and the l_j's are linear forms in six variables.

3.2. Surfaces of type (C)

We assume that $\mathrm{Quad}(S)$ is a generalized cone over the Veronese surface. Then $p_g = w+4$ and $K_S^2 = w(w+7)/2 - 1$, $w \geq 4$. By Miyaoka–Yau's inequality, we must have $w \leq 16$.

$\mathrm{Quad}(S)$ is isomorphic to the weighted projective space $\mathbb{P}(1,1,1,2,\ldots,2)$, where 2 appears $w-2$ times. The ridge R of $\mathrm{Quad}(S)$ is a linear subspace of dimension $w-3$. Blow $\mathbb{P}^{w+3}$ up along R, and let $\tilde{W}$ be the proper transform of $\mathrm{Quad}(S)$. Then $\tilde{W}$ is isomorphic to the total space of the projective bundle $\pi : \mathbb{P}(\mathcal{O}_{\mathbb{P}^2}(2) \oplus I_{w-2}) \to \mathbb{P}^2$, where I_{w-2} denotes the trivial bundle of rank $w-2$. We denote by H the tautological line bundle on $\tilde{W}$ such that $\pi_*H = \mathcal{O}(2) \oplus I_{w-2}$. We put $\Gamma = \pi^*\mathcal{O}_{\mathbb{P}^2}(1)$. Let $\{x_0, x_1, x_2\}$ be a basis for $H^0(\Gamma)$. We choose sections Z_0 of $[H - 2\Gamma]$ and Z_i of $[H]$, $1 \leq i \leq w-2$, such that they form a system of homogeneous coordinates on fibres of π. The natural map $\tilde{W} \to \text{Quad}(S)$ is obtained by $|H|$ and the divisor (Z_0) contracts to R.

By considering the projection from R, we get a net Λ on S such that $K_S = [2M+G]$ for $M \in \Lambda$, where G is the divisorial part of the inverse image of R via the canonical map. Let $\sigma : \tilde{S} \to S$ be a minimal succession of blowing-ups which eliminates $Bs\Lambda$. Let m_i be the multiplicity of a base point p_i and E_i the inverse image of p_i on $\tilde{S}$. Then we have $\sigma^*K_S \sim 2\tilde{M}+\tilde{G}$, where $\tilde{M} \in |\sigma^*M - \sum m_i E_i|$ and $\tilde{G} = \sigma^*G + 2\sum m_i E_i$. We have a holomorphic map $f : \tilde{S} \to \mathbb{P}^2$ of degree $\tilde{M}^2$ with $\tilde{M} = f^*\mathcal{O}_{\mathbb{P}^2}(1)$. If we put $E = \sum E_i$, then the canonical bundle $\tilde{K}$ of $\tilde{S}$ satisfies $\tilde{K} = \sigma^*K_S + [E]$.

Lemma 3.4. *When $\tilde{G} = 0$, σ is the identity map, $w = 7$ and $M^2 = 12$. When $\tilde{G} \neq 0$,*

$$3\tilde{M}^2 + \tilde{M}\tilde{G} \geq 4w - 2, \quad 4\tilde{M}^2 + \tilde{M}\tilde{G} \geq 5w - 2.$$

If the equality holds in one of the above inequalities, then σ is the identity map.

Proof. If $\tilde{G} = 0$, then we have $G = 0$ and σ is the identity map. Hence S is an even surface with semi-canonical bundle M. Since M^2 is even and $K_S^2 = 4M^2$, we get $w = 7$ and $M^2 = 12$. Let x_0, x_1, x_2 be a basis for $H^0(M)$. Then the products $x_i x_j$, $0 \leq i \leq j \leq 2$, are independent in $H^0(K_S)$. Since $p_g = 11$, there are additional 5 independent elements there. Using these, we can lift f to a holomorphic map of S into the total space of $\mathcal{O}_{\mathbb{P}^2}(2) \oplus I_5$. Hence we get the lifting $h : S \to \tilde{W}$ of the canonical map with $K_S = h^*H$.

If $\tilde{G} \neq 0$, we can lift f to a holomorphic map $h : \tilde{S} \to \tilde{W}$, since $\tilde{G}$ is the inverse image of R. It is a birational map of $\tilde{S}$ onto the image, and we have $\sigma^*K_S = h^*H$. We put $\zeta_i = h^*Z_i$ for $0 \leq i \leq w-2$. Then ζ_0 defines $\tilde{G}$. We identify $\{x_0, x_1, x_2\}$ as a basis for $H^0(\tilde{M})$. Since, for $j = 1, 2$, the natural map $h^* : H^0(H + j\Gamma) \to H^0(\sigma^*K_S + j\tilde{M})$ is injective, we get $h^0(\tilde{K}_S + j\tilde{M}) \geq h^0(H + j\Gamma) = (j+4)(j+3)/2 + (w-2)(j+2)(j+1)/2$. Let $C \in |j\tilde{M}|$ be a general member. Since S is regular, we have an exact sequence

$$0 \to H^0(\sigma^*K_S) \to H^0(\sigma^*K_S + j\tilde{M}) \to H^0(C, (\sigma^*K_S + j\tilde{M})|_C) \to 0.$$

Therefore, $h^0(\sigma^*K_S + j\tilde{M}) = h^0(K_C - E|_C) + w + 4$. By the Riemann–Roch theorem and Clifford's theorem, we have

$$h^0(K_C - E|_C) = \frac{1}{2}C(\sigma^*K_S + C) - \frac{1}{2}CE + h^0(E|_C) \leq \frac{j}{2}\tilde{M}((j+2)\tilde{M} + \tilde{G}) + 1$$

with equality holding only if $EC = 0$ since C is non-hyperelliptic. Hence $3\tilde{M}^2 + \tilde{M}\tilde{G} \geq 4w - 2$ and $4\tilde{M}^2 + \tilde{M}\tilde{G} \geq 5w - 2$. □

We have shown that there is a lifting $h : \tilde{S} \to \tilde{W}$ of the canonical map. Let $\tilde{X}$ be its image. Then it is the proper transform of X via $\tilde{W} \to \mathrm{Quad}(S)$. Since X has at most rational double points, so is $\tilde{X}$.

Proposition 3.5. *Assume that $w = 4$. Then $\tilde{X}$ is defined in $\tilde{W}$ by*

$$\mathrm{rank} \begin{pmatrix} x_0 & x_1 & x_2 \\ Q_0 & Q_1 & Q_2 \end{pmatrix} < 2,$$

where $Q_i \in H^0(\tilde{W}, 2H)$.

Proof. We have $21 = K_S^2 = 4M^2 + 2MG + K_S G$. By parity, $K_S G$ is a positive odd integer. Then, by the connectedness of K_S, we get $MG > 0$. Since $K_S M + M^2$ is even, we see that M^2 and MG have the same parity. Hence $(M^2, MG) = (4, 2)$, $(3, 1)$ or $(3, 3)$. By Lemma 3.4, we have $3M^2 + MG \geq 14$. Hence $M^2 = 4, MG = 2$ and $K_S G = 1$. Furthermore, σ is the identity map.

Since $K_S G = 1$, there exists an irreducible component G_0 of G satisfying $K_S G_0 = 1$. It follows that G_0 is mapped isomorphically onto a line via the canonical map of S. Hence $G_0^2 = -3$. Since the image of G_0 must be R, we have $MG_0 = 2$. In particular, we have $M(G - G_0) = G_0(G - G_0) = 0$. Since $K_S(G - G_0) = 0$ and $(G - G_0)^2 = 0$, we conclude from Hodge's index theorem that $G = G_0$.

We have $h^0(K_S + M + G) = 17$ from the exact sequence

$$0 \to H^0(K_S + M) \to H^0(K_S + M + G) \to H^0(G, \mathcal{O}_G) \to 0.$$

Hence we have a new element $\eta \in H^0(K_S + M + G)$ which is a non-zero constant on G. Since $\mathrm{Sym}^2 H^0(K_S) \to H^0(2K_S)$ is surjective, we get three relations $x_i \eta = Q_i(x, \zeta)$, $0 \leq i \leq 2$, where $Q_i \in h^* H^0(2H)$. It is easy to see that we have no further relations. Eliminating η, we see that $\tilde{X}$ is defined by the equation as in the statement. $\square$

3.3. Surfaces of type (D)

We assume that $\mathrm{Quad}(S)$ is a rational normal scroll of dimension $w \geq 4$ and $p_g \geq w + 3$.

Since $\mathrm{Quad}(S)$ is ruled by $\mathbb{P}^{w-1}$'s, it induces on S a pencil Λ of irreducible curves via the canonical map. We denote by $\rho : \hat{S} \to S$ the elimination of $\mathrm{Bs}\,\Lambda$ and by $f : \hat{S} \to \mathbb{P}^1$ the induced relatively minimal fibration. Let $\mathcal{E}$ be the saturated subsheaf of $f_* \rho^* \omega_S$ generically generated by $H^0(\mathbb{P}^1, f_* \rho^* \omega_S)$. Since Φ_{K_S} maps a general member $F \in \Lambda$ birationally onto a nondegenerate curve in $\mathbb{P}^{w-1}$, we have

$$\mathrm{rank}\{H^0(\rho^* K_S) \to H^0(\rho^* K_S|_{\hat{F}})\} = w$$

for the natural restriction map, where $\hat{F}$ is a general fibre of f. It follows that $\mathcal{E}$ is of rank w and, therefore, it is of the form

$$\mathcal{E} \simeq \bigoplus_{i=1}^{w} \mathcal{O}_{\mathbb{P}^1}(a_i), \quad a_1 \geq \cdots \geq a_w \geq 0, \quad \sum a_i = n - w + 1.$$

Furthermore, the natural sheaf homomorphism $f^*\mathcal{E} \hookrightarrow f^*f_*\rho^*\omega_S \to \rho^*\omega_S$ induces a rational map $h : \hat{S} \to \mathbb{P}(\mathcal{E})$. Since $|\rho^*K_S|$ is free from base points by Lemma 1.2, h is holomorphic and $\rho^*K_S = h^*H$, where $H = H(\mathcal{E})$ denotes a (relatively ample) tautological divisor on $\mathbb{P}(\mathcal{E})$. Note that the Picard group of $\mathbb{P}(\mathcal{E})$ is generated by H and a fiber Γ. The image of $\mathbb{P}(\mathcal{E})$ under the holomorphic map associated with $|H|$ is nothing but Quad(S) by the construction. For each i, $1 \leq i \leq w$, we have a canonical section Z_i of $[H - a_i\Gamma]$ corresponding to the natural inclusion of the i-th factor $\mathcal{O}_{\mathbb{P}^1} \hookrightarrow \mathcal{E}(-a_i)$. Then $(Z_1, \cdots, Z_w)$ form a system of homogeneous coordinates on any fiber of $\mathbb{P}(\mathcal{E}) \to \mathbb{P}^1$.

Put $\hat{X} = h(\hat{S})$. We denote by g the genus of $\hat{F}$. Note that $\hat{F}$ is a non-hyperelliptic curve, since S is a canonical surface. By Clifford's theorem, we have $(\rho^*K_S)\hat{F} \geq 2w - 1$ unless $\hat{S} = S$. Let m_i, $1 \leq i \leq N$, denote the multiplicity of the base point p_i of Λ appearing in ρ, and let E_i denote its inverse image on $\hat{S}$. Then $E_i^2 = -1$ and $E_i E_j = 0$ when $i \neq j$. We have $\rho^*F = \hat{F} + \sum_i m_i E_i$ and $K_{\hat{S}} = \rho^*K_S + [E]$, where $E = \sum_i E_i$. In particular, $F^2 = \sum_i m_i^2$ and $d := K_S F = 2g - 2 - \sum_i m_i$. We put $\zeta_i = h^*Z_i$, $1 \leq i \leq w$. Then the collection $\{\zeta_i t_0^j t_1^{a_i - j} \mid 1 \leq i \leq w, \ 0 \leq j \leq a_i\}$ forms a basis for $H^0(\rho^*K_S)$, where $\{t_0, t_1\}$ is a basis for $H^0(\hat{F})$.

Lemma 3.6. *Let the notation be as above. The ring homomorphism* $\phi : R(\hat{S}, \rho^*K_S) \to R(\hat{F}, \rho^*K_S|_{\hat{F}})$ *induced by restriction is surjective for a general fibre* $\hat{F}$.

Proof. We put $F = \rho(\hat{F})$. Then $\Phi_{\rho^*K_S}(\hat{F}) = \Phi_{K_S}(F)$ and the image of ϕ can be identified with the homogeneous coordinate ring of $\Phi_{K_S}(F) \subset \mathbb{P}^{w-1}$. Since $R(\hat{S}, \rho^*K_S)$ is generated in degree 1, so is Im(ϕ). Hence $\Phi_{K_S}(F)$ is projectively normal. This in turn implies that ϕ is surjective, because K_S induces the hyperplane bundle on F. Note also that we have shown that F is non-singular. $\square$

Lemma 3.7. $h^0(\hat{S}, \hat{F} + E) = 2$.

Proof. We have $\rho^*F \geq \hat{F} + E \geq \hat{F}$. Hence $2 = h^0(\hat{F}) \leq h^0(\hat{F} + E) \leq h^0(\rho^*F) = h^0(S, F) = 2$. $\square$

We put $L = \rho^*K_S|_{\hat{F}}$. Then $h^0(\hat{F}, L) = w$, $w \leq g$ and $h^1(L) = h^0(E|_{\hat{F}}) = w + g - 1 - d$. In particular, we have $d \leq g + w - 2$ by $h^1(L) > 0$. It follows from $h^1(\rho^*K_S) = 0$ and the cohomology long exact sequence for

$$0 \to \mathcal{O}(\rho^*K_S - \hat{F}) \to \mathcal{O}(\rho^*K_S) \to \mathcal{O}_{\hat{F}}(L) \to 0$$

that $h^1(\rho^*K_S - \hat{F}) = 0$. We have $h^0(\rho^*K_S - \hat{F}) = p_g - w$ and, by the Riemann–Roch theorem, $\chi(\rho^*K_S - \hat{F}) = (\sum m_i - d)/2 + p_g + 1$. Since $h^2(\rho^*K_S - \hat{F}) = h^0(\hat{F} + E) = 2$ by the duality theorem and Lemma 3.7, we get $h^1(\rho^*K_S - \hat{F}) = (d - \sum m_i)/2 + 1 - w$. Therefore, $w = \frac{1}{2}(d - \sum_i m_i) + 1$ and we have

$$d = g + w - 2, \quad \sum_i m_i = g - w$$

It follows from Lemma 3.6 that the image of $\hat{F}$ under the holomorphic map induced by $|\rho^*K_S|$ is non-singular. Since it factors through F, we see that F is also non-singular, implying that all the m_i's are 1. In particular, we have $F^2 = \sum_i m_i^2 = g - w$.

Lemma 3.8. $g \leq w + (w - 2)(w - 3)/2$

Proof. Since $\deg(2L) - \deg K_{\hat{F}} = 2d - (2g - 2) = d - \sum m_i > 0$, we get $h^1(2L) = 0$ and $h^0(2L) = 2d + 1 - g = g + 2w - 3$. Recall that $\text{Sym}^2 H^0(L) \to H^0(2L)$ is surjective. Therefore, there are $(w - 2)(w - 3)/2 - (g - w)$ independent quadrics through $\Phi_L(\hat{F})$. Inparticular, we get $g \leq w + (w - 2)(w - 3)/2$. □

We have $h^1(2\rho^*K_S - \hat{F}) = 0$. It follows that

$$H^0(2\rho^*K_S) \otimes H^0(\hat{F}) \to H^0(2\rho^*K_S + \hat{F})$$

is surjective by the free-pencil-trick. Since $H^0(\rho^*K_S) \otimes H^0(\rho^*K_S) \to H^0(2\rho^*K_S)$ is surjective, we see that $H^0(2\rho^*K_S + \hat{F})$ is generated by quadratic forms in $(\zeta_1, \ldots, \zeta_w)$ with coefficients in polynomials in (t_0, t_1). Namely the homomorphism $h^* : H^0(\mathbb{P}(\mathcal{E}), 2H + \Gamma) \to H^0(\hat{S}, 2\rho^*K_S + \hat{F})$ is surjective. Since $h^0(2\rho^*K_S + \hat{F}) = K_S^2 + p_g + 1 + (g + 2w - 3)$ and $h^0(\mathbb{P}(\mathcal{E}), 2H + \Gamma) = K_S^2 + p_g + 1 + w(w + 1)/2$, we have $(w - 2)(w - 3)/2 - (g - w)$ independent relative hyperquadrics Q_i in $|2H + \Gamma|$ which vanishes on $\hat{X}$. We know that they can induce all hyperquadrics through $\Phi_L(\hat{F})$. We have $h^0(3\rho^*K_S) = (3w + 1)p_g - (3w^2 + 3w + 4)/2$ and $h^0(3H) = (w + 1)(w + 2)p_g/2 - w(w + 1)(w + 2)/3$. Hence there are $w(w - 3)p_g/2 - (w - 3)(2w^2 + 3w + 4)/6$ relative cubic relations here. Since $h^0(H - \Gamma) = p_g - w$, the Q_i's can induce at most $(w - 2)(w - 3)(p_g - w)/2 - (g - w)(p_g - w)$ independent relative cubic relations in the kernel of $h^* : H^0(\mathbb{P}(\mathcal{E}), 3H) \to H^0(\hat{S}, 3\rho^*K_S)$. It follows that there are at least

$$(g - 3)(p_g - w) + (w - 3)(w - 4)(w + 1)/6$$

unknown relative cubic relations.

As to the extremal case that $g = w$, we have the following:

Theorem 3.9. *If $g = w \geq 4$, then $g = 4$. Furthermore, $\hat{X}$ is a complete intersection of a relative hyperquadric in $|2H + \Gamma|$ and a relative hypercubic in $|3H - (p_g - 5)\Gamma|$,* and

$$(a_1, a_2, a_3, a_4) = (a, a, a, 1), \ (a+1, a, a, 0), \ (a, a, a, 0),$$
$$(a+3, a+1, a, 0), \ (a+2, a+1, a, 0),$$
$$(a+1, a+1, a, 0)$$

Proof. Assume that $g = w$. We identify F as a canonical curve in $\mathbb{P}^{g-1}$. Since F is not cut out by quadrics, it is a trigonal or plane quintic curve by Enriques–Babbage–Petri theorem. Then it is well-known that there are $g - 3$ primitive cubic generators of the homogeneous ideal of F.

Let $T_j \in |3H - x_j\Gamma|$, $1 \leq j \leq g - 3$, be the relative hypercubics which induce the cubic relations satisfied on F. We may assume that $x_1 \geq x_2 \geq \cdots \geq x_{g-3} \geq 0$ by Lemma 1.6. We claim that $x_1 \leq \min\{3a_2, a_1 + 2a_3\}$. This can be seen as follows. If $x_1 > 3a_2$, then T_1 is reducible, which is absurd. If $x_1 > a_1 + 2a_3$, then T_1 is singular along the codimension 2 relative linear subspace defined by $Z_1 = Z_2 = 0$. Therefore, its restriction to Γ is a projection of a hyperquadric in $\mathbb{P}^g$. This is impossible, since F is embedded into Γ via the complete linear system $|K_F|$.

As each T_j induces $x_j + 1$ relations in $H^0(3H)$, we get $\sum_j x_j \geq (g-3)(p_g - g) + (g-3)(g-5)(g+2)/6$. It follows $x_1 \geq \deg \mathcal{E} + (g+2)(g-5)/6$. When $g > 5$, this contradicts what we have shown above. When $g = 5$, the only case to be considered is $(a_1, a_2, a_3, a_4, a_5) = (a, a, a, 0, 0)$ and $x_1 = x_2 = 3a$ for some positive integer a. Then $T_j \cap \Gamma$ is a (generalized) cone over a plane cubic curve with ridge $Z_1 = Z_2 = Z_3 = 0$. If the plane cubic curve is singular, we immediately get a contradiction, since then F would be contained in a hypercubic of projection type. Since $T_j \cap \Gamma$ are independent, it follows that $T_1 \cap T_2 \cap \Gamma$ is a union of 9 copies of $\mathbb{P}^3$. This is absurd, since $F \subset T_1 \cap T_2 \cap \Gamma$. Hence we get $g = 4$. □

We next consider the case $g > w$. Recall that Lemma 3.8 gives us a bound on g in terms of w. We also have a bound of the geometric genus as follow:

Lemma 3.10. *If $g > w$, then*

$$p_g \leq \frac{1}{2}(w+1) + \frac{g+3w-3}{w} + \frac{4(w-1)^2}{w(g-w)}$$

and $a_1 \leq (g + w - 2)/(g - w)$.

Proof. The first assertion is nothing but the inequality $(K_S F)^2 \geq K_S^2 F^2$ which follows from Hodge's index theorem. We have $(\rho^* K_S - a_1 \hat{F})\rho^* K_S \geq 0$, that is, $K_S^2 \geq a_1 K_S F$. Hence $K_S F \geq a_1 F^2$ and we get the bound of a_1. □

Since $F^2 = g - w > 0$, we must have $a_w = 0$. There is an integer k, $1 \leq k \leq w-1$, such that $a_k > 0$ and $a_i = 0$ for $i > k$. For $j < w$, put $\mathcal{E}_j := \oplus_{i=1}^j \mathcal{O}_{\mathbb{P}^1}(a_i)$ and $W_j := \mathbb{P}(\mathcal{E}_j)$. We denote by H_j the tautological bundle on W_j. The sheaf homomorphism

$$f^*\mathcal{E}_j \hookrightarrow f^*\mathcal{E} \to \rho^*\omega_S$$

induces a rational map $h_i : \hat{S} \to W_j$ over $\mathbb{P}^1$. We denote by $\hat{M}_j$ the pull-back to $\hat{S}$ of H_j. Then we have $\rho^* K_S = \hat{M}_j + [\hat{G}_j]$ with an effective divisor $\hat{G}_j$. Note that

$$\hat{G}_1 \geq \hat{G}_2 \geq \cdots \geq \hat{G}_k \geq 0.$$

Since $\hat{M}_j - a_j \hat{F}$ is nef, we have $\hat{M}_j E_i \geq a_j$ and, hence, $\hat{G}_j E_i \geq -a_j$. This implies $\hat{G}_j$ contains $a_j E_i$. Therefore $\hat{G}_j - a_j E$ is effective, and we have $\hat{G}_j \hat{F} \geq (g-w)a_j$. We have

$$\begin{aligned} g + w - 2 &= (\rho^* K_S)\hat{F} = \hat{M}_j \hat{F} + \hat{G}_j \hat{F} \\ &\geq \hat{M}_j \hat{F} + (g - w)a_j. \end{aligned}$$

Hence $\hat{M}_j \hat{D} \leq 2(w-1) - (g-w)(a_j - 1)$.

We put $L_j := \hat{M}_j|_{\hat{F}}$ and $d_j := \deg(L_j)$. It is clear that $h^0(\hat{F}, L_j) \geq j$.

Remark. Let us recall an important invariant of $\hat{F}$, the *Clifford index*. For $A \in \text{Pic}(\hat{F})$, the Clifford index of A is $\text{Cliff}(A) = \deg A - 2(h^0(A) - 1)$. Then the Clifford index of $\hat{F}$ is defined by

$$\text{Cliff}(\hat{F}) = \min\{\text{Cliff}(A) | \ h^0(A) > 1, \ h^1(A) > 1\}.$$

We have $\text{Cliff}(\hat{F}) > 0$, since $\hat{F}$ is non-hyperelliptic. Recall that L is normally generated. But L is not quadrically presented, because we need a cubic relation. It can be shown easily that there are no trisecant lines of $\Phi_L(\hat{F})$. Then, it follows from [13] that $\text{Cliff}(\hat{F}) \leq g - w + 1$. If $g > w$, then we must have $\text{Cliff}(\hat{F}) \geq 2$, since L is normally generated.

Lemma 3.11. *Suppose that $g > w$. Then $d_j \geq 2j$ for $1 < j \leq w-1$. In particular, $a_{w-1} \leq 1$.*

Proof. Since $\hat{F}$ is non-hyperelliptic, Clifford's theorem implies that $d_j \geq 2j - 1$ for $j > 1$. If $d_j = 2j - 1$ holds for $1 < j \leq w - 1$, then $h^0(L_j) = j$ and we have $h^1(L_j) = g - j \geq 2$ by the Riemann–Roch theorem, which implies $1 = \text{Cliff}(L_j) = \text{Cliff}(\hat{F})$. Since $\text{Cliff}(\hat{F}) \geq 2$ if $g > w$, we must have $d_j \geq 2j$ for $1 < j \leq w - 1$.

We assume that $k = w - 1$ and $a_{w-1} \geq 2$. Then $d_{w-1} = \hat{M}_{w-1}\hat{F} \leq 2(w-1) - (a_{w-1} - 1)(g - w) < 2(w-1)$ which is impossible. Therefore, $a_{w-1} \leq 1$. $\square$

By this lemma,

$$(a_j - 1)(g - w) \leq \begin{cases} 2(w - 1 - j) & \text{if } 1 < j \leq k, \\ 2(w - 1) & \text{if } j = 1. \end{cases}$$

Hence
$$(g-w)(p_g - w - k) \leq 2k(w-1) - (k-1)(k+2)$$

The following completes the case $w = 4$.

Proposition 3.12. *Assume that $w = 4$ and $g = 5$. Then $\hat{X}$ is defined in $\mathbb{P}(\mathcal{E})$ by*
$$\operatorname{rank} \begin{pmatrix} Z_1 & Z_2 & Z_3 \\ Q_1 & Q_2 & Q_3 \end{pmatrix} < 2$$
where $Q_i \in H^0(2H - (a_i - 1)\Gamma)$, $1 \leq i \leq 3$. Furthermore, $a_4 = 0$, $a_1 \leq 2a_2 + 1$ and $a_2 \leq 2a_3 + 1$ hold. The possible a_i's are

$$p_g = 7: \quad (1,1,1,0), \ (2,1,0,0)$$
$$p_g = 8: \quad (2,1,1,0), \ (3,1,0,0)$$
$$p_g = 9: \quad (3,1,1,0), \ (2,2,1,0)$$
$$p_g = 10: \quad (3,2,1,0)$$
$$p_g = 11: \quad (4,2,1,0), \ (3,3,1,0)$$
$$p_g = 12: \quad (5,2,1,0), \ (4,3,1,0)$$
$$p_g = 13: \quad (5,3,1,0)$$
$$p_g = 14: \quad (6,3,1,0)$$
$$p_g = 15: \quad (7,3,1,0)$$

Proof. Recall that $a_4 = 0$ holds, because $F^2 = 1$. Since $(\rho^* K_S - a_i \hat{F})E = -a_i$, we can find an element $\eta_i \in H^0(\rho^*(K_S - a_i F))$ such that $\zeta_i = \eta_i e^{a_i}$, where $e \in H^0([E])$ defines E. We can assume that $\zeta_i|_E = 0$ for $1 \leq i \leq 3$ but $\zeta_4|_E$ is a non-zero constant. Let $\eta \in H^0(\hat{K} + \hat{F}) \simeq H^0(\rho^*(K_S + F))$ be a general element which is a non-zero constant on E.

Recall that $H^0(2\rho^* K_S) \otimes H^0(\hat{F}) \to H^0(2\rho^* K_S + \hat{F})$ is surjective. We can find in $H^0(2\rho^* K_S + \hat{F})$ the collection $\{\eta\eta_i e^{a_i - 1} t_0^j t_1^{a_i - j} \mid 1 \leq i \leq 3, \ 0 \leq j \leq a_i\}$. Since $H^0(K_S) \otimes H^0(K_S) \to H^0(2K_S)$ is surjective, these sections can be expressed as a quadratic form in ζ_i's with coefficients polynomials in (t_0, t_1). Since $\hat{F}$, identified with its canonical image, is a complete intersection of three quadrics, the above relative quadric relations can be reduced to three relations. Obviously, they are of the form $\eta\eta_i e^{a_i - 1} = q_i(t, \zeta)$, $1 \leq i \leq 3$, where $q_i \in H^0(2\rho^* K_S - (a_i - 1)\hat{F})$. Multiplying e, we get three relations: $\eta\zeta_i = eq_i$. This implies that $\hat{X}$ is defined by the equation in the statement. Since the relative cubic relations thus obtained are respectively in $H^0(3H - (a_1 + a_2 - 1)\Gamma)$, $H^0(3H - (a_1 + a_3 - 1)\Gamma)$ and $H^0(3H - (a_2 + a_3 - 1)\Gamma)$, we get the inequalities for the a_i as in the proof of Theorem 3.9. Since $a_3 \leq 1$, we get the list of (a_1, a_2, a_3, a_4). □

References

[1] Ashikaga, T., and Konno, K., Algebraic surfaces of general type with $c_1^2 = 3p_g - 7$, Tôhoku Math. J. 42 (1990), 517–536.

[2] Buchsbaum, D. A., and Eisenbud, D., Algebra structures for finite free resolutions and some structure theorems for ideals in codimension 3, Amer. J. Math. 99 (1977), 447–485.

[3] Fujita, T., On the structure of polarized varieties with Δ-genera zero, J. Fac. Sci. Univ. of Tokyo 22 (1975), 103–115.

[4] Fujita, T., Defining equations for certain types of polarized varieties, in: Complex Analysis and Algebraic Geometry (W. L. Baily Jr. and T. Shioda, eds.), Iwanami Shoten Publishers/Cambridge University Press, 1977, 165–173.

[5] Green, M. L., Koszul cohomology and the geometry of projective varieties, J. Differential Geom. 19 (1984), 125–171.

[6] Green, M. L. and Lazarsfeld, R., On the projective normality of linear series on an algebraic curve, Invent. Math. 83 (1986), 73–90.

[7] Francia, P., On the base points of the bicanonical system, in: Problems in the Theory of Surfaces and Their Classification (F. Catanese, C. Ciliberto and M. Cornalba, eds.), Symposia Math. XXXII, Academic Press, 1991, 141–150.

[8] Harris, J., Curves in Projective Space, Lecture Notes, University of Montreal, 1982.

[9] Ionescu, P., On varieties whose degree is small with respect to codimension, Math. Ann. 271 (1985), 339–348.

[10] Konno, K., Algebraic surfaces of general type with $c_1^2 = 3p_g - 6$, Math. Ann. 290 (1991), 77–107.

[11] Konno, K., Even canonical surfaces with small K^2 I, III, Nagoya Math. J. 129 (1993), 115–146; ibid. 143 (1996), 1–11.

[12] Konno, K., A lower bound of the slope of trigonal fibrations, Internat. J. Math. 7 (1996), 19–27.

[13] Lange, H. and Sernesi, E., Quadrics containing a Prym-canonical curve, J. Algebraic Geom. 5 (1996), 387–399.

[14] Miyaoka, Y., On the Chern numbers of surfaces of general type, Invent. Math. 42 (1977), 225–237.

[15] Reid, M., π_1 for surfaces with small K^2, in: Algebraic Geometry, Proceedings of Summer Meeting, Copenhagen 1978, Lecture Notes in Math. 732, Springer-Verlag, Berlin 1979, 534–544.

[16] Ueno, K., Classification Theory of Algebraic Varieties and Compact Complex Spaces, Lecture Notes in Math. 439, Springer-Verlag, Berlin–Heidelberg–New York 1975.

[17] G. Xiao, Fibred algebraic surfaces with low slope, Math. Ann. 276 (1987), 449–466.

K. Konno, Department of Mathematics, Graduate School of Science, Osaka University, Machikaneyama 1-16, Toyonaka, Osaka 560-0043, Japan

E-mail: konno@math.wani.osaka-u.ac.jp

A note on Bogomolov's instability and Higgs sheaves

Adrian Langer

0. Introduction

In seventies F. A. Bogomolov discovered a remarkable inequality between Chern classes of a semistable vector bundle on a smooth complex projective surface. Slightly later Y. Miyaoka and S. T. Yau found a similar but stronger inequality between Chern classes of a smooth complex projective surface of general type. At that time there was no obvious relation between these results although Bogomolov's instability implied a weaker version of the Miyaoka–Yau inequality. Such relation was found about ten years later by C. Simpson who proved Bogomolov's inequality for stable Higgs bundles. This together with Yau's proof of stability of the tangent bundle of a manifold of general type implied the Bogomolov–Miyaoka–Yau inequality (BMY for short). In fact, in the surface case one does not need Yau's result which is necessary only if one wants to generalize the BMY inequality to higher dimensions.

The aim of this paper is to study Bogomolov's inequality and its various generalizations, e.g., to sheaves with operators, logarithmic Higgs sheaves, extensions, etc.

The BMY inequality can be generalized to log canonical surface pairs (see [La2] for the most general result in this direction). It is natural to expect that there exists Bogomolov's inequality for logarithmic Higgs sheaves. There are some partial results in this direction (see [Biq] and [LW]) but the inequality is known only in the rank 2 case (see [An]). We use algebro–geometric methods to prove Bogomolov's inequality for rank 3 semistable sheaves with logarithmic λ-connections. Logarithmic Higgs sheaves are a special case of sheaves with logarithmic λ-connections for which $\lambda = 0$.

It is desirable to have a general algebro–geometric proof of Bogomolov's inequality for Higgs bundles. Such a proof should reveal some new interesting properties of Higgs sheaves. In particular, the author of this paper expects that semistable Higgs bundles have a small number of sections if the dimension of a variety is at least two. This is no longer true in the curve case (mainly due to the absence of integrability condition). Another advantage of giving an algebro–geometric proof is that it should give some insight into the structure of Higgs bundles (or systems of Hodge bundles) corresponding to representations of the fundamental group of a variety. We present this in the rank 3 case.

It seems that one of the reasons behind the lack of Bogomolov's inequality for logarithmic Higgs sheaves is the use of analytic methods in Simpson's approach and problems with singularities. We go around some of these problems to prove a higher dimensional BMY inequality for singular varieties of general type (Theorem 5.2). This inequality was conjectured by K. Sugiyama in [Su].

In this paper we also prove Bogomolov's inequality for extensions of Higgs sheaves. This inequality was first proved in [DUW] and [BG] using analytic methods. We show a simple algebro–geometric proof of this inequality.

Later we recall a result of I. Biswas [Bis] on existence of holomorphic flat connections on manifolds with negative cotangent bundle and give a short proof of this result.

One of the points which we want to emphasize in this introduction is that the whole theory of Higgs bundles should be generalized from smooth pairs to all log canonical pairs (with $\mathbb{Q}$-divisors). The theorems in this generalized theory have much stronger implications and are useful even if one is only interested in smooth varieties (see [La2], where there are applications of such theorems, e.g., to curves on smooth surfaces). Usually, we do not state our theorems in this more general case but we point out which proofs can be actually generalized to log canonical pairs. The formulation of the results, like Bogomolov's inequality for logarithmic Higgs sheaves, is still a non-trivial problem in the higher dimensional case since one needs to define the second Chern class for reflexive sheaves in any dimension. One can go around this problem by defining only intersections of this class with $\mathbb{Q}$-divisors (for very ample divisors we can just restrict to a very general intersection surface and use Chern classes of reflexive sheaves as defined, e.g., in [La2]) but then it is not completely clear how they change, e.g., when we try to go the limit.

Finally, let us remark that although we usually work over complex numbers some of the theorems have their analogues for algebraically closed fields of positive characteristic (cf. [La3]).

The structure of the paper is as follows. In Section 1 we recall a few definitions and results used in the paper. In Section 2 we generalize Simpson's version of Bogomolov's inequality to semistable Higgs sheaves and arbitrary polarizations. We also give a counterexample to Bogomolov's inequality for semistable Higgs sheaves with non-integrable Higgs field. Section 3 contains a small generalization of Bogomolov's instability theorem which in particular says that sheaves with trivial discriminant are either always or never stable. In Section 4 we prove a logarithmic version of Bogomolov's inequality in the rank 3 case. Section 5 contains various versions of higher-dimensional generalizations of BMY inequality. In Section 6 we prove Bogomolov's inequality for extensions. In Section 7 we reprove Biswas' result and prove Bogomolov's inequality for sheaves with operators. Finally, in Section 8 we compare various stability notions for Higgs sheaves and systems of Hodge sheaves.

Notation

Throughout the paper we use the following notation.

$\Delta(\mathcal{E}) = 2 \operatorname{rk} \mathcal{E} c_2(\mathcal{E}) - (\operatorname{rk} \mathcal{E} - 1) c_1^2(\mathcal{E})$ the discriminant of $\mathcal{E}$,
$\mu(\mathcal{E})$ the slope of $\mathcal{E}$ (counted with respect to the appropriate polarization),
$\mu_{\max}(\mathcal{E}, \theta)$ the slope of the maximal destabilizing θ-invariant subsheaf of $\mathcal{E}$,
$\mu_{\min}(\mathcal{E}, \theta) = -\mu_{\max}(\mathcal{E}^*, \theta^*)$ the slope of the minimal destabilizing θ-invariant torsion–free quotient of $\mathcal{E}$,
$\widehat{\Omega}_X$ the reflexivization of the cotangent sheaf,
$\mathcal{E}_1 \widetilde{\otimes} \mathcal{E}_2$ the tensor product $\mathcal{E}_1 \otimes \mathcal{E}_2$ divided by torsion.

1. Preliminaries

Let X be a smooth projective variety defined over an algebraically closed field k. Let $\mathcal{W}$ be a torsion free sheaf on X and D a reduced normal crossing divisor.

Definition 1.1. A sheaf $\mathcal{E}$ with a $\mathcal{W}$-valued operator is a sheaf $\mathcal{E}$ with an $\mathcal{O}_X$-homomorphism $\eta \colon \mathcal{E} \to \mathcal{W} \widetilde{\otimes} \mathcal{E}$. A sheaf $\mathcal{E}$ with an Ω_X-valued ($\Omega_X(\log D)$-valued) operator θ is called a *(logarithmic) Higgs sheaf* if it satisfies the integrability condition $\theta \wedge \theta = 0$.

We also define a *system of Hodge sheaves* as a Higgs sheaf $(\mathcal{E}, \theta)$ with a decomposition $\mathcal{E} = \bigoplus \mathcal{E}^p$, such that $\theta \colon \mathcal{E}^p \to \Omega_X \otimes \mathcal{E}^{p+1}$. Similarly, one can define a *logarithmic system of Hodge sheaves*.

Definition 1.2. Let $\mathcal{E}$ be a coherent sheaf on X. Then a *logarithmic λ-connection* ∇ on $\mathcal{E}$ is a k-linear map $\nabla \colon \mathcal{E} \to \Omega_X(\log D) \otimes \mathcal{E}$ satisfying the following two conditions

(1) the Leibniz rule
$$\nabla(fs) = f\nabla(s) + \lambda df \otimes s,$$
for all local sections f and s of $\mathcal{O}_X$ and $\mathcal{E}$, respectively,

(2) the integrability condition $\nabla^2 = 0$.

If $D = 0$ then ∇ is called a *λ-connection*.

Definition 1.3. We say that a Higgs sheaf is *(slope) semistable* (or we say that $\mathcal{E}$ is θ-semistable) if for any Higgs subsheaf $(\mathcal{F}, \theta') \subset (\mathcal{E}, \theta)$ we have $\mu(\mathcal{F}) \le \mu(\mathcal{E})$. Similarly, we define the corresponding notion for systems of Hodge sheaves etc.

1.4. One can generalize all the above notions to the case of log canonical pairs. Namely, let (X, D) be a pair consisting of a normal projective variety X and a $\mathbb{Q}$-divisor $D = \sum b_i B_i$, $0 \le b_i \le 1$. Then there exists a finite map $\pi \colon Y \to X$ from a normal projective variety Y such that $\pi^* D$ is a Weil divisor. Then one can define a notion of the pull back $\widehat{\pi^* \Omega_X}(\log D)$ of logarithmic 1-forms along the $\mathbb{Q}$-divisor D (see [La1] or [La2]).

Using this sheaf it is natural to define the corresponding notions of logarithmic Higgs sheaves etc. These notions seem to be a correct generalization of Higgs sheaves but only if the pair (X, D) is log canonical. For such pairs we have the following theorems (and Bogomolov's inequality for low rank logarithmic Higgs sheaves; see Remark 4.3).

Theorem 1.5. *Let (X, D) be a log canonical pair and let $\pi : Y \to X$ be a finite map such that $\pi^* D$ is a Weil divisor. If a rank 1 reflexive sheaf $\mathcal{L}$ is contained in $\widehat{\pi^* \Omega_X^q}(\log D)$ then $\kappa(Y, \mathcal{L}) \leq q$.*

This theorem goes back to the work of G. Castelnuovo and M. de Franchis. It was reproved by F. A. Bogomolov in the case X smooth and $D = 0$ and generalized to smooth log pairs by F. Sakai and A. Sommese. The most general version works for all log canonical pairs and was proved in [La2], Theorem 4.11.

Usually, we will use this theorem in the following special case. If X is a surface, $\mathcal{O}(A) \subset \Omega_X$ and $AH \geq 0$ for some ample divisor H then $A^2 \leq 0$.

Theorem 1.6. *Let (X, D) be a log canonical surface pair and let $\pi : Y \to X$ be a finite covering such that $\pi^* D$ is a Weil divisor. Let $\mathcal{F} \subset \widehat{\pi^* \Omega_X}(\log D)$ be a rank 2 reflexive subsheaf such that $c_1 \mathcal{F}$ is pseudoeffective. Let $c_1 \mathcal{F} = P + N$ be the Zariski decomposition. Then*

$$3c_2 \mathcal{F} + \frac{1}{4} N^2 \geq (c_1 \mathcal{F})^2.$$

The above theorem was first proved by Miyaoka in the case $Y = X$ is smooth (see [Mi1], Remark 4.18). Then it was generalized in [La1], Theorem 0.1 and [La2], Theorem 5.1. For a relatively simple and short proof of this theorem we refer the reader to [La1].

2. Bogomolov's inequality for Higgs sheaves

In the sequel, X is a smooth n-dimensional complex projective variety. We also fix a collection of $(n-1)$ nef divisors $(D_1, \ldots, D_{n-1})$ such that the 1-cycle $D_1 \ldots D_{n-1}$ is not numerically trivial.

Theorem 2.1. *Let $(\mathcal{E}, \theta)$ be a $(D_1, \ldots, D_{n-1})$-semistable torsion free Higgs sheaf. Then*

$$\Delta(\mathcal{E}) D_2 \ldots D_{n-1} \geq 0.$$

Proof. If X is a surface, D_1 is ample and $\mathcal{E}$ is θ-stable the theorem follows from [Si1], Theorem 1 and Proposition 3.4. We will reduce the above statement to this special case.

First, let us generalize this to the θ-semistable case. Let $(\mathcal{E}, \theta)$ be a semistable Higgs sheaf and let $0 = \mathcal{E}_0 \subset \mathcal{E}_1 \subset \cdots \subset \mathcal{E}_m = \mathcal{E}$ be the Jordan–Hölder filtration of $(\mathcal{E}, \theta)$. Set $\mathcal{F}_i = \mathcal{E}_i/\mathcal{E}_{i-1}$, $r_i = \text{rk}\,\mathcal{F}_i$, $\mu_i = \mu(\mathcal{F}_i)$. Then by the Hodge index theorem

$$\frac{\Delta(\mathcal{E})}{r} = \sum \frac{\Delta(\mathcal{F}_i)}{r_i} - \frac{1}{r}\sum_{i<j} r_i r_j \left(\frac{c_1 \mathcal{F}_i}{r_i} - \frac{c_1 \mathcal{F}_j}{r_j}\right)^2 \geq \sum \frac{\Delta(\mathcal{F}_i)}{r_i}$$

and the last sum is nonnegative by Simpson's result.

Now the higher dimensional case with $D_1, \ldots, D_{n-1}$ ample can be reduced to the surface case by the restriction theorem (see, e.g., [Si2], Lemma 3.7). The reduction of the general statement to this special one can be found in [La3]. □

2.2. Remarks. The above theorem and Proposition 8.1 imply Bogomolov's inequality for semistable systems of Hodge sheaves (cf. [Si1], Proposition 9.8).

Let us note that Theorem 2.1 holds for normal projective varieties with singularities in codimension ≥ 3. In this case one needs to take care of an extra problem with defining appropriate Chern numbers in the inequality. It would be desirable to extend the result to reflexive sheaves on normal (log) varieties which have only log canonical singularities in codimension 2 (Note that restricting to reflexive sheaves is not really a restriction, since on smooth varieties the inequality for the reflexivization of a torsion free sheaf always implies the inequality for the sheaf.). The inequality should then be interpreted using Chern classes of reflexive sheaves on appropriate surface sections. These inequalities are better than the corresponding inequalities on the resolution of singularities (see [La1], Proposition 2.7; it is formulated for rank 2 sheaves but the proof works for any rank).

The proof of the above theorem is not satisfactory because it uses results obtained by analytic methods. It also seems to be quite difficult to prove the corresponding result in the logarithmic case. One such attempt was done in [LW], Theorem 2.7, but this result gives the desired inequality only in special cases. Another result in this direction was proved by O. Biquard ([Biq]), but it works only for a smooth divisor on a smooth manifold.

Theorem 2.1 implies the following theorem. The proof is the same as the proof of [La3], Theorem 5.1.

Theorem 2.3. *Let $(\mathcal{E}, \theta)$ be a torsion free Higgs sheaf. Then*

$$(D_1^2 D_2 \ldots D_{n-1})(\Delta(E) D_2 \ldots D_{n-1})$$
$$+ r^2(\mu_{\max}(\mathcal{E}, \theta) - \mu(\mathcal{E}))(\mu(\mathcal{E}) - \mu_{\min}(\mathcal{E}, \theta)) \geq 0,$$

where the slopes are counted with respect to $D_1 \ldots D_{n-1}$.

The above theorem implies strong restriction theorems for Higgs sheaves (cf. [La3], Theorem 5.2). Looking at the proof of the above theorem one can also see that the following proposition holds.

Proposition 2.4. *Let $(\mathcal{E}, \theta)$ be a non-semistable torsion free Higgs sheaf and assume $D_1^2 D_2 \ldots D_{n-1} > 0$. Then the equality in the inequality of Theorem 2.3 holds if and only if the Harder–Narasimhan filtration of $(\mathcal{E}, \theta)$ has exactly two factors $\mathcal{F}_1$ and $\mathcal{F}_2$, the 1-cycles $c_1(\mathcal{F}_1)D_2 \ldots D_{n-1}$ and $c_1(\mathcal{F}_2)D_2 \ldots D_{n-1}$ are numerically proportional and $\Delta(\mathcal{F}_1)D_2 \ldots D_{n-1} = \Delta(\mathcal{F}_2)D_2 \ldots D_{n-1} = 0$.*

Now we want to show that integrability condition in the definition of Higgs sheaves is indeed necessary to obtain Theorem 2.1. Note that in the rank 2 case the inequality holds without integrability condition (see Theorem 4.1).

Example 2.5. In this example we show a rank 3 vector bundle $\mathcal{E}$ with an Ω_X-valued operator θ (non-integrable), such that $\mathcal{E}$ is θ-semistable, but $\mathcal{E}$ does not satisfy Bogomolov's inequality.

Let C_1 and C_2 be smooth curves of genus $g_1, g_2 \geq 2$, respectively. Set $X = C_1 \times C_2$ and let p_i denote a projection onto C_i, $i = 1, 2$. Set $\mathcal{E}_1 = p_1^* \omega_{C_1} \otimes p_2^* \omega_{C_2}$, $\mathcal{E}_2 = p_2^* \omega_{C_2}$ and $\mathcal{E}_3 = \mathcal{O}_X$. Note that $\Omega_X = p_1^* \omega_{C_1} \oplus p_2^* \omega_{C_2}$ and $H = c_1(\mathcal{E}_1)$ is an ample divisor on X. Now we define an Ω_X-valued operator θ on $\mathcal{E} = \mathcal{E}_1 \oplus \mathcal{E}_2 \oplus \mathcal{E}_3$ by

$$\theta = \begin{pmatrix} 0 & 0 & 0 \\ i_1 & 0 & 0 \\ 0 & i_2 & 0 \end{pmatrix}$$

where $i_1 \colon \mathcal{E}_1 \hookrightarrow \Omega_X \otimes \mathcal{E}_2$ and $i_2 \colon \mathcal{E}_2 \hookrightarrow \Omega_X \otimes \mathcal{E}_3$ are obvious inclusions. Thus defined pair $(\mathcal{E}, \theta)$ is a non-integrable system of Hodge bundles.

Since $\Delta(\mathcal{E}) = 6c_2(\mathcal{E}) - 2c_1(\mathcal{E})^2 = -8(g_1 - 1)(g_2 - 1) < 0$, the promised example is provided by the pair $(\mathcal{E}, \theta)$ once we prove the following claim.

2.5.1. Claim. $\mathcal{E}$ *is θ-stable with respect to H.*

Proof. It is sufficient to show θ-semistability of $\mathcal{E}$. Indeed, if $\mathcal{E}$ is θ-semistable but not θ-stable then we can use the Jordan–Hölder filtration and Bogomolov's inequality for lower rank to get $\Delta(\mathcal{E}) \geq 0$ (cf. the proof of Theorem 2.1), a contradiction.

Let $\mathcal{F} \subset \mathcal{E}$ be a non-zero θ-invariant subsheaf of $\mathcal{E}$. Clearly, we can assume that $\mathcal{F}$ is a saturated locally free sheaf. Let us consider two cases depending on the rank of $\mathcal{F}$.

1. rk $\mathcal{F} = 1$. Since $\mathcal{E}_1 \subset \mathcal{E}$ does not contain θ-invariant subsheaves, the map $\mathcal{F} \to \mathcal{E}_2 \oplus \mathcal{E}_3$ (induced by the projection) is non-zero. So $\mu(\mathcal{F}) \leq \mu_{\max}(\mathcal{E}_2 \oplus \mathcal{E}_3) = \mu(\mathcal{E}_2) = \mu(\mathcal{E})$.

2. rk $\mathcal{F} = 2$. If the map $\mathcal{F} \to \mathcal{E}_1$, induced by the projection $\mathcal{E} \to \mathcal{E}_1$, is zero then $\mathcal{F} \hookrightarrow \mathcal{E}_2 \oplus \mathcal{E}_3$ and $\mu(\mathcal{F}) \leq \mu(\mathcal{E}_2 \oplus \mathcal{E}_3) < \mu(\mathcal{E})$. Therefore we can assume that the map $\mathcal{F} \to \mathcal{E}_1$ is non-zero. Let $\mathcal{G}$ be the kernel of this map. Then we have the induced inclusion $\mathcal{G} \hookrightarrow \mathcal{E}_2 \oplus \mathcal{E}_3$.

If the induced map $\mathcal{G} \to \mathcal{E}_3$ is non-zero then

$$\mu(\mathcal{F}) = \frac{1}{2}(c_1(\mathcal{G})H + c_1(\mathcal{F}/\mathcal{G})H) \leq \frac{1}{2}(c_1(\mathcal{E}_3)H + c_1(\mathcal{E}_1)H) = \mu(\mathcal{E}).$$

Therefore we can assume that the map $\mathcal{G} \to \mathcal{E}_3$ is zero. But $\mathcal{F} \to \mathcal{E}_3$ is non-zero, since $\theta(\mathcal{F})$ is not contained in $\Omega_X \otimes (\mathcal{E}_1 \oplus \mathcal{E}_2)$ (note that $\mathcal{G} \hookrightarrow \mathcal{E}_2$ and hence $\theta(\mathcal{G}) \hookrightarrow \Omega_X \otimes \mathcal{E}_3$). Hence $\mathcal{F}/\mathcal{G} \hookrightarrow \mathcal{E}_3$ and

$$\mu(\mathcal{F}) = \frac{1}{2}(c_1(\mathcal{G})H + c_1(\mathcal{F}/\mathcal{G})H) \leq \frac{1}{2}(c_1(\mathcal{E}_2)H + c_1(\mathcal{E}_3)H) < \mu(\mathcal{E}). \qquad \square$$

3. Bogomolov's inequality revisited

Let $\mathrm{Num}(X) = \mathrm{Pic}(X) \otimes \mathbb{R}/\sim$, where $\sim$ is an equivalence relation defined by $L_1 \sim L_2$ if and only if $L_1 A D_2 \ldots D_{n-1} = L_2 A D_2 \ldots D_{n-1}$ for all divisors A. Then we define an open cone

$$K^+ = \{D \in \mathrm{Num}(X) : D^2 D_2 \ldots D_{n-1} > 0 \text{ and } D D_1 \ldots D_{n-1} \geq 0 \text{ for all nef } D_1\}.$$

Here we prove a version of Bogomolov's instability theorem, which also works for sheaves with trivial discriminant. This will be useful in examining flat Higgs sheaves and representations of the fundamental group of a manifold (see Section 4).

The proof of this theorem is analogous to the proof of Theorem 7.3.3, [HL] (see also [Bo2], Lemma 5.3 for original but slightly different proof). We sketch the proof for the convenience of the reader.

Theorem 3.1. *Let $(\mathcal{F}, \theta)$ be a torsion free Higgs sheaf with $\Delta(\mathcal{F}) D_2 \ldots D_{n-1} \leq 0$. Then one of the following holds:*

(1) *$(\mathcal{F}, \theta)$ is $D_1 \ldots D_{n-1}$-stable for all nef $D_1 \in K^+$ and $\Delta(\mathcal{F}) D_2 \ldots D_{n-1} = 0$.*

(2) *There exists a non-trivial saturated θ-invariant subsheaf $\mathcal{F}' \subset \mathcal{F}$ such that $\xi_{\mathcal{F}',\mathcal{F}} \in \overline{K^+}$. Moreover, if $\Delta(\mathcal{F}) D_2 \ldots D_{n-1} < 0$ then $\xi_{\mathcal{F}',\mathcal{F}} \in K^+$.*

Proof. The proof is by induction on the rank of $\mathcal{F}$. Assume that $(\mathcal{F}, \theta)$ is not $D_1 \ldots D_{n-1}$-stable for some nef $D_1 \in K^+$.

Let $\mathcal{F}'$ be the maximal θ-invariant $D_1 \ldots D_{n-1}$-destabilizing subsheaf of $\mathcal{F}$. Let $\mathcal{F}''$ denote the corresponding quotient and set $r = \mathrm{rk}\,\mathcal{F}$, $r' = \mathrm{rk}\,\mathcal{F}'$ and $r'' = \mathrm{rk}\,\mathcal{F}''$. We have

$$\frac{\Delta(\mathcal{F}) D_2 \ldots D_{n-1}}{r} + \frac{rr'}{r''} \xi^2_{\mathcal{F}',\mathcal{F}} D_2 \ldots D_{n-1}$$
$$= \frac{\Delta(\mathcal{F}') D_2 \ldots D_{n-1}}{r'} + \frac{\Delta(\mathcal{F}'') D_2 \ldots D_{n-1}}{r''}.$$

If $(\mathcal{F}, \theta)$ is $D_1 \ldots D_{n-1}$-semistable then $\mathcal{F}'$ and $\mathcal{F}''$ are $D_1 \ldots D_{n-1}$-semistable (with respect to the induced Higgs fields) and $\xi^2_{\mathcal{F}',\mathcal{F}} D_2 \ldots D_{n-1} \leq 0$ by the Hodge index theorem. Since the right hand side of the above equality is non-negative by the induction assumption, we have $\xi^2_{\mathcal{F}',\mathcal{F}} D_2 \ldots D_{n-1} = \Delta(\mathcal{F}) D_2 \ldots D_{n-1} = 0$.

Since $\xi_{\mathcal{F}',\mathcal{F}} D_1 D_2 \ldots D_{n-1} = 0$, the Hodge index theorem implies that the 1-cycle $\xi_{\mathcal{F}',\mathcal{F}} D_2 \ldots D_{n-1}$ is numerically trivial, i.e., $\xi_{\mathcal{F}',\mathcal{F}} = 0$ in $\mathrm{Num}(X)$.

Therefore we can assume that $(\mathcal{F}, \theta)$ is not $D_1 \ldots D_{n-1}$-semistable. If the inequality $\xi_{\mathcal{F}',\mathcal{F}}^2 D_2 \ldots D_{n-1} \geq 0$ holds then $\mathcal{F}'$ is the required sheaf. Otherwise, one of the numbers $\Delta(\mathcal{F}')D_2 \ldots D_{n-1}$ and $\Delta(\mathcal{F}'')D_2 \ldots D_{n-1}$ is negative.

The rest of the proof is the same as in [HL], the proof of Theorem 7.3.3. We construct a subsheaf $\mathcal{G} \subset \mathcal{F}$ such that $\xi_{\mathcal{G},\mathcal{F}}$ is a positive linear combination of $\xi_{F',F}$ and some element $\xi \in K^+$. This enlarges the cone $\{D \in \overline{K^+}: D\xi D_2 \ldots D_{n-1} > 0\}$ and repeating the above procedure we get the required sheaf after a finite number of steps. □

4. Bogomolov's inequality for small rank sheaves with logarithmic λ-connections

In this section X is a smooth projective surface, D is a normal crossing divisor and H is an ample divisor on X. More generally, one can only assume that (X, D) is a log canonical surface pair (with $\mathbb{Q}$-boundary) and work in this more general case as in Theorems 1.5 and 1.6.

All the proofs in this section work in the higher dimensional case, but for the sake of simplicity we formulate the results only in the surface case. Alternatively, higher dimensional versions of Theorem 4.1 and 4.2 can be obtained from the surface case by restriction theorems (cf. the proof of Theorem 2.1). The important point here is a well known fact that the restriction of a log canonical pair to a general divisor in a base point free linear system is still log canonical.

The next theorem was proved by Anchouche in [An] (except for the second part, which follows from the proof).

Theorem 4.1 ([An], Theorem 2.2). *Let $(\mathcal{E}, \nabla)$ be a rank 2 H-semistable sheaf with (non-integrable) logarithmic λ-connection. Then $\Delta(\mathcal{E}) \geq 0$. Moreover, if $\Delta(\mathcal{E}) = 0$ then $\mathcal{E}$ is locally free and one of the following holds:*

(1) *$\mathcal{E}$ is semistable.*

(2) *Up to twist by a line bundle $\mathcal{E}$ is an extension of the structure sheaf $\mathcal{O}_X$ by a nef line bundle $\mathcal{L}$ contained in $\Omega_X(\log D)$ and such that $c_1(\mathcal{L})^2 = 0$.*

The theorem follows from Bogomolov's inequality if $\mathcal{E}$ is semistable and from Theorem 1.5 otherwise. The formulation in [An] is weaker but the proof gives a more general version stated above. In the next theorem we need to assume the integrability of ∇ (see Example 2.5).

Theorem 4.2. *Let $(\mathcal{E}, \nabla)$ be a rank 3 H-semistable sheaf with logarithmic λ-connection. Then $\Delta(\mathcal{E}) \geq 0$. Moreover, if $\Delta(\mathcal{E}) = 0$ then $\mathcal{E}$ is locally free and one of the following holds:*

(1) $\mathcal{E}$ is semistable.

(2) Up to dualization and twist by a line bundle, the sheaf $\mathcal{E}$ is an extension of the structure sheaf $\mathcal{O}_X$ by a rank 2 vector bundle $\mathcal{F}$ with nef $c_1(\mathcal{F})$ and $c_1(\mathcal{F})^2 = 3c_2(\mathcal{F})$. Moreover, either $\mathcal{F}$ is a subsheaf of $\Omega_X(\log D)$ or $c_1(\mathcal{F})^2 = 0$ and there exists a nef line bundle $\mathcal{L}$ contained in $\Omega_X(\log D)$ such that $c_1(\mathcal{L})^2 = 0$.

Proof. If $\mathcal{E}$ is semistable then the theorem follows from the usual Bogomolov's inequality, so let us assume that $\mathcal{E}$ is not H-semistable and $\Delta(\mathcal{E}) \leq 0$.

Then by Theorem 3.1 (we use this theorem only in the "usual" case) there exists a saturated subsheaf $\mathcal{F} \subset \mathcal{E}$ such that $\xi^2_{\mathcal{F},\mathcal{E}} \geq 0$ and $\xi_{\mathcal{F},\mathcal{E}}$ is pseudoeffective. Passing to the dual if necessary, we can assume that rk $\mathcal{F} = 2$. Since twisting by line bundles does not change the required inequality, we can also assume that $(\mathcal{E}/\mathcal{F})^{**} \simeq \mathcal{O}_X$.

Since $(\mathcal{E}, \nabla)$ is semistable the induced $\mathcal{O}_X$-homomorphism $\mathcal{F} \to \Omega_X(\log D) \otimes \mathcal{E}/\mathcal{F}$ is non-zero unless $\xi_{\mathcal{F},\mathcal{E}} H = 0$. In the last case the theorem follows from Theorem 4.1. Hence we can assume that the homomorphism $\varphi \colon \mathcal{F} \to \Omega_X(\log D)$, obtained from the above by inclusion $(\mathcal{E}/\mathcal{F})^{**} \hookrightarrow \mathcal{O}_X$, is non-trivial. Now we distinguish two cases.

Case 1. φ is an inclusion. Note that by assumption $c_1(\mathcal{F})$ is pseudoeffective. Therefore $3c_2(\mathcal{F}) \geq c_1(\mathcal{F})^2$ by Theorem 1.6. This implies $\Delta(\mathcal{E}) \geq 0$. If $\Delta(\mathcal{E}) = 0$ then $\mathcal{E}/\mathcal{F} \simeq \mathcal{O}_X$ and $\mathcal{F}$ provides an example of sheaf described in (2).

Case 2. φ is not an inclusion. Set $\mathcal{L} = \ker \varphi$, $\mathcal{M} = \operatorname{im} \varphi$, $L = c_1(\mathcal{L})$ and $M = c_1(\mathcal{M})$. Let us note that $c_2(\mathcal{E}) \geq c_2(\mathcal{F}) \geq LM$. Now let us distinguish two cases.

Subcase 2.1. $\mathcal{L}$ is ∇-invariant, i.e., $\nabla(\mathcal{L}) \subset \Omega_X(\log D) \otimes \mathcal{L}$. By θ-semistability of $\mathcal{E}$ we have $LH \leq \mu(\mathcal{E}) = 1/3(L+M)H$, i.e., $MH \geq 2LH$.

Hence $0 \leq (L+M)H \leq 3/2 MH$. Since $\mathcal{M} \subset \Omega_X(\log D)$, M is not big and $M^2 \leq 0$. Therefore

$$(M - 2L)^2 \geq (M - 2L)^2 + 3M^2 = 4(L+M)^2 - 12LM \geq -2\Delta(\mathcal{E}) \geq 0$$

and $M - 2L$ is pseudoeffective. If $(M - 2L)^2 > 0$ then $M - 2L$ is big. But $c_1(\mathcal{F}) = L + M$ is pseudoeffective, so $3M = 2(L+M) + (M-2L)$ is big and we get a contradiction with Theorem 1.5. Therefore $(M-2L)^2 = M^2 = \Delta(\mathcal{E}) = 0$. Similarly, we get a contradiction if $(L+M)^2 > 0$. Therefore $c_1(\mathcal{F})^2 = 0$, $c_2(\mathcal{F}) = LM = 0$ and $\mathcal{M}$ provides a line bundle described in (2). (Here we use a simple remark, that a pseudoeffective divisor A with $A^2 = 0$ is either big or nef. This immediately follows from existence of the Zariski decomposition.)

Subcase 2.2. $\mathcal{L}$ is not ∇-invariant. Since $\nabla(\mathcal{L}) \subset \Omega_X(\log D) \otimes \mathcal{F}$ we get a non-trivial homomorphism $\mathcal{L} \to \Omega_X(\log D) \otimes \mathcal{M}$ induced by ∇. At this point we need to use the integrability of ∇.

Lemma 4.3. *The induced homomorphism* $\psi \colon (\mathcal{L} \otimes \mathcal{M}^*) \oplus \mathcal{M} \to \Omega_X(\log D)$ *has rank* 1.

Proof. Locally, outside a finite number of points where $\mathcal{M}$ is not locally free, the map ψ is given by two 1-forms ω_1 and ω_2. One can check that local 2-forms $\omega_1 \wedge \omega_2$ glue together to a map $\eta: \mathcal{L} \to \Omega_X^2(\log D) \otimes \mathcal{E}/\mathcal{F}$ induced from $\nabla^2: \mathcal{E} \to \Omega_X^2(\log D) \otimes \mathcal{E}$ by the inclusion $\mathcal{L} \to \mathcal{E}$ and the surjection $\mathcal{E} \to \mathcal{E}/\mathcal{F}$. From the integrability of ∇ it follows that η is the zero map and therefore the image of ψ has rank 1. $\square$

By the lemma there exist effective divisors A_1 and A_2 such that the line bundle associated to the divisor $B = L - M + A_1 = M + A_2$ is contained in $\Omega_X(\log D)$.

If $(L + M)^2 > 0$ then $3B = L + M + A_1 + 2A_2$ is big, which is impossible by Theorem 1.5. Therefore $(L + M)^2 = 0$. Now let us note that $0 \geq \Delta(\mathcal{E}) \geq 6LM - 2(L + M)^2 = 6LM$. Hence $(L - M)^2 \geq 0$ and either $L - M$ or $M - L$ is pseudoeffective.

In the first case, since $L - M$ is not big we have $(L - M)^2 = 0$ and $\mathcal{L} \otimes \mathcal{M}^*$ is a line bundle described in (2).

In the second case, $2M = (M - L) + (M + L)$ is pseudoeffective. Since it is not big, $M - L$ can not be big and we have $(M - L)^2 = 0$. Moreover, $2M^2 = (M - L)(M + L) \geq 0$ and hence $M^2 = LM = L^2 = 0$. Clearly, $\mathcal{M}$ provides a line bundle described in (2).

In both cases $(L + M)^2 = (L - M)^2 = 0$, so $\Delta(\mathcal{E}) = 6LM = 0$.

Let us also note that in the whole Case 2 if $\Delta(\mathcal{E}) \leq 0$ then $\Delta(\mathcal{E}) = 0$, $c_1^2(\mathcal{F}) = c_2(\mathcal{F}) = 0$, $\mathcal{E} \in \mathrm{Ext}^1(\mathcal{O}_X, \mathcal{F})$ and $\mathcal{F} \in \mathrm{Ext}^1(\mathcal{O}_X(M), \mathcal{O}_X(L))$ (i.e., $\mathcal{M}$ is a line bundle) $\square$

4.3. Remark. Using Theorems 1.5 and 1.6 one can generalize Theorems 4.1 and 4.2 to all log canonical pairs.

A sheaf with λ-connection for $\lambda \neq 0$ is known to be locally free with vanishing rational Chern classes. This is no longer true for Higgs sheaves or in the logarithmic case.

4.4. Remark. Let $(\mathcal{E}, \nabla)$ be a rank 2 H-semistable sheaf with (non-integrable) logarithmic λ-connection. Then one can check that $h^0(S^{2n}\mathcal{E}(-nc_1\mathcal{E})) = o(n^3)$. This immediately implies the Bogomolov inequality. Let us also note that it is not true that $h^0(S^{2n}\mathcal{E}(-nc_1\mathcal{E})) = o(n^2)$ in the curve case (one can give easy counterexamples). This is because having a small number of section is related to Theorem 1.5.

It is an interesting question if the number of sections of semistable Higgs sheaves with trivial discriminant is in general smaller than expected.

4.5. A. Reznikov studied representations of the fundamental group of a projective manifold in $SL_2(\mathbb{C})$ (see [Re], 5.1). Similarly, Theorem 4.2 (and its proof) can be used to describe the algebro–geometric data needed to construct representations of $\pi_1(X)$ in $SL_3(\mathbb{C})$. Although Theorem 4.2 is stated only in the surface case the same proof works in higher dimensions so we really get description of representations in all dimensions.

There are three possibilities:

(1) The bundle corresponding to the representation is semistable.

(2) There exists a line bundle $\mathcal{L} \subset \Omega_X$ such that $c_1(\mathcal{L})^2 H^{n-2} = 0$, $c_1(\mathcal{L})H^{n-1} > 0$. It induces a non-trivial representation $\pi_1(X) \to \mathrm{PSL}_2(\mathbb{R})$ and an equivariant map from the universal covering of X to the upper half plane (see [Si1], Corollary 8.4).

(3) There exists a rank two vector bundle $\mathcal{F}$ in the cotangent bundle satisfying $c_1^2(\mathcal{F})H^{n-2} = 3c_2(\mathcal{F})H^{n-2} > 0$.

The last possibility seems to come from existence of an equivariant map from the universal covering of X to the 2-dimensional ball.

5. Higher-dimensional versions of the Bogomolov–Miyaoka–Yau inequality

If X is a normal $\mathbb{Q}$-Gorenstein projective variety then $\nu(X)$ denotes the numerical Kodaira dimension of X, i.e., the maximal integer k such that K_X^k is not numerically trivial.

Theorem 5.1 (Enoki, [En]). *Let X be an n-dimensional projective variety with only canonical singularities. Assume that K_X is nef and set $\nu = \min(\nu(X), n-1)$. Then for any ample divisor H the cotangent sheaf $\widehat{\Omega}_X$ is $K_X^\nu H^{n-1-\nu}$-semistable.*

Proof. The theorem is just a corollary of [En], Theorem 1.1 and the following remark. Let $f: Y \to X$ be a birational dominating morphism between normal projective varieties. Let $\mathcal{F}$ be a torsion free sheaf on Y and set $\mathcal{E} = (f_*\mathcal{F})^{**}$. Then for any nef divisor H on X (or a collection of nef divisors) $\mathcal{E}$ is H-semistable if and only if $\mathcal{F}$ is f^*H-semistable. □

Using similar method as Enoki it should be possible to prove a more general theorem stated below as a conjecture.

Conjecture 5.1' Let X be a normal n-dimensional projective variety. Assume that X has only canonical singularities in codimension 2 and that K_X is $\mathbb{Q}$-Cartier and nef. Set $\nu = \min(\nu(X), n-1)$. Then for any collection of ample divisors $D_1, \ldots, D_{n-1-\nu}$ the cotangent sheaf $\widehat{\Omega}_X$ is $K_X^\nu D_1 \ldots D_{n-1-\nu}$-semistable.

Theorem 5.2. *Let X be an n-dimensional projective variety with only canonical singularities. Assume that K_X is nef and big and let $f: Y \to X$ be any resolution of singularities. Then*

$$(2(n+1)c_2(\Omega_Y) - nc_1^2(\Omega_Y))f^*(K_X^{n-2}) \geq 0.$$

Proof. The following argument comes from [Si1]. There is a canonical system of Hodge sheaves on $\mathcal{E} = \widehat{\Omega}_X \oplus \mathcal{O}_X$ induced by the identity on the first factor and zero on the other. By Theorem 5.1 and Proposition 8.1 this system is Higgs semistable. Therefore Theorem 2.1 applied to this sheaf implies the required inequality. □

This theorem was conjectured by K. Sugiyama (see [Su], Conjecture 0.4) and claimed by S. Lu and G. Tian (see [Ti], p. 149) but to the author's knowledge the proof has never appeared. In the smooth case the theorem was proved by many people: B. Y. Chen–K. Ogiue, T. Aubin, S. T. Yau, H. Tsuji, F. Hirzebruch, etc. See [Mi3] for further references and history of the problem.

If K_X is not big then the method of proof of Theorem 5.2 gives a special case of the following result due to Y. Miyaoka:

Theorem 5.3 ([Mi2], Theorem 6.6). *Let X be a smooth n-dimensional projective variety such that K_X is nef. Then for any collection of nef divisors $D_1, \ldots, D_{n-2}$ the inequality*

$$c_2(\Omega_X) D_1 \ldots D_{n-2} \geq 0$$

holds.

However, one can expect that if the numerical Kodaira dimension is large one can get a better inequality. For simplicity the next question is formulated only in the smooth case but one can also ask it for any (non-uniruled) log canonical pair.

Question 5.4 Let X be a smooth n-dimensional projective variety such that K_X is nef. Is it true that for any $k \geq 2$ and any collection of nef divisors $D_1, \ldots, D_{n-k}$ the inequality

$$2(k+1) c_2(\Omega_X) K_X^{k-2} D_1 \ldots D_{n-k} \geq k K_X^k D_1 \ldots D_{n-k}$$

holds?

The answer is positive if $k = 2$ by the result of Y. Miyaoka (see [Mi2], Theorem 1.1 or Theorem 5.6 below) and for $k = n$ by Theorem 5.2 if K_X is big and Theorem 5.3 otherwise.

In the proof of Theorem 5.2 we mentioned that there is a natural system of Hodge bundles $(\mathcal{E} = \Omega_X \oplus \mathcal{O}_X, \theta)$. The only nontrivial saturated Hodge subsystems of this system are of the form $\mathcal{F} \oplus \mathcal{O}_X$, where $\mathcal{F}$ is a saturated subsheaf of Ω_X.

Let s_k be the rank of the maximal $K_X^{k-1} D_1 \ldots D_{n-k}$-destabilizing Hodge subsystem of $\mathcal{E}$. It is easy to see that $s_k \geq t_k + 1$, where t_k is the rank of the maximal $K_X^{k-1} D_1 \ldots D_{n-k}$-destabilizing subsheaf of Ω_X.

The following theorem generalizes Theorem 5.2 in the smooth case and is a partial answer to Question 5.4.

Theorem 5.5. *Let X be a smooth n-dimensional projective variety such that K_X is nef. Then for any $k \geq 2$ and for any collection of nef divisors $D_1, \ldots, D_{n-k}$ we have*

$$2s_k c_2(\Omega_X) K_X^{k-2} D_1 \ldots D_{n-k} \geq (s_k - 1) K_X^k D_1 \ldots D_{n-k}.$$

Proof. All the slopes in the proof are counted with respect to $K_X^{k-1} D_1 \ldots D_{n-k}$.

Let us recall that Ω_X is generically semipositive, i.e., $\mu_{\min}(\Omega_X) \geq 0$ (see [Mi2], Corollary 6.4). In particular, $\mu_{\min}(\mathcal{E}, \theta) \geq 0$. Let us also note that $s_k \mu_{\max}(\mathcal{E}, \theta) \leq d_k = K_X^k D_1 \ldots D_{n-k}$. If $d_k = 0$ then $(\mathcal{E}, \theta)$ is semistable and the inequality follows from Theorem 2.1. Therefore we can assume that $d_k > 0$. Then by Theorem 2.3 we have

$$0 \leq d_k \Delta(E) K_X^{k-2} D_1 \ldots D_{n-k} + (n+1)^2 (\mu_{\max}(\mathcal{E}, \theta) - \mu(\mathcal{E}))(\mu(\mathcal{E}) - \mu_{\min}(\mathcal{E}, \theta))$$

$$\leq d_k \Delta(E) K_X^{k-2} D_1 \ldots D_{n-k} + \left((n+1)\frac{d_k}{s_k} - d_k\right) d_k.$$

Simplifying yields the required inequality $\square$

Below we give a proof of Miyaoka's theorem [Mi2], Theorem 1.1, using a similar method. One of the reasons to give a proof is that our arguments allow us to say when equality holds (cf. [Mi2], Proposition 7.2; our argument excludes the possibility of the first case in this proposition leaving only the cases that really occur).

Theorem 5.6 ([Mi2], Theorem 1.1). *Let X be a smooth n-dimensional projective variety such that K_X is nef. Then for any collection of nef divisors $D_1, \ldots, D_{n-2}$ we have*

$$3c_2(\Omega_X) D_1 \ldots D_{n-2} \geq K_X^2 D_1 \ldots D_{n-2}.$$

Proof. Since a nef divisor is a limit of ample divisors it suffices to prove the inequality assuming D_i are very ample.

If $d = K_X^2 D_1 \ldots D_{n-2} = 0$ then the inequality follows Theorem 5.3. Therefore we can assume that $d > 0$.

Let $0 = \mathcal{E}_0 \subset \mathcal{E}_1 \subset \cdots \subset \mathcal{E}_m = (\mathcal{E}, \theta)$ be the $K_X D_1 \ldots D_{n-2}$-Harder–Narasimhan filtration of $\mathcal{E}$ (in the sense of systems of Hodge sheaves; see Section 8). Set $\mathcal{F}_i = \mathcal{E}_i / \mathcal{E}_{i-1}$, $r_i = \operatorname{rk} \mathcal{F}_i$, $\mu_i = \mu(\mathcal{F}_i)$. By generic semipositivity of the cotangent bundle all the slopes μ_i are non-negative.

By Theorem 5.5 we can assume that $r_1 \leq 2$. It follows that $r_1 = 2$ and $\mathcal{E}_1 = \mathcal{L} \oplus \mathcal{O}_X$ for some line bundle $\mathcal{L} \subset \Omega_X$. By Theorem 1.5 we have $c_1(\mathcal{L})^2 D_1 \ldots D_{n-2} \leq 0$. Therefore $\operatorname{ch}_2(\mathcal{F}_1) \leq 0$, where $\operatorname{ch}_2 = 1/2(c_1^2 - 2c_2)$ is a degree 2 component of the Chern character of the sheaf.

By Theorem 2.1 and the Hodge index theorem we have

$$2d \operatorname{ch}_2(\mathcal{F}_i) D_1 \ldots D_{n-2} \leq \left(1 - \frac{r_i - 1}{r_i}\right) d c_1^2 \mathcal{F}_i D_1 \ldots D_{n-2} \leq r_i \mu_i^2.$$

Hence using the equality $\mathrm{ch}_2(\Omega_X) = \sum \mathrm{ch}_2(\mathcal{F}_i)$ we get

$$0 \leq d\,\mathrm{ch}_2(\Omega_X)D_1 \ldots D_{n-2} \leq \sum_{i \geq 2} r_i \mu_i^2 \leq (\sum_{i \geq 2} r_i \mu_i)\mu_1 = (d - 2\mu_1)\mu_1 \leq \frac{1}{8}d^2.$$

Simplifying yields inequality

$$c_2(\Omega_X)D_1 \ldots D_{n-2} \geq \frac{7}{16} K_X^2 D_1 \ldots D_{n-2},$$

which implies the required inequality. $\square$

5.7. Remark. One can also prove the above theorem in the logarithmic case (for non-uniruled varieties) using Miyaoka's method. Actually, one can simplify Miyaoka's arguments using Theorem 4.2 in case B of the proof of [Mi2], Proposition 7.1. This argument does not say when equality holds because Bogomolov's inequality is not known for rank 4 logarithmic Higgs sheaves (which is the only obstacle).

6. Bogomolov's inequality for extensions

In this section we use the same notation as in Section 2.

6.1. Let us fix a non-negative real number τ and an extension $0 \to \mathcal{E}_1 \to \mathcal{E} \to \mathcal{E}_2 \to 0$. Assume that all the sheaves $\mathcal{E}_1$, $\mathcal{E}$ and $\mathcal{E}_2$ are torsion free.

Then for a subsheaf $\mathcal{E}'$ of $\mathcal{E}$ we set

$$\mu_\tau^Q(\mathcal{E}') = \mu(\mathcal{E}') - \tau \frac{\mathrm{rk}\,\mathcal{E}_2'}{\mathrm{rk}\,\mathcal{E}'},$$

where $\mathcal{E}_2'$ is the image of the induced map $\mathcal{E}' \to \mathcal{E}_2$.

The extension $0 \to \mathcal{E}_1 \to \mathcal{E} \to \mathcal{E}_2 \to 0$ is called τ-*semistable* if for any subsheaf $\mathcal{E}'$ of $\mathcal{E}$ we have

$$\mu_\tau^Q(\mathcal{E}') \leq \mu_\tau^Q(\mathcal{E}).$$

Let us also set

$$\mu_\tau^S(\mathcal{E}') = \mu(\mathcal{E}') + \tau \frac{\mathrm{rk}\,\mathcal{E}_1'}{\mathrm{rk}\,\mathcal{E}'},$$

where $\mathcal{E}_1'$ is the subsheaf of $\mathcal{E}'$ induced from $\mathcal{E}_1$.

An easy calculation yields the following lemma.

Lemma 6.2. *Consider the diagram*

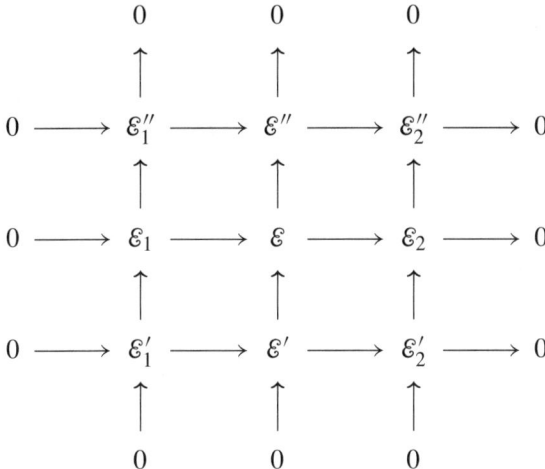

Then the following inequalities are equivalent:

(1) $\mu_\tau^S(\mathcal{E}') < \mu_\tau^S(\mathcal{E})$,

(2) $\mu_\tau^S(\mathcal{E}) < \mu_\tau^S(\mathcal{E}'')$,

(3) $\mu_\tau^Q(\mathcal{E}') < \mu_\tau^Q(\mathcal{E})$,

(41) $\mu_\tau^Q(\mathcal{E}) < \mu_\tau^Q(\mathcal{E}'')$.

In particular, τ-semistability defined by μ_τ^S and subsheaves of $\mathcal{E}$ is equivalent to τ-semistability defined by μ_τ^Q and subsheaves of $\mathcal{E}$ (see [DUW], Proposition 2.5).

The following theorem can be found in [DUW] and [BG]. For simplicity we state the results only in the usual case, but the same methods yield the results also for Higgs sheaves (Note that the methods of [BG] work only for Higgs bundles!).

Theorem 6.3 (Bogomolov's inequality for extensions). *Let* $0 \to \mathcal{E}_1 \to \mathcal{E} \to \mathcal{E}_2 \to 0$ *be a τ-semistable extension of (non-zero) torsion free sheaves for some $\tau \geq 0$. Then*

$$(D_1^2 D_2 \ldots D_{n-1})(\Delta(\mathcal{E}) D_2 \ldots D_{n-1}) + \tau^2 r_1 r_2 \geq 0.$$

Moreover, if $\tau > 0$ and the equality is attained, then we have:

(1) *The Harder–Narasimhan filtration of $\mathcal{E}$ has length 2.*

(2) *Let $\mathcal{E}'$ and $\mathcal{E}''$ denote the corresponding quotients. Then $\Delta(\mathcal{E}') D_2 \ldots D_{n-1} = \Delta(\mathcal{E}'') D_2 \ldots D_{n-1} = 0$.*

(3) *The 1-cycles $c_1(\mathcal{E}') D_2 \ldots D_{n-1}$ and $c_1(\mathcal{E}'') D_2 \ldots D_{n-1}$ are numerically proportional.*

(4) $\tau = \mu(\mathcal{E}') - \mu(\mathcal{E}'')$.

(5) *The induced maps $\mathcal{E}' \to \mathcal{E}_2$ and $\mathcal{E}_1 \to \mathcal{E}''$ are isomorphisms in codimension 0.*

Proof. Assume that $\mathcal{E}$ is not semistable and consider the maximal destabilizing subsheaf $\mathcal{E}'$ of $\mathcal{E}$. By definition of τ-semistability $\mu_\tau^Q(\mathcal{E}') \leq \mu_\tau^Q(\mathcal{E})$, i.e.,

$$\mu_{\max}(\mathcal{E}) = \mu(\mathcal{E}') \leq \mu(\mathcal{E}) - \tau\left(\frac{r_2}{r} - \frac{\operatorname{rk} \mathcal{E}_2'}{\operatorname{rk} \mathcal{E}'}\right) \leq \mu(\mathcal{E}) - \tau\left(\frac{r_2}{r} - 1\right) = \mu(\mathcal{E}) + \frac{r_1}{r}\tau.$$

Similarly, consider the quotient $\mathcal{E}''$ of $\mathcal{E}$ such that $\mu(\mathcal{E}'') = \mu_{\min}(\mathcal{E})$. Then by Lemma 6.2 $\mu_\tau^S(\mathcal{E}'') \geq \mu_\tau^S(\mathcal{E})$, i.e.,

$$\mu_{\min}(\mathcal{E}) = \mu(\mathcal{E}'') \geq \mu(\mathcal{E}) + \tau\left(\frac{r_1}{r} - \frac{\operatorname{rk} \mathcal{E}_1''}{\operatorname{rk} \mathcal{E}''}\right) \geq \mu(\mathcal{E}) + \tau\left(\frac{r_1}{r} - 1\right) = \mu(\mathcal{E}) - \frac{r_2}{r}\tau.$$

Therefore by Theorem 2.3

$$0 \leq (D_1^2 D_2 \ldots D_{n-1})(\Delta(\mathcal{E})D_2 \ldots D_{n-1}) + r^2(\mu_{\max}(\mathcal{E}) - \mu(\mathcal{E}))(\mu(\mathcal{E}) - \mu_{\min}(\mathcal{E}))$$
$$\leq (D_1^2 D_2 \ldots D_{n-1})(\Delta(\mathcal{E})D_2 \ldots D_{n-1}) + \tau^2 r_1 r_2.$$

If $\tau > 0$ and equality holds then $\mathcal{E}$ is not semistable (by Theorem 2.1), $\mu(\mathcal{E}') = \mu(\mathcal{E}) + \frac{r_1}{r}\tau$, $\mu(\mathcal{E}'') = \mu(\mathcal{E}) - \frac{r_2}{r}\tau$, and we have equality in Theorem 2.3. In particular, the Harder–Narasimhan filtration of $\mathcal{E}$ has only two quotients $\mathcal{E}'$ and $\mathcal{E}''$, the 1-cycles $c_1 \mathcal{E}' D_2 \ldots D_{n-1}$ and $c_1 \mathcal{E}'' D_2 \ldots D_{n-1}$ are numerically proportional and $\Delta(\mathcal{E}')D_2 \ldots D_{n-1} = \Delta(\mathcal{E}'')D_2 \ldots D_{n-1} = 0$ (see Proposition 2.4).

To prove the theorem it is sufficient to show that the induced maps $\mathcal{E}' \to \mathcal{E}_2$ and $\mathcal{E}_1 \to \mathcal{E}''$ are isomorphisms in codimension 0.

Since $\operatorname{rk} \mathcal{E}_2' = \operatorname{rk} \mathcal{E}'$ we have $\mathcal{E}_1' = 0$. Since $\operatorname{rk} \mathcal{E}_1'' = \operatorname{rk} \mathcal{E}''$, we have $\operatorname{rk} \mathcal{E}_2'' = 0$. □

Corollary 6.4. *Let $0 \to \mathcal{E}_1 \to \mathcal{E} \to \mathcal{E}_2 \to 0$ be a τ-polystable extension of (nonzero) torsion free sheaves for some $\tau > 0$. Then*

$$(D_1^2 D_2 \ldots D_{n-1})(\Delta(\mathcal{E})D_2 \ldots D_{n-1}) + \tau^2 r_1 r_2 = 0$$

if and only if the following conditions are satisfied:

(0) *the extension splits, i.e., $\mathcal{E} \simeq \mathcal{E}_1 \oplus \mathcal{E}_2$,*

(1) *$\mathcal{E}_1$ and $\mathcal{E}_2$ are polystable,*

(2) *$\Delta(\mathcal{E}_1)D_2 \ldots D_{n-1} = \Delta(\mathcal{E}_2)D_2 \ldots D_{n-1} = 0$,*

(3) *the 1-cycles $c_1(\mathcal{E}_1)D_2 \ldots D_{n-1}$ and $c_1(\mathcal{E}_2)D_2 \ldots D_{n-1}$ are numerically proportional,*

(4) *$\tau = \mu(\mathcal{E}_2) - \mu(\mathcal{E}_1)$.*

The above corollary should be compared with [BG], Theorem 6.2.

7. Sheaves with operators

Let X be a smooth n-dimensional projective manifold over an algebraically closed field k and let D be a reduced normal crossing divisor on X. Let $\mathcal{W}$ be a torsion free sheaf on X. As before, we fix a collection of $(n-1)$ nef divisors $(D_1, \ldots, D_{n-1})$ such that the 1-cycle $D_1 \ldots D_{n-1}$ is not numerically trivial.

The following proposition and its proof are well known and appear in many places.

Proposition 7.1. (1) *If $\mathcal{E}$ is a torsion free sheaf with a $\mathcal{W}$-valued operator η, then*

$$\mu_{\max}(\mathcal{E}) \leq \mu_{\min}(\mathcal{E}) + (r-1)[\mu_{\max}(\mathcal{W})]_+.$$

In particular, if $\mu_{\max}(\mathcal{W}) \leq 0$ and $\mathcal{E}$ is η-semistable then $\mathcal{E}$ is semistable.

(2) *If $\mathcal{E}$ is a torsion free sheaf with a logarithmic λ-connection ∇, then*

$$\mu_{\max}(\mathcal{E}) \leq \mu_{\min}(\mathcal{E}) + (r-1)[\mu_{\max}(\Omega_X(\log D))]_+.$$

In particular, if $\mu_{\max}(\Omega_X(\log D)) \leq 0$ and $\mathcal{E}$ is ∇-semistable, then $\mathcal{E}$ is semistable.

Proof. It is sufficient to prove the inequality in (1) assuming that $\mathcal{E}$ is η-semistable. Let $0 = \mathcal{E}_0 \subset \mathcal{E}_1 \subset \cdots \subset \mathcal{E}_m = \mathcal{E}$ be the usual Harder–Narasimhan filtration of $\mathcal{E}$. Then our assumption implies that all the induced maps $\mathcal{E}_i \to (\mathcal{E}/\mathcal{E}_i)\tilde{\otimes}\mathcal{W}$ are non-zero. Therefore

$$\mu(\mathcal{E}_i/\mathcal{E}_{i+1}) = \mu_{\min}(\mathcal{E}_i) \leq \mu_{\max}((\mathcal{E}/\mathcal{E}_i)\tilde{\otimes}\mathcal{W}) = \mu(\mathcal{E}_{i+1}/\mathcal{E}_i) + \mu_{\max}(\mathcal{W}).$$

Summing these inequalities we get the required inequality.

The proof of (2) is exactly the same. $\square$

As a corollary of the above proposition and Theorem 2.3 we get the following Bogomolov type inequality for sheaves with operators.

Proposition 7.2. *Assume that $\operatorname{char} k = 0$. Let $(\mathcal{E}, \eta)$ be a $D_1 \ldots D_{n-1}$-semistable torsion free sheaf with a $\mathcal{W}$-valued operator. Then*

$$4(D_1^2 D_2 \ldots D_{n-1})(\Delta(E)D_2 \ldots D_{n-1}) + (r-1)^2 r^2 ([\mu_{\max}(\mathcal{W})]_+)^2 \geq 0.$$

The following theorem was proved in [Bis]. The difference is in the second part, which was proved only in the special case when the maximal destabilizing sheaf of Ω_X is locally free. Our proof of this part is much simpler than the one in [Bis]. Let us also note that our assumptions are slightly weaker than in [Bis].

Theorem 7.3 ([Bis], Theorem A). *Assume $k = \mathbb{C}$ and let H be an ample divisor on X. The slopes below are counted with respect to $D_1 H^{n-2}$.*

(1) *Assume that $\mu_{\max}(\Omega_X) \leq 0$. Then every vector bundle $\mathcal{E}$ with an (integrable) connection admits a compatible flat connection.*

(2) *If X has Picard number one, then either $\mu_{\max}(\Omega_X) \leq 0$ or Ω_X is H-semistable.*

Proof. Let us recall that if a coherent sheaf admits an integrable $\mathbb{C}$-connection then it is locally free and all the rational Chern classes of $\mathcal{E}$ vanish (This fact is no longer true in positive characteristic). In particular, every vector bundle $\mathcal{E}$ with connection ∇ is ∇-semistable. Therefore by Proposition 7.1 $\mathcal{E}$ is $D_1 H^{n-2}$-semistable. Then Theorem 3.1 implies that $\mathcal{E}$ is H-semistable and the first part of the theorem follows from the main result of [Si1].

To prove the second part note that D_1 and H are numerically proportional and $D_1 H^{n-2}$-stability is equivalent to H-stability. Assume that $\mu_{\max}(\Omega_X) > 0$ and Ω_X is not H-semistable. Then Ω_X contains a subsheaf F of rank less then the dimension of X with ample first Chern class, contrary to Theorem 1.5. □

8. Higgs sheaves versus systems of Hodge sheaves

In this section we show that to give an algebraic proof of Bogomolov's inequality for semistable Higgs sheaves it is sufficient to prove it for stable systems of Hodge sheaves (note that the two notions of stability are different).

The proof goes as follows. From the proof of Theorem 2.1 one can see that Bogomolov's inequality for semistable Higgs sheaves follows from the one for slope stable sheaves. Such sheaves are Gieseker semistable and can be deformed through Gieseker semistable Higgs sheaves to a Gieseker semistable Higgs sheaf which underlies a system of Hodge sheaves (see [Si2], Lemma 4.1 and [Si4], Corollary 6.12). Now the claim follows from the usual reduction of semistable to stable as in the proof of Theorem 2.1 and the following proposition (which is also useful in checking Higgs semistability of Hodge systems; e.g., in the proof of Theorems 5.2 and 5.5).

Proposition 8.1. *The stability condition for a system of torsion free Hodge sheaves and the stability condition for the resulting Higgs sheaf are equivalent.*

Proof. Let us recall that a torsion free Higgs sheaf on a manifold X is the same thing as a pure sheaf of dimension $n = \dim X$ on the cotangent bundle Ω_X (see [Si4], Lemma 6.8). There is a natural fibrewise $\mathbb{C}^*$-action on Ω_X which acts on the sheaves on Ω_X in the same way as the natural $\mathbb{C}^*$-action on Higgs sheaves multiplying a Higgs field by $t \in \mathbb{C}^*$. Let us recall that systems of torsion free Hodge sheaves correspond to sheaves fixed by this $\mathbb{C}^*$-action (see [Si2], Lemma 4.1; the proof also works for torsion free sheaves).

Now let us fix a system of Hodge sheaves $\mathcal{E}$ on X. It corresponds to a pure sheaf on Ω_X, whose set theoretical support is the zero section of Ω_X.

Let us take any quotient sheaf $\mathcal{F}$ of $\mathcal{E}$ and consider the corresponding Quot scheme $\mathcal{Q}$. Since $\mathcal{E}$ is a fixed point of the $\mathbb{C}^*$-action there is an induced $\mathbb{C}^*$-action on $\mathcal{Q}$. Moreover, since $\mathcal{Q}$ is projective the orbit of $\mathcal{F}$ is a map $\mathbb{C}^* \to \mathcal{Q}$ which extends to $\mathbb{C} \to \mathcal{Q}$. The sheaf corresponding to the closed orbit over $0 \in \mathbb{C}$ is a quotient system of Hodge sheaves of $\mathcal{E}$ with the same numerical invariants as $\mathcal{F}$. □

8.2. Remark. In the case of locally free sheaves the proposition follows *a posteriori* from the results of Simpson [Si1], since both conditions are equivalent to the existence of a Hermitian–Yang–Mills metric on the underlying vector bundle. However, existence of admissible Hermitian–Yang–Mills metric for stable reflexive Higgs sheaves is difficult to prove (in the case of usual sheaves the corresponding result was proved by Bando and Siu) and in general the proposition does not seem to be known.

Acknowledgements. The problem of proving Bogomolov's inequality for Higgs bundles using algebro–geometric methods was suggested to the author by Professor M. S. Narasimhan, whom the author would like to thank. Unfortunately, as his predecessor [An], the author was unable to solve this problem in full generality.

During the preparation of this paper the author enjoyed, as a Marie Curie Research Fellow, the hospitality of the University of Warwick. The author was also partially supported by a Polish KBN grant (contract number 2P03A05022).

References

[An] B. Anchouche, Bogomolov inequality for Higgs parabolic bundles, Manuscripta Math. 100 (1999), 423–436.

[Biq] O. Biquard, Fibrés de Higgs et connexions intégrables: le cas logarithmique (diviseur lisse), Ann. Sci. École Norm. Sup. (4) 30 (1997), 41–96.

[Bis] I. Biswas, Vector bundles with holomorphic connection over a projective manifold with tangent bundle of nonnegative degree, Proc. Amer. Math. Soc. 126 (1998), 2827–2834.

[Bo1] F. A. Bogomolov, Holomorphic tensors and vector bundles on projective varieties, Izv. Akad. Nauk SSSR 42 (1978), 1227–1287; English translation: Math. USSR Izv. 13 (1979), 499–555.

[Bo2] —, Stability of vector bundles on surfaces and curves, in: Einstein Metrics and Yang–Mills Connections (T. Mabuchi and S. Mukai, eds.), Lecture Notes in Pure and Appl. Math. 145, Marcel Dekker, Inc., New York 1993, 35–49.

[Bo3] —, Stable vector bundles on projective surfaces, Mat. Sb. 185 (1994), 3–26; English translation: Sb. Math. 81 (1995), 397–419.

[BG] S. Bradlow, T. Gómez, Extensions of Higgs bundles, Illinois J. Math., to appear.

[DUW] G. Daskalopoulos, K. Uhlenbeck, R. Wentworth, Moduli of extensions of holomorphic bundles on Kähler manifolds, Comm. Anal. Geom. 3 (1995), 479–522.

[En] I. Enoki, Stability and negativity for tangent sheaves of minimal Kähler spaces, in: Geometry and analysis on manifolds, Katata/Kyoto 1987 (Toshikazu Sunada, ed.), Lecture Notes in Math. 1339, Springer-Verlag, Berlin 1988, 118–126.

[HL] D. Huybrechts, M. Lehn, The geometry of moduli spaces of sheaves, Aspects Math. 31, Vieweg, Braunschweig 1997.

[La1]   A. Langer, The Bogomolov–Miyaoka–Yau inequality for log canonical surfaces, J. London Math. Soc. 64 (2001), 327–343.

[La2]   —, Logarithmic orbifold Euler numbers of surfaces with applications, Proc. London Math. Soc., to appear.

[La3]   —, Semistable sheaves in positive characteristics, preprint 2001.

[LW]   J. Li, Y. Wang, Existence of Hermitian–Einstein metrics on stable Higgs bundles over open Kähler manifolds, Internat. J. Math. 10 (1999), 1037–1052.

[Mi1]   Y. Miyaoka, The maximal number of quotient singularities on surfaces with given numerical invariants, Math. Ann. 268 (1984), 159–171.

[Mi2]   —, The Chern classes and Kodaira dimension of a minimal variety, in: Algebraic Geometry, Sendai 1985, Adv. Stud. Pure Math. 10, Kinokuniya, Tokyo 1987, 449–476.

[Mi3]   —, Theme and variations—inequalities between Chern numbers, Sugaku Expositions 4 (1991), 157–176.

[Re]   A. Reznikov, All regulators of flat bundles are torsion, Ann. of Math. (2) 141 (1995), 373–386.

[Si1]   C. Simpson, Constructing variations of Hodge structure using Yangs–Mills theory and applications to uniformization, J. Amer. Math. Soc. 1 (1988), 867–918.

[Si2]   —, Higgs bundles and local systems, Inst. Hautes Études Sci. Publ. Math. 75 (1992), 5–95.

[Si3]   —, Moduli of representations of the fundamental group of a smooth projective variety I, Inst. Hautes Études Sci. Publ. Math. 79 (1994), 47–129.

[Si4]   —, Moduli of representations of the fundamental group of a smooth projective variety II, Inst. Hautes Études Sci. Publ. Math. 80 (1995), 5–79.

[Su]   K. Sugiyama On tangent sheaves of minimal varieties in: Kähler Metric and Moduli Spaces (T. Ochiai, ed.), Adv. Stud. Pure Math. 18-II, Academic Press Inc./Kinokuniya, Boston, MA/Tokyo 1990, 85–103.

[Ti]   G. Tian, Kähler–Einstein metrics on algebraic manifolds, in: Transcendental methods in algebraic geometry, Cetraro 1994 (F. Catanese and C. Ciliberto, eds.), Lecture Notes in Math. 1646, Springer-Verlag, Berlin 1996, 143–185.

A. Langer, Instytut Matematyki UW, ul. Banacha 2, 02–097 Warszawa, Poland
E-mail: `alan@mimuw.edu.pl`

Jets of antimulticanonical bundles on Del Pezzo surfaces of degree ≤ 2

Antonio Lanteri and Raquel Mallavibarrena

To the memory of Paolo Francia

Abstract. Let S be a complex Del Pezzo surface with $K_S^2 \leq 2$, and let $r = 4 - K_S^2$. The line bundle $L = -rK_S$ being very ample, we investigate the k-jet spannedness of $-tK_S$ for $t \geq r$. A key point is the stratification given by the rank of the evaluation map $j_2 : S \times H^0(S, L) \to J_2L$, with values in the second jet bundle of L, which puts in evidence some relevant loci related to both the intrinsic and extrinsic geometry of S. In particular, the generic 2-jet spannedness of L allows us to consider the second dual variety of (S, L), parameterizing the osculating hyperplanes to $S \subset \mathbb{P}^6$, embedded by $|L|$. Its behavior turns out to be completely different in the two cases $K_S^2 = 2$ and $K_S^2 = 1$.

Introduction

Both the definitions of k-jet ampleness and k-very ampleness for a line bundle on a smooth complex projective variety X are evolutions of that of k-spannedness, which has been the first notion introduced to formalize the concept of higher order embedding [BFS]. All these notions represent global conditions; on the contrary, k-jet spannedness represents a local one. A line bundle L on X is said to be k-jet spanned at $x \in X$ if the homomorphism $j_{k,x} : H^0(X, L) \to L \otimes \mathcal{O}_X / \mathfrak{m}_x^{k+1}$, sending every section of L to its k-th jet at x, is surjective. Here $\mathfrak{m}_x$ stands for the maximal ideal sheaf of x. Generic k-jet spannedness, when combined with the global condition of very ampleness, provides a very appropriate setting for studying the k-th osculatory behavior of a smooth variety embedded in the projective space. In fact, for $X \subset \mathbb{P}^N$ and $L = \mathcal{O}_X(1)$, the condition above simply means that the k-th osculating space to X at x is as largest as possible. In [LM3] we discussed the case of Del Pezzo surfaces S with $K_S^2 \geq 3$, studying the k-jet spannedness of their anticanonical bundle $-K_S$ and describing several properties of their second discriminant loci in a natural framework related to the second dual variety of the Veronesian surface $(\mathbb{P}^2, \mathcal{O}_{\mathbb{P}^2}(3))$. On the other hand, the k-very ampleness and k-jet ampleness properties of multiples of $-K_S$ on Del Pezzo surfaces have been investigated by several authors also for $K_S^2 = 2$ and 1 [DR], [BS2]. In this paper we look at these surfaces from the more subtle point of view of k-jet spannedness. Apart from the obvious interest in itself, this allows us to understand better and to supplement some results in those papers. In particular, trying to understand the case of the Del Pezzo surface of degree 1, left to the reader in [BS2], has been one of our motivations. In fact this case is much more interesting and intricate than the final sentence in [BS2] could lead to believe.

Let S be a Del Pezzo surface with $K_S^2 \leq 2$, let L be $-2K_S$ when $K_S^2 = 2$, and $-3K_S$ when $K_S^2 = 1$, respectively, and let $J_2 L$ be its second jet bundle. The starting point is to consider the evaluation map $j_2 : S \times H^0(S, L) \to J_2 L$ and determine the rank of $j_{2,x}$ at all points $x \in S$. Our results (Theorem (2.1) and Theorem (4.2)) describe the stratification of S given by this rank, putting in evidence some loci related to the geometry of S. E. g., for $K_S^2 = 1$ both the intrinsic and extrinsic properties (with respect to L) of the anticanonical pencil are involved. In particular, if x_0 is the base point of $|-K_S|$, it turns out that $|L - 2x_0| = |L - 3x_0|$, i. e., in the embedding $S \subset \mathbb{P}^6$ given by $|L|$, every tangent hyperplane to S at x_0 is osculating.

For $K_S^2 = 2$ we also consider the restriction of j_2 to $S \times V$, where V is a geometrically interesting vector subspace of $H^0(S, L)$ giving rise to a non-very ample linear system. The generic k-jet spannedness properties of all multiples of $-K_S$ (Corollary (2.2) and Corollary (4.3)) easily follow from the results above. It is worth comparing some of them with what follows from general results on the jet-spannedness of adjoint bundles. Actually, let $K_S^2 = 2$; then $-K_S$ being ample and spanned, the Seshadri constant $\varepsilon(-K_S, x)$ is ≥ 1 at every point $x \in S$. Hence by [La, Corollary (5.8)] we know that $-tK_S = K_S + (t+1)(-K_S)$ is $(t-2)$-jet spanned everywhere for $t \geq 2$. However this does not say that $-3K_S$ is 2-jet spanned, as in fact we obtain. Similarly, let $K_S^2 = 1$. Then, according to [La, (5.14) (i)] we know that $K_S + (k+3)(-K_S)$ is k-jet spanned at a sufficiently general point $x \in S$. In particular this says that $-4K_S$ is 2-jet spanned outside a countable set of points. In fact we show that the 2-jet spannedness of $-4K_S$ fails exactly at the singular points of the singular elements in the anticanonical pencil. By the way, this also provides a further example (e. g., see [Sh, Section 5]) of a generically 2-regular surface where the 2-regularity fails at a finite set only.

The ingredients of our analysis are the following: the restriction on the hyperosculation points of curves lying on S imposed by k-jet spannedness, which explains the failure of this condition at some points, a k-jet spannedness criterion proven in Section 1 and, above all, the local expressions of suitable sections providing a basis for $H^0(S, -tK_S)$, which allow us to compute the rank of $j_{2,x}$ at every point $x \in S$. For $K_S^2 = 1$ also the jumping sets of the ample and spanned line bundle $-2K_S$ [LPS] enter into the picture.

In particular it turns out that in both cases $K_S^2 = 2$ and 1 the line bundle L is generically 2-jet spanned. This is a good reason for considering the second dual variety of S, embedded by $|L|$ in $\mathbb{P}^6$. The fact that the second osculating space at a general point is a hyperplane defines a rational map $S \dashrightarrow S^\vee$ from S to its second dual, related to the Gauss map π_2 introduced in [LM1, p. 4832]. However the situation is totally different in the two cases. Actually, for $K_S^2 = 2$ we establish the birationality of π_2 (Theorem (5.2)). This is done by combining an improvement of [LM1, Proposition 2.4] with some results on quasi-homogeneous linear systems of plane curves due to Ciliberto and Miranda [CM]. On the other hand, for $K_S^2 = 1$, $S^\vee$ is a curve and π_2 is a fibration whose fibres (the osculating loci) are isomorphic to the elements of the anticanonical pencil (Theorem (5.3)). As far as we know, this is the first example of a generically 2-regular surface with positive second dual defect.

We would like to thank the MURST of the Italian Government (National Research Project "Geometry on algebraic varieties"), the MCYT of the Spanish Government, and our Departments for their support and for making this collaboration possible. The final stage of this research was done in the framework of the "Azione Integrata Italia–Spagna", 2001, project IT200. We are grateful to the referee for a careful reading and for helpful comments.

1. Preliminaries

Throughout all the paper we work over the complex number field. We use standard notation and terminology from algebraic geometry. We adopt the additive notation for the tensor product of line bundles, and, with a little abuse, we do not distinguish between a vector bundle and the corresponding locally free sheaf. The pull-back $\iota^*\mathcal{E}$ of a vector bundle $\mathcal{E}$ on a projective manifold X via an embedding $\iota : Y \hookrightarrow X$ is usually denoted by $\mathcal{E}_Y$.

(1.1) We specify some terminology concerned with the formalism of jets, not yet established in the literature. Let X be a smooth complex projective manifold, let $L \in \text{Pic}(X)$ and consider a vector subspace $V \subseteq H^0(X, L)$. Let $x \in X$, let $\mathfrak{m}_x$ be the ideal sheaf of x, and for every integer $k \geq 0$ consider the homomorphism

$$j_{k,x} : H^0(X, L) \to \Gamma(L \otimes \mathcal{O}_X / \mathfrak{m}_x^{k+1}),$$

sending every section $s \in H^0(X, L)$ to its k-th jet evaluated at x. Note that the range of the homomorphism above is simply $(J_k L)_x$, the fibre of the k-th jet bundle $J_k L$ at the point x. According to [BDRS], we say that: L is k-*jet spanned at x with respect to* V if $j_{k,x}|_V$, the homomorphism $j_{k,x}$ restricted to V, is surjective; L is k-*jet spanned on $\mathcal{U}$ w. r. to V* if this happens for all $x \in \mathcal{U} \subset X$. We simply say that L is *generically k-jet spanned w. r. to V* if $\mathcal{U}$ is a dense Zariski open subset of X and that L is k-*jet spanned w. r. to V* if $\mathcal{U} = X$. Moreover we always omit the expression "with respect to V" to mean that $V = H^0(X, L)$. Let $|V|$ be linear system (in general not complete, in spite of the notation) defined by V. If, in addition to the above, $|V|$ is very ample, i. e., the map $\varphi_V : X \dashrightarrow \mathbb{P}(V)$ is an embedding, we shift to the more classical terminology, saying that (X, V) $((X, L)$, if $V = H^0(X, L))$ is *(generically) k-regular* to mean that L is (generically) k-jet spanned w. r. to V. So, in accordance with [LM1], the expression "(X, L) is k-regular" means that L is very ample and k-jet spanned.

Let X be a smooth projective manifold of dimension n. Discussing the k-jet spannedness of a line bundle L on X with respect $H^0(X, L)$ instead of a subspace $V \subset H^0(X, L)$ has a clear advantage. Actually in that case the cohomological machinery allows to produce sufficient conditions adapted to the vanishing theorem one needs. Here is a very useful local k-jet spannedness criterion with respect to $H^0(X, L)$ (for $n = 2$ see [BS2] combining Lemma 4.1 with Remark 4.1).

(1.2) Lemma. *Let M be a nef line bundle on a smooth projective manifold X of dimension $n \geq 2$. Let $x \in X$, let $\sigma : \tilde{X} \to X$ be the blowing-up of X at x and let $E = \sigma^{-1}(x)$. If*

$$\sigma^*M - (k+n)E \text{ is nef and big}, \qquad (*)$$

then the line bundle $L := K_X + M$ is k-jet spanned at x. In particular, if L is very ample, M is nef and $(*)$ holds for some $x \in X$, then (X, L) is generically k regular.

Proof. In view of the exact sequence

$$0 \to L \otimes \mathfrak{m}_x^{k+1} \to L \to L \otimes \mathcal{O}_X/\mathfrak{m}_x^{k+1} \to 0,$$

to prove that $j_{k,x}$ is surjective it is enough to show that

$$H^1(X, L \otimes \mathfrak{m}_x^{k+1}) = 0.$$

Set $M := L - K_X$ and recall that $K_{\tilde{X}} = \sigma^*K_X + (n-1)E$. Then, in view of the Leray spectral sequence, we need only to prove that

$$H^1(\tilde{X}, K_{\tilde{X}} + \sigma^*M - (k+n)E) = 0.$$

But this follows from our assumptions due to the Kawamata–Viehweg vanishing theorem. □

A useful property of jet-spannedness is the following. Its proof can be extracted from [BS2, Proof of Lemma (2.2)].

(1.3) Lemma. *Let L_1, L_2 be line bundles on X and assume that L_i is k_i-jet spanned at a point $x \in X$, $i = 1, 2$. Then $L_1 + L_2$ is $(k_1 + k_2)$-jet spanned at x.*

In particular this implies that if L is the sum of k very ample line bundles on X, then (X, L) is k-regular.

As a next thing we focus on some properties of generically k-regular manifolds (X, V). Let $C \subset \mathbb{P}^N$ be an irreducible curve and let $|V|$ be the linear system cut out on C by the hyperplanes of $\mathbb{P}^N$. We say that a smooth point $x \in C$ is a *hyperosculation point of index k* if $|V - kx| = |V - (k+1)x|$. When $V = H^0(X, L)$ we write $|L - kx|$ instead of $|V - kx|$.

(1.4) Proposition. *Let $|V|$ be a very ample linear system on X, where $V \subseteq H^0(X, L)$. Assume that L is k-jet spanned at a point $x \in X$ with respect to V and let $C \subset X$ be an irreducible curve, smooth at x. Then C (in the embedding given by $|V|$) cannot admit a hyperosculation point of index $\leq k$ at x.*

Proof. By contradiction assume that C has a hyperosculation point of index $h \leq k$ at x. Since C is smooth at x we can choose local coordinates $(z, w_2, \ldots, w_n)$ on X centered at x in such a way that C is locally defined by $w_2 = \cdots = w_n = 0$ and z is a local coordinate on C. Take any section $s \in V$ and let s_C denote its restriction to C. Since x is a hyperosculation point of index h of C, in the embedding given by $|V|$, the local expansion of s_C at x has the following form

$$s_C = a_0 + a_1 z + \cdots + a_{h-1} z^{h-1} + a_{h+1} z^{h+1} + \cdots.$$

For shortness call $\eta(z)$ the expression on the right hand of the equality above, and think of s as a function of $z, w_2, \ldots, w_n$. Since C is defined by $w_2 = \cdots = w_n = 0$ we can thus write

$$s = \eta(z) + \sum_{i=2}^{n} w_i \sigma_i(z, w_2, \ldots, w_n),$$

where $\sigma_2, \ldots, \sigma_n$ are holomorphic functions in a neighborhood of x in X. But now, since $h \leq k$, computing the k-th jet $(j_k s)(x)$ of s at x, we see that the term

$$\frac{\partial^h s}{\partial z^h} = (h+1)! \, a_{h+1} z + \sum_{i=2}^{n} w_i \left(\frac{\partial^h \sigma_i}{\partial z^h} \right)$$

is zero at x. Since this holds for any $s \in V$ we conclude that the homomorphism

$$j_{k,x}|V : V \to (J_k L)_x$$

cannot be surjective. But this contradicts the assumption that L is k-jet spanned at x w. r. to V. $\square$

In particular Proposition (1.4) implies the following property of generically k regular manifolds.

(1.5) Corollary. *Assume that (X, V) is generically k-regular, k-jet spannedness holding on a dense Zariski open subset $\mathcal{U} \subseteq X$, and let $C \subset S$ be any smooth curve. Then C (in the embedding given by $|V|$) cannot admit hyperosculation points of index $\leq k$ in $\mathcal{U}$.*

We need to recall some properties of the second dual variety, in the framework of surfaces.

(1.6) Let S be a smooth surface and let L be a very ample line bundle on S whose complete linear system $|L|$ embeds S into some $\mathbb{P}^N$. Assume that (S, L) is generically 2-regular, L being 2-jet spanned on a dense Zariski open subset $\mathcal{U} \subset S$. The evaluation map $j_2 : H^0(S, L) \otimes \mathcal{O}_S \to (J_2 L)$ induces on $\mathcal{U}$ a surjective vector bundle homomorphism

$$j_{2;\mathcal{U}} : H^0(S, L) \otimes \mathcal{O}_\mathcal{U} \to (J_2 L)_\mathcal{U}.$$

Let $\mathcal{K}$ be the dual of its kernel. Let $\overline{\mathbb{P}(\mathcal{K})}$ be the closure of $\mathbb{P}(\mathcal{K}) \subset \mathcal{U} \times |L|$ inside $S \times |L|$. By definition, the second dual variety $S^\vee$ of (S, L) is the image of $\overline{\mathbb{P}(\mathcal{K})}$ via the second projection of $S \times |L|$. We denote by

$$\pi_2 : \overline{\mathbb{P}(\mathcal{K})} \to S^\vee$$

the morphism induced by this projection. Note that $\dim(\mathbb{P}(\mathcal{K})) = 2 + \text{rk}(\mathcal{K}) - 1 = 2 + h^0(L) - \text{rk}(J_2 L) - 1$. Hence

$$\dim(S^\vee) \le h^0(L) - 5,$$

with equality if and only if π_2 is generically finite. Identify $|L|$ with the dual projective space $\mathbb{P}^{N\vee}$. Then, for $(x, [s]) \in \mathcal{U} \times |L|$ with $j_2(s)(x) = 0$, we have that $\pi_2(x, [s]) = H$, the 2-osculating hyperplane to $S \subset \mathbb{P}^N$ at x such that $H \cap S = (s)_0$. Note that $(s)_0$ has a triple point at x, i.e., $H \in |L - 3x|$. We also recall [LM3, (1.4)] that the second discriminant locus $\mathcal{D}_2(S, L)$ is defined as the image of

$$\mathcal{J} := \{(x, H) \in S \times |L| \mid H \in |L - 3x|\}$$

via the second projection of $S \times |L|$. It parameterizes all elements of $|L|$ admitting some triple point. Of course $\mathcal{D}_2(S, L) \supseteq S^\vee$, with equality if (S, L) is 2-regular.

The following observation improves [LM1, Proposition (2.4)] in the very special case of surfaces in $\mathbb{P}^6$. For every $y \in S$ set $W_y := \text{Ker}(j_{2,y})$.

(1.7) Lemma. *Let (S, L) be a generically 2-regular surface with $h^0(S, L) = 7$, and assume that $\dim(S^\vee) = 2$. Let Y, Z be two Zariski closed proper subsets of S. If the homomorphism*

$$W_y \to (J_2 L)_z \qquad 1.7.1$$

induced by restricting $j_{2,z}$ to W_y is nonzero for every $y \in S \setminus Y$ and $z \in S \setminus Z$, $z \ne y$, then π_2 is birational.

Proof. The assumption that (S, L) is generically 2-regular with $h^0(S, L) = 7$ implies that the rank of the vector bundle $\mathcal{K}$ in (1.6) is one, hence the bundle projection $p : \overline{\mathbb{P}(\mathcal{K})} \to S$ is a birational morphism. Thus $p^{-1}(Y)$ and $p^{-1}(Z)$ have dimension ≤ 1, since Y and Z are Zariski closed proper subsets of S. Therefore $\pi_2(p^{-1}(Y \cup Z))$ is a proper subset of $S^\vee$, since $\dim(S^\vee) = 2$. Now suppose that π_2 is not birational, and let $\mathcal{U}$ be as in (1.6). Then for the general element $H \in S^\vee$ there are points $y, z \in \mathcal{U}$, outside Y and Z, such that H has a triple point at both y and z. But this is equivalent to saying that for such points the homomorphism (1.7.1) is zero, a contradiction. $\square$

Note that the assumption in Lemma (1.7) is equivalent to saying that $W_y \not\subseteq W_z$, and in fact our requirement means that $W_y \ne W_z$ for general y and z. Let (S, L) be a generically 2-regular surface with $h^0(S, L) = 7$. Then there is an obvious rational map $\theta : S \dashrightarrow S^\vee$ sending the general point $x \in S$ to the element of $|L|$ given by $|L - 3x|$ (representing the osculating hyperplane to $S \subset \mathbb{P}^6$ at x). As observed in the proof of Lemma (1.7), in this situation $p : \overline{\mathbb{P}(\mathcal{K})} \to S$ is a birational morphism.

Since $\pi_2 = \theta \circ p$ we see that the birationality of θ is equivalent to that of π_2. The following remark will be very useful in Section 5.

(1.8) Remark. Let (S, L) be as in Lemma (1.7) and let $y \in S$ be a general point. In view of the obvious vector bundle surjections

$$J_2 L \to J_1 L \to J_0 L$$

the homomorphisms $j_{1,z}$ and $j_{0,z}$ factor through $j_{2,z}$, for every $z \in S$. So, if the restriction of $j_{k,z}$ to W_y is nonzero for $k = 0$ or 1 then the homomorphism in (1.7.1) is nonzero a fortiori. Hence if for $m = 2$ or 1 the linear system $|L - 3y - mz|$ is strictly contained in $|L - 3y|$ for every $z \in S$ general enough, then the condition of Lemma (1.7) is satisfied and we conclude that π_2 is birational.

(1.9) Finally we collect some useful facts on the Del Pezzo surfaces which we are interested in. Recall that $S = B_{p_1,\ldots,p_t}(\mathbb{P}^2)$ is the plane blown-up at $t := 9 - K_S^2 = 7, 8$ points in general position [De, Theorem 1, p. 27]. Let $\eta : S \to \mathbb{P}^2$ be the blowing-up and let $e_i = \eta^{-1}(p_i)$ for $i = 1, \ldots, t$. Then $-K_S = \eta^* \mathcal{O}_{\mathbb{P}^2}(3) - e_1 - \cdots - e_t$. By the Riemann–Roch theorem and the Kodaira vanishing theorem, we have that $h^0(-mK_S) = 1 + \frac{1}{2}m(m+1)K_S^2$. Hence

$$h^0(-mK_S) = \begin{cases} 1 + m(m+1), & \text{for } K_S^2 = 2 \\ 1 + \frac{1}{2}m(m+1), & \text{for } K_S^2 = 1. \end{cases}$$

Let $K_S^2 = 2$. Then $|-K_S|$ is a net. Moreover, since $-K_S$ is spanned it defines a morphism $\pi : S \to \mathbb{P}^2$ representing S as a double cover of $\mathbb{P}^2$ branched along a smooth quartic curve.

Let $K_S^2 = 1$. Then $|-K_S|$ is a pencil with a single base point x_0. Moreover every curve in $|-K_S|$ is irreducible, $-K_S$ being ample. Since its arithmetic genus is 1 we thus see that any curve $\Gamma \in |-K_S|$ can admit at most one singular point, which can be either a node or a cusp. Note that x_0 is a smooth point for all curves in $|-K_S|$. Otherwise, intersecting an element $\Gamma_0 \in |-K_S|$, singular at x_0, with another element $\Gamma \in |-K_S|$, we would get the contradiction

$$1 = \Gamma^2 = \Gamma\Gamma_0 \geq \mathrm{mult}_{x_0}(\Gamma)\,\mathrm{mult}_{x_0}(\Gamma_0) \geq 2.$$

It thus follows that the general element of $|-K_S|$ is smooth, in view of the Bertini theorem. Let μ be the number of nodal curves in $|-K_S|$ and ν that of cuspidal ones. Then $\mu + 2\nu = 12$. Actually, by blowing-up S at x_0 we get a surface $\widetilde{S}$ and a fibration $\psi : \widetilde{S} \to \mathbb{P}^1$ whose general fibre is a smooth elliptic curve. Then the Euler–Poincaré characteristic of $\widetilde{S}$ is given by $e(\widetilde{S}) = \mu + 2\nu$ (e. g., see [B, Lemma VI.4]). On the other hand, by using Noether's formula we get $e(\widetilde{S}) = e(S) + 1 = 12\chi(\mathcal{O}_S) - K_S^2 + 1 = 12$.

2. Jets on Del Pezzo surfaces of degree 2

Let S be a Del Pezzo surface with $K_S^2 = 2$ and set $L := -2K_S$. Then L is very ample and the corresponding morphism embeds S as a surface of degree 8 in $\mathbb{P}^6$. In particular $-4K_S$ is 2-jet ample, by [BS2, Lemma (2.2)]. Investigating the 2-regularity of (S, L) is of some interest in itself. Actually, it seems uneasy to produce examples of generically 2-regular surfaces that are not 2-regular and such that their hyperplane bundle is 2-very ample. One is given by the general complete intersection of three quadrics in $\mathbb{P}^5$ (see [LM1, Section 5]). The pair (S, L) we are going to discuss provides another example. Of course since $L \in 2\text{Pic}(S)$, (S, L) does not contain lines; moreover, by [DR, Theorem 4] we know that L is 2-very ample. We will show that (S, L) is not 2-regular but generically 2-regular. In particular this implies that L is not 2-jet ample, which improves [BS2, Remark 5.2]. In addition, the divisor along which 2-jet spannedness fails can be described very precisely.

(2.1) Theorem. *Let S be the Del Pezzo surface with $K_S^2 = 2$, let $\pi : S \to \mathbb{P}^2$ be the double cover defined by $|-K_S|$ and let R_π be the ramification divisor of π. Set $L = -2K_S$ and for any $x \in S$ consider the map*

$$j_{2,x} : H^0(S, L) \to (J_2 L)_x.$$

Furthermore consider the vector subspace $V := \pi^*(H^0(\mathbb{P}^2, \mathcal{O}_{\mathbb{P}^2}(2)))$ of $H^0(S, L)$. Then,

$$\text{rk}(j_{2,x}) = \begin{cases} \text{rk}(j_{2,x}|V) = 6, & \text{if } x \in S \setminus R_\pi \\ \text{rk}(j_{2,x}|V) + 1 = 5, & \text{if } x \in R_\pi. \end{cases}$$

In particular L is 2-jet spanned (w. r. to V) exactly on $S \setminus R_\pi$. Moreover $-tK_S$ is 2-jet spanned for any $t \geq 3$.

Proof. Note that if $\mathcal{B}_1 = \{f, g, h\}$ is a basis for $H^0(S, -K_S)$, then

$$\mathcal{B}' = \{f^2, fg, fh, g^2, gh, h^2\}$$

is a basis for V; moreover, for any element $k \in H^0(S, L) \setminus V$, $\mathcal{B}_2 = \mathcal{B}' \cup \{k\}$ is a basis for $H^0(S, L)$. First suppose that $x \notin R_\pi$. Then π is an isomorphism around x. So we can assume that in a neighborhood U of x we have $h = 1$ and $f = u, g = v$ are local coordinates centered at x. Therefore the corresponding basis $\mathcal{B}'$ of V is locally given by $\{u^2, uv, u, v^2, v, 1, \}$. Thus we immediately see that $\text{rk}(j_{2,x}|V) = 6$ and then also $\text{rk}(j_{2,x}) = 6$, a fortiori. Now let $x \in R_\pi$. Since π is a morphism of degree 2 ramified along R_π, there are local coordinates (u, v) around x on S and (z, w) around $y = \pi(x)$ on $\mathbb{P}^2$ such that π is locally expressed by $z = u, w = v^2$, i. e., the branch divisor is locally represented by $w = 0$. Since $\{z, w, 1\}$ is the local expression of a basis of $H^0(\mathbb{P}^2, \mathcal{O}_{\mathbb{P}^2}(1))$ around y we can take $\mathcal{B}_1$ in such a way that $f = u, g = v^2, h = 1$ in a neighborhood U of x. Thus the local expression of $\mathcal{B}'$ around x is $\{u^2, uv^2, u, v^4, v^2, 1, \}$. Since the second and the fourth elements are in

m_x^3 we immediately see that $\mathrm{rk}(j_{2,x}|V) = 4$. Now let k be the element of $H^0(S, L)$ providing the equation of R_π. Note that $2R_\pi = \pi^*D$, where D is the branch divisor of π, hence $k^2 \in V$, while $k \notin V$. Since the local equation of R_π near x is $v = 0$, we have that $k_{|U} = v$ and then the local expression of $\mathcal{B}_2$ near x is the following: $\{u^2, uv^2, u, v^4, v^2, 1, v\}$. Taking into account the above discussion this shows that $\mathrm{rk}(j_{2,x}) = 5$.

Now consider $-tK_S$ for $t \geq 3$. To prove our assertion, set $M := -(t+1)K_S$ and let $\sigma : \tilde{S} \to S$ be as in Lemma (1.2), where x is any point of S. We have $\sigma^*M - 4E = \sigma^*(-(t-3)K_S) + 4(\sigma^*(-K_S) - E)$. Since $t \geq 3$ and $-K_S$ is ample, the first summand is nef and so, in order to apply Lemma (1.2) it is enough to show that $\sigma^*(-K_S) - E$ is nef and big. Let $C \subset \tilde{S}$ be any irreducible curve. If $C = E$, then $(\sigma^*(-K_S) - E)E = 1 > 0$. If $C \neq E$, then $C' = \sigma(C)$ is an irreducible curve on S and $C = \sigma^*C' - vE$, where $v = \mathrm{mult}_x(C') \geq 0$. Then

$$(\sigma^*(-K_S) - E)C = -K_S C' - v.$$

Recalling that $-K_S$ is ample and spanned with $h^0(-K_S) = 3$ we see that $|-K_S - x|$ is a pencil, whose general element D is an irreducible curve; so we can assume that D and C' are two distinct curves and then $(-K_S)C' = DC' \geq \mathrm{mult}_x(D)\mathrm{mult}_x(C') \geq v$. This shows that $\sigma^*(-K_S) - E$ is nef. On the other hand we have $(\sigma^*(-K_S) - E)^2 = K_S^2 - 1 > 0$, which proves the bigness. Then the assertion follows by using Lemma (1.2). □

As to the generic k-regularity properties of $-tK_S$ for $k \geq 3$ we have the following result. Compare this statement with the corresponding k-jet ampleness result in [BS2, p. 375].

(2.2) Corollary. *Let S be the Del Pezzo surface with $K_S^2 = 2$ and let $t \geq 3$. Then $-tK_S$ is*

i) $2[\frac{t}{3}]$-jet spanned if $t \equiv 0$ (3),

ii) $2[\frac{t}{3}]$-jet spanned, and $(2[\frac{t}{3}] + 2)$-jet spanned on $S \setminus R_\pi$ if $t \equiv 1$ (3),

iii) $(2[\frac{t}{3}] + 1)$-jet spanned, and $(2[\frac{t}{3}] + 2)$-jet spanned on $S \setminus R_\pi$ if $t \equiv 2$ (3).

Proof. We can write

$$-tK_S = \begin{cases} a(-3K_S), & \text{if } t = 3a \\ (a-1)(-3K_S) + 2(-2K_S), & \text{if } t = 3a+1 \\ a(-3K_S) + (-2K_S), & \text{if } t = 3a+2. \end{cases}$$

Then the assertion follows from Theorem (2.1) combined with Lemma (1.3). □

(2.3) Remark. Theorem (2.1) says that (S, L) is generically 2-regular, 2-regularity failing exactly on $S \setminus R_\pi$. We cannot use the same expression for (S, V) since $|V|$ does not give an embedding. Actually the morphism associated to V factors through π. Note that the second discriminant locus $\mathcal{D}_2(S, L)$ consists of the second dual variety $S_2^\vee$ of S (see Section 5), plus one component related to the ramification locus R_π. At every point $x \in R_\pi$ we have that the osculating space to $S \subset \mathbb{P}^6$ is a 4-plane. So there is a pencil of osculating hyperplane sections to S at x. This says that the additional component of $\mathcal{D}_2(S, L)$ is a family of $\mathbb{P}^1$'s parameterized by R_π.

Finally we point out that the locus R_π, where the 2-jet spannedness of L fails on S, can also be described in the following way. Let Γ be any smooth element of the net $|-K_S|$. Then $R_\pi \cap \Gamma$ consists of the four ramification points of the double cover $\Gamma \to \mathbb{P}^1$ given by $|L_\Gamma|$. So

$$R_\pi = \bigcup_{\substack{\Gamma \in |-K_S| \\ \text{smooth}}} \{x \in \Gamma \,:\, |L_\Gamma - x| = |L_\Gamma - 2x|\}.$$

3. Del Pezzo surfaces of degree 1: flexes

Let S be a Del Pezzo surface with $K_S^2 = 1$. The least positive multiple of the anti-canonical bundle being very ample is $L := -3K_S$, and the corresponding morphism embeds S as a surface of degree 9 in $\mathbb{P}^6$. Since the general element $\Gamma \in |-K_S|$ is a smooth curve of genus 1, the fact that $L\Gamma = 3$ prevents L from being 2-spanned [BS1, Corollary (1.4)]. In particular L is not 2-very ample. We can easily see that L is not even 2-jet spanned. Actually, let Γ be a smooth curve as above and consider the exact sequence

$$0 \to L + K_S \to L \to L_\Gamma \to 0. \tag{$*$}$$

Since $H^1(K_S + L) = 0$, the trace of $|L|$ on Γ coincides with $|L_\Gamma|$. On the other hand Γ is embedded by $|L|$ as a plane cubic; let $x \in S$ be a point corresponding to a flex (i.e. a hyperosculation point of index 2) of such a cubic. Then $|L_\Gamma - 2x| = |L_\Gamma - 3x|$. Then Proposition (1.4) shows that 2-jet spannedness fails at x.

To investigate further the geometry of the plane cubics $\Gamma \in |-K_S|$ embedded by $|L|$ we need explicit bases for $H^0(-mK_S)$ for $m = 2, 3$. In the next section we will need the same for $3 \leq m \leq 5$. So we insert the following remark, which will be useful several times. Recall that $h^0(-mK_S) = 1 + \frac{1}{2}m(m+1)$ by (1.9).

(3.1) Remark. Let $\mathcal{B}_1 = \{f, g\}$ be a basis of $H^0(-K_S)$. Then, for suitable elements $h \in H^0(-2K_S)$ and $k \in H^0(-3K_S)$, we can produce the following bases $\mathcal{B}_m$ for the vector spaces indicated below:

$H^0(-2K_S):$ $\quad \mathcal{B}_2 = \{f^2, fg, g^2, h\},$

$H^0(-3K_S):$ $\quad \mathcal{B}_3 = \{f^3, f^2g, fg^2, g^3, fh, gh, k\},$

$H^0(-4K_S):$ $\quad \mathcal{B}_4 = \{f^4, f^3g, f^2g^2, fg^3, g^4, f^2h, fgh, g^2h, fk, gk, h^2\},$

$$H^0(-5K_S): \quad \mathcal{B}_5 = \mathcal{B}_4 g \cup \{f^5, f^3 h, f^2 k, fh^2, hk\}.$$

Here $\mathcal{B}_4 g$ denotes the set consisting of the 11 elements of $\mathcal{B}_4$ each one multiplied by g.

Recall that the line bundle $-2K_S$ is spanned, hence for any choice of h as an element of the basis $\mathcal{B}_2$ as in Remark (3.1) we have $h(x_0) \neq 0$. Thus we can easily recover the following well known fact: $-2K_S$ gives rise to a morphism $\pi : S \to \mathbb{Q}_0 \subset \mathbb{P}^3$ exhibiting S as a double cover of the quadric cone $\mathbb{Q}_0$ branched at the vertex v and along the transverse intersection B with a cubic surface. We have $\pi^{-1}(v) = \{x_0\}$; moreover π maps two-to-one the elements of the pencil $|-K_S|$ to the generators of $\mathbb{Q}_0$. Let $\ell = \pi(\Gamma)$, where $\Gamma \in |-K_S|$, and let $x^* = \pi(x)$, for $x \in \Gamma \setminus \{x_0\}$. Let $i(B, \ell; x^*)$ be the local intersection index of B and ℓ at x^*. Note that $i(B, \ell; x^*) \leq 3$, since B is cut out on $\mathbb{Q}_0$ by a cubic surface. If $i(B, \ell; x^*) = 2$, i. e., ℓ is simply tangent to B at x^*, then Γ has a node at x; on the other hand, if $i(B, \ell; x^*) = 3$, i. e., ℓ is tangent to B at x^* but B has a hyperosculation point of index 2 at x^*, then Γ has a cusp at x. This can be immediately checked by expressing π with respect to suitable local coordinates at x and x^*. Since S is determined by B, this says that for the *general* Del Pezzo surface S with $K_S^2 = 1$ all singular curves of $|-K_S|$ are nodal.

Now we can relate the base point x_0 of $|-K_S|$ to flexes. For shortness we say that x is a flex of Γ to mean that it is a flex of Γ embedded by $|L|$, i. e., that $|L_\Gamma - 2x| = |L_\Gamma - 3x|$. Let $\Gamma \in |-K_S|$ and let $f = 0$ be its equation. Then the other element g of the basis $\mathcal{B}_1$ in Remark (3.1) vanishes along Γ only at x_0 to the first order. Consider the bases $\mathcal{B}_2$ and $\mathcal{B}_3$ introduced in Remark (3.1). Recalling that $K_S + L = -2K_S$, the exact cohomology sequence induced by $(*)$ is the following

$$0 \to H^0(-2K_S) \xrightarrow{\alpha} H^0(L) \xrightarrow{\beta} H^0(L_\Gamma) \to 0,$$

where the homomorphism α is given by $\otimes f$ and β is the restriction to Γ. Thus

$$\mathrm{Ker}\beta = H^0(-2K_S) \otimes f = \langle f^3, f^2 g, fg^2, fh \rangle,$$

which shows that

$$H^0(L_\Gamma) = \langle g_\Gamma^3, (gh)_\Gamma, k_\Gamma \rangle. \tag{**}$$

Now let $x \in \Gamma$ be any smooth point. Γ has a flex at x if and only if there is an element in $H^0(L_\Gamma)$ vanishing at x to the third order. Taking $x = x_0$ we thus see that g_Γ^3 is such an element for every $\Gamma \in |-K_S|$. Hence x_0 is a flex for all $\Gamma \in |-K_S|$; more precisely a smooth flex, recalling Remark (1.9).

Now let

$$\Phi = \bigcup_{\substack{\Gamma \in |-K_S| \\ \text{smooth}}} \{x \in \Gamma \setminus \{x_0\} : |L_\Gamma - 2x| = |L_\Gamma - 3x|\},$$

and denote by Δ the set of singular points (nodes and cusps) of the singular elements in the pencil $|-K_S|$. Recall that for an irreducible singular plane cubic, a node (a cusp) absorbs 6 (8) flexes [W, Exercise 6.4;1 p. 120]. Thus Φ parameterizes all flexes of the curves of the pencil $|-K_S|$, except possibly x_0, including those collapsed to points of Δ; so $\Delta \subset \Phi$. Recalling the double cover $\pi : S \to \mathbb{Q}_0$ discussed before we have

(3.2) Proposition. *Let $\Gamma \in |-K_S|$. The smooth flexes of Γ distinct from x_0 come in pairs via the morphism π.*

Proof. Let $x \neq x_0$ be a nonsingular point, which is a flex of Γ. Let $\{f, g\}$ be as in Remark (3.1). We can suppose that Γ is defined by f, so that $f(x) = 0, g(x) \neq 0$. Note that $|-2K_S - x|$ has codimension 1 in $|-2K_S|$, since $-2K_S$ is spanned; hence the element h of the basis $\mathcal{B}_2$ can be chosen so that it vanishes at x. On the other hand, since $x \neq x_0$ we know that $|-2K_S - 2x| \neq |-2K_S - x|$ by [LPS, Examples (1.8)(3), p. 206]. So h must vanish at x to the first order. Thus, since x is a flex, we see from (∗∗) that k can be chosen so that k_Γ vanishes at x to the third order. In other words the line defined by k in the plane $\langle \Gamma \rangle$, spanned by Γ in the embedding given by $|L|$, is the inflectional tangent at x. Now let $x' \in \Gamma$ be the conjugate of x in the involution defined by π. Since π is defined by $H^0(-2K_S)$, from the condition $\pi(x) = \pi(x')$ we get

$$((f(x'))^2 : f(x')g(x') : (g(x'))^2 : h(x')) = (0 : 0 : 1 : 0).$$

In particular $h(x') = 0$. Now consider the line defined by gh in the plane $\langle \Gamma \rangle$. It contains both x and x' since $h(x) = h(x') = 0$; moreover it contains also x_0, since $g(x_0) = 0$. We thus conclude that x' too is a flex of Γ, being collinear with x and x_0, which are flexes. □

Since the general $\Gamma \in |-K_S|$ has 9 flexes including x_0, the proof of Proposition (3.2) shows that Φ_Γ is cut out on Γ by an element of $\pi^*|\mathcal{O}_{\mathbb{P}^3}(4)|$, namely the pullback of 4 planes not containing v. In particular this shows that $\Phi_\Gamma = \deg \Phi_\Gamma = 8$, Φ meeting every smooth Γ at 8 distinct points; moreover $x_0 \notin \Phi$. Now let $\mathcal{J}_1(-2K_S)$ be the first jumping set of the ample and spanned line bundle $-2K_S$ [LPS, Section 1].

(3.3) Corollary. $(\Phi \setminus \Delta) \cap \mathcal{J}_1(-2K_S) = \emptyset$.

Proof. By [LPS, p. 206] $\mathcal{J}_1(-2K_S)$ coincides with the ramification locus of π. But this does not meet $\Phi \setminus \Delta$, by Proposition (3.2). □

Consider the obvious map $\Phi \to |-K_S| = \mathbb{P}^1$, which is a finite morphism of degree 8. Clearly Δ is contained in the ramification locus. In fact we can say more. Actually, let Γ_0 be a singular element of $|-K_S|$ and let y be its singular point; by what we said Γ_0 has a flex at x_0. Moreover, if y is a node then Γ_0 has two more flexes, say x_1, x_2. In view of the condition defining Φ and the fact that $x_0 \notin \Phi$ we have

$$\Phi \cap \Gamma_0 = \begin{cases} \{y, x_1, x_2\} & \text{if } y \text{ is a node,} \\ \{y\} & \text{if } y \text{ is a cusp.} \end{cases}$$

In the former case, the proof of Proposition (3.2) shows that the divisor supported at x_1, x_2 cut out by Φ on Γ_0 has degree 2. So we get

$$8 = \Phi\Gamma = \Phi\Gamma_0 = i(\Gamma_0, \Phi; y) + \sum_{j=1}^{2} i(\Gamma_0, \Phi; x_j) = i(\Gamma_0, \Phi; y) + 2.$$

This shows that y is a singular point of Φ (in fact a node with the same principal tangents as Γ_0). Similarly, in the latter case we get $i(\Gamma_0, \Phi, y) = 8$, hence y is a singular point of Φ as well. In fact all this says that $\Delta = \text{Sing}(\Phi)$.

4. Del Pezzo surfaces of degree 1: jets

Let (S, L) be as in Section 3. As we have seen, L is not 2-jet spanned at x_0 and at the points of Φ representing flexes. In order to compute the rank of the homomorphisms

$$j_{2,x} : H^0(S, -mK_S) \to (J_2(-mK_S))_x$$

at some special points $x \in S$ we proceed as in the proof of Theorem (2.1). To do this we will need explicit local expressions for the bases of $H^0(-mK_S)$ ($3 \le m \le 5$) described in Remark (3.1).

(4.1) Lemma. *Let S be the Del Pezzo surface with $K_S^2 = 1$, let x_0 be the base point of the pencil $|-K_S|$ and let L and Δ be defined as above. Then*

i) *The homomorphism $j_{2,x} : H^0(S, L) \to (J_2 L)_x$ has rank 3 if $x = x_0$, and rank 4 if $x \in \Delta$;*

ii) *If $x \in \Delta$, then the homomorphism $j_{2,x} : H^0(S, -4K_S) \to (J_2(-4K_S))_x$ has rank 5;*

iii) *Let $x \in \Delta$; then the homomorphism $j_{2,x} : H^0(S, -5K_S) \to (J_2(-5K_S))_x$ has rank 6.*

Proof. First look at x_0. Since any two distinct curves of the pencil are transverse at x_0, any two sections f, g, giving a basis of $H^0(-K_S)$, define local coordinates in a neighborhood U of x_0. Set $f_{|U} = w$ and $g_{|U} = z$. Since $-2K_S$ is spanned, $|-2K_S - x_0|$ has codimension 1 in $|-2K_S|$; hence looking at the basis $\mathcal{B}_2$ in Remark (3.1), we have that $h(x_0) \ne 0$. So, up to shrinking U, we can suppose that $h_{|U} = 1$. Note that also $|L - x_0|$ has codimension 1 in $|L|$; hence, looking at the basis $\mathcal{B}_3$ in Remark (3.1), we have $k(x_0) \ne 0$. Then the basis $\mathcal{B}_3$ of $H^0(S, L)$ is locally given by

$$\{w^3, w^2 z, wz^2, z^3, w, z, k_{|U}\}.$$

In particular, since the first 4 elements are in $\mathfrak{m}_{x_0}^3$ and $k(x_0) \ne 0$, we see that j_{2,x_0} has rank 3.

Now let x be a singular point of a curve $\Gamma \in |-K_S|$ and let f be a local equation of Γ around x. So f vanishes at x to the second order. Since $x \neq x_0$, we have $g(x) \neq 0$; hence, in a suitable neighborhood U of x, we can assume that $g_{|U} = 1$. Since $-2K_S$ is spanned, $|-2K_S - x|$ has codimension 1 in $|-2K_S|$; so, looking at the basis $\mathcal{B}_2$ as in Remark (3.1), we can suppose that $h(x) = 0$, since $g(x) \neq 0$. Moreover, since $x \neq x_0$ we have that $|-2K_S - 2x| \neq |-2K_S - x|$ by [LPS, Examples (1.8)(3), p. 206]. So h must vanish at x to the first order. E. g., recalling the properties of the morphism $\pi : S \to \mathbb{Q}_0 \subset \mathbb{P}^3$, we can take as h the pull-back via π of a linear form defining a hyperplane of $\mathbb{P}^3$ through $\pi(x)$, which is transverse to the generator $\pi(\Gamma)$. This means that h mod $\mathfrak{m}_x^2$ does not define a principal tangent to Γ at x. Note also that since L is very ample, the linear subsystems $|L - x|$ and $|L - 2x|$ of $|L|$ have codimension 1 and 3, respectively. So look at the basis $\mathcal{B}_3$ in Remark (3.1). Since g^3 does not vanish at x, we can choose k so that $k(x) = 0$; on the other hand, since f^3, f^2g, fg^2, fh all vanish at x to the second order, we conclude that k vanishes at x to the first order. All this shows that, in a suitable neighborhood U of x, we can assume that $h_{|U} = z$, $k_{|U} = w$ are local coordinates and

$$f_{|U} = az^2 + bzw + cw^2 + \cdots,$$

where dots represent higher order terms and $c \neq 0$, since the w-axis cannot be a principal tangent to Γ at x. Thus, looking at the local expressions of the element of the basis $\mathcal{B}_3$ of $H^0(L)$, we see that $\mathcal{B}_3$ mod $\mathfrak{m}_x^3$ is given by

$$\{0, 0, az^2 + bzw + cw^2, 1, 0, z, w\}.$$

Since $c \neq 0$, it turns out that $j_{2,x}$ has rank 4. This completes the proof of i).

Similarly, looking at the basis $\mathcal{B}_4$ of $H^0(S, -4K_S)$ in Remark (3.1), we see that $\mathcal{B}_4$ mod $\mathfrak{m}_x^3$ is given by

$$\{0, 0, 0, az^2 + bzw + cw^2, 1, 0, 0, z, 0, w, z^2\}.$$

Since $c \neq 0$ we see that $j_{2,x}$ has rank 5, which proves ii).

Finally, using the same local description as above, look at the expressions of the elements of the basis $\mathcal{B}_5$ of $H^0(S, -5K_S)$. Then $\mathcal{B}_5$ mod $\mathfrak{m}_x^3$ is given by

$$\{0, 0, 0, az^2 + bzw + cw^2, 1, 0, 0, z, 0, w, z^2, 0, 0, 0, 0, zw\}.$$

Once again, since $c \neq 0$, we see that $j_{2,x}$ has rank 6, and this proves iii). □

We have the following result.

(4.2) Theorem. *Let S be the Del Pezzo surface with $K_S^2 = 1$, let x_0 be the base point of the pencil $|-K_S|$ and let Δ, Φ be defined as above.*

(1) Let $L = -3K_S$ and consider the homomorphism $j_{2,x} : H^0(S, L) \to (J_2L)_x$. Then

$$\text{rk}(j_{2,x}) = \begin{cases} 3 & \text{if } x = x_0 \\ 4 & \text{if } x \in \Delta \\ 5 & \text{if } x \in \Phi \setminus \Delta \\ 6 & \text{if } x \notin \Phi \cup \{x_0\}. \end{cases}$$

Moreover

(2) $-4K_S$ is 2-jet spanned exactly outside Δ;

(3) $-tK_S$ is 2-jet spanned for any $t \geq 5$.

Proof. (1) Consider the pencil $|-K_S|$ and let $x \in S$. The first two assertions having been proved in Lemma (4.1), we can assume that $x \neq x_0$ and that the curve $\Gamma \in |-K_S - x|$ is smooth at x. As a basis $\mathcal{B}_1$ of $H^0(-K_S)$ we can pick $\{f, g\}$, with f defining Γ. Hence in a small neighborhood U of x we can assume that $f_{|U} = w$ gives a local coordinate, while $g_{|U} = 1$ since we know that $g(x) \neq 0$. Now look at the basis $\mathcal{B}_2$ of $H^0(-2K_S)$ as in Remark (3.1). Arguing as in the proof of Lemma (4.1) we can suppose that $h(x) = 0$. Moreover, choosing as h the pull-back via π of a linear form defining a plane transverse to the generator $\pi(\Gamma)$, up to shrinking U, we can assume that $h_{|U} = z$ provides a second local coordinate on U at x. Next consider the basis $\mathcal{B}_3$ of $H^0(L)$, produced in Remark (3.1). Note that the subspace of $H^0(L)$ of sections vanishing at x has codimension 1 and it does not contain g^3, while that of sections vanishing to the second order at x has codimension 3 and contains none of the elements fg, g^3, gh. This shows that we can choose k vanishing at x to the second order. Now recall that the restriction homomorphism $H^0(L) \to H^0(L_\Gamma)$ is surjective and that $h^0(L_\Gamma) = 3$. Since Γ is defined by $w = 0$ in U, we thus conclude that the basis of $H^0(L_\Gamma)$ in (**) of Section 3 is represented on $U \cap \Gamma$ by the three elements $1, z, k_\Gamma$.

Now assume that x is a flex of Γ; then $|L_\Gamma - 2x| = |L_\Gamma - 3x|$. On the other hand the vector subspace of $H^0(L_\Gamma)$ giving rise to $|L_\Gamma - 2x|$ is generated by k_Γ; hence we can suppose that k_Γ vanishes at x to the third order. As a consequence k has a local expansion at x of the form $k = w\mu + \cdots$, where μ is a holomorphic function of (z, w) (vanishing at the origin, since k vanishes at x to the second order) and dots denote terms of order ≥ 3. Looking at the basis $\mathcal{B}_3$ of $H^0(L)$ mod $\mathfrak{m}_x^3$, we thus see that $j_{2,x}$ has rank 5.

Finally suppose that x is not a flex of Γ. Then $|L_\Gamma - 2x| \neq |L_\Gamma - 3x|$. Since the vector subspace of $H^0(L_\Gamma)$ giving rise to $|L_\Gamma - 2x|$ is generated by k_Γ it follows that k_Γ has a local expansion at x of the form $k_\Gamma = z^2(\alpha + \cdots)$, where $\alpha \neq 0$ and dots denote terms of order ≥ 1. Thus the expansion of k at x has the form

$$k = z^2(\alpha + \cdots) + w\mu,$$

where μ is a holomorphic function of (z, w) (vanishing at the origin, since k vanishes at x to the second order), as before. Therefore the elements of the basis $\mathcal{B}_3$ of $H^0(L)$ have the following local expressions at x:

$$w^3, w^2, w, 1, zw, z, z^2(\alpha + \cdots) + w\mu,$$

with $\alpha \neq 0$. Thus, if we mod out by $\mathfrak{m}_x^3$ we immediately see that the homomorphism $j_{2,x}$ has rank 6. This shows that L is 2-jet spanned outside $\Phi \cup \{x_0\}$ and completes the proof of (1).

(2) Consider the line bundle $-4K_S$; set $M := -5K_S$ and let $\sigma : \tilde{S} \to S$ be as in Lemma (1.2), where x is any point of S. Let $C \subset \tilde{S}$ be any irreducible curve. If $C = E$, then $(\sigma^*M - 4E)E = 4 > 0$. If $C \neq E$, then $C' = \sigma(C)$ is an irreducible curve on S and $C = \sigma^*C' - \nu E$, where $\nu = \text{mult}_x(C') \geq 0$. Then

$$(\sigma^*M - 4E)C = (\sigma^*M - 4E)(\sigma^*C' - \nu E) = MC' - 4\nu = 5(-K_S)C' - 4\nu.$$

Now suppose that $x \notin \Delta$. Then there exists a smooth irreducible curve $D \in |-K_S-x|$ (in fact only one unless $x = x_0$). Now, if $C' \neq D$, then

$$(-K_S)C' \geq \text{mult}_x(D)\text{mult}_x(C') = \nu.$$

On the other hand, if $C' = D$, then $\nu = 1$ and so $(-K_S)C' = K_S^2 = 1 = \nu$. So in both cases we get $(\sigma^*M - 4E)C \geq 0$. Thus, since $(\sigma^*M - 4E)^2 = 25K_S^2 - 16 = 9$, we have that $\sigma^*M - 4E$ is nef and big and then by Lemma (1.2) we conclude that $-4K_S$ is 2-jet spanned outside Δ. Note that, for $x \in \Delta$, the argument above shows that the line bundle $\sigma^*M - 4E$ is not nef, so that Lemma (1.2) does not apply. Recalling Lemma (4.1), ii), we thus get (2).

(3) First consider the line bundle $-5K_S$. Let $x \in S$ be any point and set $M = -6K_S$. We have $\sigma^*M - 4E = 2(\sigma^*(-3K_S) - 2E)$. So, in order to apply Lemma (1.2), we have to show that $\sigma^*(-3K_S) - 2E$ is nef and big. Of course $(\sigma^*(-3K_S) - 2E)E = 2$ and $\sigma^*(-3K_S) - 2E$ is big. Now, for $C \neq E$, with the same notation as above, we get $(\sigma^*(-3K_S) - 2E)C = (-3K_S)C' - 2\nu$. Now recall that $L = -3K_S$ is very ample and $h^0(L) = 7$. If the general element D of the web $|L - 2x|$ is an irreducible curve, then we can assume that $D \neq C'$ and so $(-3K_S)C' = DC' \geq \text{mult}_x(D)\text{mult}_x(C') \geq 2\nu$. Suppose that all elements of $|L - 2x|$ are reducible and consider $D = A + B$, where $A \in |-K_S-x|$ and $B \in |-2K_S-x|$. Recall that A is an irreducible curve; moreover, if $x \neq x_0$, we can assume that B, the pull back via π of a hyperplane section of the quadric cone $\mathbb{Q}_0$ through $\pi(x)$ is irreducible. Then, if $C' \neq A$ we get

$$(-3K_S)C' = (A+B)C' \geq (\text{mult}_x(A) + \text{mult}_x(B))\, \text{mult}_x(C') \geq 2\nu.$$

On the other hand, if $x = x_0$, then A is smooth at x; hence, if $C' \neq A$ we have

$$(-3K_S)C' = 3AC' \geq 3\text{mult}_x(A)\text{mult}_x(C') \geq 3\nu.$$

Finally suppose that $C' = A$. If $x \notin \Delta$, then $\nu = 1$, hence we have again

$$(-3K_S)C' = 3K_S^2 = 3 > 2 = 2\nu.$$

Then the assertion follows by applying Lemma (1.2). Note that $\sigma^*L - 2E$ is certainly not nef if x is the double point of $C' = A$. On the other hand, $-5K_S$ is 2-jet spanned also at any point $x \in \Delta$, by Lemma (4.1), iii). Therefore $-5K_S$ is 2-jet spanned.

Finally let $t \geq 6$. Then the assertion follows from Lemma (1.3), since $-mK_S$ is very ample, hence 1-jet spanned, for $m \geq 3$. So (3) is proved. $\square$

As to the generic higher jet-spannedness properties of $-tK_S$ we have the following result.

(4.3) Corollary. *Let S be the Del Pezzo surface with $K_S^2 = 1$ and let $t \geq 5$. Then $-tK_S$ is $2[\frac{t}{5}]$-jet spanned if $t \equiv 0$ or 1 (5), and $(2[\frac{t}{5}] + 1)$-jet spanned if $t \equiv 2, 3,$ or 4 (5). Moreover, it is*

i) $(2[\frac{t}{3}] + 2)$-*jet spanned outside* $\Phi \cup \{x_0\}$ *if* $t \equiv 3$ (5),

ii) $(2[\frac{t}{3}] + 2)$-*jet spanned outside* Δ *if* $t \equiv 4$ (5).

Proof. As in the proof of Corollary (2.2) write $-tK_S$ as $a(-5K_S)$ if $t = 5a$ and as an integral convex linear combination of $-5K_S$ and $-mK_S$ if $t = 5a + m$, $m \leq 4$. Then apply Theorem (4.2) and Lemma (1.3). The only delicate point is when $m = 1$. Actually from the expression $-(5a + 1)K_S = a(-5K_S) + (-K_S)$ we conclude that $-tK_S$ is $2a$-regular except possibly at x_0. On the other hand, writing $-(5a+1)K_S = a(-4K_S) + (a + 1)(-K_S)$ and noting that $a \geq 1$ we have that the second summand is 0-jet spanned, while the first one is $2a$-jet spanned at x_0 by Theorem (4.2),(2). Therefore $-(5a + 1)K_S$ is $2a$-jet spanned everywhere. $\square$

(4.4) Remark. In analogy with Remark (2.3) we can describe the structure of the second discriminant locus $\mathcal{D}_2(S, L)$. Consider S embedded in $\mathbb{P}^6$ by $|L|$. According to Theorem (4.2), either there is only one osculating hyperplane to S at x or such hyperplanes constitute a pencil, a net or a web, respectively. Moreover $\mathcal{D}_2(S, L)$ consists of the following components: the second dual $S^\vee$ of S (see Section 5), an irreducible surface fibered over Φ, which is the closure of a scroll surface over $\Phi \setminus \Delta$, 12 planes corresponding to the points of Δ, and a $\mathbb{P}^3$, which corresponds to x_0. Due to this last component, we see that in case $K_S^2 = 1$ the second discriminant locus $\mathcal{D}_2(S, L)$ is not pure dimensional, though (S, L) is generically 2-regular.

5. Antimulticanonical models in $\mathbb{P}^6$ and their second dual varieties

In this section we consider the Del Pezzo surfaces of degree 2 and 1, focusing on the antibicanonical model in the former case and the antitricanonical model in the latter. As already observed, both are surfaces in $\mathbb{P}^6$ of degrees 8 and 9, respectively. We will refer to them as cases i) and ii) respectively.

First we deal with case i). Set $L = -2K_S$, as in Section 4. As a first thing we prove the following

(5.1) Proposition. *In case* i) *the morphism π_2 is generically finite; in particular* $\dim S^\vee = 2$.

Proof. We show that the general fibre of π_2 has dimension zero. Let $x \in S$ be a general point and suppose, by contradiction, that the only element $H \in |L - 3x|$ has the form $H = 3D + R$, with $D > 0$ and $R \geq 0$. From the equality
$$4 = 2K_S^2 = H(-K_S) = 3D(-K_S) + R(-K_S),$$
taking into account the ampleness of $-K_S$, we see that $R > 0$ and $DK_S = RK_S = -1$. In particular D and R are reduced, irreducible curves. The Hodge index theorem gives the inequality $1 = (-K_S D)^2 \geq K_S^2 D^2 = 2D^2$, hence $D^2 \leq 0$, and then the genus formula shows that D is a (-1)-curve. In the same way we conclude that R is a (-1)-curve too. We thus get
$$2 = -2K_S D = LD = HD = (3D + R)D = 3D^2 + RD = -3 + RD,$$
i. e., $RD = 5$. On the other hand, since $L^2 = 8$, we have
$$8 = H^2 = (3D + R)^2 = 9D^2 + 6DR + R^2 = -10 + 6DR,$$
which gives $DR = 3$, a contradiction. Thus π_2 is generically finite. The last assertion follows from (1.6). □

Now consider the evaluation map $j_2 : H^0(S, L) \otimes \mathcal{O}_S \to J_2 L$. By Theorem (2.1) we know that (S, L) is generically 2-regular. Let $\mathcal{U} \subseteq S$ be the dense Zariski open subset where $j_{2,x}$ is surjective.

Relying on Proposition (5.1) we use Remark (1.8), combined with a result by Ciliberto and Miranda, to show that π_2 is birational in case i). Recall that S is $\mathbb{P}^2$ blown-up at points $p_1, \ldots, p_7$ in general position, by (1.9). Thus, since $L = -2K_S$, $|L|$ corresponds to the linear system $|\mathcal{O}_{\mathbb{P}^2}(6) - 2p_1 - \cdots - 2p_7|$. Fix a point $y \in \mathcal{U} \backslash \cup_{i=1}^7 e_i$. Then by (1.9), $y = \eta^{-1}(y')$, where $y' \in \mathbb{P}^2$ is a point distinct from the p_i's. Similarly, if $z \notin \cup_{i=1}^7 e_i$ we have $z = \eta^{-1}(z')$ with $z' \in \mathbb{P}^2$, distinct from the p_i's. Moreover, we can choose y and z on S general enough so that $p_1, \ldots, p_7, y', z'$ are general. Then consider the linear subsystem $|L - 3y - 2z|$ of $|L - 3y|$. By what we said above, it corresponds to the following linear system on $\mathbb{P}^2$:
$$|\mathcal{O}_{\mathbb{P}^2}(6) - 2p_1 - \cdots - 2p_7 - 3y' - 2z'|.$$
This is a quasi-homogeneous linear system in the terminology of [CM] and, in the notation used there, it corresponds to $\mathcal{L}(6, 3, 8, 2)$. It thus follows from [CM, Theorem 8.1 and Lemma 7.1] that this linear system has the expected dimension, i. e., it is empty. On the other hand $|L - 3y| \neq \emptyset$ for $y \in \mathcal{U}$, since $\dim W_y = 1$. This says that the inclusion $|L - 3y - 2z| \subseteq |L - 3y|$ is strict. So, recalling Remark (1.8), this proves the following

(5.2) Theorem. *In case* i) *the morphism π_2 is birational. As a consequence S and $S^\vee$ are birational.*

Now consider case ii). As in Section 4, set $L = -3K_S$. Theorem (4.2) says that (S, L) is generically 2-regular, hence, for a general point $x \in S$ we have that $|L - 3x|$ consists of a single element. On the other hand, for $x \neq x_0$, also $|-K_S - x|$ consists of a single element, say Γ, and, of course, $3\Gamma \in |L - 3x|$. Moreover, according to Theorem (4.2) we have the following cases.

(1) $x \notin \Phi \cup \{x_0\}$. Clearly in this case $|L - 3x| = \{3\Gamma\}$, where $\{\Gamma\} = |-K_S - x|$.

(2) $x \in \Phi \setminus \Delta$. In this case $|L - 3x|$ is a pencil, whose elements have the form $\Gamma + C$, where $\{\Gamma\} = |-K_S - x|$ and $C \in |-2K_S - \tau|$, τ denoting the tangent vector at x determined by the inflectional tangent line to Γ, embedded by $|L|$, at x. Note that $|-2K_S - \tau|$ is in fact a pencil. Actually, since $x \in \Phi \setminus \Delta$, for every tangent vector τ at x, the linear system $|-2K_S - \tau|$ has codimension 2 in $|-2K_S|$, by Corollary (3.3).

(3) $x \in \Delta$. In this case $|L - 3x|$ is a net, whose elements have the form $\Gamma + C$, where $\{\Gamma\} = |-K_S - x| = |-K_S - 2x|$, and $C \in |-2K_S - x|$. Note that $|-2K_S - x|$ is in fact a net, since $-2K_S$ is spanned with $h^0(-2K_S) = 4$.

(4) $x = x_0$. In this very special case $|L - 3x|$ is a web whose elements have the form $\Gamma_1 + \Gamma_2 + \Gamma_3$, where $\Gamma_i \in |-K_S|$ for every i. Note that this identifies $|L - 3x|$ with the third symmetric power $(\mathbb{P}^1)^{(3)} \cong \mathbb{P}^3$.

As a consequence of this analysis we get the following

(5.3) Theorem. *In case* ii) *the morphism π_2 is a fibration with fibres isomorphic to the curves of $|-K_S|$, and $S^\vee$ is a smooth rational curve parameterizing the pencil $|-K_S|$ itself.*

As a consequence of [LM2, Lemma (2.1)] we know that if (S, L) is a generically 2-regular surface, then the union of its lines is a closed proper subset of S. Then, if $S^\vee$ is degenerate, the positive dimensional osculating locus of a general osculating hyperplane is not linear. The surface of case ii) reflects this behavior: all osculating loci are curves of arithmetic genus one, and, as far as we know, this is the first example of a generically 2-regular surface with positive second dual defect.

References

[B]   A. Beauville, Complex Algebraic Surfaces, Cambridge University Press, Cambridge 1983.

[BFS]   M. Beltrametti, P. Francia, and A. J. Sommese, On Reider's method and higher order embeddings, Duke Math. J. 58 (1989), 425–439.

[BS1]   M. Beltrametti and A. J. Sommese, On k-spannedness for projective surfaces in: Algebraic Geometry, Proc. L'Aquila 1988 (A. J. Sommese et al., eds.), Lecture Notes in Math. 1417, Springer-Verlag, Berlin 1990, 24–51.

[BS2]   —, On k-jet ampleness, in: Complex Analysis and Geometry (V. Ancona and A. Silva, eds.), Univ. Ser. Math., Plenum Press, New York 1993, 355–376.

[BDRS] M. Beltrametti, S. Di Rocco, and A. J. Sommese, On generation of jets for vector bundles, Rev. Mat. Complut. 12 (1999), 27–45.

[CM]   C. Ciliberto and R. Miranda, Degenerations of planar linear systems, J. Reine Angew. Math. 501 (1998), 191–220.

[De]   M. Demazure, Surfaces de Del Pezzo, I–V, in: Séminaire sur les singularités des surfaces, Lecture Notes in Math. 777, Springer-Verlag, Berlin 1980, 23–69.

[DR]   S. Di Rocco, k-very ample line bundles on Del Pezzo surfaces, Math. Nachr. 179 (1996), 47–56.

[LM1]   A. Lanteri and R. Mallavibarrena, Higher order dual varieties of projective surfaces, Comm. Algebra 27 (1999), 4827–4851.

[LM2]   —, Higher order dual varieties of generically k-regular surfaces, Arch. Math. 74 (2000), 1–6.

[LM3]   —, Osculatory behavior and second dual varieties of Del Pezzo surfaces, Adv. Geom. 1 (2002), 345–363.

[LPS]   A. Lanteri, M. Palleschi, and A. J. Sommese, On the discriminant locus of an ample and spanned line bundle, J. Reine Angew. Math. 477 (1996), 199–219.

[La]   R. Lazarsfeld, Lectures on linear series. With the assistance of Guillermo Fernández del Busto, in: Complex Algebraic Geometry (J. Kollár, ed.), IAS/Park City Math. Ser. 3, Amer. Math. Soc., Providence, RI, 1997, 163–219.

[Sh]   T. Shifrin, The osculatory behavior of surfaces in $\mathbb{P}^5$, Pacific J. Math. 123 (1986), 227–256.

[W]   R. J. Walker, Algebraic Curves, Dover Publications, Inc., New York 1962.

A. Lanteri, Dipartimento di Matematica "F. Enriques", Università, Via C. Saldini 50, 20133 Milano, Italy
E-mail: `lanteri@mat.unimi.it`

R. Mallavibarrena, Facultad de Matemáticas, Universidad Complutense de Madrid, Ciudad Universitaria, 28040 Madrid, Spain
E-mail: `Raquel_Mallavibarrena@mat.ucm.es`

A survey on the bicanonical map of surfaces with $p_g = 0$ and $K^2 \geq 2$

Margarida Mendes Lopes and Rita Pardini

Abstract. We give an up-to-date overview of the known results on the bicanonical map of surfaces of general type with $p_g = 0$ and $K^2 \geq 2$.

2000 Mathematics Subject Classification: 14J29

1. Introduction

Many examples of complex surfaces of general type with $p_g = q = 0$ are known, but a detailed classification is still lacking, despite much progress in the theory of algebraic surfaces. Surfaces of general type are often studied using properties of their canonical curves. If a surface has $p_g = 0$, then there are of course no such curves, and it is natural to look instead at the bicanonical system, which is not empty.

In this survey we describe the present (December 2001) state of knowledge about the bicanonical map for minimal surfaces of general type with $p_g = 0$ and $K^2 \geq 2$.

We do not consider the case $K^2 = 1$ (the so-called *numerical Godeaux* surfaces) because, in what concerns the bicanonical map, this case is special (see §2). We just remark that the numerical Godeaux surfaces are somewhat better understood than the other surfaces of general type with $p_g = 0$ and we refer to the paper [CP] and its bibliography.

This survey is organized as follows: in Section 2 we discuss the dimension of the bicanonical image and in Section 3 the base points of $|2K|$. In Section 4 we present the bounds on the degree of the bicanonical map for $K^2 \geq 2$. In Section 5 we discuss the surfaces that occur as bicanonical images, whilst in Section 6 we describe a few relevant examples. Finally in Section 7 we present some classification results and in Section 8 we present a list of open problems.

For each of the results presented we only give a very rough sketch of the proof, referring to the relevant papers for the missing details.

Acknowledgements. The present collaboration takes place in the framework of the European contract EAGER, no. HPRN-CT-2000-00099. The first author is a member of CMAF and of the Departamento de Matemática da Faculdade de Ciências da Universidade de Lisboa and the second author is a member of GNSAGA of CNR and of the Italian project PIN 2000 "Geometria sulle Varietà Algebriche".

Notation and conventions. We work over the complex numbers; all varieties are assumed to be compact and algebraic. We do not distinguish between line bundles and divisors on a smooth variety, using the additive and the multiplicative notation interchangeably. Linear equivalence is denoted by $\equiv$. The rest of the notation is standard in algebraic geometry.

2. The dimension of the bicanonical image

Let S be a minimal complex surface of general type with $p_g(S) = 0$. It is well known that:

- $q(S) = 0$,
- $1 \leq K_S^2 \leq 9$
- $P_2(S) := h^0(S, 2K_S) = 1 + K_S^2$.

We denote by $\varphi \colon S \to \mathbb{P}^{K_S^2}$ the bicanonical map of S and by Σ the image of φ. The first question one asks about the bicanonical map is what is the dimension of Σ. For $K_S^2 = 1$, one has $P_2(S) = 2$ and so Σ is a curve. For $K_S^2 \geq 2$, the answer was given by Xiao Gang in the mid-eighties:

Theorem 2.1 (Xiao Gang, [X1]). *Let S be a minimal complex surface of general type with $p_g(S) = 0$. If $K_S^2 \geq 2$ then the image of the bicanonical map of S is a surface.*

Sketch proof. We just explain the main ideas and refer the reader to [X1] for details.

By contradiction, suppose that $|2K_S|$ is composed with a pencil. Then one can show that necessarily $2K_S \equiv aF + Z$, where $|F|$ is a base point free genus 2 pencil and $a = K_S^2$. Using the very precise description of Horikawa ([Ho]) for the reducible fibres of a genus 2 fibration in terms of K_S^2 and $\chi(\mathcal{O}_S)$, one shows that for $K_S^2 \geq 3$ the surface S does not contain a genus 2 fibration, since otherwise the components of the reducible fibres would give $b_2(S)$ or more independent classes in $H^2(S, \mathbb{Z})$. For $K_S^2 = 2$, S can have a genus 2 fibration with general fibre F, but one can use the same type of argument to show that it is impossible to decompose $2K_S$ as $2K_S \equiv 2F + Z$, i.e. that $|2K_S|$ is not composed with $|F|$. $\square$

So the surfaces with $K_S^2 = 1$ and $p_g = 0$, the *numerical Godeaux*, are in a class of their own. We just mention here that there is intensive work in progress on this subject by F. Catanese and R. Pignatelli and by Y. Lee, using in particular the bicanonical fibration. As already mentioned in the introduction, for more facts on numerical Godeaux one can see the paper [CP], which has also a very complete list of references.

3. The base points of the bicanonical system

Recall that, while for $p_g(S) > 0$ the bicanonical map is defined at every point of S ([Bo], [Re], [F], [CC2], cf. [Ci]), for $p_g(S) = 0$ it is still unknown whether φ is always a morphism. For $K_S^2 \geq 5$ we have:

Theorem 3.1 (Reider, [Re]). *Let S be a minimal surface of general type with $p_g = 0$ and let $\varphi \colon S \to \mathbb{P}^{K_S^2}$ be the bicanonical map.*
If $K_S^2 \geq 5$, then φ is a morphism.

Remark 1. This is a particular case of Reider's theorem ([Re]) about adjoint systems, which only applies if $K_S^2 \geq 5$.

Remark 2. As far as we know, for all the known examples of surfaces of general type with $2 \leq K_S^2 \leq 4$ and $p_g = 0$ the bicanonical map is a morphism.

For $K_S^2 = 4$, Lin Weng ([W]) has proven that the base locus of the bicanonical system contains no -2-curve. This result has later been improved by Langer:

Theorem 3.2 (Langer, [La]). *Let S be a minimal surface of general type with $K_S^2 = 4$ and $p_g(S) = 0$. Then the system $|2K_S|$ has no fixed component.*

Still in the case $K_S^2 = 4$, F. Catanese and F. Tovena ([CT]) and D. Kotschick ([Ko]) have related the existence of base points of the bicanonical system to properties of the fundamental group of the surface. Since the statements are quite technical, we just quote here the following consequence of their results:

Theorem 3.3 (Catanese–Tovena, [CT], Kotschick, [Ko]). *Let S be a minimal surface of general type with $K_S^2 = 4$ and $p_g(S) = 0$. If $H^2(\pi_1(S), \mathbb{Z}_2) = 0$, then the bicanonical system $|2K_S|$ is base point free.*

4. The degree

Once one knows that for $K_S^2 \geq 2$ the bicanonical image of a surface S of general type with $p_g = 0$ is a surface, it is natural to look for bounds on the degree d of the bicanonical map φ.

If $K_S^2 = 2$, the bicanonical image is $\mathbb{P}^2$ and therefore $d \geq 2$. On the other hand, $(2K_S)^2 = 8$ implies $\deg \varphi \leq 8$, , with equality holding if and only if φ is a morphism. All the known examples with $K_S^2 = 2$ have $\deg \varphi = 8$.

For higher values of K_S^2 we have:

Theorem 4.1 ([M]). *Let S be a minimal complex surface of general type such that $p_g(S) = 0$, $K_S^2 \geq 3$ and let $\varphi \colon S \to \mathbb{P}^{K_S^2}$ be the bicanonical map of S. Then the degree of φ is at most 5.*
If φ is a morphism (in particular if $K_S^2 \geq 5$) then the degree of φ is at most 4.

Sketch proof (see [M] for the complete proof). Let d be the degree of φ and let m be the degree of the bicanonical image $\Sigma \subset \mathbb{P}^{K_S^2}$. Since Σ is a non-degenerate surface in $\mathbb{P}^{K_S^2}$, one has $m \geq K_S^2 - 1$. Write $|2K_S| = |M| + F$, where M and F are the moving part and the fixed part of the system, respectively. Notice that, if $F \neq 0$, then $M^2 < (2K_S)^2$, by the 2-connectedness of the bicanonical divisors. So we have $4K_S^2 \geq md$ and equality holds if and only if φ is a morphism. By an easy calculation we see that to prove the theorem it is enough to exclude the possibilities $K_S^2 = 5$, $d = 5$, and $K_S^2 = 3, d = 6$.

This is done by using the classification of surfaces of degree $n - 1$ in $\mathbb{P}^n$ (see [Nag]) to find the possibilities for Σ. Then, using the geometry of Σ, one is able to build irregular double covers of S, which in turn, by a theorem of De Franchis ([DF], see also [CC1]), yield special fibrations on S. Finally, with different "twists" for each case, the existence of such a fibration leads to a contradiction. □

Remark. As mentioned above, for all known examples of surfaces with $K_S^2 > 1$ the bicanonical map is a morphism, and so the bound 5 of the theorem above may not be effective for $K_S^2 = 3, 4$.

On the other hand, the bound 4, if φ is a morphism, is effective, as shown by the Burniat surfaces with $K_S^2 = 3, \ldots, 6$ ([Bu], [Pe], see also Example 3 of §6).

For high values of K_S^2 these bounds can be improved.

Theorem 4.2 ([MP1]). *Let S be a minimal complex surface of general type such that $p_g(S) = 0$ and let $\varphi \colon S \to \mathbb{P}^{K_S^2}$ be the bicanonical map of S. Then one has the following bounds on $d := \deg \varphi$:*

i) *if $K_S^2 = 9$, then $d = 1$;*

ii) *if $K_S^2 = 7, 8$, then $d \leq 2$;*

Sketch proof (see [MP1] for the proof). For $K_S^2 = 9$, one has $c_2(S) = 3$ and so $b_2(S) = 1$. The assertion is proven by combining Reider's theorem ([Re]) and the fact that $H^2(S, \mathbb{Q})$ is generated by the class of an ample divisor D with $D^2 = 1$.

For $K_S^2 = 7$, φ is a morphism, and therefore the degree d of φ is either 1, 2, or 4. If $d = 4$, then the bicanonical image Σ is a linearly normal surface of degree 7 in $\mathbb{P}^7$ with $p_g = q = 0$, and so it is the anti-canonical image of $\mathbb{P}^2$ blown-up at two points P, Q. Combining the information obtained from the geometry of Σ and the fact that the second Betti number of S is small ($b_2(S) = 3$), it is possible to find a contradiction, which shows that $d = 4$ does not occur.

For $K_S^2 = 8$, the technique of proof is analogous. □

Remark. The bounds in this theorem are effective, since there are examples of minimal surfaces with $p_g = 0$ and $K_S^2 = 7, 8$ for which the bicanonical map has degree 2 (see Examples 1 and 2 in Section 6).

5. The image

Another natural question that arises is finding the possibilities for the image of the bicanonical map φ, if φ is not birational. A priori, one only knows that the image of φ is a surface with $p_g = q = 0$. It turns out that it is possible to be more precise, as we will see in the next theorem.

Theorem 5.1 (Xiao Gang, [X2], [MP3]). *Let S be a minimal complex surface of general type such that $p_g(S) = 0$ and $K_S^2 \geq 2$ and let $\varphi \colon S \to \Sigma \subset \mathbb{P}^{K_S^2}$ be the bicanonical map of S. If φ is not birational, then either*

i) Σ *is a rational surface,*

or

ii) $K_S^2 = 3$, φ *is a morphism of degree 2 and $\Sigma \subset \mathbb{P}^3$ is an Enriques sextic.*

Sketch proof (see [X2] and [MP3] for the proof). If the degree d of φ is bigger than 2, then Σ is a linearly normal surface with $p_g = q = 0$ and degree lesser than or equal to $2n - 2$ in $\mathbb{P}^n$, and so a rational surface. In [X2], Xiao Gang, using double covers techniques, showed that if $d = 2$ and Σ is not rational, then $K_S^2 \leq 4$ and Σ is birationally equivalent to an Enriques surface with $K_S^2 + 4$ nodes. The Enriques surfaces with 8 nodes are classified in [MP3]. Using the knowledge of the linear systems on these surfaces, one is able to determine precisely the surfaces of general type with $p_g = 0$ and $K_S^2 = 4$ whose bicanonical map factors through a degree 2 map onto an Enriques surface. Such surfaces had been previously constructed by D. Naie ([Nai]) and their bicanonical map is of degree 4 onto a rational surface. Hence the possibility $K_S^2 = 4$ is excluded. □

6. Some examples

In this section we give a quick description of some of the known examples of surfaces with $p_g = 0$.

Example 1. *Surfaces with $K_S^2 = 8$.* All the examples known to us of surfaces S with $p_g = 0$ and $K_S^2 = 8$ are obtained by the following construction, first suggested by Beauville (cf. [Be], [Do]). One takes curves C_1, C_2 of genera g_1 and g_2, respectively, such that there exists a group G of order $(g_1 - 1)(g_2 - 1)$ that acts faithfully on both C_1 and C_2. If the quotient curves C_1/G and C_2/G are rational and the diagonal action of G on $C_1 \times C_2$ is free, then $S := (C_1 \times C_2)/G$ is a minimal surface of general type with $p_g = 0$ and $K_S^2 = 8$. By [Pa], the surfaces with these invariants and bicanonical map of degree 2 are obtained by this construction taking C_1 hyperelliptic of genus 3, and they belong to four different families. Examples with birational bicanonical map do exist. For instance, if one takes $C_1 = C_2$ to be the Fermat quintic, then it is possible to let $G = \mathbb{Z}_5^2$ act on the two copies of the curve in such a way that the diagonal action is free. The resulting surface has birational bicanonical map.

Example 2. *A surface with* $K_S^2 = 7$. This example is due to Inoue ([In]), who constructs it by taking the quotient of a complete intersection inside the product of four elliptic curves by a group isomorphic to $\mathbb{Z}_2^5$ acting freely. Alternatively, this surface can be constructed as a $\mathbb{Z}_2^2$-cover of a rational surface with 4 nodes (see [MP1]). The bicanonical map has degree 2 and the bicanonical involution belongs to the Galois group of the cover. The quotient of this surface by the bicanonical involution is a rational surface with 11 nodes.

Example 3. *Burniat surfaces.* These surfaces were discovered by Burniat ([Bu]) and studied later by Peters ([Pe]). They are obtained by taking $\mathbb{Z}_2^2$-covers of the plane branched on a configuration of lines as shown in Figure 1 in such a way that the images of the divisorial components of the fixed loci of the three nonzero elements of the Galois group of the cover are $l_{12} + m_1^1 + m_2^1$, $l_{23} + m_1^2 + m_2^2$ and $l_{13} + m_1^3 + m_2^3$. For a general choice of the lines m_j^i, the resulting surface Y is singular above the points

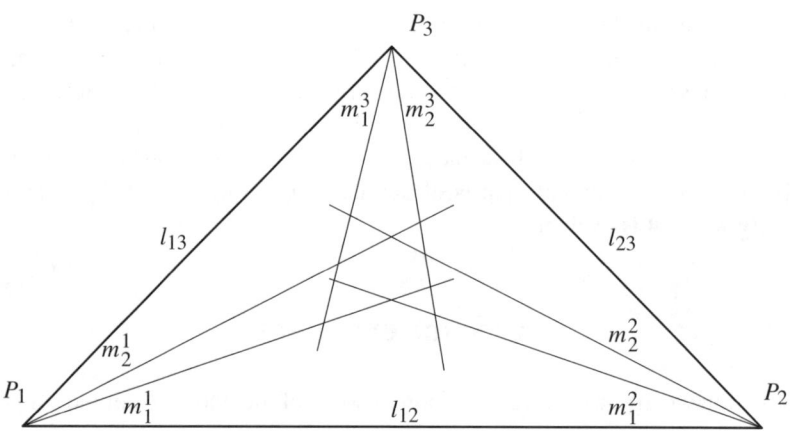

Figure 1. The branch locus of the Burniat surfaces in $\mathbb{P}^2$

P_1, P_2, P_3 and the minimal resolution of Y is obtained by taking base change with the blow-up $\hat{\mathbb{P}} \to \mathbb{P}^2$ of the plane at P_1, P_2, P_3 and then normalizing. In this way one obtains a minimal surface S with $p_g = 0$ and $K_S^2 = 6$. The bicanonical map is the composition of the induced $\mathbb{Z}_2^2$-cover $S \to \hat{\mathbb{P}}$ with the embedding of $\hat{\mathbb{P}}$ as a Del Pezzo sextic in $\mathbb{P}^6$.

Examples with the same properties and with $2 \leq K_S^2 \leq 5$ can be obtained by letting one or more subsets of 3 lines m_j^i go through the same point.

7. The limit cases

Surprisingly, the limit cases of Theorem 4.1 and Theorem 4.2, namely $K_S^2 = 7, 8$, deg $\varphi = 2$, and $K_S^2 = 6$, deg $\varphi = 4$, can be described precisely, as we will see in the next two theorems.

Theorem 7.1 ([MP4]). *Let S be a minimal complex surface of general type such that $p_g(S) = 0$ and $K_S^2 = 7, 8$. Let $\varphi \colon S \to \mathbb{P}^{K_S^2}$ the bicanonical map of S and assume that deg $\varphi = 2$. Then K_S is ample and:*

i) *there exists a fibration $f \colon S \to \mathbb{P}^1$ such that the general fibre F of f is a smooth hyperelliptic curve of genus 3 and the bicanonical involution induces the hyperelliptic involution on F;*

ii) *if $K_S^2 = 8$ then f is isotrivial and the singular fibres of f are 6 double fibres with smooth support, while if $K_S^2 = 7$ then S has 5 double fibres and exactly one fibre with reducible support.*

Remark 1. For $K_S^2 = 8$, the fact that f is an isotrivial fibration whose only singular fibres are double fibres with smooth support implies that S is one of the Beauville surfaces (see §6, Example 1). Using this fact it is possible to give a complete classification of these surfaces (see [Pa]). They belong to 4 different types, and the surfaces of each type form an irreducible connected component of the moduli space of surfaces of general type. An interesting feature of these surfaces is that they are smooth minimal models of double covers of the plane branched on a curve with certain singularities, a construction that had been suggested by Du Val ([DV], see also [Ci]). The expected number of parameters of the branch curve of this construction is negative, hence it seems very difficult prove directly its existence, that instead follows "a posteriori" from the classification of [Pa].

Remark 2. For $K_S^2 = 7$ the hyperelliptic fibration is not isotrivial and a complete classification seems out of reach. In the Inoue's surface (see §6), which is the only known example with $K_S^2 = 7$ and deg $\varphi = 2$, the unique fibre with reducible support of the fibration f is one of the double fibres. In principle one would expect this to be a special situation, hence it would be interesting to find examples where the reducible fibre is not a double fibre.

Sketch proof (see [MP4] for the proof). Consider the quotient Y of S by the bicanonical involution σ. Y is a rational surface whose only singularities are $\nu = K_S^2 + 4$ nodes, which correspond to the isolated fixed points of σ. The minimal resolution of Y is a rational surface X having $\nu \geq b_2(X) - 3$ disjoint -2-curves. Such surfaces are characterized in [DMP], where it is shown in particular that there exists a fibration $g \colon Y \to \Sigma$ with rational fibres and with $[\frac{\nu}{2}]$ double fibres. Now, using some geometrical reasoning, one shows that g pulls back to a fibration $f \colon S \to \mathbb{P}^1$ such that the general fibre of f is hyperelliptic of genus 3.

The fact that K_S is ample follows trivially for $K_S^2 = 8$ from Miyaoka's results ([Mi]) on the existence of rational curves on surfaces. For $K_S^2 = 7$, the non-existence of -2-curves on S is obtained by analyzing the structure of the unique reducible fibre of f and using the equality $b_2(S) = 3$. □

Theorem 7.2 ([MP2]). *Let S be a minimal complex surface of general type such that $p_g(S) = 0$ and $K_S^2 = 6$ and let $\varphi \colon S \to \mathbb{P}^{K_S^2}$ the bicanonical map of S. Then:*

$$\deg \varphi = 4 \text{ if and only if } S \text{ is a Burniat surface.}$$

In particular, K_S is ample.

Sketch proof (see [MP2] for the proof). The first step in the proof of this theorem consists in showing that the bicanonical image Σ of S is the non-singular Del Pezzo surface of degree 6 in $\mathbb{P}^6$. This is shown by a case by case exclusion of all the possible singular such Σ.

The second step consists in showing that the sides of the "hexagon" of -1-curves of Σ (cf. Fig. 2) are in the branch locus of φ. Using the curves in S which correspond to these sides, one produces a subgroup H of $\text{Pic}(S)$ such that $H \simeq \mathbb{Z}_2^3$.

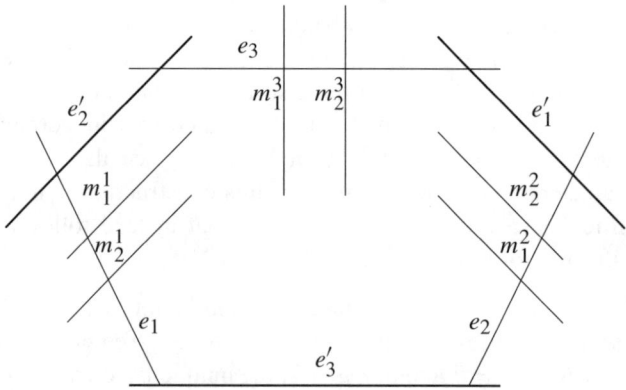

Figure 2. The branch locus of the Burniat surfaces in Σ

By studying the étale covers of S given by the nonzero elements of H, it is possible to show that the three pencils of Σ corresponding to the lines through P_1, P_2 and P_3 pull-back in S to three genus 3 hyperelliptic pencils, each having two irreducible double fibres beside the two reducible ones given by pairs of sides of the hexagon. The images of these irreducible fibres are the remaining components of the branch locus of φ (see Fig. 2).

Finally, one verifies that the bicanonical map is composed with the three involutions of S induced by the hyperelliptic pencils. It follows that the bicanonical map is a $\mathbb{Z}_2 \times \mathbb{Z}_2$-cover and S is a Burniat surface. □

Using Theorem 7.2, one can obtain very precise information on the geometry of the moduli space of surfaces with $p_g = 0$, $K_S^2 = 6$ and bicanonical map of degree 4.

Theorem 7.3 ([MP2]). *Smooth minimal surfaces of general type S with $K_S^2 = 6$, $p_g(S) = 0$ and bicanonical map of degree 4 form an unirational 4-dimensional irreducible connected component of the moduli space of surfaces of general type.*

Sketch proof (see [MP2] for the proof). By the semicontinuity of deg φ and by Theorem 4.1, the surfaces with deg $\varphi = 4$ are a closed subset of the moduli space.

Using the theory of natural deformations of abelian covers, one constructs explicitly a smooth family $\mathcal{X} \to B$ of $\mathbb{Z}_2^2$-covers of the plane blown up at three non collinear points with the following properties:

i) B is smooth and irreducible;

ii) for every $b \in B$ the fibre X_b is a Burniat surface and every Burniat surface occurs as a fibre for some $b \in B$;

iii) the family $\mathcal{X}$ is complete at every point b of B.

This shows that the Burniat surfaces are an irreducible open subset of the moduli space. Hence, in view of Theorem 7.2, the surfaces with deg $\varphi = 4$ are an irreducible open and closed subset of the moduli space. □

Remark 3. The limit cases for the degree of the bicanonical map have some common properties. First of all, by Theorem 7.1 and Theorem 7.2, they all have ample canonical class. In addition, all surfaces with $K_S^2 = 6$ and deg $\varphi = 4$, all surfaces with $K_S^2 = 8$ and deg $\varphi = 2$ (see [Pa]) and the Inoue surfaces with $K_S^2 = 7$ move in positive dimensional families, while the expected dimension of the moduli at the corresponding points is zero.

8. Some questions

Here we point out some questions that arise naturally from the results outlined in the previous sections.

Question 1 (cf. §3). Is the bicanonical map φ of surfaces with $p_g = 0$ a morphism also for $2 \leq K_S^2 \leq 4$?

Question 2 (cf. Theorem 4.1). Is there a surface with $p_g = 0$, $K_S^2 = 3$ or $K_S^2 = 4$ and deg $\varphi = 5$? Notice that for such a surface φ cannot be a morphism.

Question 3 (cf. Theorem 4.1 and Theorem 4.2). Is there is a surface with $p_g = 0$, $K_S^2 = 6$ and deg $\varphi = 3$?

Question 4. (cf. Theorem 7.2). Is it possible to characterize surfaces with $K_S^2 = 5$, $p_g = 0$ and $\deg \varphi = 4$?

Question 5. (cf. §7, Remark 2) Are the Inoue surfaces the only surfaces with $p_g = 0$, $K_S^2 = 7$ and non birational bicanonical map?

References

[Be] A. Beauville, Complex Algebraic Surfaces, Cambridge University Press, Cambridge 1983.

[Bo] E. Bombieri, Canonical models of surfaces of general type, Inst. Hautes Études Sci. Publ. Math. 42 (1973), 447–495.

[Bu] P. Burniat, Sur les surfaces de genre $P_{12} > 0$, Ann. Mat. Pura Appl. (4) 71 (1966), 1–24.

[CC1] F. Catanese, C. Ciliberto, On the irregularity of cyclic coverings of algebraic varieties, in: Geometry of Complex Projective Varieties, Cetraro 1990 (A. Lanteri, M. Palleschi, D. C. Struppa, eds.), Mediterranean Press, Rende 1993, 89–116.

[CC2] F. Catanese, C. Ciliberto, Surfaces with $p_g = q = 1$, Sympos. Math. XXXII (1991), 49–79.

[CP] F. Catanese, R. Pignatelli, On simply connected Godeaux surfaces, in: Complex analysis and algebraic geometry (T. Peternell and F.-O. Schreyer, eds.), Walter de Gruyter, Berlin 2000, 117–153.

[CT] F. Catanese, F. Tovena, Vector bundles, linear systems and extensions of π_1, in: Complex algebraic varieties (K. Hulek et al., eds.), Lecture Notes in Math. 1507, Springer-Verlag, Berlin 1992, 51–70.

[Ci] C. Ciliberto, The bicanonical map for surfaces of general type, Proc. Sympos. Pure Math. 62 (1997), 57–84.

[DF] M. De Franchis, Sugl'integrali di Picard relativi a una superficie doppia, Rend. Circ. Mat. Palermo XX (1905), 331–334.

[Do] I. Dolgachev, Algebraic surfaces with $p = p_g = 0$, in: Algebraic surfaces (C.I.M.E. 1977, Varenna), Liguori, Napoli 1981, 97–215.

[DMP] I. Dolgachev, M. Mendes Lopes, R. Pardini, Rational surfaces with many nodes, Compositio Math. 132 (2002), 349–363.

[DV] P. Du Val, On surfaces whose canonical system is hyperelliptic, Canadian J. Math. 4 (1952), 204–221.

[F] P. Francia, On the base points of the bicanonical system, Sympos. Math. XXXII (1991), 141–150.

[Ho] E. Horikawa, On algebraic surfaces with pencils of curves of genus 2, in: Complex Analysis and Algebraic Geometry. A collection of papers dedicated to K. Kodaira, Cambridge University Press, Cambridge 1977, 79–90.

[In] M. Inoue, Some new surfaces of general type, Tokyo J. Math. 17 (1994), 295–319.

[Ko]   D. Kotschick, On the pluricanonical maps of Godeaux and Campedelli surfaces, Internat. J. Math. 5 (1994), 53–60.

[La]   A. Langer, Pluricanonical systems on surfaces with small K^2, Internat. J. Math. 11 (2000), 379–392.

[M]   M. Mendes Lopes, The degree of the generators of the canonical ring of surfaces of general type with $p_g = 0$, Arch. Math. 69 (1997), 435–440.

[MP1] M. Mendes Lopes, R. Pardini, The bicanonical map of surfaces with $p_g = 0$ and $K^2 \geq 7$, Bull. London Math. Soc. 33 (2001), 265–274.

[MP2] M. Mendes Lopes, R. Pardini, A connected component of the moduli space of surfaces of general type with $p_g = 0$, Topology 40 (2001), 977–991.

[MP3] M. Mendes Lopes, R. Pardini, Enriques surfaces with eight nodes, Math. Z., to appear.

[MP4] M. Mendes Lopes, R. Pardini, The bicanonical map of surfaces with $p_g = 0$ and $K^2 \geq 7$, II, Bull. London Math. Soc., to appear.

[Mi]   Y. Miyaoka, The maximal number of quotient singularities on surfaces with given numerical invariants, Math. Ann. 268 (1973), 159–171.

[Nag] M. Nagata, On rational surfaces I, Mem. Coll. Sci., Univ. Kyoto, Ser. A, Vol. XXXII, Mathematics 3 (1960), 351–370.

[Nai] D. Naie, Surfaces d'Enriques et une construction de surfaces de type général avec $p_g = 0$, Math. Z. 215 (1994), 269–280.

[Pa]   R. Pardini, The classification of double planes of general type with $K^2 = 8$ and $p_g = 0$, J. Algebra, to appear.

[Pe]   C. Peters, On certain examples of surfaces with $p_g = 0$ due to Burniat, Nagoya Math. J. 166 (1977), 109–119.

[Re]   I. Reider, Vector bundles of rank 2 and linear systems on algebraic surfaces, Ann. of Math. 127 (1988), 309–316.

[W]   L. Weng, A result on bicanonical maps of surfaces of general type, Osaka J. Math. 32 (1995), 467–473.

[X1]   G. Xiao, Finitude de l'application canonique des surfaces de type général, Bull. Soc. Math. France 113 (1985), 23–51.

[X2]   G. Xiao, Degree of the bicanonical map of a surface of general type, Amer. J. Math. 112 (1990), 713–737.

M. Mendes Lopes, CMAF, Universidade de Lisboa, Av. Prof. Gama Pinto, 2, 1649-003 Lisboa, Portugal
E-mail: mmlopes@lmc.fc.ul.pt

R. Pardini, Dipartimento di Matematica, Università di Pisa, Via Buonarroti 2, 56127 Pisa, Italy
E-mail: pardini@dm.unipi.it

The antibirational involutions of the plane and the classification of real del Pezzo surfaces

*Francesco Russo**

Introduction

Real algebraic curves and real algebraic surfaces were studied and classified extensively during the last century: quartic plane curves and their bitangents beginning with Plücker [11] and Zeuthen [19], the complete intersections of a quadric cone with a cubic surface and their tritangent planes by Klein [6] and Comessatti [2]; for cubic surfaces Schläfli [12] gave a classification later extended by B. Segre to arbitrary fields[13], while for complete intersection of two quadrics there was a list by C. Segre [15], [14].

All these results are connected with real del Pezzo surfaces. Classically del Pezzo surfaces were defined as linearly normal smooth surfaces of degree n in $\mathbb{P}^n$ and in 1887 del Pezzo [4] showed that there are no such surfaces for $n > 9$. Over the complex field and for $3 \leq n \leq 9$ del Pezzo surfaces form – with the exception of the quadric surface embedded in $\mathbb{P}^8$ – a single series $X^9, X^8, \ldots, X^3$, such that X^n is always the projection of X^{n+1} from a point of itself. The original series was enlarged to include the surfaces X^2 and X^1. In modern language del Pezzo surfaces are defined as (smooth) surfaces with ample anticanonical bundle and they appear in the classification theory of surfaces and 3-folds as basic examples (Mori fiber spaces).

Early work of Enriques [5] on del Pezzo surfaces over arbitrary fields used direct geometrical methods, later put in a modern context by Manin [9], [10]. Recently the classification of real del Pezzo surfaces was considered by Wall [18] by combining topological methods with the study of involutions in Weyl groups; by Silhol [17] by using Galois cohomology and minimal model theory; and by Kollár in [7], where real del Pezzo surfaces appear as a final step in the minimal model program to classify real surfaces which are rational over $\mathbb{C}$.

Here we return on the subject and show how the ideas introduced by Comessatti in the fundamental papers [2], [3] can be applied to get a classification of real del Pezzo

---

*The author has been supported by the CNPq grant 300761/97 and by PRONEX-ALGA (Algebra Comutativa e Geometria Algebrica).

surfaces in an elementary and unitary way by using projective methods and also how they furnish a description of the geometry of this class of surfaces: number of real lines, topological form of the real part, minimal models over $\mathbb{R}$.

Comessatti studied real algebraic surfaces, which are rational over $\mathbb{C}$, by representing on the complex projective plane the conjugation naturally associated to them and by defining in this way an antibirational involution of the plane. Comessatti's point of view was essentially birational while here we will study a biregular classification. For smooth real del Pezzo surfaces the antibirational involutions associated are not defined in at most 8 points in general position with respect to lines, conics and cubics with a node and have distinct base points.

The classification of antiprojective and antiquadratic involutions of the plane is applied to obtain a well known theorem of Enriques–Manin for the field of real numbers, to get the classification of real del Pezzo surfaces of degree greater or equal to 5 and a full description of their geometry (Proposition 2.1, Corollaries 2.2, 2.3, 2.4).

Examples of real structures on complex del Pezzo surfaces are constructed by composing the classical involutions of the plane (de Jonquiéres, Geiser, Bertini) with complex conjugation. One obtains examples of real del Pezzo surfaces minimal over $\mathbb{R}$ in degree 4 and 2 (de Jonquiéres), in degree 2 (Geiser) and in degree 1 (Bertini) (see Sections 3, 4 and 5). Moreover it turns out that de Jonquiéres antibirational involutions are naturally associated to minimal conic bundles (see Section 3).

The classification of real del Pezzo surfaces of degree less or equal to 4 is achieved by showing that in this range the ones minimal over $\mathbb{R}$ are isomorphic to those described above. By analyzing these explicit descriptions one immediately recovers the number of real lines and the topology of the real part of these surfaces. The analysis of the geometry is non-trivial especially in degree 1 and 2, whose treatment is perhaps the most interesting and original part of the paper. Moreover in lower degrees the geometry is very different from the complex case (see for example Corollary 5.3). These results of Section 4 and 5 give also simple answers to some questions posed in [18] and [17] and complete the picture of the last sections of [7].

As suggested by Comessatti in the footnote on page 47 of [2], we deduced from the classification of real del Pezzo surfaces of degree 2, respectively of degree 1 or 4, all the results about the geometry of real quartic plane curves: real bitangents and topological types; about real complete intersection sextic space curves: real tritangent planes and topological types; about pencil of smooth real quadrics in $\mathbb{P}^4$. This is done systematically in the last two sections (see Corollary 4.3, 4.4 and 5.3). We point out that usually the topology of real quartic plane curves and real sextic complete intersection space curves is used to deduce the topology of real del Pezzo surfaces of degree 2 and 1 (see [7]) while here we bypass it by using the notion of *associated surface* introduced by Comessatti and in order to count the tritangent plane to a sextic curve as above we use the notions of *tritangent plane of the first or second kind* (see [2] and Sections 4 and 5).

Most of the results presented here are somehow implicit in Comessatti's papers [2] and [3], even if not stated in the form presented below and they were not proved in this way also in the modern literature. In particular in our approach we do not use any result coming from Galois cohomology as in [17], [1]. The computation of the number of real lines for del Pezzo surfaces is deduced in a geometrical way and does not use the theory of Weyl groups as in [18] (see especially Corollary 5.3). The classification of minimal real del Pezzo surfaces of degree 2 and 1, Propositions 4.1 and 5.1, follows from an easy argument of projective geometry. This last question was left as an open problem in [17]. In conclusion we can say that most of the results contained here are more or less known but the point of view and the arguments used are elementary and perhaps original.

By combining some minimal model theory over the complex field, for example Mori's theory, with the results presented here one could deduce the classification of antibirational involutions of the plane generalizing the well known theorem of Bertini. The answer is that, modulo conjugation by Cremona transformations, there are only the following types: antiprojectivities, antiquadratic involutions without fixed points, de Jonquiéres, Geiser and Bertini antibirational involutions (see [2] for the statement of this result and also [1] or [7] Corollary 3.4).

We end by remarking that in recent years there was a renewed interest in real algebraic geometry especially due to the work of Kollár, see the recent paper [8] and the numerous references cited there; moreover the paper [1] contains an account written in modern terms of the contributions of Comessatti in real algebraic geometry, focusing especially on the birational classification of real rational surfaces and on the number of connected components of their real parts.

Acknowledgements. I wish to thank Ciro Ciliberto for his kind advice, for his interest and especially for many discussions and suggestions on various subjects along last years.

Writing these notes I was also influenced and helped by János Kollár's lectures given at the Summer School Levico '97 and collected in [7].

1. Notation and definitions

For any scheme $X_\mathbb{R}$ over $\mathbb{R}$, let $X_\mathbb{C} := X_\mathbb{R} \times_{\mathrm{Spec}(\mathbb{R})} \mathrm{Spec}(\mathbb{C})$ be the *complexification* of $X_\mathbb{R}$. Let $\alpha : \mathrm{Spec}(\mathbb{C}) \to \mathrm{Spec}(\mathbb{C})$ be the morphism induced by complex conjugation and let $\overline{\sigma} : X_\mathbb{C} \to X_\mathbb{C}$ be the automorphism $\mathbb{I}_{X_\mathbb{R}} \times \alpha$. Since $\overline{\sigma}^2 = \mathbb{I}_{X_\mathbb{C}}$ we call $\overline{\sigma}$ an *antiinvolution* on $X_\mathbb{C}$. The bar is justified by the fact that $\overline{\sigma}$ is not a $\mathbb{C}$-automorphism but it becomes only after composing by α. A pair $(X_\mathbb{C}, \overline{\sigma})$ consisting of an automorphism of $X_\mathbb{C}$ which becomes a $\mathbb{C}$-automorphism after composition with α and such that $\overline{\sigma}^2 = \mathbb{I}$ will be called a *real structure* on the scheme $X_\mathbb{C}$. If $X_\mathbb{C}$ is of finite type over $\mathbb{C}$ and if it is quasi-projective over $\mathbb{C}$, then there exists a unique separated scheme $X_\mathbb{R}$ of finite type over $\mathbb{R}$, such that $X_\mathbb{R} \times_{\mathrm{Spec}(\mathbb{R})} \mathrm{Spec}(\mathbb{C}) \simeq X_\mathbb{C}$ and such that this isomorphism identifies the given antiinvolution of $X_\mathbb{C}$ with the natural antiinvolution on $X_{\mathbb{R}} \times \operatorname{Spec}(\mathbb{C})$ described above. If $(X_{\mathbb{C}}, \overline{\sigma})$ and $(X_{\mathbb{C}}, \overline{\sigma}')$ are two different real structures on the $\mathbb{C}$-scheme $X_{\mathbb{C}}$, they correspond to $\mathbb{R}$-isomorphic $\mathbb{R}$-schemes if and only if there exists a $\mathbb{C}$-automorphism ϕ of $X_{\mathbb{C}}$ such that $\overline{\sigma} = \phi^{-1} \circ \overline{\sigma}' \circ \phi$. In this case the two real structures $(X_{\mathbb{C}}, \overline{\sigma})$ and $(X_{\mathbb{C}}, \overline{\sigma}')$ are said to be *equivalent*. In the projective case, the study of equivalence classes of real structures on $X_{\mathbb{C}}$ is equivalent to the study of $\mathbb{R}$-isomorphism classes of $\mathbb{R}$-schemes for which the complexification is isomorphic over $\mathbb{C}$ to $X_{\mathbb{C}}$.

For any separated scheme $X_{\mathbb{R}}$ of finite type over $\mathbb{R}$, we indicate by $X(\mathbb{R})$ (respectively $X(\mathbb{C})$) the set of real points (respectively of complex points) of $X_{\mathbb{R}}$ with Euclidean topology. Then $\mathbb{P}_{\mathbb{R}}^n$ is projective n-space as a scheme over $\mathbb{R}$ and $\mathbb{P}^n(\mathbb{R})$ is denoted by $\mathbb{RP}^n$.

If $C \subset X_{\mathbb{C}}$ is a subscheme, then $\overline{\sigma}(C) \subset X_{\mathbb{C}}$ is again a subscheme. Then $C + \overline{\sigma}(C) \subset X_{\mathbb{C}}$ can be viewed as a subscheme of $X_{\mathbb{R}}$, i.e. there is a unique subscheme $C_{\mathbb{R}} \subset X_{\mathbb{R}}$ such that $C_{\mathbb{R}} \times_{\operatorname{Spec}(\mathbb{R})} \operatorname{Spec}(\mathbb{C}) = C + \overline{\sigma}(C)$.

Let us discuss the birational morphisms, defined over $\mathbb{R}$, between smooth geometrically irreducible real surfaces. They can be factorized into elementary transformations of two kinds: the *blowing up of a closed real point* or the *blowing up of two conjugate closed points*. In fact on the complexification we can factorize the morphisms into blowing ups of closed points and the two described above are the only ones defined over $\mathbb{R}$. The blowing up of conjugate points has no effect on the set of real points, i.e. if $Y = \operatorname{Bl}_{p_1, p_2}(X)$, with $\overline{\sigma}(p_1) = p_2$, then $Y(\mathbb{R})$ is homeomorphic to $X(\mathbb{R})$. If we blow up a closed real point p, then $\operatorname{Bl}_p(X)(\mathbb{R})$ is homeomorphic to $\mathbb{RP}^2 \# X(\mathbb{R})$, the connected sum with $\mathbb{RP}^2$. If X and Y are birational over $\mathbb{R}$, then $X(\mathbb{R})$ and $Y(\mathbb{R})$ have the same number of connected components.

As in the complex case, a smooth geometrically irreducible real surface $X_{\mathbb{R}}$ is said to be *minimal over $\mathbb{R}$* if every real birational morphism from $X_{\mathbb{R}}$ into a smooth real surface is an isomorphism. This is clearly equivalent to the fact that on $X_{\mathbb{C}}$ there are no *real* (-1)-curves, i.e. (-1)-curves $C \subset X_{\mathbb{C}}$ such that $\overline{\sigma}(C) = C$ or couples of disjoint conjugate (-1)-curves, i.e. $C_1, C_2 \subset X_{\mathbb{C}}$ are (-1)-curves such that $\overline{\sigma}(C_1) = C_2$ and $(C_1 \cdot C_2) = 0$. We recall that the *Picard number of $X_{\mathbb{R}}$*, $\rho(X_{\mathbb{R}})$, is the dimension of the vector space of 1-cycles on $X_{\mathbb{R}}$ with real coefficients modulo numerical equivalence.

Let us introduce the natural generalization of this setting to rational maps and define the notion of *antibirational involution of the plane*. If $\phi : \mathbb{P}_{\mathbb{C}}^2 \dashrightarrow \mathbb{P}_{\mathbb{C}}^2$ is a $\mathbb{C}$-rational map, then $\overline{\phi} := \phi \circ \alpha$ is called an *antirational map of the plane*. If ϕ is birational, $\overline{\phi}$ is said an *antibirational transformation of the plane*.

We are mainly interested in the antibirational transformations of the plane which are involutory, i.e. if $\overline{\phi} := \phi \circ \alpha$, then $\phi^{-1} = \alpha \circ \phi \circ \alpha$. These transformations are called the *antibirational involutions of the plane*. If ϕ is a real birational involution of the plane, i.e. $\alpha \circ \phi = \phi \circ \alpha$ and $\phi^2 = \mathbb{I}_{\mathbb{P}_{\mathbb{C}}^2}$, then $\overline{\phi} = \phi \circ \alpha$ is an antibirational involution of the plane.

Let us now suppose that $X_{\mathbb{R}}$ is a *geometrically rational* real surface, i.e. $X_{\mathbb{C}}$ is birational over $\mathbb{C}$ to $\mathbb{P}_{\mathbb{C}}^2$. If $\pi : X_{\mathbb{C}} \dashrightarrow \mathbb{P}_{\mathbb{C}}^2$ is a birational isomorphism, then $\pi \circ \overline{\sigma} \circ \pi^{-1}$ in an antibirational involution of the plane, where $\overline{\sigma}$ is the natural antiinvolution on

$X_{\mathbb{C}}$ associated to the real structure induced by $X_{\mathbb{R}}$. By abuse of notation we indicate by $\bar{\sigma} : \mathbb{P}^2_{\mathbb{C}} \dashrightarrow \mathbb{P}^2_{\mathbb{C}}$ also the antibirational involution $\pi \circ \bar{\sigma} \circ \pi^{-1}$. This is justified by the fact that $\pi \circ \bar{\sigma} \circ \pi^{-1}$ represents on the plane the real structure $\bar{\sigma} : X_{\mathbb{C}} \to X_{\mathbb{C}}$. By definition, the fundamental points of the antibirational involution $\bar{\sigma} : \mathbb{P}^2_{\mathbb{C}} \dashrightarrow \mathbb{P}^2_{\mathbb{C}}$ are contained in the fundamental points of π^{-1}.

To two different birational representations on the plane π_1 and π_2 as above, there will correspond two representation of the conjugation $\bar{\sigma}$, $\bar{\sigma}_1$ and $\bar{\sigma}_2$, conjugate (in the sense of group theory) under the Cremona transformation $\pi_1 \circ \pi_2^{-1}$. If $X_{\mathbb{R}}$ and $X_{\mathbb{R}}'$ are geometrically rational surfaces, $\mathbb{R}$-birational, then $\bar{\sigma}$ and $\bar{\sigma}'$ are conjugate by $\psi \in Bir(\mathbb{P}^2_{\mathbb{C}})$, i.e. $\bar{\sigma}' = \psi \circ \bar{\sigma} \circ \psi^{-1}$, with ψ a Cremona transformation of the plane. Hence the classification of $\mathbb{R}$-birational equivalence classes of geometrically rational real surfaces is equivalent to the classification of equivalence classes (in the above sense) of antibirational involutions of the plane. This was Comessatti's point of view (see [2], [3] and also [1]).

To clarify the above definitions we classify varieties $X_{\mathbb{R}}$ such that their complexification $X_{\mathbb{C}}$ is isomorphic to $\mathbb{P}^n_{\mathbb{C}}$. We have to determine the equivalence classes of real structures on $\mathbb{P}^n_{\mathbb{C}}$, i. e. according to C. Segre or Comessatti the equivalence classes of involutory antiprojectivities modulo projective transformations. An (*involutory*) *antiprojectivity* (or *projective antiinvolution*) is a transformation $\bar{\sigma} : \mathbb{P}^n_{\mathbb{C}} \to \mathbb{P}^n_{\mathbb{C}}$ such that $\bar{\sigma} \circ \alpha$ is a $\mathbb{C}$-automorphism of $\mathbb{P}^n_{\mathbb{C}}$ (and such that $\bar{\sigma}^2 = \mathbb{I}_{\mathbb{P}^n_{\mathbb{C}}}$ for the involutory ones), where from now on by abusing notations we will indicate by $\alpha : \mathbb{P}^n_{\mathbb{C}} \to \mathbb{P}^n_{\mathbb{C}}$ also the antiinvolution induced by $\mathbb{I}_{\mathbb{P}^n_{\mathbb{R}}} \times \alpha$, i.e. conjugation on $\mathbb{P}^n_{\mathbb{C}}$. Since $\bar{\sigma} \circ \alpha$ is a $\mathbb{C}$-automorphism of $\mathbb{P}^n_{\mathbb{C}}$, we have $\bar{\sigma} = \omega \circ \alpha$, with $\omega : \mathbb{P}^n_{\mathbb{C}} \to \mathbb{P}^n_{\mathbb{C}}$ a projectivity and α complex conjugation on $\mathbb{P}^n_{\mathbb{C}}$.

The following proposition solves the problem.

Proposition 1.1 ([2], [16]). *If n is even, then every antiprojective involution of $\mathbb{P}^n_{\mathbb{C}}$ is equivalent to complex conjugation on $\mathbb{P}^n_{\mathbb{C}}$. In particular, if $X_{\mathbb{C}} \simeq \mathbb{P}^n_{\mathbb{C}}$, n even, then $X_{\mathbb{R}}$ is $\mathbb{R}$-isomorphic to $\mathbb{P}^n_{\mathbb{R}}$.*

If n is odd and if an antiprojective involution of $\mathbb{P}^n_{\mathbb{C}}$ has a fixed point, then it is equivalent to complex conjugation. In particular, if $X_{\mathbb{C}} \simeq \mathbb{P}^n_{\mathbb{C}}$, n odd, and if $X(\mathbb{R}) \neq \emptyset$, then $X_{\mathbb{R}}$ is $\mathbb{R}$-isomorphic to $\mathbb{P}^n_{\mathbb{R}}$.

Proof. Let n be even and let $\bar{\sigma}_1, \bar{\sigma}_2$ be two antiprojective involutions of $\mathbb{P}^n_{\mathbb{C}}$. Fix $\frac{n+2}{2}$ generic couples of points homologous for $\bar{\sigma}_1$, $p_1, p_1', \ldots p_{\frac{n+2}{2}}, p'_{\frac{n+2}{2}}$, and analogously $\frac{n+2}{2}$ generic couples $q_1, q_1', \ldots, q_{\frac{n+2}{2}}, q'_{\frac{n+2}{2}}$ homologous for $\bar{\sigma}_2$. Let ω be the projective transformation of $\mathbb{P}^n_{\mathbb{C}}$ defined by $\omega(p_i) = q_i$, $\omega(p_i') = q_i'$ with $i = 1, \ldots, \frac{n+2}{2}$. Then $\bar{\sigma}_2 \circ \omega = \omega \circ \bar{\sigma}_1$, because $\bar{\sigma}_2 \circ \omega \circ \bar{\sigma}_1^{-1} \circ \omega^{-1}$ is a projectivity with $n+2$ generic fixed points.

If n is odd and $\bar{\sigma}_1, \bar{\sigma}_2$ have a fixed point, we can take $\frac{n+1}{2}$ generic couples of fixed points homologous for $\bar{\sigma}_1$ and $\bar{\sigma}_2$ and the fixed point to define ω as above. □

Remark 1. For $n = 1$ one easily sees that there are only two distinct equivalence classes of involutory antiprojectivities:

$$z' = \pm \frac{1}{\overline{z}}$$

and two equivalence classes of real curves $z \cdot \overline{z} = \pm 1$, $\mathbb{P}_{\mathbb{R}}^1$ and the conic without real points.

Let us now consider smooth real surfaces $X_{\mathbb{R}}$ such that $X_{\mathbb{C}} \simeq \mathbb{P}_{\mathbb{C}}^1 \times \mathbb{P}_{\mathbb{C}}^1$. The quadric hypersurface of equation $(x_1^2 + \cdots + x_r^2 - x_{r+1}^2 - \cdots - x_{r+s}^2 = 0) \subset \mathbb{P}^{r+s-1}$ will be denoted by $Q^{r,s}$.

Proposition 1.2. *Let $X_{\mathbb{R}}$ be a smooth real surface such that $X_{\mathbb{C}} \simeq \mathbb{P}_{\mathbb{C}}^1 \times \mathbb{P}_{\mathbb{C}}^1$. Then $X_{\mathbb{R}}$ is isomorphic to one of the following:*

$X_{\mathbb{R}}$	$\rho(X_{\mathbb{R}})$	$X(\mathbb{R})$
$Q^{2,2} \cong Q^{2,1} \times Q^{2,1}$	2	$S^1 \times S^1$
$Q^{4,0} \cong Q^{3,0} \times Q^{3,0}$	2	$\emptyset$
$Q^{3,0} \times Q^{2,1}$	2	$\emptyset$
$Q^{3,1}$	1	S^2

Proof. Let $\overline{\sigma} : X_{\mathbb{C}} \to X_{\mathbb{C}}$ be the antiinvolution associated to $X_{\mathbb{R}}$. Then either $\overline{\sigma}$ sends each complex ruling into itself inducing on it an antiprojective involution or it interchanges the 2 rulings of $\mathbb{P}_{\mathbb{C}}^1 \times \mathbb{P}_{\mathbb{C}}^1$. In the first case the two projections $\pi_i : \mathbb{P}_{\mathbb{C}}^1 \times \mathbb{P}_{\mathbb{C}}^1 \to \mathbb{P}_{\mathbb{C}}^1$ are defined over $\mathbb{R}$ giving the 3 possible splittings, according to the different antiinvolution $\overline{\sigma}$ induces on each ruling. In the second case $\mathcal{O}_{\mathbb{P}_{\mathbb{C}}^1 \times \mathbb{P}_{\mathbb{C}}^1}(1, 1)$ is defined over $\mathbb{R}$ and gives an embedding of $X_{\mathbb{R}}$ as the quadric surface in $\mathbb{P}_{\mathbb{R}}^3$ of equation $x_1 x_4 - x_2 x_3 = 0$, which is projectively equivalent over $\mathbb{R}$ to the quadric $Q^{3,1}$. In the last case $X(\mathbb{R})$ is homeomorphic to S^2. $\square$

Let us introduce the definition of smooth real del Pezzo surface.

Definition 1. A smooth geometrically irreducible real surface $X_{\mathbb{R}}$ is said a *smooth real del Pezzo surface* if $-K_X$ is ample. A smooth real del Pezzo surface $X_{\mathbb{R}}$ is said a *minimal real del Pezzo surface* if $\rho(X_{\mathbb{R}}) = 1$.

The classification of smooth complex del Pezzo surfaces is classical and can be summarized in the following:

Proposition 1.3. *Let $X_{\mathbb{C}}$ be a smooth complex del Pezzo surface (i.e. a smooth complex surface such that $-K_X$ is ample). Then either $X_{\mathbb{C}} \simeq \mathrm{Bl}_{p_1,\ldots,p_r}(\mathbb{P}_{\mathbb{C}}^2)$, $r \leq 8$, where no 3 points are collinear, no 6 of them lie on a conic, and there is no cubic through 7 of them having a double point at the 8th, or $X_{\mathbb{C}} \simeq \mathbb{P}_{\mathbb{C}}^1 \times \mathbb{P}_{\mathbb{C}}^1$. In particular, $K_X^2 \leq 9$ and we have either $K_X^2 = 9 - r$ or $K_X^2 = 8$.*

In Section 1 we classified smooth real surfaces $X_\mathbb{R}$, such that $X_\mathbb{C} \simeq \mathbb{P}^2_\mathbb{C}$, proving that $X_\mathbb{R} \simeq \mathbb{P}^2_\mathbb{R}$ (Proposition 1.1). Equivalently, if $X_\mathbb{R}$ is a smooth real del Pezzo surface of degree 9, then $X_\mathbb{R} \simeq \mathbb{P}^2_\mathbb{R}$. By the above result, we have that if $X_\mathbb{R}$ is a smooth del Pezzo surface of degree 8, then $X_\mathbb{C} \simeq \mathbb{P}^1_\mathbb{C} \times \mathbb{P}^1_\mathbb{C}$ or $X_\mathbb{C} \simeq \mathrm{Bl}_p(\mathbb{P}^2_\mathbb{C})$. Then $X_\mathbb{R}$ is isomorphic over $\mathbb{R}$ to one of the surfaces listed in Proposition 1.2 or $X_\mathbb{R} \simeq \mathrm{Bl}_p(\mathbb{P}^2_\mathbb{R})$, $p \in \mathbb{R}\mathbb{P}^2$ (see Corollary 2.4 below).

Since we will be interested in the classification of smooth real del Pezzo surfaces $X_\mathbb{R}$, by Proposition 1.2 we can suppose that it is given a standard plane representation of $X_\mathbb{C} \simeq \mathrm{Bl}_{p_1,\ldots,p_r}(\mathbb{P}^2_\mathbb{C}) \xrightarrow{\pi} \mathbb{P}^2_\mathbb{C}$ as the blowing of the plane in $1 \leq r \leq 8$ points, general with respect to lines, conics and cubics with a node. The homaloidal nets defining the antibirational involutions of the plane $\overline{\sigma}$ associated to (and defining) antiinvolutions on $X_\mathbb{C}$ are centered at most in $p_1, \ldots, p_r$ and have no infinitely near base points. Let us indicate by $C^n[p_1^{n_1}, \ldots, p_r^{n_r}]$ a homaloidal net of curves of degree $n \geq 2$ with distinct base points of multiplicity $n_1 \geq n_2 \geq \cdots \geq n_r \geq 1$.

We will see that some simple remarks on linear systems of plane curves with distinct base points–together with the fact that if $\pi : X_\mathbb{R} \to X'_\mathbb{R}$ is a birational morphism between smooth real surfaces with $X_\mathbb{R}$ del Pezzo, then $X'_\mathbb{R}$ is also del Pezzo– will lead us to the classification of smooth real del Pezzo surfaces.

If X is a real smooth algebraic surface, we will indicate by $X(a, 2b)$ the smooth real surface obtained by blowing up a real closed points and b couples of closed conjugate points on X. With these definitions we have $\mathbb{P}^2_\mathbb{R}(2, 0) \simeq Q^{2,2}(1, 0)$ and $\mathbb{P}^2_\mathbb{R}(0, 2) \simeq Q^{3,1}(1, 0)$

2. Antiquadratic involutions of the plane and real del Pezzo surfaces of degree $d = K^2_X \geq 5$

Let us begin with the first non trivial examples of antibirational involutions of the plane: the ordinary antiquadratic involutions, i.e. antiquadratic involutions with distinct base points. These are naturally associated to the birational representation of antiinvolution on surfaces $X_\mathbb{R}$ such that $X_\mathbb{C} \simeq \mathbb{P}^1_\mathbb{C} \times \mathbb{P}^1_\mathbb{C}$.

Example 1 (Antiquadratic involutions, [2]). An ordinary antiquadratic involution $\overline{\sigma} : \mathbb{P}^2_\mathbb{C} \dashrightarrow \mathbb{P}^2_\mathbb{C}$ is associated to a net of conics with three distinct non-collinear base points. Let p_1, p_2, p_3 be the points where $\overline{\sigma}$ is not defined. There exist transformations of two different kinds: those in which every vertex of the triangle $\{p_1, p_2, p_3\}$ corresponds to the opposite edge (let us call these ones *standard*) or the ones in which a vertex, let us suppose p_1, is sent into the opposite edge $\langle p_2, p_3 \rangle$ and the other two $p_i, i = 2, 3$, into the edge $\langle p_1, p_i \rangle$, which passes through it (analogous to the inversion with respect to a circle and examples of de Jonquiéres antiinvolutions discussed below). In the second case the transformations have always fixed points, are projectively equivalent and they will be discussed in detail in Example 3 below.

From now on let us suppose that $\bar\sigma$ is a standard antiquadratic involution. Then $\bar\sigma$ induces an antiprojective involution in each pencil of lines through the fundamental points p_i, $i = 1, 2, 3$. These antiprojective involutions can have or not fixed lines. If at least two pencils have fixed lines, then $\bar\sigma$ has fixed elements, at least the point p of intersection of two fixed lines belonging to the two different pencils. It follows that the three pencils have fixed elements, at least the lines $\langle p, p_j \rangle$, $j = 1, 2, 3$.

Let us construct examples of antiquadratic involutions with fixed points and without fixed points. Let $q \in X_{\mathbb{C}} \simeq Q \simeq \mathbb{P}_{\mathbb{C}}^1 \times \mathbb{P}_{\mathbb{C}}^1 \subset \mathbb{P}_{\mathbb{C}}^3$ be a point such that $\bar\sigma(q) = q' \in Q \setminus (l \cup l')$, where l and l' are the two lines of the quadric Q passing through q and $\bar\sigma : Q \to Q$ is one of the four real structures on the quadric Q described in Proposition 1.2. Let $\pi : Q \dashrightarrow \mathbb{P}_{\mathbb{C}}^2$ be the projection from the point q, let $p_2 = \pi(l)$, $p_3 = \pi(l')$ and let $p_1 = \pi(q')$. Then clearly $\mathrm{Bl}_{q,q'}(Q) \simeq \mathrm{Bl}_{p_1, p_2, p_3}(\mathbb{P}_{\mathbb{C}}^2)$ and $\bar\sigma$ induces an antiinvolution $\bar\sigma'$ on $\mathrm{Bl}_{q,q'}(Q) \simeq \mathrm{Bl}_{p_1, p_2, p_3}(\mathbb{P}_{\mathbb{C}}^2)$ such that $\bar\sigma'(E) = E'$, where E and E' are the exceptional curves of the blow-up. Equivalently we have an antibirational involution of the plane $\bar\sigma : \mathbb{P}_{\mathbb{C}}^2 \dashrightarrow \mathbb{P}_{\mathbb{C}}^2$ such that $\bar\sigma(p_1) = \langle p_2, p_3 \rangle$, which is clearly an antiquadratic involution: if L is a general line in $\mathbb{P}_{\mathbb{C}}^2$, then $\pi^{-1}(L) = C$ is a conic through q; the curve $\bar\sigma(C)$ is a conic passing through q' and $\pi(C)$ is a conic through p_1 which clearly passes through p_2 and p_3 since it intersects the lines l and l' in one point. The rulings of the quadrics are mapped by π into the pencil of lines through p_2 and p_3. Let us remark that the ruling to which l belongs is mapped into the pencil of lines through p_3 and the ruling to which l' belongs into the pencil of lines through p_2.

If $\bar\sigma$ corresponds to the real ruled quadric $Q^{2,2} \simeq Q^{2,1} \times Q^{2,1}$, we constructed an example of antiquadratic involution with fixed points and such that in the pencils of lines through p_2 and p_3 there are fixed lines. If $\bar\sigma$ corresponds to $Q^{4,0} \simeq Q^{3,0} \times Q^{3,0}$ (or to $Q^{3,0} \times \mathbb{P}_{\mathbb{R}}^1$) we have an example of a standard antiquadratic involution without fixed points. In both cases the antiquadratic involution is standard because each pencil of lines through p_2 and p_3 is mapped into itself. If $\bar\sigma$ corresponds to $Q^{3,1}$ then the pencil of lines through p_2 is transformed into the pencil of lines through p_3 (and viceversa) and the corresponding antiquadratic involution is not standard but of de Jonquiéres type (see Example 3).

We now prove that, modulo projective transformations of the plane, there are only two distinct classes of standard antiquadratic involutions: with fixed points or without fixed points. In other words there exists only the two types described above and the two surfaces $Q^{4,0} \simeq Q^{3,0} \times Q^{3,0}$ and $Q^{3,0} \times \mathbb{P}_{\mathbb{R}}^1$ give, modulo projective equivalence, the same antiquadratic involution, i.e. $Q^{4,0}(0, 2) \simeq (Q^{3,0} \times \mathbb{P}_{\mathbb{R}}^1)(0, 2)$.

Let $\bar\sigma : \mathbb{P}_{\mathbb{C}}^2 \dashrightarrow \mathbb{P}_{\mathbb{C}}^2$ be a standard antiquadratic involution of the plane having no fixed points (respectively with fixed points) and centered in the three distinct points p_1, p_2, p_3. If $\bar\sigma' : \mathbb{P}_{\mathbb{C}}^2 \dashrightarrow \mathbb{P}_{\mathbb{C}}^2$ is another antiquadratic involution of the same type centered in p_1', p_2', p_3', then $\bar\sigma$ and $\bar\sigma'$ can be transformed into each other by a projective transformation of the plane ω, i.e. $\bar\sigma = \omega^{-1} \circ \bar\sigma' \circ \omega$.

First we can suppose $p_i = p_i'$, $i = 1, 2, 3$, and let $\bar\sigma_i, \bar\sigma_i'$ be the antiprojective involutions $\bar\sigma$, respectively $\bar\sigma'$, induce in the pencil of lines through p_i. There is a projective transformation in the pencil of lines through p_1 sending $\bar{\sigma}_1$ into $\bar{\sigma}'_1$, fixing the rays $\langle p_1, p_2 \rangle$ and $\langle p_1, p_3 \rangle$ (which are homologous in $\bar{\sigma}_1$ and $\bar{\sigma}'_1$). In fact, two antiprojective involutions on $\mathbb{P}^1_{\mathbb{C}}$ given in affine equations by $z' = \frac{k}{\bar{z}}$, $z' = \frac{k'}{\bar{z}}$ ($k, k' \in \mathbb{R}$), having in common the couple 0 and ∞ and without fixed points (k, k' negative, or with fixed points: k, k' positive) are sent into each other by the projective transformation $z' = mz$ ($m\bar{m} = \frac{k'}{k}$) having the fixed points 0 and ∞. There also exists a projective transformation in the pencil of lines through p_2 sending $\bar{\sigma}_2$ into $\bar{\sigma}'_2$, fixing the rays $\langle p_2, p_1 \rangle$ and $\langle p_2, p_3 \rangle$. These two projective transformation of the pencils define a projective transformation of the plane (which fixes the triangle p_1, p_2, p_3) sending $\bar{\sigma}$ into $\bar{\sigma}'$.

When an antiquadratic involution, standard or not, has fixed points, we can reduce it to complex conjugation $\alpha : \mathbb{P}^2_{\mathbb{C}} \to \mathbb{P}^2_{\mathbb{C}}$ by a quadratic transformation centered in a fixed point and in the 2 points p_2, p_3 (notations as above), i.e. there exists $T : \mathbb{P}^2_{\mathbb{C}} \dashrightarrow \mathbb{P}^2_{\mathbb{C}}$ quadratic transformation such that $\alpha = T^{-1} \circ \bar{\sigma} \circ T$. This amounts to represent the real structure associated to $Q^{2,2}$, or to $Q^{3,1}$, by projecting from a real point. If you prefer, $Q^{2,2}(1, 0) \simeq \mathbb{P}^2_{\mathbb{R}}(2, 0)$ and $Q^{3,1}(1, 0) \simeq \mathbb{P}^2_{\mathbb{R}}(0, 2)$.

The above discussion can be summarized in the following proposition.

Proposition 2.1 ([2]). *Every ordinary antiquadratic involution of the plane is projectively equivalent to one of the following:*

1. *a de Jonquiéres antiinvolution of order 2;*
2. *a standard antiquadratic involution with fixed points;*
3. *a standard antiquadratic involution without fixed points.*

Furthermore the first two cases can be reduced by a quadratic transformation to conjugation, i.e. there exists $T : \mathbb{P}^2_{\mathbb{C}} \dashrightarrow \mathbb{P}^2_{\mathbb{C}}$ ordinary quadratic transformation such that $\alpha = T^{-1} \circ \bar{\sigma} \circ T$.

Here is a corollary, which was proved by Enriques for del Pezzo surfaces X_k defined over any (infinite) field k (see [5]).

Corollary 2.2 ([5], [10]). *Let $X_{\mathbb{R}}$ be a real del Pezzo surface of degree $d = K_X^2 \geq 5$ with $X(\mathbb{R}) \neq \emptyset$. Then $X_{\mathbb{R}}$ is $\mathbb{R}$-birational to $\mathbb{P}^2_{\mathbb{R}}$.*

Proof. If $X_{\mathbb{C}} \simeq \mathbb{P}^1_{\mathbb{C}} \times \mathbb{P}^1_{\mathbb{C}}$, we are done by projecting from a real point. Otherwise, since $d = 9 - r$, we have $r \leq 4$ so that $\bar{\sigma}$ is an antiprojective or an antiquadratic involution with fixed points. The first one can be reduced by a projective transformation to complex conjugation, the latter by a quadratic transformation. Hence for every real structure $\bar{\sigma}$ with fixed points on the complexification $X_{\mathbb{C}}$ of a real surface as above, there exists a birational map $\phi : X_{\mathbb{C}} \dashrightarrow \mathbb{P}^2_{\mathbb{C}}$ such that $\phi \circ \bar{\sigma} = \alpha \circ \phi$, i.e. ϕ is defined over $\mathbb{R}$ and $X_{\mathbb{R}}$ is $\mathbb{R}$-birational to $\mathbb{P}^2_{\mathbb{R}}$. □

Another corollary is the classification of real structures on the complex surfaces $\text{Bl}_{p_1,\ldots,p_r}(\mathbb{P}^2_{\mathbb{C}})$, $1 \leq r \leq 4$.

Corollary 2.3. *There do not exist real smooth del Pezzo surfaces $X_\mathbb{R}$ of degree $d = K_X^2 \geq 5$, which are minimal over $\mathbb{R}$ and for which $X_\mathbb{C} \simeq \mathrm{Bl}_{p_1,\ldots,p_r}(\mathbb{P}_\mathbb{C}^2)$, $1 \leq r \leq 4$.*

Proof. The antibirational involutions $\overline{\sigma}$ associated to a real structure on an $X_\mathbb{C}$ as above are antiprojective involutions or antiquadratic involutions. In the first case we can suppose that $X_\mathbb{R}$ dominates $\mathbb{P}_\mathbb{R}^2$, while an antiquadratic involution carries at least one couple of disjoint conjugate (-1)-curves on $X_\mathbb{C}$. □

Corollary 2.4. *Let $X_\mathbb{R}$ be a smooth real del Pezzo surface of degree $d = K_X^2 \geq 5$ such that $X_\mathbb{C} \simeq \mathrm{Bl}_{p_1,\ldots,p_r}(\mathbb{P}_\mathbb{C}^2)$, $0 \leq r \leq 4$. Then $X_\mathbb{R}$ is isomorphic to one of the following:*

d	$X_\mathbb{R}$	$\rho(X_\mathbb{R})$	$X(\mathbb{R})$	# real lines
9	$\mathbb{P}_\mathbb{R}^2$	1	$\mathbb{R}\mathbb{P}^2$	0
8	$\mathbb{P}_\mathbb{R}^2(1,0)$	2	$\#2\mathbb{R}\mathbb{P}^2$	1
7	$\mathbb{P}_\mathbb{R}^2(2,0)$	3	$\#3\mathbb{R}\mathbb{P}^2$	3
7	$\mathbb{P}_\mathbb{R}^2(0,2)$	2	$\mathbb{R}\mathbb{P}^2$	1
6	$\mathbb{P}_\mathbb{R}^2(3,0)$	4	$\#4\mathbb{R}\mathbb{P}^2$	6
6	$\mathbb{P}_\mathbb{R}^2(1,2)$	3	$\#2\mathbb{R}\mathbb{P}^2$	2
6	$Q^{3,1}(0,2)$	2	S^2	0
6	$Q^{2,2}(0,2)$	3	$S^1 \times S^1$	0
6	$(Q^{3,0} \times Q^{2,1})(0,2) \cong (Q^{3,0} \times Q^{3,0})(0,2)$	3	$\emptyset$	0
5	$\mathbb{P}_\mathbb{R}^2(4,0)$	5	$\#5\mathbb{R}\mathbb{P}^2$	10
5	$\mathbb{P}_\mathbb{R}^2(2,2)$	4	$\#3\mathbb{R}\mathbb{P}^2$	4
5	$\mathbb{P}_\mathbb{R}^2(0,4)$	3	$\mathbb{R}\mathbb{P}^2$	2

Proof. Remember that $Q^{2,2}(1,b) \simeq \mathbb{P}_\mathbb{R}^2(2,b)$, that $Q^{3,1}(1,b) \simeq \mathbb{P}_\mathbb{R}^2(0,b+2)$ and that by Corollary 2.3 the only real del Pezzo surfaces minimal over $\mathbb{R}$ of degree $d \geq 5$ are either $\mathbb{P}_\mathbb{R}^2$ or the 4 surfaces whose complexification is $\mathbb{P}_\mathbb{C}^1 \times \mathbb{P}_\mathbb{C}^1$. The isomorphism in the table is a consequence of the fact that, modulo projective transformations, there exists only one equivalence class of antiquadratic involutions without fixed points. The number of real lines is easily deduced from the description of the antiinvolutions. □

3. De Jonquiéres antiinvolutions and real del Pezzo surfaces of degree 4 and 3

Let us associate to the classical examples of (real) birational involution of the plane with distinct general base points (de Jonquiéres, Geiser, Bertini) an antibirational involution of the plane $\overline{\sigma}$ representing a conjugation $\overline{\sigma}$ on the complex del Pezzo surface $X_\mathbb{C}$ obtained by blowing up the fundamental points of $\overline{\sigma}$.

Definition 2. A morphism $f : X_\mathbb{R} \to B_\mathbb{R}$ from a real surface $X_\mathbb{R}$ to a smooth real curve $B_\mathbb{R}$ is said a *conic bundle* if every fiber is isomorphic to a plane conic. It is said a *minimal conic bundle* if $\rho(X_\mathbb{R}) = 2$.

Example 2 (de Jonquiéres). Let Γ^n be a (real) plane curve of degree n having an $(n-2)$-fold point in the (real) point p_0, $n \geq 2$. Let $p_1, \ldots, p_{2n-2}$ be the $2n - 2$ (real) points of contact of tangents from p_0 to Γ^n. When dealing with del Pezzo surfaces, we will suppose the p_i's in general position and clearly $n \leq 4$.

Let us define on $\mathbb{P}_\mathbb{C}^2 \setminus \bigcup_{i=1}^{2n-2} \langle p_0, p_i \rangle$ a *(real) de Jonquiéres involution* by taking as a pair of the involution any pair of points p, q such that p and q lie on a line through p_0 and separate harmonically the 2 residual intersections of this line with Γ^n. For $n = 2$ this is a generalization of the inversion with respect to a circle.

If $X_\mathbb{C} = \mathrm{Bl}_{p_0,\ldots,p_{2n-2}}(\mathbb{P}_\mathbb{C}^2) \xrightarrow{\pi} \mathbb{P}_\mathbb{C}^2$ is the blow up of $\mathbb{P}_\mathbb{C}^2$ at the p_i's, the above involution extends to an involution $\tau : X_\mathbb{C} \to X_\mathbb{C}$ such that the strict transform $l_{0,i}$ of $\langle p_0, p_i \rangle$ is sent into $\pi^{-1}(p_i) = E_i$ and $\pi^{-1}(p_0) = E_0$ into the unique curve $C_0 = C^{n-1}[p_0^{n-2}, p_i^1]$ of degree $n - 1$, having an $(n-2)$-fold point in p_0 and passing simply through the p_i's (C_0 is the first polar to Γ^n).

The birational involution of the plane $\tilde{\tau} = \pi \circ \tau \circ \pi^{-1}$ is a de Jonquiéres involution of order n, determined by the homaloidal net $C^n[p_0^{n-1}, p_i^1]$ and such that $\tilde{\tau}(p_0) = C_0$, $\tilde{\tau}(p_i) = \langle p_0, p_i \rangle$. In particular for $n = 2$ we have a net of conics through three general points. For $n = 3$ the involution τ can be seen also as the involution associated to the ramified double cover $\psi : X_\mathbb{C} \to \mathbb{P}_\mathbb{C}^1 \times \mathbb{P}_\mathbb{C}^1$ defined in the following way. Take the pencil of lines through p_0 and the pencil of conics through $p_1, \ldots, p_4$ (we suppose $p_0, \ldots, p_4$ in general position). The strict transforms of these pencils give the morphism ψ, which is then ramified along a smooth curve C of type $(2, 2)$. The 4 singular *conics* of each pencils are mapped into lines of the corresponding ruling, which are tangent to C. The involution τ sends each irreducible component of a singular *conic* into the other one. The inverse image of C on $X_\mathbb{C}$ is a smooth curve, which projects on $\mathbb{P}_\mathbb{C}^2$ to the curve Γ^3, going simply through p_0 and tangent to $\langle p_0, p_i \rangle$ in the points p_i.

Let us now suppose the curve Γ^n and the p_i's real. If, as always, $\alpha : \mathbb{P}_\mathbb{C}^2 \to \mathbb{P}_\mathbb{C}^2$ is complex conjugation, let $\overline{\sigma} = \tilde{\tau} \circ \alpha = \alpha \circ \tilde{\tau}$. Then $\overline{\sigma}$ defines an antibirational involution of the plane of order n, having fundamental points in the p_i's and as fundamental curves the $2n - 2$ lines $\langle p_0, p_k \rangle, k = 1, \ldots, 2n - 2$, and the curve C_0. Furthermore $\overline{\sigma}(p_0) = C_0, \overline{\sigma}(p_i) = \langle p_0, p_i \rangle$. The antibirational involution $\overline{\sigma}$ induces an antiinvolution $\overline{\sigma}$ on $X_\mathbb{C}$ and hence a real structure on $X_\mathbb{C}$. For $n = 2$ and the three points non-collinear we obtain an example of non-standard antiquadratic involution and clearly all such antiquadratic involutions are projectively equivalent.

The pencil of lines through p_0 is transformed into itself by $\overline{\sigma}$ and the total transform on $X_\mathbb{C}$ of this pencil defines a structure of conic bundle $\psi : X_\mathbb{C} \to \mathbb{P}_\mathbb{C}^1$; thus $\psi \circ \overline{\sigma} = \overline{\omega} \circ \psi$, with $\overline{\omega}$ the antiinvolution on $\mathbb{P}_\mathbb{C}^1$ induced by $\overline{\sigma}$. In other words if $X_\mathbb{R}$ is the real smooth surface defined by $(X_\mathbb{C}, \overline{\sigma})$, ψ comes from an $\mathbb{R}$-morphism $\psi_\mathbb{R} : X_\mathbb{R} \to B_\mathbb{R}$, with $B_\mathbb{R}$ a smooth real rational curve. The strict transforms $l_{0,i}$ of the $2n - 2$ lines

$\langle p_0, p_i \rangle$ together with the corresponding exceptional divisors E_i are the $2n-2$ singular fibers of ψ and lie over *real* points q_i of $B_{\mathbb{R}}$ ($q_i = \psi(l_{0,i}) = \psi(E_i) = \overline{\omega} \circ \psi(l_{0,i}) = \overline{\omega}(q_i)$) for every $i = 1, \ldots, 2n-2$, and we can suppose $B_{\mathbb{R}}$ isomorphic to $\mathbb{P}^1_{\mathbb{R}}$.

Let us show that $\rho(X_{\mathbb{R}}) = 2$, i.e. that $\psi_{\mathbb{R}} : X_{\mathbb{R}} \to B_{\mathbb{R}}$ is a minimal conic bundle. More precisely we prove that $Pic(X_{\mathbb{R}}) = \langle K_{X_{\mathbb{R}}}, F \rangle$, where F is the class of a fiber of $\psi_{\mathbb{R}}$. Let $M \in Pic(X_{\mathbb{C}})$ be a divisor such that $\overline{\sigma}(M) = M$. Then $(M \cdot F) = 2l$, $l \in \mathbb{Z}$. In fact $(M \cdot F) = (M \cdot l_{0,i} + E_i) = 2(M \cdot E_i)$ because $\overline{\sigma}(l_{0,i}) = E_i$. Hence $M' = M + lK_X$ is such that $(M' \cdot F) = 0$ and clearly $\overline{\sigma}(M') = M'$. From the description of each singular fiber and once again from the fact that $\overline{\sigma}(l_{0,i}) = E_i$ we deduce that the restriction of M' to each irreducible component of every fiber of ψ has degree 0, so that for each fiber F we have $h^0(M'_{|F}) = 1$ and $h^1(M'_{|F}) = 0$. This yields that $\psi_*(M')$ is a line bundle and that $M' \simeq \psi^*(\psi_*(M'))$, from which the claim follows.

The $2n-2$ points q_i divide $B(\mathbb{R}) \sim \mathbb{RP}^1$ into $2n-2$ closed intervals and around a point q_i, $\psi_{\mathbb{R}}$ has local equation $x^2 + y^2 = \epsilon$, which implies $X(\mathbb{R}) \sim \amalg_{n-1} S^2$, the disjoint union of $(n-1)$ copies of S^2. Since $n \geq 2$ we have $2(n-1) \geq 2$ singular fibers and $X_{\mathbb{C}}$ has even degree $K_X^2 = 8 - 2(n-1)$

Let us remark that $\text{Bl}_p(\mathbb{P}^2_{\mathbb{R}}) \to \mathbb{P}^1_{\mathbb{R}}$ is a minimal conic bundle which is not minimal over $\mathbb{R}$ and that in this case $Pic(\text{Bl}_p(\mathbb{P}^2_{\mathbb{R}})) \otimes \mathbb{Q} = \langle K_{X_{\mathbb{R}}}, F \rangle$. This would be the case $n = 1$ not treated here.

If $n = 2$, then the (-1)-curves $l_{1,2}$ and E_0 are conjugate and disjoint. By contracting them we obtain a real del Pezzo surface X' of degree 8 with $X'(\mathbb{R}) \sim S^2$, i.e. $X_{\mathbb{R}}$ is isomorphic to $Q^{3,1}(0,2)$ and it is not minimal over $\mathbb{R}$.

If $n = 3$ and if the points are in general position, $\overline{\sigma}$ sends also the pencil of conics through $p_1, \ldots, p_4$ into itself and induces on it an antiprojective involution with fixed points and another structure of conic bundle. Hence the totality of the 16 lines of $X_{\mathbb{C}}$ is divided into 8 couples of lines, intersecting in one point. In conclusion for every (-1)-curve on $X_{\mathbb{C}}$ we have $(C \cdot \overline{\sigma}(C)) = 1$ and $X_{\mathbb{R}}$ is minimal over $\mathbb{R}$. Then $X(\mathbb{R}) \sim S^2 \amalg S^2$ and the monodromy interchanges the two components. A surface of this kind will be indicated by $\mathbb{D}_4$, where $\mathbb{D}$ stands for de Jonquiéres.

More generally we will indicate by $\mathbb{D}_d$ the real surface associated to a real de Jonquiéres antibirational involution of order n centered at $2n-1$ points real points, where $d = K_X^2 = 8 - 2(n-1)$. If $n = 4$ and the points are general, the surface $\mathbb{D}_2$ is also a real del Pezzo surface which is minimal over $\mathbb{R}$ as we will show at the end of Example 4 in Section 5.

On the contrary, let us suppose we have a minimal conic bundle $\psi : X_{\mathbb{R}} \to \mathbb{P}^1_{\mathbb{R}}$ with $2m > 0$ singular fibers. This implies $K_X^2 = 8 - 2m$. By a suitable contraction of the (-1)-curves in the fibers of $\psi : X_{\mathbb{C}} \to \mathbb{P}^1_{\mathbb{C}}$, we can represent $X_{\mathbb{C}}$ as the blowing-up of $\mathbb{P}^2_{\mathbb{C}}$ at $2m+1$ points $p_0, \ldots, p_{2m}$ and the fibers of ψ as the pencil of lines through p_0. Moreover the singular fibers of ψ are the strict transforms of the lines $\langle p_0, p_i \rangle$, $i = 1, \ldots, 2m$. If $X_{\mathbb{R}}$ is a del Pezzo surface, then the points are in general position with respect to lines, conics and cubics with a node. It is then easy to see that the antiinvolution $\bar{\sigma} : X_{\mathbb{C}} \to X_{\mathbb{C}}$ is represented on $\mathbb{P}^2_{\mathbb{C}}$ by a de Jonquiéres antibirational involution of order $m+1$ and that modulo a projective transformation we can suppose the points p_j, $j = 0, \ldots, 2m$ real.

We show in the next proposition that every degree 4 smooth real del Pezzo surface minimal over $\mathbb{R}$ is isomorphic to a surface of type $\mathbb{D}_4$.

Proposition 3.1. *Let $X_{\mathbb{R}}$ be a smooth real del Pezzo surface of degree 4 which is minimal over $\mathbb{R}$. Then $X_{\mathbb{R}}$ is isomorphic to a del Pezzo surface of type $\mathbb{D}_4$. Moreover smooth real del Pezzo surfaces of type $\mathbb{D}_4$ form a connected family.*

Proof. For every (-1)-curve $C \subset X_{\mathbb{C}}$ we have $1 \geq (C \cdot \bar{\sigma}(C)) > 0$, where the second inequality follows from the minimality of $X_{\mathbb{R}}$.

Then it is immediate to see that $|C + \bar{\sigma}(C)|$ is a linear system without base points, of selfintersection 0 and of dimension 1 (C and $\bar{\sigma}(C)$ are (-1)-curves on $X_{\mathbb{C}}$). This linear system defines a morphism $\phi : X_{\mathbb{C}} \to \mathbb{P}^1_{\mathbb{C}}$ with reduced and connected fibers having 4 singular fibers ($K_X^2 = 4$). If $\phi^{-1}(p) = F_1 + F_2$ is a reducible fiber, then $\bar{\sigma}(F_1) = F_2$ by minimality.

Let us represent the pencil of *conics* $|C + \bar{\sigma}(C)|$ on the plane as the pencil of lines through p_0, where $X_{\mathbb{C}} = \text{Bl}_{p_0,\ldots,p_5}(\mathbb{P}^2_{\mathbb{C}})$. The 4 singular fibers are mapped into the lines $\langle p_0, p_i \rangle$, $i = 1, \ldots, 4$. If $\bar{\sigma}$ is the corresponding representation of the conjugation, then we have $\bar{\sigma}(p_i) = \langle p_0, p_i \rangle$, which implies $\bar{\sigma}(p_0) = C^2[p_k^1] = C_0$: $1 = (E_0 \cdot l_{0,i}) = (\bar{\sigma}(E_0) \cdot E_i)$.

If $\tilde{\tau}$ is the de Jonquiéres involution centered in $p_0, \ldots, p_5$, then $\bar{\omega} = \tilde{\tau} \circ \bar{\sigma}$ is an antiprojectivity since it transforms lines into lines. Moreover $\bar{\omega}(p_k) = p_k$ for every $k = 0, \ldots 4$. It is involutory because $\bar{\omega}^2$ is a projectivity with five general fixed points, i.e. the identity. After reducing $\bar{\omega}$ to complex conjugation, we can suppose the p_k's real and $\bar{\sigma} = \tilde{\tau} \circ \alpha$, with $\tilde{\tau}$ the real de Jonquiéres involution generated by the p_k's.

The surfaces of type $\mathbb{D}_4$ form a connected family since they are in one-to-one correspondence with the blow-up of $\mathbb{P}^2_{\mathbb{R}}$ in 5 general real points and these surfaces form a connected family. □

Corollary 3.2. *Let $X_{\mathbb{R}}$ be a smooth real del Pezzo surface of degree $d = K_X^2 = 4$. Then $X_{\mathbb{R}}$ is isomorphic to one of the following:*

$X_{\mathbb{R}}$	$\rho(X_{\mathbb{R}})$	$X(\mathbb{R})$	# real lines
$\mathbb{P}^2_{\mathbb{R}}(5,0)$	6	$\#6\mathbb{RP}^2$	16
$\mathbb{P}^2_{\mathbb{R}}(3,2)$	5	$\#4\mathbb{RP}^2$	8
$\mathbb{P}^2_{\mathbb{R}}(1,4)$	4	$\#2\mathbb{RP}^2$	4
$Q^{3,1}(0,4)$	3	S^2	0
$Q^{2,2}(0,4)$	4	$S^1 \times S^1$	0
$(Q^{3,0} \times Q^{3,0})(0,4)$	4	$\emptyset$	0
$\mathbb{D}_4$	2	$S^2 \sqcup S^2$	0

Proof. By Proposition 3.1 we know that either $X_{\mathbb{R}}$ is isomorphic to $\mathbb{D}_4$ or it dominates a real del Pezzo surface of degree $d' \geq 5$ and the result easily follows. The number of real lines is computed as follows: if $X(\mathbb{R})$ is orientable, then there are no real lines; in the other cases we can easily compute the number because $\bar{\sigma}$ is an antiprojectivity. □

Corollary 3.3. *Let $X_{\mathbb{R}}$ be a smooth real del Pezzo surface of degree $d = K_X^2 = 3$. Then $X_{\mathbb{R}}$ is isomorphic to one of the following:*

$X_{\mathbb{R}}$	$\rho(X_{\mathbb{R}})$	$X(\mathbb{R})$	# real lines
$\mathbb{P}_{\mathbb{R}}^2(6,0)$	7	$\#7\mathbb{RP}^2$	27
$\mathbb{P}_{\mathbb{R}}^2(4,2)$	6	$\#5\mathbb{RP}^2$	15
$\mathbb{P}_{\mathbb{R}}^2(2,4)$	5	$\#3\mathbb{RP}^2$	7
$\mathbb{P}_{\mathbb{R}}^2(0,6)$	4	$\#\mathbb{RP}^2$	3
$\mathbb{D}_4(1,0)$	3	$\mathbb{RP}^2 \amalg S^2$	3

Proof. The surface $X_{\mathbb{C}}$ has 27 (-1)-curves and hence $\bar{\sigma}$ has to fix one of them. This means that $X_{\mathbb{R}}$ is the blow-up of a real del Pezzo surface of degree 4 in a real point, leading to the above cases. In the first 4 cases $\bar{\sigma}$ is an antiprojectivity and we can easy deduce the number of real lines. In the last case there exists a plane representation $X_{\mathbb{C}} = \mathrm{Bl}_{p_0,\ldots,p_5}(\mathbb{P}_{\mathbb{C}}^2) \xrightarrow{\pi} \mathbb{P}_{\mathbb{C}}^2$ such that $\bar{\sigma}$ is a real de Jonquiéres antibirational involution centered in $p_0, \ldots, p_4$ and such that $\bar{\sigma}(p_5) = p_5$, where p_5 is the center of the blow-up on $\mathbb{D}_4$. It follows that the real lines on $X_{\mathbb{C}}$ are $E_5 = \pi^{-1}(p_5)$, the strict transform of the line $\langle p_0, p_5 \rangle$ and of the conic through $p_1, \ldots, p_5$. Let us remark that the surfaces of type $\mathbb{D}_4(1,0)$ (or $\mathbb{D}_4$) are unirational over $\mathbb{R}$ but not rational over $\mathbb{R}$. □

Remark 2. It is well known that a smooth cubic hypersurface $X_k \subset \mathbb{P}_k^n$ defined over a field k is unirational over k if it contains a k-line, i.e. a line defined over k. Hence a smooth real cubic surface is unirational over $\mathbb{R}$. On the other hand, if two real smooth surfaces are $\mathbb{R}$-birational, then their real parts have the same number of connected components. As a consequence the surfaces of type $\mathbb{D}_4(1, 0)$ (or $\mathbb{D}_4$) are unirational over $\mathbb{R}$ but not rational over $\mathbb{R}$.

4. Geiser antiinvolutions and degree 2 real del Pezzo surfaces

We define through Geiser involution a real structures on complex del Pezzo surfaces of degree 2, such that the resulting real surfaces are minimal del Pezzo surfaces.

Example 3 (Geiser). The Geiser involution of the plane is generated by cubic curves through seven fixed points $p_1, \ldots, p_7$ in general position: a pair of the involution is a pair of points which form with the p_i's a set of nine associated points, i.e. the nine points are the intersections of the pencil of cubics that eight of them generate.

This is well defined on $\mathbb{P}_\mathbb{C}^2 \setminus \bigcup_{i=1}^7 K_i$, where K_i is the cubic through $p_1, \ldots, p_7$ having a double point in p_i. It uniquely extends to a well defined involution τ on $X_\mathbb{C} = \mathrm{Bl}_{p_1,\ldots,p_7}(\mathbb{P}_\mathbb{C}^2) \xrightarrow{\pi} \mathbb{P}_\mathbb{C}^2$ such that $\tau(K_i) = \pi^{-1}(p_i) = E_i$, where K_i also denotes the strict transform on $X_\mathbb{C}$.

Let $\tilde{\tau} = \pi \circ \tau \circ \pi^{-1}$ be the birational involution of the plane described above; since $\tilde{\tau}(p_i) = K_i$ for every i, the homaloidal net associated to $\tilde{\tau}$ is given by curves having triple points in the p_i's and it is then of type $C^8[p_i^3]$, curves of degree 8 with seven triple points in the p_i's.

Equivalently, τ can be seen to be the involution associated to the ramified double covering $\psi_{|-K_X|} : X_\mathbb{C} \to \mathbb{P}_\mathbb{C}^2$. The covering ψ is ramified along a smooth quartic curve D and since $E_i + K_i = -K_X$, the curves E_i and K_i are mapped into lines of $\mathbb{P}_\mathbb{C}^2$, which are bitangent to D ($(E_i \cdot K_i) = 2$) and we have $\tau(E_i) = K_i$. Furthermore τ sends the strict transforms $l_{i,k}$ of the lines $\langle p_i, p_k \rangle$ into the strict transforms $C_{i,k}$ of the conics passing through the five points p_j, $j \neq i$, $j \neq k$ and $\tau(l_{i,k}) = C_{i,k}$ since $l_{i,k} + C_{i,k} = -K_X$.

Let us suppose the p_i's real and define the antibirational involution of the plane $\bar{\sigma} = \tilde{\tau} \circ \alpha = \alpha \circ \tilde{\tau}$. It induces a real structure $\bar{\sigma}$ on $X_\mathbb{C}$ which interchanges the elements of the 28 couples of (-1)-curves $\{E_i, K_i\}$, $\{l_{i,k}, C_{i,k}\}$, the totality of the 56 lines on $X_\mathbb{C}$. In particular, for every (-1)-curve $C \subset X_\mathbb{C}$, we have $C + \bar{\sigma}(C) = -K_X$, $(C \cdot \bar{\sigma}(C)) = 2$. This gives $Pic(X_\mathbb{R}) \otimes \mathbb{Q} = \langle -K_X \rangle$, i.e. $X_\mathbb{R}$ is a minimal real del Pezzo surface of degree 2. Surfaces of this kind will be indicated by $\mathbb{G}_2$, where $\mathbb{G}$ stands for Geiser.

We also verify that the surfaces $\mathbb{D}_2$ are minimal over $\mathbb{R}$. If $\bar{\sigma}$ is the corresponding antibirational de Jonquiéres involution of order 4, then, after renumbering as in Example 3, we have $\bar{\sigma}(E_0) = K_0$, $\bar{\sigma}(l_{0,i}) = E_i$, $\bar{\sigma}(K_i) = C_{0,i}$, $\bar{\sigma}(l_{i,k}) = C_{i,k}$, $i = 1, \ldots, 6$, $i < k$. Hence for every (-1)-curve $C \subset X_\mathbb{C}$ we have $(C \cdot \bar{\sigma}(C)) = 1$ or 2.

Let us now introduce the definition of *associated* surface to a degree 2 smooth real del Pezzo surface and find the topological type of surfaces of type $\mathbb{G}_2$.

Since for every real structure on a complex surface $X_\mathbb{C}$ we have $\bar{\sigma}(K_{X_\mathbb{C}}) = K_{X_\mathbb{C}}$, $K_{X_\mathbb{R}}$ gives to a smooth real del Pezzo surface $X_\mathbb{R}$ of degree 2 a structure of double cover of $\mathbb{P}_\mathbb{R}^2$, ramified along a smooth quartic real plane curve $D_\mathbb{R}$. If $f(x, y) = 0$ is a local equation of $D_\mathbb{R}$, with f real polynomial, we can suppose that the double cover $X_\mathbb{R}$ has local equation $z^2 = f(x, y)$. Then the smooth real del Pezzo surface $X_\mathbb{R}'$ of degree 2 defined by the local equation $z^2 = -f(x, y)$ is said *associated* to $X_\mathbb{R}$ (see [2], p. 50).

The two surfaces are clearly isomorphic over $\mathbb{C}$ and hence we can suppose $X_\mathbb{C} = \mathrm{Bl}_{p_1,\ldots,p_7}(\mathbb{P}_\mathbb{C}^2) \xrightarrow{\pi} \mathbb{P}_\mathbb{C}^2$. If $\bar{\sigma}$ and $\bar{\sigma}'$ are the corresponding antibirational involution of the plane, since $\tilde{\tau}$ is the birational involution of the double cover $\psi : X_\mathbb{C} \to \mathbb{P}_\mathbb{C}^2$ ramified along D, we have

$$\bar{\sigma} \circ \bar{\sigma}' = \bar{\sigma}' \circ \bar{\sigma} = \tilde{\tau}.$$

Since $D_\mathbb{R}$ is an even degree smooth real plane curve, each connected component of the real part $D(\mathbb{R})$ of $D_\mathbb{R}$ is homotopically trivial and disconnects $\mathbb{RP}^2$ into two connected components, one homeomorphic to a disk, the other one to a Möbius strip. Hence $D(\mathbb{R})$ divides the plane into connected regions, one of which is not orientable and on which f has the same sign.

If $X_\mathbb{R}$ is a surface of type $\mathbb{G}_2$, then $\overline{\sigma} = \widetilde{\tau} \circ \alpha$, implying that $X'_\mathbb{R}$ is isomorphic to the blowing-up of $\mathbb{P}^2_\mathbb{R}$ in seven real points, $\mathrm{Bl}_{p_1,\ldots,p_7}(\mathbb{P}^2_\mathbb{R}) \xrightarrow{\pi} \mathbb{P}^2_\mathbb{R}$. From $X'(\mathbb{R}) \sim \#8\mathbb{RP}^2$, one deduces that $f(x, y)$ is negative on the non-orientable component and $D(\mathbb{R})$ has 4 connected components. This finally gives $X(\mathbb{R}) \sim \sqcup 4S^2$.

We are in position to prove that minimal real del Pezzo surfaces of degree 2 are isomorphic to surfaces of type $\mathbb{G}_2$. This answers a question posed by Silhol (see [17], p. 130).

Proposition 4.1. *Let $X_\mathbb{R}$ be a smooth minimal real del Pezzo surface of degree 2. Then $X_\mathbb{R}$ is isomorphic to a surface of type $\mathbb{G}_2$. Furthermore, minimal smooth real del Pezzo surfaces of degree 2 form a connected family.*

Proof. Let $X_\mathbb{C} = \mathrm{Bl}_{p_1,\ldots,p_7}(\mathbb{P}^2_\mathbb{C}) \xrightarrow{\pi} \mathbb{P}^2_\mathbb{C}$. For every (-1)-curve $C \subset X_\mathbb{C}$, $C + \overline{\sigma}(C)$ is defined over $\mathbb{R}$ and hence $C + \overline{\sigma}(C) = -rK_X$. Thus $-2r = -rK_X^2 = (K_X \cdot C + \overline{\sigma}(C)) = -2$, gives $C + \overline{\sigma}(C) = -K_X$ and $(C \cdot \overline{\sigma}(C)) = 2$. Since the 28 couples of (-1)-curves $\{C_j, C'_j\}$ lying over the complex bitangents to D have the properties $(C_j \cdot C'_j) = 2$, $(C_h \cdot C_l) \leq 1$ and $(C_h \cdot C'_l) \leq 1$ if $h \neq l$, we have $\overline{\sigma}(p_i) = K_i$ and $\overline{\sigma}(\langle p_i, p_k \rangle) = C_{i,k}$. The homaloidal net associated to $\overline{\sigma}$ is then of type $C^8[p_i^3]$ and, if $\widetilde{\tau}$ is the Geiser involution generated by the p_i's, then $\overline{\omega} = \widetilde{\tau} \circ \overline{\sigma}$ is an antiprojectivity such that $\overline{\omega}(p_i) = p_i$. It is involutory because $\overline{\omega}^2$ is a projectivity with 7 general fixed points. After reducing $\overline{\omega}$ to complex conjugation we can suppose the p_i's real and $\overline{\sigma} = \widetilde{\tau} \circ \alpha$, with $\widetilde{\tau}$ the real Geiser involution generated by the p_i's.

The connectedness of the family follows from the fact that there is a one-to-one correspondence with the blowing-ups of $\mathbb{P}^2_\mathbb{R}$ in 7 real points and these latter form a connected family. □

To conclude the classification of real del Pezzo surfaces of degree 2 we prove that those minimal over $\mathbb{R}$, which are not minimal del Pezzo surfaces, are isomorphic to surfaces of type $\mathbb{D}_2$.

Proposition 4.2. *Let $X_\mathbb{R}$ be a smooth real del Pezzo surface of degree 2 minimal over $\mathbb{R}$. Then $X_\mathbb{R}$ is either isomorphic to a surface of type $\mathbb{G}_2$ or to a surface of type $\mathbb{D}_2$.*

Proof. Since $X_\mathbb{R}$ is minimal over $\mathbb{R}$, we have $(C \cdot \sigma(C)) > 0$ for every (-1)-curve $C \subset X_\mathbb{C}$. On the other hand, we also have $(C \cdot \overline{\sigma}(C)) \leq 2$. Hence either $(C \cdot \overline{\sigma}(C)) = 2$ for every (-1)-curve $C \subset X_\mathbb{C}$ and applying the same arguments as in the proof of Proposition 4.1 one deduces that $X_\mathbb{R}$ is a minimal del Pezzo surface of type $\mathbb{G}_2$ or there exists a (-1)-curve $C \subset X_\mathbb{C}$ such that $(C \cdot \overline{\sigma}(C)) = 1$. In the last case it is immediate that $|C + \overline{\sigma}(C)|$ defines a structure of minimal conic bundle on $X_\mathbb{R}$.

To conclude that $X_\mathbb{R}$ is a surface of type $\mathbb{D}_2$ we argue as in the proof of Proposition 3.1. Let us represent the pencil of *conics* $|C + \overline{\sigma}(C)|$ on the plane as the pencil of lines through p_0, where $X_\mathbb{C} = \mathrm{Bl}_{p_0,\ldots,p_6}(\mathbb{P}^2_\mathbb{C})$. The 6 singular fibers are mapped into the lines $\langle p_0, p_i \rangle$, $i = 1, \ldots, 6$. If $\overline{\sigma}$ is the corresponding representation of the conjugation of $X_\mathbb{C}$, then we have $\overline{\sigma}(p_i) = \langle p_0, p_i \rangle$ from which it follows $\overline{\sigma}(p_0) = C^3[p_0^2, p_i^1] = C_0$. If $\widetilde{\beta}$ is the de Jonquiéres involution centered in $p_0, \ldots, p_6$, then $\overline{\omega} = \widetilde{\beta} \circ \overline{\sigma}$ is an antiprojectivity and $\overline{\omega}(p_k) = p_k$ for every $k = 0, \ldots 6$. It is then involutory because $\overline{\omega}^2$ is a projectivity with seven general fixed points, i.e. the identity. Then after reducing $\overline{\omega}$ to complex conjugation, we can suppose the p_k's real and $\overline{\sigma} = \widetilde{\beta} \circ \alpha$, with $\widetilde{\beta}$ the real de Jonquiéres involution generated by the p_k's. $\square$

Let us indicate by $\mathbb{D}_4(2,0)_0^2$, respectively by $\mathbb{D}_4(2,0)_1^1$, the smooth real del Pezzo surface of degree 2 obtained by blowing-up 2 real points on the same connected component of $\mathbb{D}_4$, respectively one real point on each connected component of $\mathbb{D}_4$.

Corollary 4.3. *Let $X_\mathbb{R}$ be a smooth real del Pezzo surface of degree $d = K_X^2 = 2$. Then $X_\mathbb{R}$ is isomorphic to one of the following:*

$X_\mathbb{R}$	$\rho(X_\mathbb{R})$	$X(\mathbb{R})$	# real lines
$\mathbb{P}^2_\mathbb{R}(7,0)$	8	$\#8\mathbb{RP}^2$	56
$\mathbb{P}^2_\mathbb{R}(5,2)$	7	$\#6\mathbb{RP}^2$	32
$\mathbb{P}^2_\mathbb{R}(3,4)$	6	$\#4\mathbb{RP}^2$	16
$\mathbb{P}^2_\mathbb{R}(1,6)$	5	$\#2\mathbb{RP}^2$	8
$\mathbb{D}_4(2,0)_1^1$	4	$\mathbb{RP}^2 \sqcup \mathbb{RP}^2$	8
$\mathbb{D}_4(2,0)_0^2$	4	$\#2\mathbb{RP}^2 \sqcup S^2$	8
$Q^{3,1}(0,6)$	4	S^2	0
$Q^{2,2}(0,6)$	5	$S^1 \times S^1$	0
$(Q^{3,0} \times Q^{3,0})(0,6)$	5	$\emptyset$	0
$\mathbb{D}_4(0,2)$	3	$S^2 \sqcup S^2$	0
$\mathbb{D}_2$	2	$\sqcup 3 S^2$	0
$\mathbb{G}_2$	1	$\sqcup 4 S^2$	0

Proof. We know all real del Pezzo surfaces of degree $d' \geq 2$ which are minimal over $\mathbb{R}$ and hence we obtain the description of $X_\mathbb{R}$. The number of real lines is deduced from the corresponding description of $\overline{\sigma}$. Surfaces with $X(\mathbb{R})$ orientable contain no real line, while if $\overline{\sigma}$ is an antiprojectivity the number of real lines is easily deduced. Let us consider then surfaces of type $\mathbb{D}_4(2,0)$. If $X_\mathbb{C} = \mathrm{Bl}_{p_0,\ldots,p_6}(\mathbb{P}^2_\mathbb{C}) \xrightarrow{\pi} \mathbb{P}^2_\mathbb{C}$, then we can assume that $\overline{\sigma}$ is a de Jonqiuéres antiinvolution centered in the real points $p_0, \ldots, p_4$ and that $\overline{\sigma}(p_5) = p_5$, $\overline{\sigma}(p_6) = p_6$. Then the following are the unique (-1)-curves which are fixed by $\bar{\sigma}$: the lines $\langle p_0, p_5 \rangle$ and $\langle p_0, p_6 \rangle$ together with their transforms by the Geiser involution, the conics $C_{0,5}$ and $C_{0,6}$ (notations as in Example 4); $E_5 = \pi^{-1}(p_5)$ and $E_6 = \pi^{-1}(p_6)$ and the cubics K_6 and K_7. □

Corollary 4.4 ([11], [19]). *There are 6 topological types of degree 4 smooth real plane curves $D_{\mathbb{R}}$:*

$D(\mathbb{R})$	# real bitangents	$X_{\mathbb{R}}$	$X_{\mathbb{R}}'$
○○○○	28	$\mathbb{G}_2$	$\mathbb{P}^2_{\mathbb{R}}(7, 0)$
○○○	16	$\mathbb{D}_2$	$\mathbb{P}^2_{\mathbb{R}}(5, 2)$
○○	8	$\mathbb{D}_4(0, 2)$	$\mathbb{P}^2_{\mathbb{R}}(3, 4)$
○	4	$Q^{3,1}(0, 6)$	$\mathbb{P}^2_{\mathbb{R}}(1, 6)$
∅	4	$(Q^{3,0} \times Q^{3,0})(0, 6)$	$\mathbb{D}_4(2, 0)^1_1$
⊙	4	$Q^{2,2}(0, 6)$	$\mathbb{D}_4(2, 0)^2_0$

If d is the number of ovals of $D(\mathbb{R})$ which are not contained in another oval, then $D(\mathbb{C})$ has $4 + 2d \cdot (d - 1)$ real bitangents.

Proof. To deduce the number of ovals of $D(\mathbb{R})$ and their mutual position we can restrict for example to real del Pezzo surfaces of degree 2 with $X(\mathbb{R})$ not orientable.

The number of real bitangents is deduced by looking at the action of $\bar{\sigma}$ on the couples of (-1)-curves lying over the 28 complex bitangent to $D(\mathbb{C})$. Since to every quartic real plane curve there correspond two real del Pezzo surfaces of degree 2, which are associated (see Example 4), one containing real lines, the other one without real lines, to count the number of real bitangents we can look at the action of $\bar{\sigma}$ on surfaces containing real lines.

If C is a real line and if τ is the corresponding Geiser involution on $X_{\mathbb{C}}$, then $\tau(C)$ is a real line. Remember that for every (-1)-curve $C \subset X_{\mathbb{C}}$, $-K_X = C + \tau(C)$ and hence $C + \tau(C) = -K_X = \sigma(-K_X) = \bar{\sigma}(C) + \bar{\sigma}(\tau(C)) = C + \bar{\sigma}(\tau(C))$. The curves C and $\tau(C)$ are mapped into a real line of $\mathbb{P}^2_{\mathbb{R}}$, which is bitangent to $D(\mathbb{C})$. The number of real bitangents is half the number of real lines on the corresponding degree 2 real del Pezzo surface and the last part of the corollary is easily verified. □

5. Bertini antiinvolutions and degree 1 real del Pezzo surface

We describe a structure of minimal real del Pezzo surface on the blow-up of $\mathbb{P}^2_{\mathbb{C}}$ in eight real points by composing the Bertini involution generated by these points with complex conjugation.

Example 4 (Bertini). The Bertini involution of the plane is generated by the sextics with eight double points in general position $p_1, \ldots, p_8$: a pair of the involution is a pair of points p, q such that, if C is the cubic through p and the p_i's and D any curve in the linear systems of sextics through p and having double points in the p_i's, $D \cap C = \{2p_1, \ldots, 2p_8, p, q\}$.

This defines an involution on $\mathbb{P}^2_\mathbb{C} \setminus (\bigcup_{i=1}^{8} D_i \cup r)$, where the D_i's are the sextics having double points in p_k, $k \neq i$ and a triple point in p_i and r is the nine point of intersection of two cubics passing through $p_1, \ldots, p_8$.

It uniquely extends to a well defined involution τ on $X_\mathbb{C} = \mathrm{Bl}_{p_1,\ldots,p_8}(\mathbb{P}^2_\mathbb{C}) \xrightarrow{\pi} \mathbb{P}^2_\mathbb{C}$ such that $\tau(D_i) = \pi^{-1}(p_i) = E_i$, where D_i denotes also the strict transform of D_i on $X_\mathbb{C}$.

Let $\tilde{\tau} = \pi \circ \tau \circ \pi^{-1}$ be the birational involution of the plane associated to τ. Since $\tilde{\tau}(p_i) = D_i$ for every i, the homaloidal net associated to $\tilde{\tau}$ is given by curves having points of multiplicity 6 in the p_i's and it is then of type $C^{17}[p_i^6]$.

Equivalently, τ can be seen as the involution associated to the ramified double cover of a quadric cone Q given by $\psi_{|-2K_X|} : X_\mathbb{C} \to Q \subset \mathbb{P}^3_\mathbb{C}$; ψ is ramified along a smooth sextic curve Γ, which is the complete intersection of Q with a smooth cubic surface of $\mathbb{P}^3_\mathbb{C}$ (and along the vertex). Since $E_i + D_i = -2K_X$, these curves are mapped into conics on Q, which are tritangent to Γ ($(E_i \cdot C_i) = 3$).

On $X_\mathbb{C}$ there are 240 (-1)-curves, divided into 120 couples $\{A_i, A'_i\}$, such that $A_i + A'_i = -2K_X$ and $(A_i \cdot A'_i) = 3$: these are the 28 couples given by the lines $\langle p_i, p_k \rangle = l_{i,k}$ together with the quintics $l'_{i,k}$ passing simply through p_i and p_k and having double points in the other 6 points p_j, $j \neq i$, $j \neq k$; the 56 couples formed by the conics $C_{i,j,k}$ passing through the 5 points different from p_i, p_j, p_k together with the quartics $C'_{i,j,k}$ having double points in p_i, p_j, p_k and simple points in the other 5 points; the 28 couples formed by the cubics $e_{i,k}$, having a double point in p_i and passing through the points different from p_k, together with the cubics $e_{k,i}$.

These 120 couples of curves are mapped by ψ into 120 planes tritangent to Γ (or equivalently into 120 conics on Q tritangent to Γ) and $\tau(A_i) = A'_i$ for every $i = 1, \ldots, 120$.

Let us now suppose the p_i's real and define the antibirational involution of the plane $\overline{\sigma} = \tilde{\tau} \cup \alpha - \alpha \circ \tilde{\tau}$. It induces a real structure σ on $X_\mathbb{C}$: For every (-1)-curve $C \subset X_\mathbb{C}$, we have $C + \overline{\sigma}(C) = -2K_X$, $(C \cdot \overline{\sigma}(C)) = 3$ so that $Pic(X_\mathbb{R}) \otimes \mathbb{Q} = \langle -K_X \rangle$, i.e. the smooth real del Pezzo surface $X_\mathbb{R}$, given by the real structure $(X_\mathbb{C}, \overline{\sigma})$ is a minimal real del Pezzo surface of degree 1. Surfaces of this kind will be indicated by $\mathbb{B}_1$, where $\mathbb{B}$ stands for Bertini.

To find the topological type of these surfaces we have to make some preliminary remarks on the components of the ramification smooth real curve $\Gamma_\mathbb{R}$ of degree 6 of the double cover structure given by $|-2K_{X_\mathbb{R}}|$. For more details one can consult [2], p. 49). The cone Q is real, has real points and $\Gamma_\mathbb{R}$ is the complete intersection of Q with a cubic surface not passing through the vertex of the cone.

Then $\Gamma(\mathbb{R})$ can have connected components of two kinds: components cut by a line of the ruling in an even number of points (2 or 0), called *ovals* (null homotopic), and components cut by a line of the ruling in an odd number of points (3 or 1), called *big-circles* (homotopically equivalent to a plane section not going through the vertex).

Hence $\Gamma(\mathbb{R})$ contains either a big-circle and eventually some oval or 3 big-circles and no ovals. If there are at least 3 ovals, they lie in one of the two regions determined by the big circle. Otherwise, taking a plane through 3 points on three distinct ovals lying in different regions, we have six point of intersection with these components and further intersections with the big-circle and Γ would have degree greater or equal to 8. There are no nested ovals because some line of the ruling would intersect Γ in 4 points.

Then as in the degree 2 case we can define the notion of *associated surface* to a degree 1 real del Pezzo surface of degree 1. If we have 3 big-circles, 1 big-circle and no ovals, one big-circle and one oval in each region determined, the surface $X_\mathbb{R}$ and the *associated* surface $X'_\mathbb{R}$ are *autoassociated*, i.e. isomorphic over $\mathbb{R}$.

If we have 1 big-circle and $k \geq 2$ ovals in one region, the configuration is not symmetric and the two associated covers of Q ramified along $\Gamma_\mathbb{R}$ are not isomorphic; one has topological type $\#(2k+1)\mathbb{RP}^2$, the associated one $\mathbb{RP}^2 \amalg kS^2$.

The associated surface to a minimal del Pezzo surface of type $\mathbb{B}_1$ is the blow-up of $\mathbb{P}^2_\mathbb{R}$ in 8 real points and has topological type $\#9\mathbb{RP}^2$; thus $\mathbb{B}_1(\mathbb{R}) \sim \mathbb{RP}^2 \amalg 4S^2$.

Let us show that smooth minimal real del Pezzo surfaces of degree 1 are all of this kind, answering to a question posed by Silhol (see [17], p. 130).

Proposition 5.1. *Let $X_\mathbb{R}$ be a minimal smooth real del Pezzo surface of degree 1. Then $X_\mathbb{R}$ is isomorphic to a del Pezzo surface of type $\mathbb{B}_1$. Moreover, the family of minimal real del Pezzo surfaces of degree 1 is connected.*

Proof. If $X_\mathbb{R}$ is a minimal real del Pezzo surface of degree 1, then we have $X_\mathbb{C} = \text{Bl}_{p_1,\ldots,p_8}(\mathbb{P}^2_\mathbb{C}) \xrightarrow{\pi} \mathbb{P}^2_\mathbb{C}$. Since for every (-1)-curve $C \subset X_\mathbb{C}$, $C + \overline{\sigma}(C)$ is defined over $\mathbb{R}$, we have $C + \overline{\sigma}(C) = -rK_X, r \in \mathbb{R}$. Thus $-r = -rK_X^2 = (K_X \cdot C + \overline{\sigma}(C)) = -2$ gives $C + \overline{\sigma}(C) = -2K_X$ and $(C \cdot \overline{\sigma}(C)) = 3$ for every (-1)-curve $C \subset X_\mathbb{C}$. Since $(A_i \cdot A_k) \leq 2$ and $(A_i \cdot A'_j) \leq 2$, if $i \neq j$, we have $\overline{\sigma}(A_i) = A'_i$ for every $i = 1,\ldots,120$ (notations as in Example 5). Furthermore, $\overline{\sigma}(p_j) = D_j$, $j = 1,\ldots,8$, where $\overline{\sigma}$ is the corresponding antibirational involution of the plane. The associated homaloidal net is then of type $C^{17}[p_i^6]$ and if $\widetilde{\tau}$ is the Bertini involution generated by the p_k's, $\overline{\omega} = \widetilde{\tau} \circ \overline{\sigma}$ is an antiprojectivity. Since $\overline{\omega}(p_k) = p_k$ it is involutory, because $\overline{\omega}^2$ has 8 general fixed points. Then reducing $\overline{\omega}$ to complex conjugation we see that $X_\mathbb{R}$ is a surface of type $\mathbb{B}_1$. □

To conclude the classification of real del Pezzo surfaces of degree 1 we need the following lemma.

Lemma 5.2. *A smooth real del Pezzo surface of degree 1 minimal over $\mathbb{R}$ is a minimal real del Pezzo surface.*

Proof. We have $3 \geq (C \cdot \overline{\sigma}(C)) > 0$ for every (-1)-curve $C \subset X_\mathbb{C}$ and it is sufficient to prove that there are no (-1)-curves for which $(C \cdot \overline{\sigma}(C)) = 1$ or 2. In fact if $(C \cdot \overline{\sigma}(C)) = 3$ for every (-1)-curve $C \subset X_\mathbb{C}$ we can argue as in the previous proposition.

Real del Pezzo surfaces 309

If $(C \cdot \overline{\sigma}(C)) = 1$, then $|C + \overline{\sigma}(C)|$ would define a structure of minimal conic bundle with an even number $2m$ of singular fibers and $K_X^2 = 8 - 2m$ would be even.

If $(C \cdot \overline{\sigma}(C)) = 2$, the divisor $D = K_X + C + \overline{\sigma}(C)$ would be $\overline{\sigma}$ invariant, linearly equivalent to an effective divisor ($h^0(D) \geq 1$) and such that $D^2 = (D \cdot K_X) = -1$, i.e. there would be a real (-1)-curve on $X_{\mathbb{R}}$. □

We concluded the classification of smooth real del Pezzo surfaces of degree 1 since we have a complete description of all real del Pezzo surfaces which are minimal over $\mathbb{R}$.

Corollary 5.3. *Let $X_{\mathbb{R}}$ be a smooth real del Pezzo surface of degree $d = K_X^2 = 1$. Then $X_{\mathbb{R}}$ is isomorphic to one of the following:*

$X_{\mathbb{R}}$	$\rho(X_{\mathbb{R}})$	$X(\mathbb{R})$	# real lines
$\mathbb{P}^2_{\mathbb{R}}(8, 0)$	9	$\#9\mathbb{R}\mathbb{P}^2$	240
$\mathbb{P}^2_{\mathbb{R}}(6, 2)$	8	$\#7\mathbb{R}\mathbb{P}^2$	126
$\mathbb{P}^2_{\mathbb{R}}(4, 4)$	7	$\#5\mathbb{R}\mathbb{P}^2$	60
$\mathbb{P}^2_{\mathbb{R}}(2, 6)$	6	$\#3\mathbb{R}\mathbb{P}^2$	26
$\mathbb{P}^2_{\mathbb{R}}(0, 8)$	5	$\#\mathbb{R}\mathbb{P}^2$	8
$\mathbb{D}_4(3, 0)^1_2$	5	$\#2\mathbb{R}\mathbb{P}^2 \sqcup \mathbb{R}\mathbb{P}^2$	24
$\mathbb{D}_4(3, 0)^0_3$	5	$\#3\mathbb{R}\mathbb{P}^2 \sqcup S^2$	24
$\mathbb{D}_4(1, 2)$	4	$\mathbb{R}\mathbb{P}^2 \sqcup S^2$	6
$\mathbb{D}_2(1, 0)$	3	$\mathbb{R}\mathbb{P}^2 \sqcup 2S^2$	4
$\mathbb{G}_2(1, 0)$	2	$\mathbb{R}\mathbb{P}^2 \sqcup 3S^2$	2
$\mathbb{B}_1$	1	$\mathbb{R}\mathbb{P}^2 \sqcup 4S^2$	0

Correspondingly there are 7 topological types of degree 6 smooth real complete intersection curves $\Gamma_{\mathbb{R}}$ on a cylinder with real points, not passing through the vertex:

$\Gamma(\mathbb{R})$	# real tritangent planes	$X_{\mathbb{R}}$	$X_{\mathbb{R}}'$
1 *big circle* + 4 *ovals*	120	$\mathbb{B}_1$	$\mathbb{P}^2_{\mathbb{R}}(8, 0)$
1 *big circle* + 3 *ovals*	64	$\mathbb{G}_2(1, 0)$	$\mathbb{P}^2_{\mathbb{R}}(6, 2)$
1 *big circle* + 2 *ovals*	32	$\mathbb{D}_2(1, 0)$	$\mathbb{P}^2_{\mathbb{R}}(4, 4)$
1 *big circle* + 1 *oval*	16	$\mathbb{D}_4(1, 2)$	$\mathbb{P}^2_{\mathbb{R}}(2, 6)$
1 *big circle* + 0 *oval*	8	$\mathbb{P}^2_{\mathbb{R}}(0, 8)$	$\mathbb{P}^2_{\mathbb{R}}(0, 8)$
1 *big circle* + 1 + 1 *ovals*	24	$\mathbb{D}_4(3, 0)^0_3$	$\mathbb{D}_4(3, 0)^0_3$
3 *big circles*	24	$\mathbb{D}_4(3, 0)^1_2$	$\mathbb{D}_4(3, 0)^1_2$

Proof. We will deduce simultaneously the number of real lines and the number of tritangent planes on the corresponding curve $\Gamma_\mathbb{R}$. To this aim let us recall some definitions introduced by Comessatti([2], p. 39). If a tritangent plane is real, then the corresponding couple $\{A_i, A'_i\}$ is fixed by $\overline{\sigma}$. If $\overline{\sigma}$ interchanges the curves, the *real tritangent plane* is said to be *of the first kind* for $(X_\mathbb{C}, \overline{\sigma})$; if $\overline{\sigma}$ fixes each element of the couple, then the *real tritangent plane* is said to be *of the second kind* for $(X_\mathbb{C}, \overline{\sigma})$.

It follows that the number of real lines on $(X_\mathbb{C}, \overline{\sigma})$ is the double of the number of real tritangent planes of the second kind and that every tritangent plane which is of the first kind for $(X_\mathbb{C}, \overline{\sigma})$ becomes of the second kind on the associated surface $(X_\mathbb{C}, \overline{\sigma}')$ and viceversa. To determine the number of real lines on real del Pezzo surfaces of degree 1 it is enough to count the number of real tritangent planes of the second kind (or of the first kind) on surfaces $X_\mathbb{R}$ and the number of real tritangent planes of the corresponding curve $\Gamma_\mathbb{R}$. Let notations be as in Example 5.

If $X_\mathbb{R}$ is $\mathbb{B}_1$, then $\overline{\sigma}$ is a Bertini antibirational involution. All the 120 couples are fixed and the corresponding 120 real tritangent planes are of the first kind and there are no real lines. On $\mathbb{P}^2_\mathbb{R}(8,0)$ we have 240 real lines.

If $X_\mathbb{R}$ is $\mathbb{G}_2(1,0)$, we can assume that $\overline{\sigma}$ is a Geiser antibirational involution centered in the seven real points $p_1, \ldots, p_7$ and that $\overline{\sigma}(p_8) = p_8$. Then $\overline{\sigma}$ fixes the following couples: a couple point-sextic $\{E_8, D_8\}$; 7 couples line-quintic $\{l_{i,8}, l'_{i,8}\}$, $i = 1, \ldots, 7$; 35 couples conic-quartic $\{C_{k,l,m}, C'_{k,l,m}\}, k, l, m = 1, \ldots, 7$; 21 couples cubic-cubic $\{e_{i,k}, e_{k,i}\}, i, k = 1, \ldots, 7$; we have 64 real tritangent planes, one of which is of the second kind corresponding to the 2 real lines E_8, K_8. On $\mathbb{P}^2_\mathbb{R}(6,2)$ there are $63 \cdot 2 = 126$ real lines.

If $X_\mathbb{R}$ is $\mathbb{D}_2(1,0)$, in a suitable representation we can assume that $\overline{\sigma}$ is an antibirational involution of order 4 and given by the homaloidal net of curves with one triple point p_1 and 6 simple points $p_2, \ldots, p_7$ and that the point p_8 is fixed. Let us recall that $\overline{\sigma}$ maps the points $p_2, \ldots, p_7$ to the lines $l_{1,2}, \ldots, l_{1,7}$. The couples fixed by $\overline{\sigma}$ are: 1 couple point-sextic $\{E_8, D_8\}$; 1 couple line-quintic $\{l_{1,8}, l'_{1,8}\}$; 15 couples conic-quartic $\{C_{1,l,m}, C'_{1,l,m}\}, l, m = 2, \ldots, 7$; 15 couples cubic-cubic $\{e_{i,k}, e_{k,i}\}$, $i, k = 2, \ldots, 7$; we have 32 real tritangent planes, 2 of which of the second kind corresponding to the 4 real lines $E_8, D_8, l_{1,8}, l'_{1,8}$. On $\mathbb{P}^2_\mathbb{R}(4,4)$ there are $30 \cdot 2 = 60$ real lines.

If $X_\mathbb{R}$ is $\mathbb{D}_4(1,2)$, respectively one of the two surfaces of type $\mathbb{D}_4(3,0)$, we can assume that $\overline{\sigma}$ is a de Jonquiéres antibirational involution of order 3 centered in the real points $p_1, \ldots, p_5$ and that $\overline{\sigma}(p_6) = p_6$, $\overline{\sigma}(p_7) = p_8$, respectively $\overline{\sigma}(p_k) = p_k$, $k = 6, 7, 8$. In the first case are fixed: 1 couple point-sextic $\{E_8, D_8\}$; 1 couple line-quintic $\{l_{1,8}, l'_{1,8}\}$; 7 couples conic-quartic $\{C_{1,l,m}, C'_{1,l,m}\}, l, m = 2, \ldots, 5$; 7 couples cubic-cubic $\{e_{i,k}, e_{k,i}\}, i, k = 2, \ldots, 5$; we have 16 real tritangent planes, 3 of which of the second kind corresponding to the 6 real lines $E_8, D_8, l_{1,8}, l'_{1,8}$, and $C_{1,6,7}, C'_{1,6,7}$. On $\mathbb{P}^2_\mathbb{R}(2,6)$ we have $13 \cdot 2 = 26$ real lines. In the second case are fixed: 3 couples point-sextic $\{E_i, D_i\}, i = 6, 7, 8$; 3 couples line-quintic $\{l_{1,i}, l'_{1,i}\}, i = 6, 7, 8$; 9 couples conic-quartic $\{C_{1,l,m}, C'_{1,l,m}\}, l, m = 2, \ldots, 5$ or $6, 7, 8$; 9 couples cubic-

cubic $\{e_{i,k}, e_{k,i}\}$, $i, k = 2, \ldots, 5$ or $6, 7, 8$; we have 24 real tritangent planes, 12 of which of the second kind corresponding to the 24 real lines E_i, D_i, $i = 6, 7, 8$, $l_{1,i}, l'_{1,i}$; $i = 6, 7, 8$; $C_{1,l,m}$, $C'_{1,l,m}$, $l, m = 6, 7, 8$ and $\{e_{i,k}, e_{k,i}\}$, $i, k = 6, 7, 8$. Each surface is autoassociated.

If $X_{\mathbb{R}}$ is the autoassociated surface $\mathbb{P}^2_{\mathbb{R}}(0, 8)$, we can assume that $\overline{\sigma}$ is an antiprojectivity and that $\overline{\sigma}(p_1) = p_2, \overline{\sigma}(p_3) = p_4, \overline{\sigma}(p_5) = p_6, \overline{\sigma}(p_7) = p_8$. Then there are 8 couples fixed by $\overline{\sigma}$: 4 couples line-quintic $\{l_{i,i+1}, l'_{i,i+1}\}$, $i = 1, 3, 5, 7$; 4 couples cubic-cubic $\{e_{i,i+1}, e_{i+1,i}\}$, $i = 1, 3, 5, 7$; we have 8 real tritangent planes, 4 of which are of the first kind corresponding to the 8 real lines $\{l_{i,i+1}, l'_{i,i+1}\}$, $i = 1, 3, 5, 7$. $\square$

References

[1] C. Ciliberto, C. Pedrini, Annibale Comessatti and real algebraic geometry, Rend. Circ. Mat. Palermo Suppl. 36 (1994), 71–102.

[2] A. Comessatti, Fondamenti per la geometria sopra le superficie razionali dal punto di vista reale, Math. Ann. 73 (1912), 1–72.

[3] A. Comessatti, Sulla connessione delle superficie razionali reali, Ann. Mat. 23 (1914), 215–283.

[4] P. del Pezzo, Sulle superficie dell' n-esimo ordine immerse nello spazio a n dimensioni, Rend. Circ. Mat. Palermo 12 (1887), 241–271.

[5] F. Enriques, Sulle irrazionalità da cui può farsi dipendere la risoluzione di una equazione algebrica $f(x, y, z) = 0$ con funzioni razionali di due parametri, Math. Ann. 49 (1897), 1–23.

[6] F. Klein, Über Realitätsverhältnisse bei der einem beliebigen Geschlecht zugehörigen Normalkurve der ϕ, Math. Ann. 42 (1893), 1–29.

[7] J. Kollár, Real Algebraic surfaces, e-print `alg-geom/9712003`.

[8] J. Kollár, Which are the simplest algebraic varieties ?, Bull. Amer. Math. Soc. 38 (2001), 409–433.

[9] Y. I. Manin, Rational surfaces over perfect fields, Inst. Hautes Études Sci. Publ. Math. 30 (1966), 55–114.

[10] Y. I. Manin, Cubic forms, North-Holland, 1974 and 1986.

[11] J. Plücker, Theorie der algebraischen Kurven, Bonn, 1839.

[12] L. Schläfli, On the distribution of surfaces of the third order into species, Philos. Trans. Roy. Soc. London, 153 (1863), 193–241.

[13] B. Segre, The non-singular cubic surfaces, Clarendon Press, 1942.

[14] C. Segre, Studio sulle quadriche in uno spazio lineare ad un numero qualunque di dimensioni, Mem. Reale Acc. Torino 36 (1883), 3–86.

[15] C. Segre, Étude des differents surfaces du 4e ordre à conique double ou cuspidale, etc, Math. Ann. 24 (1884), 313–444.

[16] C. Segre, Le rappresentazioni reali delle forme complesse e gli enti iperalgebrici, Math. Ann. 40 (1892), 413–467.

[17] R. Silhol, Real algebraic surfaces, Lecture Notes in Math. 1392, Springer-Verlag, Berlin, 1989

[18] C. T. C. Wall, Real forms of smooth real del Pezzo surfaces, J. Reine Angew. Math. 375 (1987), 47–66.

[19] H. G. Zeuthen, Sur les différentes formes des courbes du quatrième ordre, Math. Ann. 7 (1874), 410–432.

F. Russo, Departamento de Matematica, Universidade Federal de Pernambuco, 50670-901 Recife–PE, Brasil

E-mail: `frusso@dmat.ufpe.br`

Letters of a bi-rationalist
IV. Geometry of log flips

Vyacheslav V. Shokurov *

Flops and flips first appeared in mathematics as geometrical constructions:

(1) during Fano's modification of a 3-fold cubic into a Fano 3-fold $X_{14} \subset \mathbb{P}^9$ [10, Theorem 4.6.6];

(2) Atiyah's flop: one of his first papers [3] in 1958 treated the simultaneous resolution of the surface ODP, and was the initial stimulus for Brieskorn's simultaneous resolution of Du Val singularities[1];

(3) Kulikov's perestroikas [15, Modifications in 4.2–3];

(4) Francia's flip (see Example 3 below);

(5) Reid's pagodas [18];

(6) semistable flips [28], [12], [22];

(7) Kawamata's nonsingular 4-fold flip [13];

(8) geometrical 4-fold flips [11]; and

(9) the Thaddeus principle [27], [5].

However in general, for higher dimensions, one can hardly imagine an effective and explicit *geometric* construction (for instance, a chain of certain blow-ups and blow-downs) for *flips*, even for log ones, except for very special situations with extra structures, e.g., as in (6) (8), and for moduli spaces as in (9). On the other hand, we hope that the log flips exist and this can be established in a more formal and algebraic way.

---

*Partially supported by NSF grants DMS-9800807 and DMS-0100991.

[1]Thanks to Miles Reid for this historical remark. He added also "Possibly a little later, Moishezon (and Hironaka) were using the same kind of thing to construct algebraic spaces (minischemes) that were not varieties.

However, as I said in my Old Person's View, one can trace the idea back through Zariski and Kantor and Cremona, even as far back as papers of Beltrami in 1863 and Magnus in 1837 referred to in Hilda Hudson's bibliography – these papers study the standard monoidal involution $\mathbb{P}^3 - \to \mathbb{P}^3$ given by $(x, y, z, t) \mapsto (1/x, 1/y, 1/z, 1/t)$, which flops the 6 edges of the coordinate tetrahedron".

Recently, this was done for the log flips up to dimension 4 [26, Corollary 1.8]. Since these flips were obtained without the use of any classification or concrete geometry of them, it is worthwhile in the aftermath to get some of the aforementioned geometrical facts. This is a goal of the note which we pursue in a more general situation. For the convenience of the reader, we put the list of notation and terminology at the end of the letter.

Definition. A birational transform $X- \to X^+/Z$ between two birational contractions $f: X \to Z$ and $f^+: X^+ \to Z$ of normal algebraic varieties is called a (directed) *D-quasi-flip/Z* or, shortly, *-qflip/Z*, for a Weil $\mathbb{R}$-divisor D on X, when there is a semiample/Z $\mathbb{R}$-Cartier divisor D^+ on X^+ such that $f_*^+ D^+ \sim_\mathbb{R} f_* D$.

A *D*-qflip/Z can be given in a *log form* or, shortly, in *lf*, that is, in terms of log structures on X/Z and X^+/Z, namely, there are Weil $\mathbb{R}$-divisors B and B^+ on X and X^+ respectively such that:

- $f_*^+ B^+ = f_* B$; and

- $D = K + B$ and $D^+ = K_{X^+} + B^+$.

The qflip is a *log* one if in addition:

- B and B^+ are boundaries; and

- pairs (X, B) and (X^+, B^+) are log canonical.

Note that up to an $\mathbb{R}$-linear equivalence of D and/or D^+ we can assume that $f_*^+ D^+ = f_* D$ in the definition. Then any *D*-qflip is a qflip in lf for some B and B^+ (but maybe not a log qflip). (We always take all canonical divisors K, K_{X^+}, etc. on modifications of X given by the same differential form, or by the same bi-divisor [23, Example 1.1.3].) Any *D*-flip is a *D*-qflip with D^+ as the birational transform of D. The inverse holds when X^+/Z is small and D^+ is ample/Z; for example, by Monotonicity below, for any qflip in lf, X^+/Z is small if X/Z is small, $-D = -(K+B)$ is nef/Z, and (X, B) is terminal in codimension 2, that is, $a(Y) > 1$ for any subvariety $Y \subset X$ of codimension ≥ 2 in notation below. Thus a log qflip is a log flip under the last conditions, and with ample $D^+ = K_{X^+} + B^+/Z$. Log qflips, with nonsmall X^+/Z and a boundary B^+, are naturally induced by log flips on the reduced part of B (cf. the proof of [26, Special termination 2.2]).

Even always assuming that the characteristic of base field k is 0, we expect that most of the results and statements below hold without such an assumption, e.g., the following generalizations of [14, Lemma 5-1-17] and Monotonicity [19, (2.13.3)] – our basic tools.

Lemma. *Let $X- \to X^+/Z$ be a D-qflip for an $\mathbb{R}$-Cartier divisor D on X such that:*

1. *X/Z is a D-**contraction**, that is, $-D$ is numerically ample/Z [23, Section 5], and*

2. X/Z is a nonisomorphism.

Then
$$^+c \le d + 1$$
where

- $d = d(X/Z)$ is the **minimal** dimension of the irreducible components of the exceptional locus E of X/Z; and

- ^+c is the **minimal** codimension in X^+ of the irreducible components of the **rational transform** ^+E of E in X^+/Z.

If E has the **pure** dimension d, that is, each irreducible component of E has the dimension d, then ^+c can be taken as the **maximal** codimension.

Warning 1. In general, ^+E is quite different from the exceptional locus $E^+ = E(X^+/Z)$. However $E^+ \subseteq {}^+E$ whenever $X^- \to X^+$ is an isomorphism on $X \setminus E$; for instance, the latter holds for the D-flips of D-contractions but not for all qflips.

Remark 1. The minimal dimensions and codimensions can be replaced in the dual form of the lemma by maximal ones, namely, $c \le {}^+d + 1$ in its *maximal* form. The lemma itself, in the maximal form, is dual to its symmetric statement $c \le {}^+d + 1$ in the *minimal* form holding for the same E and ^+E.

Note also that taking the minimal dimension and maximal codimension we consider only nonempty components, in particular, such (co-)dimensions are defined only for nonempty subvarieties. This explains Condition 2.

Proof. After a birational contraction of X^+/Z given by D^+ we can assume that X^+/Z is $-D^+$-contraction; this change only increases ^+c.

Then the ampleness of $p_1^*(-D) + p_2^* D^+$ on $X \times_Z X^+/Z$ and [14, Proof of Lemma 5-1-17] imply that $X \times_Z X^+$ is divisorial over Z (see also [21, Negativity 1.1]). Moreover, for each irreducible component Y of E and its rational image $^+Y \subseteq {}^+E \subset X^+$,
$$\dim Y + \dim {}^+Y \ge \dim Y \times_Z {}^+Y = \dim X - 1.$$
That gives the required inequality.

The last statement for the pure d follows from the maximal case mentioned in Remark 1. □

Monotonicity. Let $(X, B) \dashrightarrow (X^+, B^+)/Z$ be a qflip in lf with nef $-(K + B)/Z$. Then, for each prime bi-divisor P of X,
$$a(X^+, B^+, P) \ge a(X, B, P).$$
Moreover,
$$a(X^+, B^+, P) > a(X, B, P)$$

for each P with center$_X$ $P \subseteq E$ when $-(K+B)$ is numerically ample/Z; the equivalent inclusion is center$_{X^+}$ $P \subseteq {}^+E$.

Proof. As for [19, (2.13.3)]. □

Let $(X/Z, B)$ be a log pair with a boundary B such that:

(BIR) $f : X \to Z$ is a *birational* contraction which we always consider locally over some fixed point in Z; and

(WLF) the pair is a *weak log canonical Fano* contraction, that is, (X, B) is log canonical, and $-(K+B)$ is nef/Z;

it is said to be a *log canonical Fano* contraction or *log contraction* when $-(K+X)$ is numerically ample/Z. Note that $-(K+B)$ is big/Z by (BIR). The log pairs include, in particular, the birational *log contractions* of LMMP (Log Minimal Model Program), which are birational contractions X/Z of extremal faces numerically negative/Z with respect to $K+B$ [23, 5.1.1b]. However in this letter we do not touch fibred contractions [2].

The most fundamental questions in geometry concern *dimensions*. In our situation they are

- $n = \dim X$; and

- the *minimal* dimension $d = d(X/Z)$ for the exceptional locus E of X/Z.

Other more modern numerical invariants:

- the *(log) length* $l = l(X/Z, B)$ of $(X/Z, B)$, that is the minimal $-(K+B.C)$ for generic curves C in the covering families of *contracted* locus E (this is the exceptional locus whenever X/Z is birational, and $E = X$ otherwise) of X/Z; and

- the *m.l.d. (minimal log discrepancy)* $a = a(E) = a(X, B, E)$ of (X, B) in E, that is the minimal log discrepancy $a(X, B, P)$ at prime bi-divisors P having the center *in* E; the latter means that center$_X$ $P \subseteq E$.

It is known that the dimension d depends on the length [24, Theorem], and on the singularities [4, Théorème 0]. Sometimes a more subtle interaction occurs.

Example 1. Suppose that (X, B) has only canonical (terminal) singularities in codimension 2, and a curve C is an irreducible component of the exceptional locus E of projective X/Z. Then the *existence of log flips in dimension $n \geq 3$ in the formal/$\bar{k}$, or analytic category when $\bar{k} = \mathbb{C}$, implies that $(K+B.C) \geq -1$ (respectively, > -1)*. More precisely, for $n \geq 2$, we can assume just $a(C) \geq 1$ (respectively > 1). In other words, this means that the length l of the contraction is ≤ 1 (respectively < 1) whenever $d = 1$ and $a(C) \geq 1$ (respectively > 1). One can drop the existence of the log flip in dimension $n \leq 4$.

We verify that $l \leq 1$ in the canonical case; the terminal case is similar (cf. [7, Lemma 3.4]). Indeed, suppose that $l > 1$ and $a(C) \geq 1$. Then over a small neighborhood of $f(C)$ in the classical complex topology for $\bar{k} = \mathbb{C}$ (or formally over arbitrary algebraically closed $\bar{k}$), there exists a rather generic hyperplane section H/Z that intersects C transversely in a single point. In addition, changing the contraction over such a neighborhood we can assume that $E = C$. So, locally/Z, $(X/Z, B + H)$ is again a log pair under (BIR) and (WLF). Since the exceptional locus $E = C$ is a *proper* subvariety in X, the new m.l.d. $a := a(X, B + H, E) = a(X, B, E) \geq 1$, too.

On the other hand, according to our assumptions, there exists a flip $X - \to X^+/Z$ with respect to $K + B + H$. Actually it is also the flip for $K + B$ and is the $(-H)$-flip [26, Corollary 3.4]. So, the flip transform ^+H of H is the birational transform H^+ of H and numerically negative/Z on the exceptional locus E^+ of the flipped contraction X^+/Z. Hence $E^+ \subseteq {}^+E \subset H^+$, and $E^+ = {}^+E$ unless $n = 2$ with $E^+ = \emptyset$ (cf. Warning 1). Moreover, ^+E has the minimal codimension $c^+ = 2$ by the lemma, and the m.l.d. $^+a = a(X^+, B^+ + H^+, {}^+E) \leq 1$; it is enough to establish the latter for $n = 2$, when it is well-known [23, Example 4.2.1]. But this contradicts to the assumption $a \geq 1$ because $a < {}^+a \leq 1$ by Monotonicity.

Remark 2. In the last paragraph we proved a little bit more. Let $(X/Z, B)$ be a purely log terminal pair with the reduced divisor H. Then each flip of $(X/Z, B)$ with $a \geq 1$ gives the flip on $(H/f(H), B_H)$ where B_H is given by the adjunction. (So, then $d \geq 2$ by the lemma when $H \sim_{\mathbb{R}} -h(K + B - H)$, with $h \in \mathbb{R}$, is numerically ample and $\not\equiv 0/Z$ as in the example.) Therefore, for a purely log terminal and canonical in codimension 2 pair $(X/T, B)$, LMMP with only *flipping* contractions $(X/Z/T, B)$ induces LMMP on $(H/f(H), B_H)$ (cf. the proof of [26, Special Termination 2.3]). Moreover, the same holds for any chain of *birational* contractions in LMMP for $(X/T, B)$ unless one of them contracts a component of Supp B.

Advertisement 1. A generalization and applications of the improvement in Remark 2 will be treated in one of the following letters.

In the 3-dimensional terminal case with $B = 0$, Example 1 implies the Benveniste result [4, Théorème 0][2]; now without linear systems arguments. But it looks difficult to apply his arguments in higher dimensions; even in dimension 4. Deformation arguments in any dimension gives the weaker inequality $l < 2$ in Example 1. In general, $d > l/2$ even for more difficult singularities [24, Theorem]. On the other hand, we expect

Conjecture 1. Under conditions (BIR) and (WLF), suppose that X/Z is *projective*. Then $d \geq a - 1$ ($>$ in the *log Fano* case), or, equivalently, $d \geq \lceil a - 1 \rceil$.

---

[2] In general the strict inequality in the theorem fails in presence of canonical singularities along curve C, e.g., when C is obtained by the contraction along the second factor of a surface $\mathbb{P}^1 \times \mathbb{P}^1$ in a nonsingular 3-fold X with the normal bundle $\pi_1^* \mathcal{O}_{\mathbb{P}^1}(-1) \oplus \pi_2^* \mathcal{O}_{\mathbb{P}^1}(-2)$ where $\pi_i : \mathbb{P}^1 \times \mathbb{P}^1 \to \mathbb{P}^1$ is the projection on i-th factor.

In addition, if $d = \lceil a - 1 \rceil$, $E \neq \emptyset$, and $(X/Z, B)$ is a *log contraction*, that is, X/Z is a *D*-contraction for $D = K + B$, then, for any log qflip $X- \to X^+/Z$, the transform ^+E satisfies the following properties:

(CDM) $^+c = d + 1$;

(NSN) each irreducible component of ^+E of the minimal codimension $d + 1$ is *nonsingular* as a scheme point of X^+; and

(PDM) if d is the *pure dimension*, then ^+E is also of the pure codimension $d + 1$.

Moreover, for the log flip $X- \to X^+/Z$, $^+E = E^+$ is the exceptional locus of X^+/Z.

It is enough to establish the conjecture when $E \neq \emptyset$. Otherwise we put $d = -\infty$ and $a = -\infty$ as it used to be.

Example 2. In particular, if X is nonsingular and $B = 0$ then $a = n - d$ and Conjecture 1 implies that $d \geq (n - 1)/2$ (respectively $d > (n - 1)/2$. This is Wiśniewski's inequality [29, Theorem 1.1, p. 147] in a *single* formula).

Perhaps there is a relation between the length l and the m.l.d. a. I am not sure. But definitely, $l/2 \geq a - 1$ does not hold always.

Example 3. Francia's flip corresponds to the contraction, which is obtained from the relative model Y after contraction of its plane E in [8, Section 2], and it has $a = 3/2$ and $l = 1/2$. So, the above inequality $d > l/2$ is less sharp than that of in Conjecture 1.

Conjecture 1 can be derived from the conjecture on existence of log flips [23, Conjecture 5.1.2] and another conjecture on the m.l.d. [20, Problem 5a]:

Conjecture 2. For any (scheme) point $P \in X$ the m.l.d. $a(X, B, P) \leq \text{codim } P$, with = holds only when P is nonsingular in X and $B = 0$ near P. Taking hyperplane sections, it is enough to prove that $a(X, B, P) \leq \dim X$ and the = case for closed points.

Moreover, the nonsingularity of X still holds when we replace codim P by codim $P - \varepsilon$ with any $0 \leq \varepsilon < 1$. Equivalently, $\lceil a(X, B, P) \rceil \leq \text{codim } P$, where = holds only when P is nonsingular in X.

Warning 2. In the conjecture we still assume that B is a boundary!

Conjecture 2 was proven for codim $P \leq 3$ (after a $\mathbb{Q}$-factorialization, follows from [23, Corollary 3.3] with a final step by Markushevich [16, Theorem 0.1]). The stronger form with $\lceil \ \rceil$ follows from the weaker one, the covering trick and LMMP. From LMMP, we need only the existence of $\mathbb{Q}$-factorializations. (It is expected that the next case with $a(X, 0, P) = \text{codim } P - 1$ corresponds to the higher dimensional cDV singularities; cf. [1, Proposition 3.3].)

Since the m.l.d. measures the singularity, it is natural to expect that it decreases under the specialization that is stated in Ambro's conjecture [1, Conjecture 0.2]. It implies Conjecture 2, and is proved up to dimension 3 [1, Theorem 0.1] and for toric varieties by [1, Theorem 4.1]. The former gives again codim $P \leq 3$ and the latter gives the toric case.

Example 4. Conjecture 2 holds for toric varieties with invariant B.

Theorem. *The existence of **log flips in dimension n** and Conjecture 2 in dimension m implies Conjecture 1 in dimension n for any $d \leq m - 1$.*

Actually, log flips can be weaken to log qflips. More precisely, it is enough to have the existence of log qflips for log contractions $(X/Z, B)$ instead of log flips.

Proof. According to the closing remark in Conjecture 1, we can assume that $E \neq \emptyset$. In particular, $n \geq 2$.

We can suppose also that $a > 0$. Otherwise, $a = 0$ and $d \geq 1 > -1 = a - 1 = \lceil a - 1 \rceil$.

Since X/Z is projective, after perturbation of B we can assume that X/Z is a log contraction. Note that taking quite a small perturbation, which increases B and decreases a, we preserve $\lceil a - 1 \rceil$. If the original B gave the log contraction we do not change B.

Now we can apply the lemma. Let $X \dashrightarrow X^+/Z$ be a log qflip of $(X/Z, B)$ with a boundary B^+ on X^+. Then by the lemma there is an irreducible component $Y \subseteq {}^+E$ such that codim $Y \leq d + 1$. Hence Conjecture 2 implies that $a(X^+, B^+, Y) \leq d + 1$, and Monotonicity

$$a = a(X, B, E) \leq a(X^+, B^+, Y) \leq d + 1$$

gives the required inequality.

Under additional assumptions in Conjecture 1, if codim $Y \leq d$ then, according to the same inequalities, $a \leq d$ and $\lceil a - 1 \rceil \leq d - 1 < d$. This proves (CDM) because, according to these assumptions, $d = \lceil a - 1 \rceil$ or $\lceil a \rceil = \lceil a(X^+, B^+, Y) \rceil = d + 1$. So, (NSN) follows from Conjecture 2. The last statement in the lemma implies (PDM). If $X \dashrightarrow X^+/Z$ is the log flip, then $E^+ = {}^+E$ (cf. Warning 1). Indeed, for small X/Z, the inverse transform is the anti-flip. Otherwise $a \leq 1, d = n - 1 = \lceil u - 1 \rceil \leq 0$, and $n \leq 1$, which contradicts to our assumptions. □

Corollary 1. *Conjecture 1 holds for the toric contractions X/Z, with only canonical singularities and $B = 0$, which are numerically negative with respect to K.*

A more general case we consider in one of our future letters (see Advertisement 2). It would be interesting to know whether the combinatorics behind this statement were known. In particular, how important is the projectivity in this statement (cf. Question 1)?

Proof. Immediate by Example 4 and the existence of toric flips [17, Theorem 0.2]. We do not need to perturb $B = 0$ since K itself is numerically negative. □

Corollary 2. *The theorem holds without Conjecture 2 for all $d \leq m = 2$.*

Proof. Immediate by [23, Corollary 3.3] and [16, Theorem 0.1]. □

The main inequality $d \geq a - 1$ in Conjecture 1 can be established for rather high dimensions d without flips.

Example 5. For all $d \geq (n - 1)/2$, Conjecture 1 follows form Conjecture 2 for $m \leq (n + 1)/2$; in particular, up to $n = 6$ we can drop Conjecture 2. Indeed, let Y be an irreducible component of E then $a \leq \text{codim } Y \leq (n + 1)/2 \leq d + 1$. Moreover, $d = \lceil a - 1 \rceil$ only if $d = (n - 1)/2$, the dimension is pure and E is nonsingular in each of its irreducible components as a scheme point; the additional statements (CDM), (NSN) and (PDM) in Conjecture 1 hold by the lemma and our hypothesis (cf. the proof of the theorem). Otherwise $d \geq n/2$ is integral, and $\lceil a \rceil \leq n/2 = d < d + 1$.

Corollary 3. *In dimensions $n \leq 6$ one can drop Conjecture 2 in the theorem.*

Proof. In Example 5 it was proven for all $d \geq (n - 1)/2$ because $m = (n + 1)/2 \leq (6 + 1)/2 = 7/2$. For $d \leq 2 < (n - 1)/2 \leq (6 - 1)/2 = 5/2$, the corollary was proven in Corollary 2. □

Corollary 4. *In dimension $n \leq 4$ one can drop both conjectural hypotheses in the theorem, namely, the existence of log flip and Conjecture 2.*

Proof. Immediate by Corollary 3 because the log flips exists. The latter was proven in [26, Corollary 1.8] when (X, B) is Kawamata log terminal. The other log flips also exist due to [26, Special termination 2.3] and the local log semiampleness (cf. [23, Conjecture 2.6], and see the proof of [23, Log Flip Theorem 6.13]).

Actually, by Example 5 it is enough to consider the case with pure $d = 1$. Then $a \leq 2$. Indeed, otherwise after a strict log terminal resolution we can assume that each reduced component H in B is $\mathbb{Q}$-Cartier, and intersects properly the curves C of E. Indeed, we can construct the log terminal resolution/X using Kawamata log terminal flips and [26, Special termination 2.3] as in [21, Reduction 6.5] (cf. the proof of [26, Proposition 10.6]). This is not an isomorphism only over a finite set of points in C because X is nonsingular in the generic points of C by our assumption. This is impossible by the Kawamata log terminal case (cf. Example 1) because C is contractible at least in the formal or analytic category, $a > 2$ can only be increased after the construction, and now the flip in C exists (see [26, Remark 1.12]). □

Example 6 (Minimal contractions). Under conditions (BIR) and (WLF), suppose that X/Z is projective and dimension $d = \lceil a - 1 \rceil$ is *pure* (or maximal). Then a log pair $(X/Z, B)$ will be called a *minimal (log) contraction* (respectively, when $B \neq 0$).

In particular, a minimal log contraction with $d = a - 1$ is possible only when $(X/Z, B)$ is a 0-log pair, that is, $K + B \equiv 0/Z$ in our situation. This follows from a more subtle version of $>$ in Monotonicity under $\not\equiv 0/Z$ by [21, Negativity 1.1], or LMMP including the log termination and our theorem. If X/Z is projective under hypotheses of the theorem (cf. its proof) there exists a nonidentical directed flop

$X \dashrightarrow X^+/Z$ with the transform ^+E of pure codimension and satisfying the properties (CDM), (NSN), and (PDM) of Conjecture 1 (perturb B by D as a negative to a polarization). In particular, such a flop is unique when X is $\mathbb{Q}$-factorial and X/Z is *formally extremal*, that is, the *formal* (in the formal or analytic category) relative Picard number of X/Z is 1. Moreover, for dimension $d = (n-1)/2 = a - 1$, it is expected that $(X/Z, B)$ is nonsingular in the irreducible components of E as scheme points, $f(E)$ is a *closed point* (take a general hyperplane section of $f(E)$ when $\dim f(E) \geq 1$), and each directed flop (qflop) $X \dashrightarrow X^+/Z$ should be "symmetric", that is, the exceptional $E^+ = {}^+E$ of pure dimension $d = (n-1)/2$ with the nonsingular irreducible components of E^+ as scheme points (cf. Questions 1 and 2 below). The same follows from LMMP for any projective flop $(X^+/Z, B)$ of $(X/Z, B)$ as a composition of directed ones.

For instance, if X is nonsingular and $B = 0$, then it is expected that $d \geq (n-1)/2$ (cf. Example 2). So, by Conjecture 2 $(X/Z, 0)$ is minimal only when $d = a - 1 = (n-1)/2$ and the dimension is pure. Again by Conjecture 2 it is expected that any directed or/and projective flop X^+/Z is nonsingular with the same number of irreducible components of E^+ as for E (the number of exceptional prime divisors over E or E^+ with the log discrepancy $a = (n+1)/2$) whenever X^+/Z is *formally $\mathbb{Q}$-factorial* (in the formal or analytic category; cf. Question 2 below). The latter should hold for any nontrivial flop when X/Z is formally extremal. In this case the flop is unique when it exists and will be called *minimal formally extremal*. The LMMP implies that each of the directed and projective flops to X^+/Z is a composition of (formally) extremal contractions and flops; only such flops are enough when both X and X^+/Z are (formally) $\mathbb{Q}$-factorial.

An elementary example with nonsingular X, $B = 0$, and $E = \mathbb{P}^d$ belongs to the toric geometry. Its invariant divisors are $(n+1)/2 = d + 1$ numerically negative D_i^- (intersecting E up to the linear equivalence by a negative to its hyperplane; their intersection in X is E itself) and $(n-1)/2 + 1 = d + 1$ numerically positive D_j^+ (intersecting E in hyperplane sections in a general position – an anti-canonical divisor in total). The construction of such a contraction X/Z see in [9, Example 3.12.2(iii)] with $r = (n+1)/2 = d+1$, and $a_1 = \cdots = a_r = 1$. This contraction and its flop are *formally or analytically toric* (cf. a conjecture after the proof of [25, Theorem 6.4]) and *semistable* (cf. (6)); in this situation the latter means that there exists a nonsingular hypersurface H passing through E and $\equiv 0/Z$. Thus they induce a flop on this hypersurface as in Example 7 below with E/pt. Moreover, it is a symplectic one (cf. Question 4). Each toric contraction $(X/Z, B)$ is algebraically (analytically or formally) log i-symplectic for any $0 \leq i \leq n = \dim X$ with B as invariant divisor (take a general linear combination of invariant i-forms $\wedge^i (dz_j)/z_j$). In our situation, we can take 3-form (3-symplectic structure)

$$\frac{d(f_1^- f_1^+)}{f_1^- f_1^+} \wedge \left(\sum_{2 \leq i,j \leq d+1} \frac{d(f_i^-) d(f_j^+)}{f_i^- f_j^+} \right)$$

where f_i^- and f_j^+ are respectively local sections/Z of $\mathcal{O}_X(-D_i^-)$ and $\mathcal{O}_X(-D_j^+)$. Then after a generic perturbation of products $f_1^- f_1^+$ and $f_i^- f_j^+$ into g_{11}, with $H = \{g_{11} = 0\}$ vanishing on $\mathbb{P}^d$, and g_{ij} (the latter does not vanishing on $\mathbb{P}^d$ at all), the 3-form

$$\frac{dg_{11}}{g_{11}} \wedge \left(\sum_{2 \leq i,j \leq d+1} \frac{d(f_i^-)d(f_j^+)}{g_{ij}} \right)$$

induces by its residue a symplectic 2-form on H. Such flops will be called *induced toric*.

Note that, according to A. Borisov (a private communication) the *top* m.l.d. in dimension n for the toric *isolated* $\mathbb{Q}$-factorial singularities P is $a(P) = n/2 < (n+1)/2$. The latter is the mld for the minimal contraction of nonsingular X with $B = 0$. Actually, *for any toric isolated $\mathbb{Q}$-Gorenstein singularity P, $a(P) \leq (n+1)/2$, and $a(P) = (n+1)/2$ only for the above toric contraction.* For such a toric singularity $P \in Z$, a toric projective $\mathbb{Q}$-factorialization X/Z is small birational, with the exceptional locus $E = f^{-1}P$, and crepant; it exists by [17, Theorem 0.2] because we can assume that $a(P) = a(Z, 0, P) > 1$ (otherwise $a(P) \leq 1 < (n+1)/2$ for $n \geq 2$). Thus $a(P) = a(Z, 0, P) = a = a(X, 0, E)$. We can suppose also that X is nonsingular and $d \geq 1$ since otherwise, by the above $\mathbb{Q}$-factorial case, $a \leq n/2$. If, in addition, it is extremal then, by Example 2, the arguments in the proof of our Theorem, Example 4, and the existence of any D-flip (flop) [26, Example 3.5.1], $d \geq (n-1)/2$ (cf. Corollary 1), and $a \leq n - d \leq (n+1)/2$. The case $a = (n+1)/2$ is possible only when $d = (n-1)/2$, and this is a minimal contraction. Since the latter is toric and extremal, E is a projective nonsingular toric variety with only ample invariant divisors, that is, $E = \mathbb{P}^d$ (the Fano variety of the maximal index $d+1$), and this is the above contraction. Suppose now that the relative Picard number of X/Z is 2. Then X/Z has two extremal contractions/Z (it is known in the toric geometry, and follows from the Cone Theorem [14, Theorem 4-2-1] after a perturbation of $B = 0$). They are small and, by the induction on dimension n, give $a \leq n/2$ if one of them does not contract the exceptional locus into a point. Thus, if $a \geq (n+1)/2$, we can assume that both are again the above contractions of $E_1 = \mathbb{P}^d \neq E_2 = \mathbb{P}^d$ with $d = (n-1)/2$, and, by the extremal properties of both contractions $E_1 \cap E_2$ is at most a point. Since E is connected and all numerically negative invariant divisors on E_i/Z give E_i in intersection, there is one of them, say D_i, which is numerically positive/Z on the other E_{3-i}. According to the previous description $D_1 + D_2 \equiv 0/Z$, and their intersection $D_1 \cap D_2$ gives the induction in dimension n. However, such a contraction is impossible when $n = 3$. Indeed, for $n = 3$, there is no invariant divisor D which is numerically negative/Z on two curves in E, because these curves form a connected exceptional sublocus of E in D and each of its components is a (-1)-curve. In this situation, each E_i is an intersection of two pairwise distinct invariant divisors. So, $E = E_1 \cup E_2$ because each invariant divisor passing through a curve of E is negative on this curve only. This gives a contradiction because 4 invariant divisors passing through one of curves E_i pass through the intersection point $E_1 \cap E_2$. In particular, we prove that $a \leq n/2$ when the relative Picard number is 2. The same holds for higher Picard numbers by the last case and the induction when we consider a contraction of 2 dimensional face of the Kleiman-Mori cone for X/Z. This completes our proof and explains what are the minimal toric contractions.[3]

So, in dimension $n = 3$, Atiyah's flops (2-3) and their contractions are the only *nonsingular toric minimal* ones.

Other elementary examples of minimal semistable flops are Reid's pagodas and their possible higher dimensional generalizations (cf. La Torre Pendante [11, 8.12]). For $n \geq 5$, the induced flop on H can be different from the toric induced (or symplectic) one. All nonsingular minimal flops and their contractions for 3-fold are semistable (that is, their contractions Z have the cDV type). They are *absolutely* extremal, that is, formally/$\bar{k}$ or analytically when $\bar{k} = \mathbb{C}$, if and only if $E = \mathbb{P}^1$ is irreducible, and the flops are pagodas – (1-3) (5) above. Other minimal 3-fold flops are their composite over $\bar{k}$, e.g., modifications of Kulikov's model for K3 surfaces semistable degenerations.

Question 1. Does Conjecture 1 hold for nonprojective X/Z? Or at least in the nonsingular case of Example 6? It is interesting and nontrivial for $n \geq 4$ because in dimension 3 each small X/Z is at least formally or analytically projective.

Question 2 (cf. Example 6). What do minimal contractions of nonsingular X with $n \geq 5$ look like? Their flops? Absolutely extremal? Are they still semistable? Their combinatorics? Is E irreducible and normal ($= \mathbb{P}^d$) for absolutely extremal contraction X/Z?

Example 7 (Almost minimal contractions). Under conditions (BIR) and (WLF), the next important class having pure $d = a$ is *almost minimal*.

Suppose also that X is nonsingular and $B = 0$ then, for such a contraction, $d = a = n/2$, and the dimension is pure, but either 0-log or non 0-log pairs $(X/Z, 0)$ are possible, e.g., Mukai's flop and Kawamata's flip for $n = 4$. By the theorem it is expected that $f(E) \leq 1$, and $f(E)$ is a nonsingular curve in $P \in f(E)$ only when $(X/Z, 0)$ is locally over Z a 0-log pair with absolutely extremal X/Z near P; then it is expected to be a minimal nonsingular contraction over the transversal hyperplane sections of $f(E)$ through P. The directed or projective flop is fibred. In particular, this should be a nonsingular (nonsymplectic even locally over the contraction, for example, with $\mathbb{P}^{d-1}$-fibration because then the lines have $2d - 4 + 1 = 2d - 3 = n - 3 < n - 2$ parameters and this contradicts to Ran's estimation [6, Lemma 2.3]) flop.

The case with the point $f(E)$ is more complicated. First, suppose that $(X/Z, 0)$ is a non 0-log pair (this is the *minimal* case *among* the *non 0-contractions*; cf. Question 3 below). Then in dimension 4 the MMP implies that the flip $X - \to X^+$ exists, and X^+/Z is nonsingular with a curve $^+E = E^+$. The absolutely extremal components

---
[3] A. Borisov knows a pure combinatorial proof of this fact.

of such flips are Kawamata's flips (7). One can hope for a similar picture in higher dimensions.

Now suppose that $f(E) = pt.$ is a closed point, and $(X/Z, 0)$ is a 0-log pair. Such contractions appear as the toric ones induced in Example 6 and also as the *nonsingular* birational *symplectic* contractions with pure dimension $\leq n/2$ for E (then the pure dimension of E is $n/2$ and as we know E/pt). In these cases, for $n \geq 4$, one can expect the existence of a nontrivial nonsingular "symmetric" direct and/or projective flop. It should have $E^+ = {}^+E$ of the pure dimension $d = n/2$ with the same number of irreducible components as E (the number exceptional prime divisors over E or over E^+ with the log discrepancy $a = n/2$) in the formal or analytic case. For instance, conjecturally the Mukai flop is the only nonsingular almost minimal symplectic flop in dimension $n = 4$ (cf. Question 4 below).

But there are also singular flops. For instance, in dimension 4, an extremal flop $X{-}{-}{\to} X^+/Z$ which transforms $E = \mathbb{P}^2$ into ${}^+E = E^+ = \mathbb{P}^1$ with a single simple singularity having the m.l.d. $= 2$.

In dimension 2, the almost minimal contractions $(X/Z, 0)$, which are 0-log pairs, are the minimal resolutions of Du Val singularities $P = f(E) \in Z$. They are always symplectic, but toric only of type $\mathbb{A}_*$, and, by Example 6, induced toric only of type $\mathbb{A}_1$ (correspond to the ordinary double singularity).

Question 3. Is the nonsingular non 0-log pair case with $d = n/2$ in higher dimensions similar to dimension $n = 4$? In particular, is the flip X^+ always nonsingular? $E = \mathbb{P}^d$ for formally or analytically extremal contractions?

Question 4. Classify the nonsingular birational almost minimal 0-log pairs $(X/Z, 0)$ with $f(E) = pt.$ and their flops. In particular, such nonsingular symplectic flops. Is any such flop induced toric when $n \geq 4$ (see Example 6)?

Example 8 (More minimal contractions). A *nonidentical* contraction with the m.l.d. $a \leq 1$ is never minimal. For the next (terminal) segment $a \in (1, 2]$, the contraction is minimal only when $d = 1$ and E has only curve components C. Moreover, X/Z is *small* for $n \geq 3$. In particular, in dimension 3, $E = C$ is a curve.

Such contractions for terminal 3-folds with $B = 0$ appear in the MMP as 0-log pairs when $a = 2$ and non 0-log pairs with $a = (m+1)/m$ where $m \geq 2$ is the index of K in $E = C$. So, $a \leq 3/2$ in the latter case. In addition, Francia's flip of Example 3 corresponds to (the only one with difficulty 1) an extremal terminal minimal log contraction $(X/Z, 0)$ with $a = 3/2$ or of the index 2; it also has the maximal length $l = 1/2$ among the index 2 contractions (cf. Example 1). For 3-folds with locally complete intersection singularities, we have only 0-log pairs with terminal Gorenstein singularities and their flops that are well-known.

Flops similar to the latter in dimension $n = 4$ are still not classified (even the absolutely extremal amongst them; cf. Question 4). However flips of some (maybe nonminimal) terminal Gorenstein contractions are explicitly known (8); they have $d = 2$ by Example 1.

Advertisement 2. An opposite class of *maximal* contractions and its application to the termination of log flips will be given in one of our future letters.

List of notation and terminology

$a = a(E) = a(X, B, E)$, the m.l.d. of (X, B) *in* the exceptional locus E

$a(Y) = a(X, B, Y)$, the *m.l.d.* (*minimal log discrepancy*) of (X, B) *in* subvariety Y, that is, the minimal log discrepancy $a(X, B, P)$ at the prime bi-divisors P having the center *in* Y; the latter means that $\text{center}_X P \subseteq Y$; this assumes that $Y \neq \emptyset$ (otherwise we can put $a(Y) = -\infty$; cf. Conjecture 1)

$a(X, B, P)$, for a prime bi-divisor P (prime divisors on some model of X [23, p. 2668]), the log discrepancy of (X, B) or $K + B$ at P [21, p. 98]; P is considered here as its general or scheme point but not as a subvariety

B, a Weil $\mathbb{R}$-divisor on X; usually a boundary (except for Monotonicity), that is, all its multiplicities $0 \leq b_i \leq 1$, and $K + B$ is $\mathbb{R}$-Cartier

B^+, a Weil $\mathbb{R}$-divisor on X^+ such that $f_*^+ B^+ = f_* B$; usually a boundary (except for Monotonicity); for the log flips, the birational transform of B

(BIR), the condition on p. 316 which we assume afterwards, e.g., in Conjecture 1

^+c, the minimal codimension in X^+ of the irreducible components of the rational transform ^+E of E in X^+/Z; the codimension is *pure* when all the irreducible components are of the same codimension

$\text{center}_X P$, for a prime bi-divisor P (prime divisors on some model of X [23, p. 2668]), its center in X [23, p. 2669]; P is considered here as a subvariety

(CDM), the property on p. 318 under additional assumptions in Conjecture 1

codim P, for a scheme point $P \in X$, its codimension $\dim X - \dim P$ in X, e.g., codim $P = \dim X$ if and only if P is a closed point in X

$d = d(X/Z)$, the minimal dimension of the irreducible components of the exceptional locus E of X/Z; this assumes that $E \neq \emptyset$ (otherwise we can put $d = -\infty$; cf. Conjecture 1); the dimension is *pure* when all the irreducible components are of the same dimension

D, a Weil $\mathbb{R}$-divisor D on X

D^+, a semiample/Z $\mathbb{R}$-Cartier divisor on X^+ [23, Definition 2.5] such that $f_*^+ D^+ \sim_\mathbb{R} f_* D$ (see Definition on p. 314); for the log flips, D^+ is the birational transform of D [21, p. 98]

$E = E(X/Z)$, the exceptional locus E of X/Z, that is, the union of contractible curves

$E^+ = E(X^+/Z)$, the exceptional locus of X^+/Z; thus E means here to be exceptional; in general, ^+E is quite different from the rational transform ^+E

^+E, the *rational* or *complete birational transform* of E in X^+/Z (see ^+Y)

$f : X \to Z$, a birational contraction of normal algebraic varieties over k; sometimes we use such contractions in the analytic or formal category (cf. Example 1), and most of the statements work in them

$f^+ : X^+ \to Z$, its qflip or log flip

K, K_{X^+}, canonical divisors respectively on X and X^+ given by the same differential form, or by the same bi-divisor [23, Example 1.1.3]

$l = l(X/Z, B)$, the *(log) length* of log pair $(X/Z, B)$; see p. 316

LMMP, the log minimal model program and its conjectures [23, 5.1]

$n = \dim X$, the dimension of X

(NSN), the property on p. 318 under additional assumptions in Conjecture 1

(PDM), the property on p. 318 under additional assumptions in Conjecture 1

pt., a closed point

(WLF), the condition on p. 316 which we assume afterwards, e.g., in Conjecture 1

^+Y, the *rational* or *complete birational transform* of Y in X^+/Z; it is defined for any subvariety $Y \subseteq X$ and any rational map $g : X-\to Y$ as $g(Y) = \psi \circ \phi^{-1}(Y)$, where

$$g = \psi \circ \phi^{-1} : X \xleftarrow{\phi} W \xrightarrow{\psi} Y$$

with a birational contraction ϕ, and independent on the decomposition

$X-\to X^+/Z$, either a D-qflip$/Z$, or log qflip (see Definition on p. 314), or a log flip [23, p. 2684]

$(X, B)-\to (X^+, B^+)/Z$, in Monotonicity, a qflip in lf (see Definition on p. 314) with possibly nonboundaries B and B^+

$(X/Z, B)$, a log pair which usually satisfies (BIR) and (WLF) on p. 316

a 0-log pair is a log pair $(X/Z, B)$ such that (X, B) is log canonical and $K + B \equiv 0/Z$ (cf. [26, Remark 3.27, (2)])

$\sim_\mathbb{R}$, the $\mathbb{R}$-linear equivalence [23, Definition 2.5]

References

[1] Ambro, F., On minimal log discrepancies, Math. Res. Letters 6 (1999), 573–580.

[2] Andreatta, M., Wiśniewski, J., A view on contractions of higher dimensional varieties, Proc. Sympos. Pure Math. 62 (1997), 153–183.

[3] Atiyah, M. F., On analytic surfaces with double points, Proc. Roy. Soc. London. Ser. A 247 (1958), 237–244.

[4] Benveniste, X., Sur les cône des 1-cycles effectifs en dimension 3, Math. Ann. 272 (1985), 257–265.

[5] Boden, H., Hu, Y., Variations of moduli of parabolic bundles, Math. Ann. 301 (1995), 539–559.

[6] Burns, D., Yi Hu, Tie Luo, HyperKähler manifolds and birational transformations in dimension 4, preprint.

[7] Cheltsov, I., Park, J., Generalized Eckardt points, submitted for publication, e-print: Math.AG/0003121.

[8] Francia, P., Some remarks on minimal models. I, Compositio Math. 40 (1980), 301–313.

[9] Iskovskikh, V. A., Birational rigidity of Fano hypersurfaces in the framework of Mori theory, Russian Math. Surveys 56 (2001), 207–291.

[10] Iskovskikh, V. A., Prokhorov Yu.G., Fano varieties, Encyclopaedia Math. Sci. 47, Springer-Verlag, Berlin 1999.

[11] Kachi, Y., Flips from 4-folds with isolated complete intersection singularities, Amer. J. Math. 120 (1998), 43–102.

[12] Kawamata, Y., Crepant blowing-up of 3-dimensional canonical singularities and its application to degenerations of surfaces, Ann. of Math. 127 (1988), 93–163.

[13] Kawamata, Y., Small contractions of four dimensional algebraic manifolds, Math. Ann. 284 (1989), 595–600.

[14] Kawamata, Y., Matsuda, K., Matsuki K., Introduction to the Minimal Model Problem, in: Algebraic Geometry, Sendai 1985 (T. Oda, ed.), Adv. Stud. Pure Math. 10, Kinokuniya Book Store/North Holland, Tokyo/Amsterdam 1987, 283–360.

[15] Kulikov, Vik. S., Degenerations of K3 surfaces and Enriques surfaces, Math. SSSR Izv. 11 (1977), 957–989

[16] Markushevich, D., Minimal discrepancy for a terminal cDV singularity is 1, J. Math. Sci. Univ. Tokyo 3 (1996), 445-456.

[17] Reid, M., Decomposition of toric morphisms, in: Arithmetic and Geometry II (M. Artin and J. Tate, eds.), Progr. Math. 36, Birkhäuser, Basel 1983, 395–418.

[18] Reid, M., Minimal models of canonical 3-folds, in: Algebraic Varieties and Analytic Varieties (S. Iitaka, ed.), Adv. Stud. Pure Math. 1, Kinokuniya Book Store/North Holland, Tokyo/Amsterdam 1983, 131–180.

[19] Shokurov, V. V., The nonvanishing theorem, Math. USSR Izv. 26 (1986), 591–604.

[20] Shokurov, V. V., Problems about Fano varieties, in: Birational Geometry of Algebraic Varieties: Open problems. The XXIIIrd International Symposium, Division of Mathematics, The Taniguchi Foundation, August 22 – August 27, 1988, 30–32.

[21] Shokurov, V. V., 3-fold log flips, Russian Acad. Sci. Izv. Math. 40 (1993), 95–202.

[22] Shokurov, V. V., Semistable 3-fold flips, Russian Acad. Sci. Izv. Math. 42 (1994), 371–425.

[23] Shokurov, V. V., 3-fold log models, J. Math. Sci. 81 (1996), 2667–2699.

[24] Shokurov, V. V., Anticanonical boundedness for curves, appendix to Nikulin, V. V., The diagram method for 3-folds and its application to the Kähler cone and Picard number of Calabi-Yau 3-folds, in: Higher-dimensional complex varieties, Trento 1994 (M. Andreatta, T. Peternell, eds.), Walter de Gruyter, Berlin 1996, 321–328.

[25] Shokurov, V. V., Complements on surfaces, J. Math. Sci. 102 (2000), 3876–3932.

[26] Shokurov, V. V., Prelimiting flips, preprint, Baltimore–Moscow (available on http://www.maths.warwick.ac.uk/ miles/Unpub/Shok/pl.ps, 2002, 235pp.

[27] Thaddeus, M., Geometric invariant theory and flips, J. Amer. Math. Soc. 9 (1996), 691–723.

[28] Tsunoda, S., Degenerations of surfaces, Algebraic Geometry, Sendai 1985 (T. Oda, ed.), Adv. Stud. Pure Math. 10, Kinokuniya Book Store/North Holland, Tokyo/Amsterdam 1987, 755–764.

[29] Wiśniewski, J., On contractions of extremal rays of Fano manifolds, J. Reine. Angew. Math. 417 (1991), 141–157.

V. V. Shokurov, Department of Mathematics, Johns Hopkins University, Baltimore, MD 21218, U.S.A.

E-mail: shokurov@math.jhu.edu

A method for tracking singular paths with application to the numerical irreducible decomposition

Andrew J. Sommese, Jan Verschelde and Charles W. Wampler

Abstract. In the numerical treatment of solution sets of polynomial systems, methods for sampling and tracking a path on a solution component are fundamental. For example, in the numerical irreducible decomposition of a solution set for a polynomial system, one first obtains a "witness point set" containing generic points on all the irreducible components and then these points are grouped via numerical exploration of the components by path tracking from these points. A numerical difficulty arises when a component has multiplicity greater than one, because then all points on the component are singular. This paper overcomes this difficulty using an embedding of the polynomial system in a family of systems such that in the neighborhood of the original system each point on a higher multiplicity solution component is approached by a cluster of nonsingular points. In the case of the numerical irreducible decomposition, this embedding can be the same embedding that one uses to generate the witness point set. In handling the case of higher multiplicities, this paper, in concert with the methods we previously proposed to decompose reduced solution components, provides a complete algorithm for the numerical irreducible decomposition. The method is applicable to tracking singular paths in other contexts as well.

2000 Mathematics Subject Classification:Primary 65H10, 14Q99; Secondary 68W30

Key words and phrases. Component of solutions, embedding, interpolation, irreducible component, irreducible decomposition, generic point, homotopy continuation, numerical algebraic geometry, multiplicity, path following, polynomial system, sampling, singularity.

1. Introduction

Our main motivation for developing a method to track singular paths is to complete our numerical algorithms for computing the irreducible decomposition of the solution set of a polynomial system of equations. Conventional path-tracking methods are limited to nonsingular paths; they are sufficient for handling solution components of multiplicity one, but cannot cope with components having multiplicity greater than one. With the addition of a path-tracking algorithm applicable to these components, they can be treated in all other respects in the same manner as the components of multiplicity one. Moreover, the new path tracker can be applied in other similar contexts as well, even those in which the path to be followed is defined by functions that are merely analytic rather than polynomial.

Since it is our main motivation, we begin with a brief description of the irreducible decomposition problem. Let

$$f(x) := \begin{bmatrix} f_1(x_1, \ldots, x_N) \\ \vdots \\ f_n(x_1, \ldots, x_N) \end{bmatrix}, \qquad (1)$$

be a system of polynomial equations in $\mathbb{C}^N$, where for simplicity we assume that not all of the f_i are identically zero. For this system, the solution set, $Z = V(f) := \{x \in \mathbb{C}^N \mid f(x) = 0\}$, can be written as the union

$$Z := \bigcup_{i=0}^{N-1} Z_i := \bigcup_{i=0}^{N-1} \bigcup_{j \in \mathcal{I}_i} Z_{ij} \qquad (2)$$

where Z_i is the union of all i-dimensional irreducible components Z_{ij} of Z, and the index sets $\mathcal{I}_i$ are finite and possibly empty. Note that Z is the reduction of the possibly nonreduced algebraic set $f^{-1}(0)$ defined by the equations $f_1, \ldots, f_n$. The irreducible decomposition problem is to determine all of the irreducible components, Z_{ij}, meaning, at a minimum, to enumerate the components and provide a set membership test for each of them.

In [16], we presented algorithms to numerically describe and manipulate the irreducible decomposition. In that article we gave an implementation of the algorithm for the components Z_{ij} of $V(f)$, which occur with multiplicity one as components of $f^{-1}(0)$. In this article we present an implementation for the remaining components of $V(f)$ that have multiplicity at least two as components of $f^{-1}(0)$.

Components of multiplicity at least two specialize in the classical case of a one variable polynomial to roots of multiplicity at least two. Beyond the difficulties present in the classical case, there are a number of non-classical difficulties. For example, perturbations of the system can in some situations cause the whole component to disappear, or to change dimension.

One consequence of our new algorithm is the first implemented homotopy method of finding exactly the isolated solutions of a polynomial system of n polynomials in n variables. All previous implementations deliver a finite set of solutions containing the isolated solutions, but also possibly containing non-isolated solutions. Though previous implementations picked out the nonsingular isolated solutions from this set, they gave little or no information about whether a singular solution was isolated or not.

Computing a numerical irreducible decomposition depends on an algorithm for continuing from a generic point on a solution component to nearby points on the component. In [16] and the related papers [17, 18, 19, 20], widely spaced points on a component are found by tracking paths along linear slices of the component. As we shall describe further below, the numerical exploration of a component by path tracking allows one to classify witness points on each component, thereby determining the degree of the component and creating an efficient membership test for the component.

The essential difficulty, which this paper overcomes, is that standard predictor-corrector path tracking fails on components of multiplicity greater than one. This is because every point on such a set is singular: the Jacobian matrix of the polynomial system evaluated at a point on the set has rank smaller than the co-dimension of the set. In other words, the null space of the Jacobian matrix includes directions that are not in the tangent space of the set. Indeed, in extreme cases the Jacobian matrix can become identically zero, thus yielding no information about the tangent space of the underlying set. To predict a new point along the path, standard path trackers use the Jacobian matrix to compute a tangent vector along the path, so they cannot handle singular sets. These trackers also use the Jacobian matrix to compute corrections to the prediction via Newton's method. Here the ill-conditioning of the Jacobian matrix in the vicinity of the singular solution set slows convergence and limits the final accuracy.

Besides the numerical irreducible decomposition, other problems can create the need to track singular paths. For example, as in [7, 9], suppose we have a family of polynomial systems, $f(x, q) = 0$, that depend analytically on some parameters $q \in \mathbb{C}^M$. For some generic parameters q_0, we may use continuation to find a set of solution points that contains the generically isolated solutions of $f(x, q_0) = 0$, some of which may be multiple roots. One may obtain such a set for another set of parameters q_1 by tracking solution paths in the homotopy $f(x, tq_0 + (1-t)q_1) = 0$. With standard path tracking, one could only apply this technique to the generically nonsingular solutions. The approach of this paper removes this restriction.

Although the above discussion has posed the problem in terms of polynomial systems, the method applies more generally to systems of analytic equations. We make no essential use of the polynomial nature of the equations. As prediction and correction both act locally, only the analytic properties of functions of several complex variables are germane.

The method we propose is related to previous work [10, 11, 12] in computing singular solutions of polynomial and analytic systems of equations. For polynomial systems, related recent work on dealing with components of solutions of multiplicity at least two is described in [3] and in [5, 6]. A semi-numerical approach to restore the quadratic convergence of Newton's method by deflation can be found in [13] and [14]. The main purpose of this paper is to outline algorithms that complement our numerical decomposition method in [16]. In Section 5, we present a numerical experiment to illustrate feasibility on a small example.

Acknowledgments. We gratefully acknowledge the support of this work by Volkswagen-Stiftung (RiP-program at Oberwolfach). The first author thanks the Duncan Chair of the University of Notre Dame and National Science Foundation. This material is based upon work supported by the National Science Foundation under

Grant No. 0105653. The second author thanks the Department of Mathematics of the University of Illinois at Chicago and National Science Foundation. This material is based upon work supported by the National Science Foundation under Grant No. 0105739. The third author thanks General Motors Research and Development for their support.

2. Singular path tracking

In this section we present a general statement of the singular path-tracking problem, and explain our solution to the problem. In Section 3, we explain in more detail how this technique is used as part of a larger algorithm for computing a numerical irreducible decomposition.

Our approach is to meld together the approach we used for path tracking on a reduced component of the solution set of $f(x) = 0$ with the classical homotopy continuation approach to isolated singular points of polynomial systems.

2.1. Path tracking on components

First we recall the bare-bones framework for path tracking on reduced components. We assume we have a solution x_0 of $f(x) = 0$ which is a general point of some reduced k-dimensional component Z_{kj} of $f^{-1}(0)$. Indeed, the inductive procedures of [16, 17, 18, 19] lead to such points. Moreover those approaches lead to a new system of equations

$$F(x) := \begin{bmatrix} F_1(x_1, \ldots, x_N) \\ \vdots \\ F_{N-k}(x_1, \ldots, x_N) \\ L_1(x_1, \ldots, x_N) \\ \vdots \\ L_k(x_1, \ldots, x_N) \end{bmatrix}, \qquad (3)$$

where

1. Z_{kj} is a reduced component of the solution set of the equations $F_1 = \mathbf{0}$,..., $F_{N-k} = \mathbf{0}$; and

2. the L_i are generic linear functions whose set of common zeroes is a generic k-dimensional linear space L transverse to Z_{kj} and containing x_0.

An appropriate homotopy, chosen so that L varies within a $(k+1)$-dimensional linear space L', gives a smooth path on the irreducible reduced curve $L' \cap Z_{kj}$.

In the situation where we have a nonreduced k-dimensional component, we again have the same sort of system F, plus the generic linear spaces L and L', except that the k-dimensional irreducible component Z_{kj} and the irreducible curve $L' \cap Z_{kj}$ are not reduced.

2.2. Singular isolated solutions

Next recall how an isolated solution x_0 of multiplicity at least two of a system $f(x) = \mathbf{0}$ is handled classically. We assume the system is square, i.e., $n = N$, since we usually reduce to such systems at the cost of possibly increasing the multiplicity ν of x_0 when $\nu > 1$. The system is embedded in a larger system $f(x, t) = \mathbf{0}$ of N equations in $N + 1$ unknowns such that $f(x) = f(x, 0)$. In this case it follows from the upper semicontinuity of the dimension of fibers of an analytic map, that the zero set of $f(x, t) = \mathbf{0}$ is a one-dimensional analytic set C in a neighborhood of $(x_0, 0)$. The system $f(x, t) = \mathbf{0}$ is chosen so that the projection map of C down to the $\mathbb{C}$ under the projection from $\mathbb{C}^N \times \mathbb{C}$ to the second factor is ν-sheeted from a neighborhood of $(x_0, 0)$ down to a neighborhood of 0. This setup gives smooth maps $x_1(t), \ldots, x_\nu(t)$ from $(0, t_0]$ to C for some real positive t_0 such that the $x_i(t)$ are a cluster of distinct smooth isolated solutions of $f(x, t) = \mathbf{0}$ and $\lim_{t \to 0} x_i(t) \to x_0$. It is a useful fact [11] that the centroid of the cluster, $(x_1(t) + \cdots + x_\nu(t))/\nu$ is holomorphic.

2.3. Singular paths

Now let us explain how we meld these two setups together. As noted at the start of this section, we will explain in more detail in Section 3 how this technique is used as part of a larger algorithm for computing a numerical irreducible decomposition. Let

$$g(v, s, t) := \begin{bmatrix} g_1(v_1, \ldots, v_N, s, t) \\ \vdots \\ g_N(v_1, \ldots, v_N, s, t) \end{bmatrix}, \qquad (4)$$

be a system of N functions analytic in $\mathbb{C}^{N+2}$. We call v the variables and (s, t) the parameters, and we are interested in solutions of the equations $g(v, s, t) = \mathbf{0}$ for the variables given the values of the parameters. Note that in contrast to (1), this system is assumed to be square: it has the same number of functions as variables. We call a point (v, s, t) *singular* if the $N \times N$ Jacobian matrix of partial derivatives with respect to v is less than full rank at the point. In practice

1. we are interested in the in isolated solutions of $g(v, 0, 0) = \mathbf{0}$ or the irreducible components of $g(v, s, 0)$;

2. the analytic set $g(v, s, 0) = \mathbf{0}$ contains a curve, which would let us do the continuation if it was reduced; and

3. the variable t plays the same role as in the classical procedure sketched for the isolated solution.

Let us make this precise by listing conditions under which our path-tracking algorithm applies. First,

1. the solution set of $g(v, s, 0) = \mathbf{0}$ is an irreducible one-dimensional curve C in a neighborhood of v_0, whose reduction is isomorphic to a neighborhood of $0 \in \mathbb{C}$ under the projection $\mathbb{C}^N \times \mathbb{C} \times \{0\} \to \mathbb{C}$ given by $(v, s, t) \to s$.

This in particular implies that v_0 is an isolated solution of the system $g(v, 0, 0) = \mathbf{0}$. We denote by $v_0(s)$ the point on C that goes to s under the above projection.

Next we assume that

2. the multiplicity ν of C in the analytic set defined by $g(v, s, 0) = \mathbf{0}$ is the same as the multiplicity ν of v_0 as an isolated solution of $g(v, 0, 0) = \mathbf{0}$.

Using the above conditions, we know by semicontinuity of dimension of fibers of analytic maps that the solution set of $g(v, s, t) = \mathbf{0}$ is a pure dimensional two-dimensional analytic set S in a neighborhood of $(v_0, 0, 0)$. We further assume that,

3. for any fixed s in a neighborhood of $[0, 1]$, the projection $(v, s, t) \to t$ gives a proper ν-sheeted finite mapping of a neighborhood of $(v_0, s, 0)$ in the reduction of $C_s := S \cap (\mathbb{C}^N \times \{s\} \times \mathbb{C})$ to a neighborhood of $0 \in \mathbb{C}$.

In the situation of Section 3, we know this for $s = 0$, and by genericity of construction, for a Zariski open set of $s \in \mathbb{C}$, which by again using the genericity of the construction can be assumed to contain $[0, 1]$.

By the above condition there must exist a real positive t_0 such that

4. for all $s \in [0, 1]$, we have smooth maps $v_1(s, t), \ldots, v_\nu(s, t)$ from $\{s\} \times (0, t_0]$ to C_s such that the $v_i(s, t)$ are a cluster of distinct smooth isolated solutions of $g(v, s, t) = \mathbf{0}$ and $\lim_{t \to 0} v_i(s, t) \to v_0(s)$.

We emphasize some consequences of the above for the cluster $\mathcal{C} = \{v_1, \ldots, v_\nu\}$ of ν distinct starting points that satisfy $g(v, 0, t_0) = \mathbf{0}$ for a positive real $t_0 < \delta$ for a sufficiently small δ. For a sufficiently small δ

5. the centroid of the cluster, $(v_1(s, t) + \cdots + v_\nu(s, t))/\nu$, is an analytic function of (s, t) in some neighborhood of $[0, 1] \times [0, \delta]$; and

6. for fixed (s, t) in this neighborhood and with $t \neq 0$, the cluster points $v_i(s, t)$ are nonsingular isolated solutions of $g(v, s, t) = \mathbf{0}$.

Item 1 in this list simply asserts that for $t = 0$, we have a well-defined solution path parameterized by s. We wish to track this path as s goes from zero to one on the reals and accurately compute the endpoint of this path. However, for $\nu > 1$, this solution path is a multiple solution and is therefore singular. Fortunately, for t in the vicinity of zero, item 4 asserts that the solution component of $g(v, s, 0) = \mathbf{0}$ containing v_0 is part of a solution component of $g(v, s, t) = \mathbf{0}$, which is smooth for $s \in [0, 1]$ and $t \in (0, t_0]$ for some small positive t_0. It may not be apparent how such an embedding $g(v, s, t)$ of a given system $g(v, 0, 0)$ is constructed, but we can often arrange it as a consequence of the homotopies we use in solving systems of polynomial equations. It is important to note that the choice of a v_0 on a given component for which this procedure works will depend on the given construction of the system $g(v, s, t) = \mathbf{0}$. A particular instance, the numerical irreducible decomposition problem, is discussed in the next section.

3. Numerical irreducible decomposition

In this section, we review the algorithm from [16] and explain how it can lead to a kind of singular path-tracking problem. Improvements to the algorithm given in [17, 18, 19] also depend on path tracking and hence require singular path tracking to handle components of multiplicity greater than one. After a quick overview of these algorithms, we show how the particular systems arising in this context can be reformulated as in the foregoing Section 2.

3.1. Overview of decomposition algorithms

Let $\widehat{W} = \{\widehat{W}_i \mid i = 0, \ldots, N-1\}$ denote the witness point superset produced by the routine **WitnessGenerate** of [16]. We refer the reader to [16] for the construction of $\widehat{W}_i$, but note that geometrically $\widehat{W}_i$ is a subset of the intersection of Z with a generic linear space L_{N-i} of dimension $N - i$, which contains $Z_i \cap L_{N-i}$. As explained in detail in [16, 21], we model generic objects, e.g., coefficients of linear equations, by using random number generators. Each $\widehat{W}_i$ contains generic points on each i-dimensional component, plus additional junk points which will be filtered out, i.e., points lying on irreducible components of dimension greater than i. In particular, we showed in [16] how to do a breakup of each $\widehat{W}_i$ of $\widehat{W}$ as $\widehat{W}_i := \left(\cup_{j \in \mathcal{I}_i} W_{ij}\right) \cup J_i$, where

1. W_{ij} consists of $\deg Z_{ij}$ generic points of Z_{ij} each occurring ν_{ij} times, where ν_{ij} is a positive integer. Moreover, $\nu_{ij} \geq \mu_{ij}$, where μ_{ij} is the multiplicity of Z_{ij} in the possibly nonreduced algebraic set $f^{-1}(\mathbf{0})$, and $\nu_{ij} = 1$ if and only if $\mu_{ij} = 1$; and

2. $J_i \subset \cup_{k>i} Z_k$.

In [16, 17], this classification is accomplished by the construction of filtering polynomials

$$\{ p_{ij} \mid 0 \leq i \leq N-1 \, ; \, j \in \mathcal{I}_i \}. \tag{5}$$

The construction involves the operation of sampling the component containing a given generic point. The samples are used to interpolate and construct a filtering polynomial. The structure of the underlying set of a component of multiplicity at least two does not differ from the corresponding set for a multiplicity one component. If the component of multiplicity at least two could be sampled, the interpolation and further steps of the algorithm of [16] can be dealt with using the same routines with no change. To have a stable interpolation method, it is necessary to sample the irreducible component at widely separated points. In [16] we showed how to implement this sampling for multiplicity one components using homotopy continuation. The difficulty is that for a component of multiplicity at least two, this would amount to tracking paths singular at every point.

In [18], the classification of the witness points into components is effected by finding points that are connected by monodromy loops. These loops must be numerically tracked, and any witness points that appear with multiplicity greater than one will require a singular path-tracking algorithm. The monodromy method does not stand alone: in many instances an additional check is necessary to confirm that the monodromy classification is correct. In [19], it is shown that the computation of the linear trace is sufficient. As above, this is computed by interpolating points found by path tracking. Path tracking is fundamental to all these operations, and if we can succeed to accurately and reliably track paths on components of higher multiplicity, then these components can be treated exactly as the multiplicity-one components.

3.2. Witness point generation

In [15], witness points are generated using a cascade of homotopies, one dimension at a time. In case the number of equations n differs from the number of variables N, we add $N - n$ random hyperplanes to an underdetermined system, or $n - N$ extra "slack" variables to an overdetermined system. In the notation below we thus can take N as the maximum of N and n. The witness point set for dimension i is constructed with the homotopy

$$H(x, z, \lambda, t) := \begin{bmatrix} f_1(x_1, \ldots, x_N) + \gamma_{1,1} z_1 + \cdots + \gamma_{1,i+1} z_{i+1} \\ \vdots \\ f_N(x_1, \ldots, x_N) + \gamma_{N,1} z_1 + \cdots + \gamma_{N,i+1} z_{i+1} \\ z_1 + \lambda_{1,0} + \lambda_{1,1} x_1 + \cdots + \lambda_{1,N} x_N \\ \vdots \\ z_i + \lambda_{i,0} + \lambda_{i,1} x_1 + \cdots + \lambda_{i,N} x_N \\ z_{i+1} + t(\alpha_0 + \alpha_1 x_1 + \cdots + \alpha_N x_N) \end{bmatrix}. \qquad (6)$$

The arguments in H are as follows:

1. $x \in \mathbb{C}^N$ are the original variables from the system

$$f(x) = 0, \quad x = (x_1, x_2, \ldots, x_N);$$

2. $z \in \mathbb{C}^{i+1}$ are auxiliary variables in the embedding of the component, $z = (z_1, z_2, \ldots, z_{i+1})$;

3. $t \in (0, 1]$ is the continuation parameter, going from one to zero;

4. $\lambda \in \mathbb{C}^{(i) \times (N+1)}$ is a matrix of generic coefficients for i hyperplanes.

In addition, both $\gamma \in \mathbb{C}^{N \times (i+1)}$ and $\alpha \in \mathbb{C}^{N+1}$ are generic parameters. The hyperplanes, defined by λ, slice any i-dimensional components down to isolated points; these are the witness points for the components. The hyperplanes also meet components of dimension greater than i, so some endpoints of the homotopy may lie on these sets. If so, these are the "junk" points J_i mentioned above.

In this homotopy we have smooth paths, indexed by k, of the form $(x_k(t), z_k(t))$: $(0, 1] \to \mathbb{C}^N \times \mathbb{C}^{i+1}$, defined by $H(x_k(t), z_k(t), \lambda, t) = \mathbf{0}$. The witness point superset (which consists of both the proper witness points and the junk points) for dimension i is given by the endpoints of those paths that have all slack variables equal to zero; that is, those which are of the form

$$\lim_{t \to 0} (x_k(t), z_k(t)) = (w_k, \mathbf{0}).$$

Since the slack variables are zero, one sees from (6) that w_k is both a solution point of $f = \mathbf{0}$ and lies on the linear slice defined by λ.

3.3. Sampling components

In the homotopy above, the matrix λ defines a linear slice that cuts out the witness points. To classify the witness points and otherwise explore the components, one may move λ in a secondary homotopy and track the movements of the witness points. A common example is a linear homotopy going from $s = 0$ to $s = 1$ between the original slice $\lambda^{(1)}$ and a new slice $\lambda^{(2)}$ as

$$H(w, \mathbf{0}, (1-s)\lambda^{(1)} + s\lambda^{(2)}, 0) = \mathbf{0}, \tag{7}$$

where H is as defined in (6). For nonsingular components, this can be tracked by conventional methods to give a new sample on the component. These methods fail, however, when w is on a multiple component, because of the singularity of the Jacobian matrix for all points on the component.

This leads us back to the singular path-tracking problem as defined in §2. We must begin with the cluster of ν points that approach w as $t \to 0$ and use the doubly parameterized homotopy

$$H(x, z, (1-s)\lambda^{(1)} + s\lambda^{(2)}, t) = \mathbf{0}. \tag{8}$$

This is of the same form as (4), with variables $v = (x, z)$. The witness point generator tracks all solution paths, so we are assured that the cluster is complete. The existence of a one-dimensional path to track, is known to be true by the method of constructing of the witness point set, and because the slicing coefficients, $\lambda^{(1)}$ and $\lambda^{(2)}$ are chosen generically.

4. The algorithm

Before presenting the new algorithm for tracking singular paths, it is useful to sketch out existing predictor-corrector algorithms for tracking nonsingular paths. Our singular path tracker builds on the same framework, but we replace both the predictor and the corrector with new routines that work for singular paths.

4.1. Nonsingular tracking algorithm

A generic path-tracking algorithm proceeds as follows [2], (see also [1, 8]). In our homotopies, we may assume that the path parameter, s, is strictly increasing, that is, the path has no turning points.

- **Given:** System of full-rank equations, $g(v, s) = 0$, initial point v_0 at $s_0 = 0$ such that $g(v_0, 0) \approx 0$, and initial step length h.
- **Find:** Sequence of points (v_i, s_i), $i = 1, 2, \ldots$, along the path such that $g(v_i, s_i) \approx 0$, $s_{i+1} > s_i$, terminating with $s_n = 1$. Return a high-accuracy estimate of v_n.
- **Procedure:**
    - **Loop:** For $i = 1, 2, \ldots$
        1. **Predict:** Predict solution (u, s') such that $||(u, s') - (v_{i-1}, s_{i-1})|| \approx h$ with $s' > s_{i-1}$.
        2. **Correct:** In the vicinity of (u, s'), attempt to find a corrected solution (w, s'') such that $g(w, s'') \approx 0$.
        3. **Update:** If correction step was successful, update $(v_i, s_i) = (w, s'')$. Increment i.
        4. **Adjust:** Adjust the step length h.
    - **Terminate:** Terminate when $s_i = 1$.
    - **Refine endpoint:** At $s_i = 1$, correct v_i to high accuracy.

There are many possible choices for the implementation of each step. Some useful choices are as follows.

- The simplest predictor is just $u = v_{i-1}$, but it is much better to use a prediction along the tangent direction as $u = v_i - \alpha(\partial g/\partial v)^{-1}(\partial g/\partial s)$, where α is calculated to give the desired step length. Higher-order predictions can also be used.

- The step length can be measured as a weighted two-norm or simply as $s' - s_{i-1}$, in which case $s' = s_{i-1} + h$.

- A common corrector strategy is to hold s constant, that is, $s'' = s'$, and compute $\boldsymbol{w}$ by Newton's method, allowing a fixed number of iterations. The correction is deemed successful if Newton's method converges within a pre-specified path tracking tolerance within the allowed number of iterations.

- A good step length adjustment strategy is to cut the step length in half on failure of the corrector and to double it if several successive corrections at the current step size have been successful.

- Near the end of the path-tracking interval, the step length is adjusted to land exactly on $s = 1$.

By keeping the number of iterations in the corrector small (no larger than three) and the path tracking tight, all intermediate points are kept close to the exact path, minimizing any chance that a solution will jump tracks. However, to save computation time, the path-tracking tolerance is generally looser than that used in the final refinement at $s = 1$.

4.2. Singular tracking algorithm

When the path to be tracked is singular, Newton's method cannot be used effectively for correction. In the formulation of §2, this can be overcome by considering the cluster of nonsingular points that approach the singular path as $t \to 0$.

For our new corrector, we turn to methods that have been proposed for the accurate estimation of a singular endpoint, known as the "endgame." The simplest approach is to detect a cluster of paths approaching each other as the paths are tracked as close as possible to $t = 0$. In [10], it was shown that the centroid of such a cluster is an analytic function of t, which can be extrapolated to $t = 0$ to get an accurate approximation of the solution. More advanced "endgames" are proposed in [4, 11, 12, 22]. All of these depend, directly or indirectly, on a breakup of the ν roots into subsets of cardinality $c_1, \ldots, c_k$ with $c_1 + \cdots + c_k = \nu$, each subset forming a cycle (see [12]). For each cycle, there exists a fractional power series

$$\boldsymbol{v}(t) = \boldsymbol{v}_0 + \boldsymbol{a}_1 t^{1/c_i} + \boldsymbol{a}_2 t^{2/c_i} + \cdots \qquad (9)$$

which can be used for accurate estimation of $\boldsymbol{v}(0)$.

In [12], the notion of an "endgame operating range" was introduced. This simply recognizes that on the one hand, the fractional power series (9), is convergent only inside a certain radius, ρ_1, while on the other hand, ill-conditioning prevents accurate numerical accuracy inside a second radius, ρ_0, dependent on the precision of arithmetic used. If the precision is sufficient so that $\rho_0 < \rho_1$, the solution can be sampled in the annular region $\rho_0 < |t| < \rho_1$ to estimate the endpoint as the constant term in the power series.

In the singular path-tracking problem, the endgame operating range changes as s advances along the path. Our algorithm attempts to move the cluster of points $\mathcal{C}(s, t)$ forward in s, adjusting t at every step to stay within the endgame operating range. For intermediate values of s, the endgame can use loose tolerances — just enough to verify that t is within range. Upon reaching $s = 1$, the endgame is applied a final time with tight tolerance to compute an accurate endpoint.

We begin by specifying the singular path tracker at a high level, similar to above, and then describe the basic options for implementing the predictor and corrector steps for the singular case. For a cluster of points $\mathcal{C} = \{v_1, \ldots, v_n\}$, let $g(\mathcal{C}, s, t) = \mathbf{0}$ be an abbreviation for $g(v_i, s, t) = \mathbf{0}, i = 1, \ldots, n$. For simplicity, the algorithm to follow assumes that step length is measured by the change in s rather than by some more general notion of arc length.

- **Given:** System of equations, $g(v, s, t) = \mathbf{0}$, initial cluster $\mathcal{C}_0$, such that $g(\mathcal{C}_0, 0, t_0) \approx \mathbf{0}$, meeting conditions (1,2,3) in §2. Also, initial step length h.

- **Find:** Sequence of clusters $(\mathcal{C}_i, s_i, t_i)$, $i = 1, 2, \ldots$, along the path such that $g(\mathcal{C}_i, s_i, t_i) \approx \mathbf{0}$ meeting the same cluster conditions, with $s_{i+1} > s_i$, terminating with $s_n = 1$. Return the final cluster and a high-accuracy estimate of the path point at $s = 1$.

- **Procedure:**

    - **Loop:** For $i = 1, 2, \ldots$
        1. **Predict:** Predict cluster $(\mathcal{U}, s', t')$ with $s' = s_{i-1} + h$ and $t' = t_{i-1}$.
        2. **Correct:** In the vicinity of $\mathcal{U}$, attempt to find a corrected cluster $\mathcal{W}$ such that $g(\mathcal{W}, s', t') \approx \mathbf{0}$.
        3. **Recondition:** If correction is successful, play a singular endgame in t' to verify the cluster conditions and compute an approximation to the point $\lim_{t' \to 0} \mathcal{W}(s', t')$, where $g(\mathcal{W}(s', t'), s', t') = \mathbf{0}$. If the endgame is successful, then
            * **Adjust t:** Pick a new t_i in the endgame operating zone.
            * **Update:** Set $s_i = s'$ and generate the corresponding cluster $\mathcal{C}_i$. Increment i.
        4. **Adjust h:** Adjust the step length h.
    - **Terminate:** Terminate when $s_i = 1$.
    - **Refine endpoint:** Play the endgame at $s = 1$ to compute the final path point to high accuracy.

Note that this algorithm assumes that the initial cluster has already been properly reconditioned. If not, the main loop should be preceded by an initial reconditioning.

In contrast to the nonsingular path tracker, this time the correction step does not attempt to compute a point on the path. Instead, it serves to compute the nonsingular cluster of points for a nonzero $t = t'$ after a step forward in s. The endgame in the reconditioning step now plays the role formerly played by the corrector: it computes the path point. In doing so successfully, it verifies that the path is being followed to sufficient precision. It also builds a local model of the solution cluster's dependence on t, so that t_i can be safely placed in the endgame operating range.

Since the paths of the cluster points are all nonsingular for small enough nonzero t, prediction and correction can proceed as in the nonsingular path tracker. In particular, we may use a first-order predictor and Newton's method for correction. For added protection against path crossing in the predictor-corrector step, one could check that the distances between the corrected cluster points $\mathcal{W}$ is greater than their Newton residuals. Adjustment of the step length proceeds as in the nonsingular tracker, except that a step is now successful only if both the corrector and the reconditioner succeed.

A simple version of reconditioning is to monitor the condition number while tracking the cluster paths versus t with s held constant. Along these paths, determine when the maximal condition number of the points in the cluster hits a predetermined value. The condition number can be estimated either from the Jacobian matrix of partial derivatives or as the inverse of the distance between nearest points in the cluster. Note that if the initial condition number is already past the mark, t must be increased. If it is far too high, the correction step may fail, and the step length will then be decreased.

A more sophisticated version of reconditioning is to use a fit to the fractional power series (9), to estimate the path point. By monitoring the estimate as t is decreased, one may verify that the sequence is converging accurately. This is safer than the condition number criterion, which can fail if the specified condition number is lower than the condition number at the outer edge of the endgame operating range (the radius of convergence of the power series). Furthermore, detection of convergence at larger t, where the condition number is smaller, will allow the procedure to advance more quickly.

The final refinement may also take advantage of the fractional power series to compute a high-order estimate of the path point. A simple version, though, is to track t towards zero until the cluster radius is smaller than a specified tolerance, then take the centroid of the cluster as the estimate.

These comments notwithstanding, our current level of implementation is very basic. In the numerical experiments to follow, we use the condition number criterion for reconditioning. We also only recondition at the end ($s = 1$) to prepare the cluster for further tracking. We used 10^5 as a practical value for double-precision arithmetic, as it still leaves about 8 digits of accuracy in the cluster points.

5. A numerical experiment

The algorithms outlined above have been implemented with the aid of the path-following routines in PHCpack [23]. Recently the package has been upgraded with a module (see [20] for an overview) to numerically decompose solution sets of polynomial systems into irreducible components.

We tested the implementation on a rational normal curve of degree four, of multiplicity two:

$$\begin{cases} x_1^2 - x_2 = 0 \\ x_2^2 - x_3 = 0 \\ (x_3 - x_4)^2 = 0 \end{cases} \tag{10}$$

We can read off the solution set: $(x_1, x_1^2, x_1^4, x_1^4)$, for any $x_1 \in \mathbb{C}$. The system is underdetermined: three equations in four unknowns. We use the following embedding of the system in (10):

$$\begin{cases} x_1^2 - x_2 + \gamma_{1,1} z_1 = 0 \\ x_2^2 - x_3 + \gamma_{2,1} z_1 = 0 \\ (x_3 - x_4)^2 + \gamma_{3,1} z_1 = 0 \\ z_1 - t = 0 \\ z_1 + \lambda_{1,0} + \lambda_{1,1} x_1 + \lambda_{1,2} x_2 + \lambda_{1,3} x_3 + \lambda_{1,4} x_4 = 0 \end{cases} \tag{11}$$

where the γ and λ constants are randomly generated complex numbers. The variable z_1 is an added "slack" variable. Clearly, for $z_1 = 0$ and $t = 0$, we obtain generic points on the curve. In particular, we have four clusters of two points each. The last equation is a generic slice, so that the limits of the solution points as $t \to 0$ are witness points for the set.

For the purposes of this numerical test, we use the interpolation procedure of [16] to determine the degree of the space curve. Sampling is accomplished by a sequence of homotopies to randomly generated slices, $\lambda^{(1)}, \lambda^{(2)}, \ldots, \lambda^{(n)}$, each homotopy having the form of (8). The samples are projected onto a randomly-oriented plane, $\mathbb{C}^2$, and fit with polynomials ranging in degree from 1 to 4. Extra samples serve as independent test points to see when the interpolant fits the entire component.

In final refinement, we use 64 decimal places. With this precision, the magnitude of the cluster radius ranges between 10^{-34} and 10^{-33}. We did not recondition except at the endpoints, so t was held constant during each leg of the homotopy. However, the condition numbers of the paths were monitored: they ranged between 10^5 and 10^6. The calculations are summarized in Table 1. At degree four, the interpolant fits the test points to full precision.

In Table 1 we observe a steady worsening of the condition of the interpolation problem as the degree rises. For degrees higher than four, one would need to sharpen the clusters with more than 64 decimal places as working precision or implement a more sophisticated endgame. Alternatively, instead of ordinary linear projections, one can (if the linear span of the component permits) project from a point on the component.

degree of the interpolator	number of sampled clusters	magnitude of condition number	magnitude of residual at test points
1	3	3	−6
2	7	8	−10
3	12	13	−20
4	16	21	−∞

Table 1. Summary of incremental interpolation to determine the degree of a space curve. For every degree of the interpolator, we list the total number of sampled clusters, the magnitude of the condition number (3 stands for 10^3) of the interpolation problem and the magnitude of the residual at some test points.

These central projections [17] reduce the degree of the interpolating polynomial and thus improve the numerical conditioning of the problem.

Complementary to using central projections is the application of extrapolation techniques to improve the centroid of the cluster. For this example, extrapolation is used on the fractional power series (9) with cycle number $c = 2$ at values 10^{-20}, 10^{-32}, and 10^{-44} for t in the homotopy (11). The working precision is 64 decimal places. This second-order extrapolator achieves a cluster radius of magnitude in the range between 10^{-45} and 10^{-44}. In comparison, if we approximate the centroid of the cluster just by taking the average of the solution vectors at $z_1 = 10^{-72}$, we obtain a radius in the range between 10^{-34} and 10^{-33}.

6. Conclusions

We have proposed a tracking algorithm for singular paths. The system of equations defining the singular path are embedded in a perturbed system having a cluster of nearby nonsingular paths. These can be handled effectively with nonsingular path-tracking techniques. A singular endgame can then be applied to the cluster to estimate points on the original singular path. A bare-bones version of the tracker is shown to be effective on a simple, fourth-degree curve having multiplicity two.

Section 4 indicates several possible enhancements to the safety and efficiency of the algorithm. Chief among these is to use a higher-order singular endgame instead of just the centroid of the cluster, which is correct only to first-order in t. Higher-order approximations will decrease the error propagation caused by the multiplicity and thus increase the numerical stability. To get a point on a singular path having cycle number c accurate up to d decimal places, a rough rule of thumb is that with first-order approximation we need to take t as low as 10^{-cd}, whereas with extrapolation of order c, we can stop t at 10^{-d}. This behavior is confirmed in the experiment in Section 5.

This algorithm, although ripe for further improvement, serves to show feasibility for extending our techniques for irreducible decomposition to components of multiplicity greater than one. In this sense, the irreducible decomposition algorithm is now complete, although many improvements are still possible.

References

[1] E. L. Allgower and K. Georg, Numerical Continuation Methods, an Introduction, Springer Ser. Comput. Math. 13, Springer-Verlag, Berlin 1990.

[2] E. L. Allgower and K. Georg, Numerical Path Following, in: Techniques of Scientific Computing, Part 2 (P. G. Ciarlet and J. L. Lions, eds.), Handb. Numer. Anal. 5, North-Holland, Amsterdam 1997, 3–203.

[3] A. Galligo and D. Rupprecht, Semi-numerical determination of irreducible branches of a reduced space curve, in: Proceedings of the 2001 International Symposium on Symbolic and Algebraic Computation (B. Mourrain, ed.), ACM 2001, 137–142.

[4] B. Huber and J. Verschelde, Polyhedral end games for polynomial continuation, Numer. Algorithms 18 (1998), 91–108.

[5] G. Lecerf, Une alternative aux méthodes de réécriture pour la résolution des systèmes algébriques, PhD Thesis, École Polytechnique, France, 2001.

[6] G. Lecerf, Quadratic Newton iteration for systems with multiplicity, J. FoCM 2 (2002), 247–293.

[7] T. Y. Li and X. Wang, Nonlinear homotopies for solving deficient polynomial systems with parameters, SIAM J. Numer. Anal. 29 (1992), 1104–1118.

[8] A. Morgan, Solving polynomial systems using continuation for engineering and scientific problems, Prentice-Hall, Englewood Cliffs, NJ, 1987.

[9] A. P. Morgan and A. J. Sommese, Coefficient-parameter homotopy continuation, Appl. Math. Comput. 29 (1989), 123–160; Errata: Appl. Math. Comput. 51 (1992), 207.

[10] A. P. Morgan, A. J. Sommese, and C. W. Wampler, Computing singular solutions to nonlinear analytic systems, Numer. Math. 58 (1991), 669–684.

[11] A. P. Morgan, A. J. Sommese, and C. W. Wampler, Computing singular solutions to polynomial systems, Adv. Appl. Math. 13 (1992), 305–327.

[12] A. P. Morgan, A. J. Sommese, and C. W. Wampler, A power series method for computing singular solutions to nonlinear analytic systems, Numer. Math. 63 (1992), 391–409.

[13] T. Ojika, Modified deflation algorithm for the solution of singular problems. I. A system of nonlinear algebraic equations, J. Math. Anal. Appl. 123 (1987),199–221.

[14] T. Ojika, S. Watanabe, and T. Mitsui, Deflation algorithm for the multiple roots of a system of nonlinear equations, J. Math. Anal. Appl. 96 (1983), 463–479.

[15] A. J. Sommese and J. Verschelde, Numerical homotopies to compute generic points on positive dimensional algebraic sets, J. Complexity 16 (2000), 572–602.

[16] A. J. Sommese, J. Verschelde, and C. W. Wampler, Numerical decomposition of the solution sets of polynomial systems into irreducible components, SIAM J. Numer. Anal. 38 (2001), 2022–2046.

[17] A. J. Sommese, J. Verschelde, and C. W. Wampler, Numerical irreducible decomposition using projections from points on the components, in: Symbolic Computation: Solving Equations in Algebra, Geometry, and Engineering (E. L. Green, S. Hoşten, R. C. Laubenbacher, and V. Powers, eds.), Contemp. Math. 286, Amer. Math. Soc., Providence, RI, 2001, 37–51.

[18] A. J. Sommese, J. Verschelde, and C. W. Wampler, Using monodromy to decompose solution sets of polynomial systems into irreducible components, in: Application of Algebraic Geometry to Coding Theory, Physics and Computation (C. Ciliberto, F. Hirzebruch, R. Miranda, and M. Teicher, eds.), Proceedings of a NATO Conference, February 25–March 1, 2001, Eilat, Israel, NATO Sci. Ser. II, Math. Phys. Chem. 36, Kluwer Academic Publishers, Dordrecht 2001, 297–315.

[19] A. J. Sommese, J. Verschelde, and C. W. Wampler, Symmetric functions applied to decomposing solution sets of polynomial systems, to appear in SIAM J. Numer. Anal.

[20] A. J. Sommese, J. Verschelde, and C. W. Wampler, Numerical irreducible decomposition using PHCpack. in: Mathematics and Visualization (M. Joswig and N. Takayama, eds.), Proceedings of Dagstuhl seminar no. 01421, to appear.

[21] A. J. Sommese and C. W. Wampler, Numerical algebraic geometry, in: The Mathematics of Numerical Analysis (J. Renegar, M. Shub, and S. Smale, eds.), Lectures in Appl. Math. 32, Proceedings of the AMS-SIAM Summer Seminar in Applied Mathematics, Park City, Utah, July 17–August 11, 1995, Park City, Utah, Amer. Math. Soc., Providence, RI, 1996, 749–763.

[22] M. Sosonkina, L. T. Watson, and D. E. Stewart, Note on the end game in homotopy zero curve tracking, ACM Trans. Math. Software 22 (1996), 281–287.

[23] J. Verschelde, Algorithm 795: PHCpack: A general-purpose solver for polynomial systems by homotopy continuation, ACM Trans. Math. Software 25 (1999), 251–276, software available at http://www.math.uic.edu/~jan.

A. J. Sommese, Department of Mathematics, University of Notre Dame, Notre Dame, IN 46556-4618, U.S.A.
E-mail: sommese@nd.edu
URL: http://www.nd.edu/~sommese

J. Verschelde, Department of Mathematics, Statistics, and Computer Science, University of Illinois at Chicago, 851 South Morgan (M/C 249), Chicago, IL 60607-7045, U.S.A.
E-mail: jan@math.uic.edu jan.verschelde@na-net.ornl.gov
URL: http://www.math.uic.edu/~jan

C. W. Wampler, General Motors Research and Development, Mail Code 480-106-359, 30500 Mound Road, Warren, MI 48090-9055, U.S.A.
E-mail: Charles.W.Wampler@gm.com

Lectures of the Conference

K. Ueno: Abelian conformal field theory under semi-stable degeneration

L. Bădescu: On the problem of extending formal functions in algebraic geometry

M. Gross: Geometric approaches to mirror symmetry

A. Beauville: On the Chow ring of a K3 surface

A. Langer: Maruyama's conjecture

Y. Kawamata: On Francia's flip

M. Reid: 3-fold flips from 1980 to 2001

M. Andreatta: On manifolds whose tangent bundle contains an ample locally free subsheaf

J. Wisniewski: Rational curves and classification of higher dimensional varieties

V. Guletskiĭ: The Chow motive on a Godeaux surface

K. Konno: Quadric hull of canonical surfaces

M. Mendes Lopes: Rational surfaces with many nodes

A. Corti: Fano 3-folds and Weighted Grassmannians

M. Mella: Special projective varieties

L. Ein: Multiplier ideals and applications

V. Shokurov: A.c.c. of m.l.d.'s implies the log termination

S. Verra: Quaternionic Pryms and Hodge classes

Y. Lee: The family of curves in the bicanonical pencil of Godeaux surfaces

T. de Fernex: Birational transformations of prime order of the projective plane

Y. Miyaoka: Rational curves of small degrees and applications

J. Murre: Chow–Kuenneth decomposition for elliptic modular varieties

C. Ciliberto: On the bicanonical map for surfaces of general type: Francia's legacy

List of Contributors

Lucian Bădescu, Institute of Mathematics of the Romanian Academy and Department of Mathematics, University of Bucarest, P. O. Box 1-764, 70700 Bucharest, Romania.
E-mail address: lbadescu@stoilow.imar.ro

Luca Barbieri-Viale, Dipartimento di Metodi e Modelli Matematici, Università degli Studi di Roma "La Sapienza", Via A. Scarpa 16, 00161, Roma, Italy.
E-mail address: barbieri@dmmm.uniroma1.it

Arnaud Beauville, Laboratoire J.A. Dieudonné, Université de Nice, Parc Valrose, 06108 Nice Cedex 2, France.
E-mail address: beauvill@math.unice.fr

Giuseppe Borrelli, Via Volci 49, 00052 Cerveteri (Roma), Italy.
E-mail address: borrelli@mat.uniroma3.it

Fabrizio Catanese, Lehrstuhl Mathematik VIII, Universität Bayreuth, 95440 Bayreuth, Germany.
E-mail address: Fabrizio.Catanese@uni-bayreuth.de

Ciro Ciliberto, Dipartimento di Matematica, Università di Roma Tor Vergata, Via della Ricerca Scientifica, 00133 Roma, Italy.
E-mail address: cilibert@mat.uniroma2.it

Alberto Conte, Dipartimento di Matematica, Università di Torino, Via Carlo Alberto 10, 10023 Torino, Italy.
E-mail address: conte@dm.unito.it

Alessio Corti, DPMMS, CMS, Wilberforce Road, Cambridge CB3 0WB, England.
E-mail address: A.Corti@dpmms.cam.ac.uk

Tommaso de Fernex, Department of Mathematics, Statistics, and Computer Sciences, University of Illinois at Chicago, 851 South Morgan, Chicago, IL 60607-7045, U.S.A.
E-mail address: defernex@math.uic.edu

Lawrence Ein, Department of Mathematics, Statistics, and Computer Sciences, University of Illinois at Chicago, 851 South Morgan, Chicago, IL 60607-7045, U.S.A.
E-mail address: ein@math.uic.edu

Vladimir Guletskiĭ, Institute of Mathematics, Surganova 11, Minsk 220072, Belarus.
E-mail address: guletskii@im.bas-net.by

Yujiro Kawamata, Department of Mathematical Sciences, University of Tokyo, Komaba, Meguro, Tokyo, 153-8914, Japan.
E-mail address: kawamata@ms.u-tokyo.ac.jp

Kazuhiro Konno, Department of Mathematics, Graduate School of Sciences, Osaka University, Machikaneyama 1-16, Toyonaka, Osaka, 560–0043, Japan.
E-mail address: konno@math.wani.osaka-u.ac.jp

Adrian Langer, Instytut Matematyki UW, Ul. Banacha 2, 02-097 Warsawa, Poland.
E-mail address: alan@mimuw.edu.pl

Antonio Lanteri, Dipartimento di Matematica "F. Enriques", Via C. Saldini 50, 20133 Milano, Italy.
E-mail address: lanteri@mat.unimi.it

Raquel Mallavibarrena, Departamento de Algebra, Facultad de Matemáticas, Universidad Complutense de Madrid, Ciudad Universitaria, 28040 Madrid, Spain.
E-mail address: raquel@mat.ucm.es

Margarida Mendes Lopes, CMAF, Universidade de Lisboa, Av. Prof. Gama Pinto, 2, 1649-003 Lisboa, Portugal.
E-mail address: mmlopes@lmc.fc.ul.pt

Marina Marchisio, Dipartimento di Matematica, Università di Torino, Via Carlo Alberto 10, 10023 Torino, Italy.
E-mail address: marchisio@dm.unito.it

Jacob Murre, Department of Mathematics, University of Leiden, P. O. Box 9512, 2300 RA Leiden, The Netherlands.
E-mail address: murre@math.leidenuniv.nl

Rita Pardini, Dipartimento di Matematica, Università di Pisa, Via Buonarroti 2, 56127 Pisa, Italy.
E-mail address: pardini@dm.unipi.it

Claudio Pedrini, Dipartimento di Matematica, Università di Genova, Via Dodecaneso 35, 16146, Genova, Italy.
E-mail address: pedrini@dima.unige.it

Miles Reid, Mathematical Institute, University of Warwick, Coventry CV4 7AL, England.
E-mail address: Miles@Maths.Warwick.Ac.UK

Francesco Russo, Departamento de Matematica, Universidade Federal de Pernambuco, 50670-901 Recife-Pe, Brasil.
E-mail address: frusso@dmat.ufpe.br

Frank-Olaf Schreyer, Mathematik und Informatik Universität des Saarlandes, Im Stadtwald, 66123 Saarbrücken, Germany.
E-mail address: schreyer@math.uni-sb.de

Vyacheslav V. Shokurov, Department of Mathematics, Johns Hopkins University, Baltimore, MD-21218, U.S.A.
E-mail address: shokurov@math.jhu.edu

Andrew J. Sommese, Department of Mathematics, University of Notre Dame, Notre Dame, IN 46556-4618, U.S.A.
E-mail address: sommese@nd.edu

Jean Verschelde, Department of Mathematics, Statistics, and Computer Sciences, University of Illinois at Chicago, 851 South Morgan (M/C 249), Chicago, IL 60607-7045, U.S.A.
E-mail address: jan@math.uic.edu

Charles Wampler, General Motors Research & Development, Mail Code 480-106-359, 30500 Mount Road, Warren, MI 48090-9055, U.S.A.
E-mail address: Charles.W.Wampler@gm.com

List of Participants

A. Albano (Torino, albano@dm.unito.it)

A. Alzati (Milano, Alberto.Alzati@mat.unimi.it)

M. Andreatta (Trento, andreatt@science.unitn.it)

C. Arezzo (Parma, claudio.arezzo@ipruniv.cce.unipr.it)

D. Arezzo (Genova, arezzo@dima.unige.it)

M. A. Barja (Barcellona, barja@ma1.upc.es)

A. Beauville (Nice, beauvill@math.unice.fr)

L. Bădescu (Bucarest, lbadescu@stoilow.imar.ro)

C. Bartocci (Genova, bartocci@dima.unige.it)

M. C. Beltrametti (Genova, beltrame@dima.unige.it)

M. Bertin (Paris, marie.bertin@parisfree.com)

G. Borrelli (Roma, borrelli@matrm3.mat.uniroma3.it)

G. Canonero (Genova, canonero@dima.unige.it)

V. Catalisano (Genova, catalisa@dima.unige.it)

M. P. Cavaliere (Genova, cavalier@dima.unige.it)

C. Ciliberto (Roma, cilibert@mat.uniroma2.it)

L. Chiantini (Siena, chiantini@unisi.it)

N. Chiarli (Torino, chiarli@polito.it)

E. Colombo (Milano, colombo@mat.unimi.it)

A. Conte (Torino, conte@alpha01.dm.unito.it)

A. Corti (Cambridge, A.Corti@dpmms.cam.ac.uk)

T. de Fernex (Chicago, defernex@math.uic.edu)

A. Del Centina (Ferrara, cen@dns.unife.it)

L. Ein (Chicago, ein@math.uic.edu)

M. L. Fania (L'Aquila, fania@univaq.it)

P. Frediani (Pavia, dimat.unipv.it)

A. Gimigliano (Bologna, gimiglia@dm.unibo.it)

S. Greco (Torino, greco@polito.it)

M. Gross (Warwick, mgross@maths.warwick.ac.uk)

V. Guletskiĭ (Minsk, guletskii@im.bas-net.by)

Y. Kawamata (Tokyo, kawamata@ms.u-tokyo.ac.jp)

K. Konno (Osaka, konno@math.wani.osaka-u.ac.jp)

A. Langer (Varsaw, alan@mimuw.edu.pl)

A. Lanteri (Milano, lanteri@mat.unimi.it)

M. Lattarulo (Genova, lattarul@dima.unige.it)

Y. Lee (Seoul, ynlee@ccs.sogang.ac.kr)

H. Maeda (Tokyo, hmaeda@mse.waseda.ac.jp)

M. Manaresi (Bologna, manaresi@dm.unibo.it)

M. Marchisio (Torino, marchisio@alpha01.dm.unito.it)

M. G. Marinari (Genova, marinari@dima.unige.it)

M. Mella (Ferrara, mll@dns.unife.it)

M. Mendes Lopes (Libona, mmlopes@lmc.fc.ul.pt)

Y. Miyaoka (Tokyo, miyaoka@kurims.kyoto-u.ac.jp)

J. Murre (Leiden, murre@math.leidenuniv.nl)

G. Occhetta (Milano, occhetta@mat.unimi.it)

P. Oliverio (Cosenza, oliverio@unical.it)

A. Oneto (Genova, oneto@dima.unige.it)

F. Orecchia (Napoli, orecchia@matna2.dma.unina.it)

D. Panizzolo (Trento, panizza@science.unitn.it)

R. Pardini (Pisa, pardini@dm.unipi.it)

List of Participants

C. Pedrini (pedrini@dima.unige.it)

R. Pignatelli (Göttingen, pignatel@uni-math.gwdg.de)

F. Pioli (Trieste, pioli@sissa.it)

L. Ramella (Genova, ramella@dima.unige.it)

M. Reid (Warwick, miles@maths.warwick.ac.uk)

C. Ronconi (Padova, ronconi@dmsa.unipd.it)

M. E. Rossi (Genova, rossim@dima.unige.it)

F. Russo (Recife, russo@dmat.ufpe.br)

P. Salmon (Bologna, salmon@dm.unibo.it)

M. E. Serpico (Genova, serpico@dima.unige.it)

V. Shokurov (Baltimore, shokurov@math.jhu.edu)

R. Strano (Catania, sstrano@dmi.unict.it)

G. Tamone (Genova, tamone@dima.unige.it)

A. Tironi (Milano, atironi2001@yahoo.it)

F. Tonoli (Göttingen, tonoli@uni-math.gwdg.de)

C. Turrini (Milano, turrini@mat.unimi.it)

K. Ueno (Kyoto, ueno@kusm.kyoto-u.ac.jp)

L. Ugaglia (Milano, ugo@socrates.mat.unimi.it)

G. Valla (Genova, valla@dima.unige.it)

L. Van Geemen (Milano, geemen@mat.unimi.it)

A. Verra (Roma, verra@matrm3.mat.uniroma3.it)

J. Wisniewski (Varsaw, jarekw@duch.mimuw.edu.pl)

K. Yamaki (Kyoto, yamaki@kusm.kyoto-u.ac.jp)